Intelligent Techniques for Warehousing and Mining Sensor Network Data

Alfredo Cuzzocrea
University of Calabria, Italy

INFORMATION SCIENCE REFERENCE

Hershey · New York

Director of Editorial Content:	Kristin Klinger
Senior Managing Editor:	Jamie Snavely
Assistant Managing Editor:	Michael Brehm
Publishing Assistant:	Sean Woznicki
Typesetter:	Kurt Smith, Sean Woznicki
Cover Design:	Lisa Tosheff
Printed at:	Yurchak Printing Inc.

Published in the United States of America by
Information Science Reference (an imprint of IGI Global)
701 E. Chocolate Avenue
Hershey PA 17033
Tel: 717-533-8845
Fax: 717-533-8661
E-mail: cust@igi-global.com
Web site: http://www.igi-global.com/reference

Library of Congress Cataloging-in-Publication Data

Intelligent techniques for warehousing and mining sensor network data / Alfredo Cuzzocrea, editor.
    p. cm.
  Includes bibliographical references and index.

  Summary: "This book focuses on the relevant research theme of warehousing and mining sensor network data, specifically for the database, data warehousing and data mining research communities"--Provided by publisher.

  ISBN 978-1-60566-328-9 (hardcover) -- ISBN 978-1-60566-329-6 (ebook) 1.
Sensor networks. 2. Data mining. 3. Information retrieval. 4. Computer
storage devices. I. Cuzzocrea, Alfredo, 1974-
  TK7872.D48.I48 2010
  681'.2--dc22
                              2009043965

British Cataloguing in Publication Data
A Cataloguing in Publication record for this book is available from the British Library.

Table of Contents

Section 1
Warehousing and OLAPing Sensor Network Data

Detailed Table of Contents

Section 1
Warehousing and OLAPing Sensor Network Data

Chapter 1

    Marcos M. Campos, Oracle Data Mining Technologies
    Boriana L. Milenova, Oracle Data Mining Technologies

Warehousing and analytics of sensor network data is an area growing in relevance as more and more sensor data are collected and made available for analysis. Applications that involve processing of streaming sensor data require efficient storage, analysis, and monitoring of data streams. Traditionally, in these applications, RDBMSs have been confined to the storage stage. While contemporary RDBMSs were not designed to handle stream-like data, the tight integration of sophisticated analytic capabilities into the core database engine offers a powerful infrastructure that can more broadly support sensor network applications. Other useful components found in RDBMs include: extraction, transformation and load (ETL), centralized data warehousing, and automated alert capabilities. The combination of these components addresses significant challenges in sensor data applications such as data transformations, feature extraction, mining model build and deployment, distributed model scoring, and alerting/messaging infrastructure. This chapter discusses the usage of existing RDBMS functionality in the context of sensor network applications.

Chapter 2

    Alfredo Cuzzocrea, ICAR-CNR, Italy and University of Calabria, Italy
    Filippo Furfaro, University of Calabria, Italy
    Elio Masciari, ICAR-CNR, Italy
    Domenico Saccà, University of Calabria, Italy

Sensor networks represent a leading case of data stream sources coming from real-life application scenarios. Sensors are non-reactive elements which are used to monitor real-life phenomena, such as live weather conditions, network traffic etc. They are usually organized into networks where their readings are transmitted using low level protocols. A relevant problem in dealing with data streams consists in the fact that they are intrinsically multi-level and multidimensional in nature, so that they require to be analyzed by means of a multi-level and a multi-resolution (analysis) model accordingly, like OLAP, beyond traditional solutions provided by primitive SQL-based DBMS interfaces. Despite this, a significant issue in dealing with OLAP is represented by the so-called curse of dimensionality problem, which consists in the fact that, when the number of dimensions of the target data cube increases, multidimensional data cannot be accessed and queried efficiently, due to their enormous size. Starting from this practical evidence, several data cube compression techniques have been proposed during the last years, with alternate fortune. Briefly, the main idea of these techniques consists in computing compressed representations of input data cubes in order to evaluate time-consuming OLAP queries against them, thus obtaining approximate answers. Similarly to static data, approximate query answering techniques can be applied to streaming data, in order to improve OLAP analysis of such kind of data. Unfortunately, the data cube compression computational paradigm gets worse when OLAP aggregations are computed on top of a continuously flooding multidimensional data stream. In order to efficiently deal with the curse of dimensionality problem and achieve high efficiency in processing and querying multidimensional data streams, thus efficiently supporting OLAP analysis of such kind of data, this chapter proposes novel compression techniques over data stream readings that are materialized for OLAP purposes. This allows the authors to tame the unbounded nature of streaming data, thus dealing with bounded memory issues exposed by conventional DBMS tools. Overall, this chapter introduces an innovative, complex technique for efficiently supporting OLAP analysis of multidimensional data streams.

Chapter 3

    Hector Gonzalez, University of Illinois at Urbana-Champaign, USA
    Jiawei Han, University of Illinois at Urbana-Champaign, USA
    Hong Cheng, University of Illinois at Urbana-Champaign, USA
    Tianyi Wu, University of Illinois at Urbana-Champaign, USA

Massive Radio Frequency Identification (RFID) datasets are expected to become commonplace in supply-chain management systems. Warehousing and mining this data is an essential problem with great potential benefits for inventory management, object tracking, and product procurement processes. Since RFID tags can be used to identify each individual item, enormous amounts of location-tracking data are generated. Furthermore, RFID tags can record sensor information such as temperature or humidity. With such data, object movements can be modeled by movement graphs, where nodes correspond to locations, and edges record the history of item transitions between locations and sensor readings recorded during the transition. This chapter shows the benefits of the movement graph model in terms of compact representation, complete recording of spatio-temporal and item level information, and its role in facilitating multidimensional analysis. Compression power, and efficiency in query processing are gained by organizing the model around the concept of gateway nodes, which serve as bridges connecting different regions of graph, and provide a natural partition of item trajectories. Multi-dimensional analysis is provided by a graph-based object movement data cube that is constructed by merging and collapsing nodes and edges according to an application-oriented topological structure.

Chapter 4

S. Orlando, Università Ca' Foscari di Venezia, Italy
A. Raffaetà, Università Ca' Foscari di Venezia, Italy
A. Roncato, Università Ca' Foscari di Venezia, Italy
C. Silvestri, Università Ca' Foscari di Venezia, Italy

This chapter discusses how data warehousing technology can be used to store aggregate information about trajectories of mobile objects, and to perform OLAP operations over them. To this end, the authors define a data cube with spatial and temporal dimensions, discretized according to a hierarchy of regular grids. Tbe authors analyse some measures of interest related to trajectories, such as the number of distinct trajectories in a cell or starting from a cell, the distance covered by the trajectories in a cell, the average and maximum speed and the average acceleration of the trajectories in the cell, and the frequent patterns obtained by a data mining process on trajectories. The authors focus on some specialised algorithms to transform data, and load the measures in the base cells. Such stored values are used, along with suitable aggregate functions, to compute the roll-up operations. The main issues derive, in this case, from the characteristics of input data, i.e., trajectory observations of mobile objects, which are usually produced at different rates, and arrive in streams in an unpredictable and unbounded way. Finally, this chapter also discusses some use cases that would benefit from such a framework, in particular in the domain of supervision systems to monitor road traffic (or movements of individuals) in a given geographical area.

Section 2
Mining Sensor Network Data

Chapter 5

Alec Pawling, University of Notre Dame, USA
Ping Yan, University of Notre Dame, USA
Julián Candia, Northeastern University, USA
Tim Schoenharl, University of Notre Dame, USA
Greg Madey, University of Notre Dame, USA

This chapter considers a cell phone network as a set of automatically deployed sensors that records movement and interaction patterns of the population. The authors discuss methods for detecting anomalies in the streaming data produced by the cell phone network. The authors motivate this discussion by describing the Wireless Phone Based Emergency Response (WIPER) system, a proof-of-concept decision support system for emergency response managers. This chapter also discusses some of the scientific work enabled by this type of sensor data and the related privacy issues. The authors describe scientific studies that use the cell phone data set and steps they have taken to ensure the security of the data. This chapter describes the overall decision support system and discuss three methods of anomaly detection that the authors have applied to the data.

Chapter 6

Pedro Pereira Rodrigues, LIAAD - INESC Porto L.A. & University of Porto, Portugal

João Gama, LIAAD - INESC Porto L.A. & University of Porto, Portugal

Luís Lopes, CRACS - INESC Porto L.A. & University of Porto, Portugal

Knowledge discovery is a wide area of research where machine learning, data mining and data warehousing techniques converge to the common goal of describing and understanding the world. Nowadays applications produce infinite streams of data distributed across wide sensor networks. This ubiquitous scenario raises several obstacles to the usual knowledge discovery work flow, enforcing the need to develop new techniques, with different conceptualizations and adaptive decision making. The current setting of having a web of sensory devices, some of them enclosing processing ability, represents now a new knowledge discovery environment, possibly not completely observable, that is much less controlled by both the human user and a common centralized control process. This ubiquitous and fast-changing scenario is nowadays subject to the same interactions required by previous static and centralized applications. Hence the need to inspect how different knowledge discovery techniques adapt to ubiquitous scenarios such as wired/wireless sensor networks. This chapter explores different characteristics of sensor networks which define new requirements for knowledge discovery, with the common goal of extracting some kind of comprehension about sensor data and sensor networks, focusing on clustering techniques which provide useful information about sensor networks as it represents the interactions between sensors. This network comprehension ability is related with sensor data clustering and clustering of the data streams produced by the sensors. A wide range of techniques already exists to assess these interactions in centralized scenarios, but the seizable processing abilities of sensors in distributed algorithms present several benefits that shall be considered in future designs. Also, sensors produce data at high rate. Often, human experts need to inspect these data streams visually in order to decide on some corrective or proactive operations. Visualization of data streams, and of data mining results, is therefore extremely relevant to sensor data management, and can enhance sensor network comprehension, and should be addressed in future works.

Chapter 7

Yang Zhang, University of Twente, The Netherlands

Nirvana Meratnia, University of Twente, The Netherlands

Paul Havinga, University of Twente, The Netherlands

Raw data collected in wireless sensor networks are often unreliable and inaccurate due to noise, faulty sensors and harsh environmental effects. Sensor data that significantly deviate from normal pattern of sensed data are often called outliers. Outlier detection in wireless sensor networks aims at identifying such readings, which represent either measurement errors or interesting events. Due to numerous shortcomings, commonly used outlier detection techniques for general data seem not to be directly applicable to outlier detection in wireless sensor networks. This chapter reports on the current state-of-the-art on outlier detection techniques for general data, provides a comprehensive technique-based taxonomy for these techniques, and highlight their characteristics in a comparative view. Furthermore, the authors address challenges of outlier detection in wireless sensor networks, provide a guideline on requirements that suitable outlier detection techniques for wireless sensor networks should meet, and will explain why general outlier detection techniques do not suffice.

Section 3
Clustering Sensor Network Data

Chapter 8

Elena Baralis, Politecnico di Torino, Italy
Tania Cerquitelli, Politecnico di Torino, Italy
Vincenzo D'Elia, Politecnico di Torino, Italy

After the metaphor "the sensor network is a database," wireless sensor networks have become an important research topic in the database research community. Sensing technologies have developed new smart wireless devices which integrate sensing, processing, storage and communication capabilities. Smart sensors can programmatically measure physical quantities, perform simple computations, store, receive and transmit data. Querying the network entails the (frequent) acquisition of the appropriate sensor measurements. Since sensors are battery-powered and communication is the main source of power consumption, an important issue in this context is energy saving during data collection. This chapter thoroughly describes different clustering algorithms to efficiently discover spatial and temporal correlation among sensors and sensor readings. Discovered correlations allow the selection of a subset of good quality representatives of the whole network. Rather than directly querying all network nodes, only the representative sensors are queried to reduce the communication, computation and power consumption costs. Experiments with different clustering algorithms show the adaptability and the effectiveness of the proposed approach.

Chapter 9

Stefano Lodi, University of Bologna, Italy
Gabriele Monti, University of Bologna, Italy
Gianluca Moro, University of Bologna, Italy
Claudio Sartori, University of Bologna, Italy

This work proposes and evaluates distributed algorithms for data clustering in self-organizing ad-hoc sensor networks with computational, connectivity, and power constraints. Self-organization is essential in environments with a large number of devices, because the resulting system cannot be configured and maintained by specific human adjustments on its single components. One of the benefits of in-network data clustering algorithms is the capability of the network to transmit only relevant, high level information, namely models, instead of large amounts of raw data, also reducing drastically energy consumption. For instance, a sensor network could directly identify or anticipate extreme environmental events such as tsunami, tornado or volcanic eruptions notifying only the alarm or its probability, rather than transmitting via satellite each single normal wave motion. The efficiency and efficacy of the methods is evaluated by simulation measuring network traffic, and comparing the generated models with ideal results returned by density-based clustering algorithms for centralized systems.

Section 4
Query Languages and Query Optimization Techniques for Warehousing and Mining Sensor Network Data

Chapter 10

Shi-Kuo Chang, University of Pittsburgh, USA
Gennaro Costagliola, Università di Salerno, Italy
Erland Jungert, Swedish Defense Research Agency, Sweden
Karin Camara, Swedish Defense Research Agency, Sweden

Sensor data fusion imposes a number of novel requirements on query languages and query processing techniques. A spatial/temporal query language called ΣQL has been proposed to support the retrieval of multimedia information from multiple sources and databases. This chapter investigates intelligent querying techniques including fusion techniques, multimedia data transformations, interactive progressive query building and ΣQL query processing techniques using sensor data fusion. The authors illustrate and discuss tasks and query patterns for information fusion, provide a number of examples of iterative queries and show the effectiveness of ΣQL in a command-action scenario.

Chapter 11

Mark Roantree, Dublin City University, Ireland
Alan F. Smeaton, Dublin City University, Ireland
Noel E. O'Connor, Dublin City University, Ireland
Vincent Andrieu, Dublin City University, Ireland
Nicolas Legeay, Dublin City University, Ireland
Fabrice Camous, Dublin City University, Ireland

One of the more recent sources of large volumes of generated data is sensor devices, where dedicated sensing equipment is used to monitor events and happenings in a wide range of domains, including monitoring human biometrics and behaviour. This chapter proposes an approach and an implementation of semi-automated enrichment of raw sensor data, where the sensor data can come from a wide variety of sources. The authors extract semantics from the sensor data using their XSENSE processing architecture in a multi-stage analysis. The net result is that the authors transform sensor data values into XML data so that well-established XML querying via XPATH and similar techniques, can be followed. The authors then propose to distribute the XML data on a peer-to-peer configuration and the authors show, through simulations, what the computational costs of executing queries on this P2P network, will be. The authors validate their approach through the use of an array of sensor data readings taken from a range of biometric sensor devices, fitted to movie-watchers as they watched Hollywood movies. These readings were synchronised with video and audio analysis of the actual movies themselves, where movie highlights were automatically detected, which the authors try to correlate with observed human reactions. The XSENSE architecture is used to semantically enrich both the biometric sensor readings and the outputs of video analysis, into one large sensor database. This chapter thus presents and validate a scalable means of semi-automating the semantic enrichment of sensor data, thereby providing a means of large-scale sensor data management which is a necessary step in supporting data mining from sensor networks.

Section 5
Intelligent Techniques for Efficient Sensor Network Data Warehousing and Mining

Chapter 12

Sotiris Nikoletseas, University of Patras, Greece
Olivier Powell, University of Geneva, Switzerland
Jose Rolim, University of Geneva, Switzerland

Geographic routing is becoming the protocol of choice for many sensor network applications. Some very efficient geographic routing algorithms exist, however they require a preliminary planarization of the communication graph. Planarization induces overhead which makes this approach not optimal when lightweight protocols are required. On the other hand, georouting algorithms which do not rely on planarization have fairly low success rates and either fail to route messages around all but the simplest obstacles or have a high topology control overhead (e.g. contour detection algorithms). This chapter describes the GRIC algorithm which was designed to overcome some of those limitations. The GRIC algorithm is the first lightweight and efficient on demand (i.e. all-to-all) geographic routing algorithm which does not require planarization, has almost 100% delivery rates (when no obstacles are added), and behaves well in the presence of large communication blocking obstacles.

Chapter 13

David J. Yates, Bentley University, USA
Jennifer Xu, Bentley University, USA

This research is motivated by data mining for wireless sensor network applications. The authors consider applications where data is acquired in real-time, and thus data mining is performed on live streams of data rather than on stored databases. One challenge in supporting such applications is that sensor node power is a precious resource that needs to be managed as such. To conserve energy in the sensor field, the authors propose and evaluate several approaches to acquiring, and then caching data in a sensor field data server. The authors show that for true real-time applications, for which response time dictates data quality, policies that emulate cache hits by computing and returning approximate values for sensor data yield a simultaneous quality improvement and cost saving. This "win-win" is because when data acquisition response time is sufficiently important, the decrease in resource consumption and increase in data quality achieved by using approximate values outweighs the negative impact on data accuracy due to the approximation. In contrast, when data accuracy drives quality, a linear trade-off between resource consumption and data accuracy emerges. The authors then identify caching and lookup policies for which the sensor field query rate is bounded when servicing an arbitrary workload of user queries. This upper bound is achieved by having multiple user queries share the cost of a sensor field query.

Finally, the authors discuss the challenges facing sensor network data mining applications in terms of data collection, warehousing, and mining techniques.

Section 6
Intelligent Techniques for Advanced Sensor Network Data Warehousing and Mining

Chapter 14

Qingchun Jiang, Oracle Corporation, USA
Raman Adaikkalavan, Indiana University, USA
Sharma Chakravarthy, University of Texas, Arlington, USA

Event processing in the form of ECA rules has been researched extensively from the situation monitoring viewpoint to detect changes in a timely manner and to take appropriate actions. Several event specification languages and processing models have been developed, analyzed, and implemented. More recently, data stream processing has been receiving a lot of attention to deal with applications that generate large amounts of data in real-time at varying input rates and to compute functions over multiple streams that satisfy quality of service (QoS) requirements. A few systems based on the data stream processing model have been proposed to deal with change detection and situation monitoring. However, current data stream processing models lack the notion of composite event specification and computation, and they cannot be readily combined with event detection and rule specification, which are necessary and important for many applications. In this chapter, the authors discuss a couple of representative scenarios that require both stream and event processing. The authors then summarize the similarities and differences between the event and data stream processing models. The comparison clearly indicates that for most of the applications considered for stream processing, event component is needed and is not currently supported. And conversely, earlier event processing systems assumed primitive (or simple) events triggered by DBMS and other applications and did not consider computed events. By synthesizing these two and combining their strengths, the authors present an integrated model – one that will be better than the sum of its parts. This chapter discusses the notion of a semantic window, which extends the current window concept for continuous queries, and stream modifiers in order to extend current stream computation model for complicated change detection. The authors further discuss the extension of event specification to include continuous queries. Finally, this chapter demonstrates how one of the scenarios discussed earlier can be elegantly and effectively modeled using the integrated approach.

Chapter 15

Biswajit Panja, Morehead State University, USA
Sanjay Kumar Madria, Missouri University of Science and Technology, USA

In sensor networks, the large numbers of tiny sensor nodes communicate remotely or locally among themselves to accomplish a wide range of applications. However, such a network poses serious security protocol design challenges due to ad hoc nature of the communication and the presence of constraints such as limited energy, slower processor speed and small memory size. To secure such a wireless network, the efficient key management techniques are important as existing techniques from mobile ad hoc networks assume resource-equipped nodes. There are some recent security protocols that have been proposed for sensor networks and some of them have also been implemented in a real environment. This chapter provides an overview of research in the area of key management for sensor networks mainly focused on using a cluster head based architecture. First, the authors provide a review of the existing security protocols based on private/public key cryptography, Kerberos, Digital signatures and IP security. Next, the authors investigate some of the existing work on key management protocols for sensor networks along with their advantages and disadvantages. Finally, some new approaches for providing key management, cluster head security and dynamic key computations are explored.

Preface

This book focuses on the relevant research theme of warehousing and mining sensor network data, which is attracting a lot of attention from the Database, Data Warehousing and Data Mining research communities. With this main idea in mind, this book is oriented to fundamentals and theoretical issues of sensor networks as well as sensor network applications, which have become of relevant interest for next-generation intelligent information systems. Sensor network applications are manifolds: from environmental data collection/management to alerting/alarming systems, from intelligent tools for monitoring/managing IP networks to novel RFID-based applications etc.

Sensor network data management poses new challenges that are outside the scope of capabilities of conventional DBMS, where data are represented and managed according to a tuple-oriented approach. As an example, DBMS expose a limited memory that is not compatible with the prominent unbounded-memory requirement of sensor network data, which, ideally, originate an unbounded data flow. In this respect, collecting and querying sensor network data is questioning, and it cannot be accomplished via conventional DBMS-inspired methodologies. Also, time is completely neglected in DBMS, whereas it plays a leading role in sensor network data management.

Under a broader view, sensor network data are a specialized class of data streams, which can be defined as intermittent sources of information. The above-mentioned issues become, in consequence of this, the guidelines for the design and development of next-generation Data Stream Management Systems (DSMS), which can be reasonably intended as the next challenge for data management research. Therefore, under another perspective, warehousing and mining sensor network data, and, more generally, data streams can be viewed as methodologies and techniques on top of DSMS, oriented to extend data-intensive capabilities of such systems. The same happened for conventional DBMS, with OLAP and Data Mining tools.

Warehousing and mining sensor network data research can also be roughly indented as the application of traditional warehousing and mining techniques developed in the context of DBMS for relational data as well as non-conventional data (e.g., textual data, raw data, XML data etc) to novel scenarios drawn by sensor networks. Despite this, models and algorithms developed in conventional data warehousing and mining technologies cannot be applied "as-they-are" to the novel context of sensor network data management, as the former are not suitable to requirements of sensor data, such as: time-oriented processing, multiple-rate arrivals, unbounded memory, single-pass processing etc. From this, it follows the need for designing and developing models and algorithms able to deal with previously-unrecognized characteristics of sensor network intelligent information systems, thus overcoming actual limitations of data warehousing and data mining systems and platforms.

Based on these motivations and pursuing these aims, this book covers a broad range of topics: data warehousing models for sensor network data, intelligent acquisition techniques for sensor network data,

ETL processes over sensor network data, advanced techniques for processing sensor network data, efficient storage solutions for sensor network data, collecting sensor network data, querying sensor network data, query languages for sensor network data, fusion and integration techniques for heterogeneous sensor network data. cleaning techniques over sensor network data, mining sensor network data, frequent item set mining over sensor network data, intelligent mining techniques over sensor network data, mining outliers and deviants over sensor network data, discovery of complex knowledge patterns from sensor network data, privacy preserving issues of warehousing and mining sensor network data etc.

The main mission of this book is represented by the achievement of a high-quality publication on fundamentals, state-of-the-art techniques and future trends of warehousing and mining sensor network data research. Themes proposed by this book are viable since, traditionally, data methodologies play a leading role in the research community. In turn, this is due to the fact that data processing issues are orthogonal issues for a broad range of next-generation systems and applications, among which we re-call: distributed databases, data warehouses, data mining tools, information systems, knowledge-based systems etc. In this respect, themes proposed by this book have a plus-value as they are focused on a very interesting application field such as sensor network data management, which can be reasonably considered as one of the most relevant research themes presently.

Therefore, this book expands the sensor networks research field by putting the basis for novel research trends in the context of warehousing and mining sensor network data, via addressing topics that are, at now, rarely investigated, such as data mining query languages for sensor data. Indeed, the most important unique characteristic of this book is represented by its interdisciplinarity across different research fields spanning from traditional DBMS to Data Warehousing and Data Mining, all concerned with the innovative research theme of sensor network data management.

This book consists of fifteen chapters organized in six major sections. The first section, titled "Warehousing and OLAPing Sensor Network Data", focuses the attention on models, techniques and algorithms for warehousing and OLAPing several kinds of sensor network data, from conventional ones to RFID data, location-based sensor data and streaming mobile object observations. The second section, titled "Mining Sensor Network Data", moves the attention on several mining sensor network data issues, such as anomaly detection in streaming sensor data, knowledge discovery from sensor network data in order to improve the quality of sensor network comprehension tasks, and outlier detection in wireless sensor networks. The third section, titled "Clustering Sensor Network Data", is related to clustering techniques and algorithms for sensor network data, with particular emphasis over applications in specialized contexts, such as intelligent acquisition techniques for sensor network data and peer-to-peer data clustering in self-organizing sensor networks. The fourth section, titled "Query Languages and Query Optimization Techniques for Warehousing and Mining Sensor Network Data", focuses the attention on query methodologies for sensor networks, particularly on intelligent query techniques for sensor network data fusion and optimization approaches for query activities embedded in Data Mining tasks over peer-to-peer sensor networks. The fifth section, titled "Intelligent Techniques for Efficient Sensor Network Data Warehousing and Mining", moves the attention on intelligent techniques devoted to improve the performance of warehousing and mining sensor network data, such as geographic routing of sensor data in the presence of voids and obstacles, and sensor field resource management approaches aiming at improving Data Mining tasks over sensor networks. Finally, the sixth section, titled "Intelligent Techniques for Advanced Sensor Network Data Warehousing and Mining", is related to advanced aspects of warehousing and mining sensor network data, such as the synergy between event and stream processing for complex applications, and dynamic security key management schemes in sensor networks.

In the following, chapters of the section "Warehousing and OLAPing Sensor Network Data" are summarized.

In the first chapter, titled "Integrated Intelligence: Separating the Wheat from the Chaff in Sensor Data", Marcos M. Campos and Boriana L. Milenova investigate the issue of warehousing and analytics of sensor network data, which is an area growing in relevance as more and more sensor data are collected and made available for analysis. Applications that involve processing of streaming sensor data require efficient storage, analysis, and monitoring of data streams. Traditionally, in these applications, RDBMSs have been confined to the storage stage. While contemporary RDBMSs were not designed to handle stream-like data, the tight integration of sophisticated analytic capabilities into the core database engine offers a powerful infrastructure that can more broadly support sensor network applications. Other useful components found in RDBMs include: extraction, transformation and load (ETL), centralized data warehousing, and automated alert capabilities. The combination of these components addresses significant challenges in sensor data applications such as data transformations, feature extraction, mining model build and deployment, distributed model scoring, and alerting/messaging infrastructure. Based on these motivations, chapter "Integrated Intelligence – Separating the Wheat from the Chaff in Sensor Data" discusses the usage of existing RDBMS functionality in the context of sensor network applications.

In the second chapter, titled "Improving OLAP Analysis of Multidimensional Data Streams via Efficient Compression Techniques", Alfredo Cuzzocrea, Filippo Furfaro, Elio Masciari and Domenico Saccà consider multidimensionality issues of data streams, and propose efficient compression techniques for improving OLAP analysis of multidimensional data streams. Authors state that a relevant problem in dealing with data streams consists in the fact that they are intrinsically multi-level and multidimensional in nature, so that they require to be analyzed by means of a multi-level and a multi-resolution (analysis) model accordingly, like OLAP, beyond traditional solutions provided by primitive SQL-based DBMS interfaces. Despite this, a significant issue in dealing with OLAP is represented by the so-called curse of dimensionality problem, which consists in the fact that, when the number of dimensions of the target data cube increases, multidimensional data cannot be accessed and queried efficiently, due to their enormous size. Starting from this practical evidence, several data cube compression techniques have been proposed during the last years, with alternate fortune. Briefly, the main idea of these techniques consists in computing compressed representations of input data cubes in order to evaluate time-consuming OLAP queries against them, thus obtaining approximate answers. Similarly to static data, approximate query answering techniques can be applied to streaming data, in order to improve OLAP analysis of such kind of data. Unfortunately, the data cube compression computational paradigm gets worse when OLAP aggregations are computed on top of a continuously flooding multidimensional data stream. In order to efficiently deal with the curse of dimensionality problem and achieve high efficiency in processing and querying multidimensional data streams, thus efficiently supporting OLAP analysis of such kind of data, this chapter proposes novel compression techniques over data stream readings that are materialized for OLAP purposes. This allows us to tame the unbounded nature of streaming data, thus dealing with bounded memory issues exposed by conventional DBMS tools. Overall, chapter "Improving OLAP Analysis of Multidimensional Data Streams via Efficient Compression Techniques" introduces an innovative, complex technique for efficiently supporting OLAP analysis of multidimensional data streams.

In the third chapter, titled "Warehousing RFID and Location-Based Sensor Data", Hector Gonzalez, Jiawei Han, Hong Cheng and Tianyi Wu focus the attention on the problem of efficiently warehousing RFID and location-based sensor data. Authors recognize that RFID datasets are expected to become commonplace in supply-chain management systems. Warehousing and mining this data is an essential problem with great potential benefits for inventory management, object tracking, and product procurement processes. Since RFID tags can be used to identify each individual item, enormous amounts of location-tracking data are generated. Furthermore, RFID tags can record sensor information such as temperature or humidity. With such data, object movements can be modeled by movement graphs, where nodes correspond to locations, and edges record the history of item transitions between locations and sensor readings recorded during the transition. In this chapter, benefits of the movement graph model in terms of compact representation, complete recording of spatio-temporal and item level information, and its role in facilitating multidimensional analysis are demonstrated. Compression power, and efficiency in query processing are gained by organizing the model around the concept of gateway nodes, which serve as bridges connecting different regions of graph, and provide a natural partition of item trajectories. Moreover, in chapter "Warehousing RFID and Location-Based Sensor Data" multi-dimensional analysis is provided by a graph-based object movement data cube that is constructed by merging and collapsing nodes and edges according to an application-oriented topological structure.

In the fourth chapter, titled "Warehousing and Mining Streams of Mobile Object Observations", Salvatore Orlando, Alessandra Raffaetà, Alessandro Roncato and Claudio Silvestri study the problem of warehousing and mining streams of mobile object observations, by discussing how data warehousing technology can be used to store aggregate information about trajectories of mobile objects, and to perform OLAP operations over them. To this end, authors define a data cube with spatial and temporal dimensions, discretized according to a hierarchy of regular grids. Authors analyze some measures of interest related to trajectories, such as the number of distinct trajectories in a cell or, starting from a cell, the distance covered by the trajectories in a cell, the average and maximum speed and the average acceleration of the trajectories in the cell, and the frequent patterns obtained by a data mining process on trajectories. Furthermore, authors focus on some specialized algorithms to transform data, and load the measures in the base cells. Such stored values are used, along with suitable aggregate functions, to compute the roll-up operations. In this case, as author observe, main issues derive from characteristics of input data, i.e. trajectory observations of mobile objects, which are usually produced at different rates, and arrive in streams in an unpredictable and unbounded way. Finally, chapter "Warehousing and Mining Streams of Mobile Object Observations" also discusses some use cases that would benefit from the proposed framework, in particular in the domain of supervision systems to monitor road traffic (or movements of individuals) in a given geographical area.

In the following, chapters of the section "Mining Sensor Network Data" are summarized.

In the fifth chapter, titled "Anomaly Detection in Streaming Sensor Data", Alec Pawling, Ping Yan, Julián Candia, Tim Schoenharl and Greg Madey consider a cell phone network as a set of automatically deployed sensors that records movement and interaction patterns of the target population. Authors discuss methods for detecting anomalies in streaming data produced by the cell phone network, and motivate this discussion by describing the Wireless Phone Based Emergency Response (WIPER) system, a proof-of-concept decision support system for emergency response managers. Authors also discuss some of the scientific work enabled by this type of sensor data and the related privacy issues, and describe scientific studies that use the cell phone data set and steps they have taken to ensure the security of the data. Finally, chapter "Anomaly Detection in Streaming Sensor Data" describes the overall decision support system and discusses three methods of anomaly detection that can be applied to the data.

In the sixth chapter, titled "Knowledge Discovery for Sensor Network Comprehension", Pedro Pereira Rodrigues, João Gama and Luís Lopes explore different characteristics of sensor networks that define new requirements for knowledge discovery, with the common goal of extracting some kind of comprehension about sensor data and sensor networks, focusing on clustering techniques that provide useful information about sensor networks via representing interactions between sensors. This network comprehension ability is related with sensor data clustering and clustering of the data streams produced by the sensors. A wide range of techniques already exists to assess these interactions in centralized scenarios, but the processing abilities of sensors in distributed algorithms present several benefits that shall be considered in future designs. Also, sensors produce data at high rate. Often, human experts need to inspect these data streams visually in order to decide on some corrective or proactive operations. Therefore, chapter "Knowledge Discovery for Sensor Network Comprehension" asserts that visualization of data streams, and of data mining results, is extremely relevant to sensor data management, and can enhance sensor network comprehension, thus it should be addressed in future works.

In the seventh chapter, titled "Why General Outlier Detection Techniques Do Not Suffice for Wireless Sensor Networks", Yang Zhang, Nirvana Meratnia and Paul Havinga start from recognizing that raw data collected in wireless sensor networks are often unreliable and inaccurate due to noise, faulty sensors and harsh environmental effects. Sensor data that significantly deviate from normal patterns of sensed data are often called outliers. Outlier detection in wireless sensor networks aims at identifying such readings, which represent either measurement errors or interesting events. Due to numerous shortcomings, commonly-used outlier detection techniques for general data seem not to be directly applicable to outlier detection in wireless sensor networks. In this chapter, authors report on the current state-of-the-art on outlier detection techniques for general data, provide a comprehensive technique-based taxonomy for these techniques, and highlight their characteristics in a comparative view. Furthermore, chapter "Why General Outlier Detection Techniques do not Suffice for Wireless Sensor Networks?" addresses challenges of outlier detection in wireless sensor networks, provides a guideline on requirements that suitable outlier detection techniques for wireless sensor networks should meet, and explains why general outlier detection techniques do not suffice.

In the following, chapters of the section "Clustering Sensor Network Data" are summarized.

In the eighth chapter, titled "Intelligent Acquisition Techniques for Sensor Network Data", Elena Baralis, Tania Cerquitelli and Vincenzo D'Elia investigate the issue of querying sensor networks, which entails the (frequent) acquisition of appropriate sensor measurements. Since sensors are battery-powered and communication is the main source of power consumption, an important issue in this context is energy saving during data collection. This chapter thoroughly describes different clustering algorithms to efficiently discover spatial and temporal correlation among sensors and sensor readings. Discovered correlations allow the selection of a subset of good quality representatives of the whole network. Rather than directly querying all network nodes, only the representative sensors are queried to reduce the communication, computation and power consumption costs. Finally, chapter "Intelligent Acquisition Techniques for Sensor Network Data" presents several experiments with different clustering algorithms demonstrating the adaptability and the effectiveness of the proposed approach.

In the ninth chapter, titled "Peer-to-Peer Data Clustering in Self-Organizing Sensor Networks", Stefano Lodi, Gabriele Monti, Gianluca Moro and Claudio Sartori propose and evaluate distributed algorithms for data clustering in self-organizing ad-hoc sensor networks with computational, connectivity, and power constraints. Self-organization is essential in environments with a large number of devices, because the resulting system cannot be configured and maintained by specific human adjustments on its single components. One of the benefits of in-network data clustering algorithms is the capability of the network to transmit only relevant, high level information, namely models, instead of large amounts of raw data, also reducing drastically energy consumption. For instance, a sensor network could directly identify or anticipate extreme environmental events such as tsunami, tornado or volcanic eruptions notifying only the alarm or its probability, rather than transmitting via satellite each single normal wave motion. In chapter "Peer-To-Peer Data Clustering in Self-Organizing Sensor Networks", the efficiency and efficacy of the methods is evaluated by simulation measuring network traffic, and comparing the generated models with ideal results returned by density-based clustering algorithms for centralized systems.

In the following, chapters of the section "Query Languages and Query Optimization Techniques for Warehousing and Mining Sensor Network Data" are summarized.

In the tenth chapter, titled "Intelligent Querying Techniques for Sensor Data Fusion", Shi-Kuo Chang, Gennaro Costagliola, Erland Jungert and Karin Camara focus the attention on sensor network data fusion, which imposes a number of novel requirements on query languages and query processing techniques. A spatial/temporal query language called ΣQL has been proposed by the same authors previously, in order to support the retrieval of multimedia information from multiple sources and databases. In this chapter, authors investigate intelligent querying techniques including fusion techniques, multimedia data transformations, interactive progressive query building and ΣQL query processing techniques using sensor data fusion. Furthermore, chapter "Intelligent Querying Techniques for Sensor Data Fusion" illustrates and discusses tasks and query patterns for information fusion, provides a number of examples of iterative queries and shows the effectiveness of ΣQL in a command-action scenario.

In the eleventh chapter, titled "Query Optimisation for Data Mining in Peer-to-Peer Sensor Networks", Mark Roantree, Alan F. Smeaton, Noel E. O'Connor, Vincent Andrieu, Nicolas Legeay and Fabrice Camous move the attention on sensor devices, which represent one of the more recent sources of large volumes of generated data in sensor networks where dedicated sensing equipment is used to monitor events and happenings in a wide range of domains, including monitoring human biometrics and behavior. In this chapter, authors propose an approach and an implementation of semi-automated enrichment of raw sensor data, where the sensor data can come from a wide variety of sources. Also, authors extract semantics from the sensor data using the proposed XSENSE processing architecture in a multi-stage analysis. Sensor data values are thus transformed into XML data so that well-established XML querying via XPATH and similar techniques can be followed. In this respect, authors propose to distribute XML data on a peer-to-peer configuration and show, through simulations, what the computational costs of executing queries on this P2P network, will be. Authors validate the proposed approach through the use of an array of sensor data readings taken from a range of biometric sensor devices, fitted to movie-watchers as they watched Hollywood movies. These readings were synchronized with video and audio analysis of the actual movies themselves, where movie highlights were automatically detected, in order to correlate these highlights with observed human reactions. XSENSE architecture is used to semantically enrich both the biometric sensor readings and the outputs of video analysis, into one large sensor database. Chapter "Query Optimisation for Data Mining in Peer-to-Peer Sensor Networks" thus presents and validates a scalable means of semi-automating the semantic enrichment of sensor data, thereby providing a means of large-scale sensor data management which is a necessary step in supporting data mining from sensor networks.

In the following, chapters of the section "Intelligent Techniques for Efficient Sensor Network Data Warehousing and Mining" are summarized.

In the twelfth chapter, titled "Geographic Routing of Sensor Data around Voids and Obstacles", Sotiris Nikoletseas, Olivier Powell and Jose Rolim start from recognizing that geographic routing is becoming the protocol of choice for many sensor network applications. Some very efficient geographic routing algorithms exist, however they require a preliminary planarization of the communication graph.

Planarization induces overhead which makes this approach not optimal when lightweight protocols are required. On the other hand, georouting algorithms which do not rely on planarization have fairly low success rates and either fail to route messages around all but the simplest obstacles or have a high topology control overhead (e.g. contour detection algorithms). In order to fulfill this gap, chapter "Geographic Routing of Sensor Data around Voids and Obstacles" describes the GRIC algorithm, the first lightweight and efficient on demand (i.e., all-to-all) geographic routing algorithm which does not require planarization, has almost 100% delivery rates (when no obstacles are added), and behaves well in the presence of large communication blocking obstacles.

In the thirteenth chapter, titled "Sensor Field Resource Management for Sensor Network Data Mining", David J. Yates and Jennifer Xu motivate their research by data mining for wireless sensor network applications. Authors consider applications where data is acquired in real-time, and thus data mining is performed on live streams of data rather than on stored databases. One challenge in supporting such applications is that sensor node power is a precious resource that needs to be managed as such. To conserve energy in the sensor field, authors propose and evaluate several approaches to acquiring, and then caching data in a sensor field data server. Authors show that for true real-time applications, for which response time dictates data quality, policies that emulate cache hits by computing and returning approximate values for sensor data yield a simultaneous quality improvement and cost saving. This "win-win" is because when data acquisition response time is sufficiently important, the decrease in resource consumption and increase in data quality achieved by using approximate values outweighs the negative impact on data accuracy due to the approximation. In contrast, when data accuracy drives quality, a linear trade-off between resource consumption and data accuracy emerges. Authors then identify caching and lookup policies for which the sensor field query rate is bounded when servicing an arbitrary workload of user queries. This upper bound is achieved by having multiple user queries share the cost of a sensor field query. Finally, chapter "Sensor Field Resource Management for Sensor Network Data Mining" discusses the challenges facing sensor network data mining applications in terms of data collection, warehousing, and mining techniques.

In the following, chapters of the section "Intelligent Techniques for Advanced Sensor Network Data Warehousing and Mining" are summarized.

In the fourteenth chapter, titled "Event/Stream Processing for Advanced Applications", Qingchun Jiang, Raman Adaikkalavan and Sharma Chakravarthy state that event processing in the form of ECA rules has been researched extensively from the situation monitoring viewpoint to detect changes in a timely manner and to take appropriate actions. Several event specification languages and processing models have been developed, analyzed, and implemented. More recently, data stream processing has been receiving a lot of attention to deal with applications that generate large amounts of data in real-time at varying input rates and to compute functions over multiple streams that satisfy quality of service (QoS) requirements. A few systems based on the data stream processing model have been proposed to deal with change detection and situation monitoring. However, current data stream processing models lack the notion of composite event specification and computation, and they cannot be readily combined with event detection and rule specification, which are necessary and important for many applications. In this chapter, authors discuss a couple of representative scenarios that require both stream and event processing, and then summarize the similarities and differences between the event and data stream processing models. The comparison clearly indicates that for most of the applications considered for stream processing, event component is needed and is not currently supported. And conversely, earlier event processing systems assumed primitive (or simple) events triggered by DBMS and other applications and did not consider computed events. By synthesizing these two and combining their strengths, authors present an integrated model – one that will be better than the sum of its parts. Authors discuss the notion of a semantic window, which extends the current window concept for continuous queries, and stream modifiers in order to extend current stream computation model for complicated change detection. Authors further discuss the extension of event specification to include continuous queries. Finally, chapter "Event/Stream Processing for Advanced Applications" demonstrates how one of the scenarios discussed earlier can be elegantly and effectively modeled using the integrated approach.

Finally, in the fifteenth chapter, titled "A Survey of Dynamic Key Management Schemes in Sensor Networks", Biswajit Panja and Sanjay Kumar Madria observe that, in sensor networks, the large numbers of tiny sensor nodes communicate remotely or locally among themselves to accomplish a wide range of applications. However, these networks pose serious security protocol design challenges due to ad hoc nature of the communication and the presence of constraints such as limited energy, slower processor speed and small memory size. To secure such a wireless network, efficient key management techniques are important as existing techniques from mobile ad hoc networks assume resource-equipped nodes. There are some recent security protocols that have been proposed for sensor networks and some of them have also been implemented in a real environment. This chapter provides an overview of research in the area of key management for sensor networks mainly focused on using a cluster head based architecture. First, authors provide a review of the existing security protocols based on private/public key cryptography, Kerberos, Digital signatures and IP security. Next, authors investigate some of the existing work on key management protocols for sensor networks along with their advantages and disadvantages. Finally, chapter "A Survey of Dynamic Key Management Schemes in Sensor Networks" explores some new approaches for providing key management, cluster head security and dynamic key computations.

Overall, this book represents a solid research contribution to state-of-the-art studies and practical achievements in warehousing and mining sensor network data, and puts the basis for further efforts in this challenging scientific field that will more and more play a leading role in next-generation Database, Data Warehousing and Data Mining research.

Alfredo Cuzzocrea
Editor

Acknowledgment

The editor wishes to thank all the authors for their insights and excellent contributions to this book. The editor would like to also acknowledge the invaluable help of all people involved in the review process of this book, without whose support the project could not have been successfully completed. Most of authors of the chapters included in this book also served as referees for chapters by other authors. Constructive and comprehensive reviews, which have involved two rigorous review rounds, have determined the success of this book. Complete list of reviewers is reported next.

The editor is also grateful to all the staff at IGI Global whose contribution to the whole process from the initial idea of the book to the final publication has been invaluable.

Alfredo Cuzzocrea
Editor

Section 1
Warehousing and OLAPing
Sensor Network Data

Chapter 1
Integrated Intelligence:
Separating the Wheat from the Chaff in Sensor Data

Marcos M. Campos
Oracle Data Mining Technologies

Boriana L. Milenova
Oracle Data Mining Technologies

ABSTRACT

Warehousing and analytics of sensor network data is an area growing in relevance as more and more sensor data are collected and made available for analysis. Applications that involve processing of streaming sensor data require efficient storage, analysis, and monitoring of data streams. Traditionally, in these applications, RDBMSs have been confined to the storage stage. While contemporary RDBMSs were not designed to handle stream-like data, the tight integration of sophisticated analytic capabilities into the core database engine offers a powerful infrastructure that can more broadly support sensor network applications. Other useful components found in RDBMs include: extraction, transformation and load (ETL), centralized data warehousing, and automated alert capabilities. The combination of these components addresses significant challenges in sensor data applications such as data transformations, feature extraction, mining model build and deployment, distributed model scoring, and alerting/messaging infrastructure. This chapter discusses the usage of existing RDBMS functionality in the context of sensor network applications.

INTRODUCTION

Sensor data analysis has become an area of growing importance as rapid advances in sensor technology have produced a variety of powerful and low-cost sensor arrays. Such sensors can be used to collect information from many domains, including satellite remote sensing, surveillance, computer network security, and health management. With the continuing expansion of these domains of interest and the increasing volume and complexity of the collected information, the effective and efficient storage, analysis, and monitoring of streaming sensor data requires a novel engineering approach. In addition to the challenges of handling enormous quantities of data, the requirements for sensor information

DOI: 10.4018/978-1-60566-328-9.ch001

processing applications usually place a premium on sophisticated analytic methods, performance, scalability, data integrity, and data security. Due to the specialized nature of many sensor driven applications, many engineering endeavors result in ad-hoc prototypes that are narrow in scope and are not transferable. Integration and re-use of pre-existing technology is also problematic.

It would clearly be advantageous to establish common engineering standards and to leverage a computational platform with analytical capabilities within an integrated framework. While databases were not originally designed to support streaming data, modern databases, with their capabilities of enabling mission critical applications, distributed processing, and integrated analytics, can significantly facilitate application development. Given the data-centric nature of sensor data analysis, an analytic data warehouse can provide a powerful infrastructure that allows for fast prototyping and efficient implementation. Tight coupling between data integration and analysis is inherent to an analytic data warehouse environment. This represents an important advantage over standalone applications that typically incur costly (and often prolonged) creation of infrastructure.

The following sections highlight important concepts in stream data processing, describe a database-centric approach for instrumenting applications dedicated to sensor data integration, analysis, and monitoring, and offer specific SQL query examples for the usage of individual components. An emphasis is placed on the benefits of leveraging an integrated platform with built-in analytics versus adopting an approach that relies on external components. The integrated methodology is illustrated with examples from two domains - network intrusion detection and satellite data imagery. These examples expand upon earlier work in (Campos & Milenova, 2005; Milenova & Campos, 2005). To make it concrete, the different aspects of the proposed architecture, as well as the examples, are described using an implementation based on Oracle's database technology.

BACKGROUND

Streaming sensor data is set apart from more traditional types of data by two important characteristics: its massive volume and its dynamic, distributed, and heterogeneous nature. Most applications that address domains with streaming data, strive to accumulate the data to some extent (often placing emphasis on recent data) and archive it in a, possibly off-site, data warehouse. Access to archived data can be prohibitively expensive and therefore can hinder analytical efforts. In addition to achieving some level of storage and retroactive analysis of continuous and unbounded data, applications processing sensor data are often required to perform online monitoring of the data stream and must be capable of real-time pattern detection and decision making.

Increased interest in this area has spurred the development of data-processing techniques that specialize in handling streaming data. Good surveys of research efforts in stream processing can be found in (Golab & Ozsu, 2003; Gaber et al., 2005; Garofalakis et al., 2002). One focus of exploration is on one-pass methods for constructing concise, but reasonably accurate, summaries/synopses of the data. These summaries could potentially provide approximate answers to user queries while guaranteeing reasonable quality of approximation. Popular methods for generating approximations include random samples, histograms, sketches, and wavelets. More complex techniques have also been considered – for example, one-pass data mining algorithms could extract information from the data and can later be used to describe and characterize (e.g., clustering (Guha et al., 2003), Singular Value Decomposion (Brand, 2006)) or monitor (e.g., classification (Hulten et al., 2001; Aggarwal et al, 2004)) the data stream. Recent developments in commercial RDBMSs towards tight integration of analytics with data warehousing and query processing offer new opportunities for leveraging the summary-based approach to achieve decreased storage, insightful analysis, and efficient monitoring of the data stream. This chapter will explore these topics.

Other efforts have focused on stream query languages, streaming operators, and continuous query processing and optimization (Babu & Widom, 2001; Babcock et al., 2002). While these efforts are very instructive, their effect on real-world streaming applications is very limited. Applications cannot easily leverage such new advances and are faced with the constraints of existing infrastructure. In the context of real world application implementation, the most noteworthy recent development is the introduction of a continuous query capability in the Oracle RDBMS (Witkowski et al., 2007). This feature allows monitoring of real-time changes to a query as the result of changes to the underlying tables referenced in the query. This feature will be illustrated later in the *ETL* section.

Analytic Data Warehousing for Sensor Data

An Analytic Data Warehouse (ADW) is a data warehouse with analytic capabilities, including integrated statistical functions, data mining, and OLAP. Incorporating analytics into a data warehouse provides many advantages, such as: performance, scalability, data integrity, and security over performing data analysis outside the data warehouse. The rest of this section describes how an ADW can be used as a platform for sensor data applications.

An ADW-based architecture (Figure 1) for sensor data applications typically includes the following major components:

- Sensor arrays
- Extraction, transformation and load (ETL) of sensor data
- Centralized data warehousing
- Analytic module
- Alerts, visualization, and reports

Sensor arrays produce streams of data that need to be integrated, analyzed, and sometimes monitored. Sensor data are processed and loaded into a centralized data repository. Other relevant features can also be generated during the ETL stage (e.g., spatial features for satellite imagery data). All the required analytical methods are integral to the database infrastructure – no data movement is required. The stored data can be used for data mining model generation. The generated models can undergo scheduled distribution and deployment across different database instances. These models can then be used for monitoring the incoming sensor data. The database can issue alerts when anomalous activity is detected. The models and the stored sensor data can be further investigated using database reporting and analysis tools (e.g., via OLAP).

The key aspect to the described data flow is that processing is entirely contained within the database. With the exception of the sensor array, all other components can be found in modern RDBMSs. The following sections will describe each component.

Sensor Arrays

Sensor arrays can contain a single type or multiple types of sensors. Sensors can be either local or distributed, using a remote connection to the system. Data from sensor arrays can be also complemented with other domain relevant data. The examples in this chapter are based on two types of sensors – hyperspectral sensor arrays used in satellite imagery and network (cyber) sensor arrays that monitor computer network traffic. Network sensors typically filter and reassemble TCP/IP packets in order to extract high-level connection features.

In a distributed architecture, lightweight sensors are often preferred since they are the only component that needs to run on the remote system. An ADW-based system also favors the lightweight sensor approach since all computa-

Figure 1. ADW-based architecture for sensor data applications

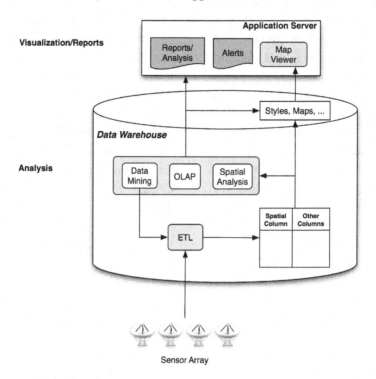

tionally intensive tasks (feature extraction, model generation, and detection) take place in the RDBMS. The ADW approach does not imply that the system operates in a non-distributed fashion with a single point of failure and computational bottleneck. On the contrary, modern databases emphasize distributed architecture capabilities, including sophisticated failover mechanisms to ensure uninterruptible system uptime.

ETL

Given the unbounded nature of raw stream data, the feature extraction stage is of critical importance for making data storage and subsequent data mining feasible. Native SQL and user-defined functions offer a high degree of flexibility and efficiency when extracting key pieces of information from a data stream. Useful SQL capabilities include math, aggregate, and analytic functions. For example, windowing functions can be used

to compute aggregates over time intervals or sets of rows. The following code snippet shows a windowing analytic function that computes the number of http connections to a given host during the last 5 seconds:

```
SELECT count(*) OVER
    (ORDER BY time_stamp
    RANGE INTERVAL '5'
    SECOND PRECEDING) as http_cnt
FROM connection_data
WHERE dest_host = 'myhost'
AND service = 'http';
```

In the context of ETL on streams, it can be very relevant to monitor a change in a measurement rather than the measurement itself. For example, Oracle's Continuous Query (CQ) feature allows monitoring the difference in a query result over time and alerting the user when changes occur. The following code sample shows a continuous query that will produce a notification when the current sensor measurement exceeds the previous value by a certain threshold. The output of this query is appended to a table. Alternatively, the result can be posted in a notification queue. The queue alert mechanism will be discussed later in the *Alerts* section.

```
CREATE CONTINUOUS QUERY sensor_change_cq
    COMPUTE ON COMMIT
    DESTINATION change_table
SELECT location, time_stamp, measurement
FROM (SELECT location, time_stamp, mea-
surement,
    LAG(measurement, 1) OVER (PARTITION
BY location
        ORDER BY time) prev_measurement
    FROM sensor_measurements)
WHERE prev_measurement - measurement >
100;
```

The ETL stage can also include more sophisticated data mining approaches that extract information from the data stream. It is possible to embed, through SQL queries, mining models that monitor, filter, summarize, or de-noise the sensor data. More details on this integration are provided in the *Analytical Module* section.

Data Warehousing

Using an RDBMS as a centralized data repository offers significant flexibility in terms of data manipulation. Without replicating data, database views or materialized views can capture different slices of the data (e.g., data over a given time interval, data for a specific location). Such views can be used directly for mining model generation and data analysis. A good overview of data warehousing in the context of analytical applications can be found in (Stackowiak et al., 2007).

OLAP technology provides an efficient means for ad hoc data analysis. In the context of data streams, advanced OLAP methods can leverage

data compression to further improve efficiency (Margaritis et al., 2001). A new feature in Oracle RDBMS allows that materialized views be defined over OLAP cubes (Smith & Hobbs, 2008). This can significantly improve materialized views management and performance.

Another technology that is very relevant to some sensor network applications is spatial analysis. Tight synergy with core database functionality is again highly desirable. Examples of the interoperability between data mining and spatial components in the context of processing satellite sensor data are given in the satellite imagery case study.

Analytic Module

This module carries out analysis of the data stream that can be implemented with machine learning/data mining algorithms. It can also perform OLAP and spatial analysis. The output of the analysis can be captured in reports that are displayed in the reporting module. The data mining component combines automated model generation and distribution, as well as real-time and offline monitoring.

Data mining techniques that have been used in the context of sensor data include maximum likelihood classifiers, neural networks, decision trees, and support vector machines (SVM). Modern RDBMSs offer, to different degrees, robust and effective implementations of data mining algorithms that are fully integrated with core database functionality. The incorporation of data mining eliminates the necessity of data export outside the database, thus enhancing data security. Since the model representation is native to the database, no special treatment is required to ensure interoperability.

In order to programmatically operationalize the model generation process, data mining capabilities can be accessed via APIs (e.g., JDM standard Java API, PL/SQL data mining API in Oracle RDBMS). Specialized GUIs can be also easily developed as entry points, building upon the available API infrastructure (e.g., Oracle Data Miner, SPSS Clementine, Inforsense). Such GUI tools enable interactive data exploration and initial model investigation.

In ADW, model distribution is greatly simplified since models are not only stored in the database but are also executed in it as well. Models can be periodically updated by scheduling automatic builds. The newly generated models can then be automatically deployed to multiple database instances.

Sensor data applications implemented within a modern RDBMS framework can transparently leverage the grid computing infrastructure available for the database. Grids make possible pooling of available servers, storage, and networks into a flexible on-demand computing resource capable of achieving scalability and high availability. An example of a grid computing infrastructure is Oracle's Real Application Clusters (RAC) architecture. RAC allows a single Oracle database to be accessed by concurrent database instances running across a group of independent servers (nodes).

An application built on top of RAC can successfully leverage server load balancing (distribution of workload across nodes) and client load balancing (distribution of new connections among nodes). Transparent application and connection failover mechanisms are also available, thus ensuring uninterruptible system uptime. A grid-enabled RDBMS system needs to be seamlessly integrated with a scheduling infrastructure. Such an infrastructure enables scheduling, management, and monitoring of model build and deployment jobs. An example of a scheduling system meeting the above requirements is Oracle Scheduler.

Detection can be performed either real-time or offline. Real-time detection (and alarm generation) is essential for the instrumentation of many sensor data applications. In the context of ADW, an effective real-time identification mechanism can be implemented by leveraging the parallelism and

scalability of a modern RDBMS. This removes the need for a system developer to design and implement such infrastructure. Inside the database, detection can be tightly integrated, through SQL, into the ETL process itself. The benefits of integrating analytics within core database query language are illustrated with examples using Oracle's PREDICTION SQL operator.

Audit network data are classified as anomalous or not by a data mining model. The model scoring is part of a database INSERT statement:

```
INSERT INTO predictions
    (id, prediction)
VALUES(10001,
       PREDICTION(
       model1 USING
       'tcp' AS protocol_type,
       'ftp' AS service,
       ...
       'SF' AS flag,
       27 AS duration));
```

In addition to real-time detection, it is often useful to perform offline scoring of stored sensor data. This provides an assessment of model performance, characterizes the type network activity, and assists in the discovery of unusual patterns. Having detection cast as an SQL operator allows other database features to be leveraged. The following SQL code snippet shows how to create a functional index on the probability of a case being an anomaly:

```
CREATE INDEX anomaly_prob_idx
   ON audit_data
      (PREDICTION_PROBABILITY(
      model2,
      0
      USING *));
```

The functional index optimizes query performance on the audit data table when filtering or sorting on anomaly probability is desired.

The following query, which returns all cases in audit_data with probability of being an anomaly greater than 0.5, will have better performance if the anomaly_prob_idx index is used:

```
SELECT *
FROM audit_data
WHERE PREDICTION_PROBABILITY(
    model2,
    0
    USING *) > 0.5;
```

The SQL PREDICTION operators also allow for the combination of multiple data mining models that are scored either serially or in parallel, thus enabling hierarchical and cooperative detection approaches. Models can be built on different types of sensor data or different timeframes, can have different scope (localized vs. global detectors), and can use different data mining algorithms. The next example shows a hypothetical use case where two models perform parallel cooperative detection. The query returns all cases where either model1 or model2 indicate an anomaly with probability higher than 0.4:

```
SELECT *
FROM audit_data
WHERE PREDICTION_PROBABILITY(
    model1,
    'anomaly'
    USING *) > 0.4
OR PREDICTION_PROBABILITY(
    model2,
    'anomaly'
    USING *) > 0.4;
```

Alternatively, one can consider sequential (or "pipelined") cooperative detection, where the results of one model influence the predictions of another model. In this case, when the model2 classifies a case as an anomaly with probability greater than 0.5, model1 will attempt to identify the type of anomaly:

```
SELECT id, PREDICTION(
    model1 USING *)
FROM audit_data
WHERE PREDICTION_PROBABILITY(
    Model2,
    0 USING *) > 0.5;
```

Alerts

If a mining model monitoring a data stream detects certain types of events (e.g., anomalies), it needs to generate an alarm, notify interested parties, and possibly initiate a response. Database triggers are powerful mechanisms that initiate a predefined action when a specific condition is met. The following statement shows the trigger definition where an anomaly results in posting of a message to a queue for handling anomaly notifications:

```
CREATE TRIGGER alert_trg
    BEFORE INSERT ON
        predictions
    FOR EACH ROW
WHEN (new.prediction <> 'normal')
BEGIN
DBMS_AQ.ENQUEUE(
    'anomaly_notify_queue',
    ...);
END;
```

Distributed systems that operate in a loosely-coupled and autonomous fashion and require operational immunity from network failures need the ability to handle asynchronous communications. A publish-subscribe messaging infrastructure provides such a framework where detectors act as publishers that send alerts without explicitly specifying recipients. The subscribers (users/applications) receive only messages that they have registered an interest in. The decoupling of senders and receivers is achieved via a queuing mechanism. Each queue represents a subject or a channel. An example of such infrastructure can be found in Oracle's DBMS_AQ PL/SQL package. This mechanism is also leveraged by the Oracle's continuous query feature to communicate query results.

Active publication of information to end-users in an event-driven manner complements the more traditional pull-oriented approaches to accessing information that are also available. Alerts can be also delivered via a diverse range of channels (e.g., e-mails, cell phone messages).

Visualization and Reports

Using a database as the platform for sensor data applications facilitates the generation of data analysis results and reports. Collected sensor data, predictions, as well as model contents, can be inspected either directly using queries or via higher level reporting and visualization tools (e.g., Discoverer, Oracle Reports). Analysis results can also be leveraged by a large set of report and interface development tools. For example, web portal creation tools offer infrastructure for the development of application 'dashboards'. This allows circumvention of a lengthy application development process and provides standardized and easily customized report generation and delivery mechanisms (e.g., Oracle Business Intelligence Enterprise Edition (OBIEE), Business Objects Crystal Reports, IBM Cognos Business Intelligence).

Case Studies

Two case studies are described. The first one uses hyperspectral satellite image data. It illustrates the database mining capabilities and the synergy of data mining and spatial analysis. The second case study uses network intrusion data. It describes an online network monitoring application that takes advantage of analytic components as well as of the scheduling, alerting, and visualization infrastructure available in the database.

Satellite Imagery

This example illustrates ADW's integrated approach in the context of hyperspectral satellite image processing and analysis. The focus is on the data mining and spatial analysis modules in Figure 1 and their interoperability. The spectral bands for an image are stored in a database table. A spatial column encodes the coordinate information. Data mining and spatial analysis can be performed directly on this table within the RDBMS. Unsupervised (clustering) and supervised (classification) mining approaches are described. The data mining results are further combined with spatial functionality.

The purpose of this application is multifold:

- Create annotated maps that faithfully represent the underlying features in the image;
- Achieve significant reduction in data storage (data annotations create by clustering or classification models can be stored instead of the original hyperspectral data);
- Use developed models to monitor evolving land usage and urbanization or to estimate crop yields.

ADW functionality can be leveraged for rapid prototyping in data analysis and modeling with the ultimate goal of fast delivery of reliable applications.

The hyperspectral image data has 224 spectral bands and was acquired from the AVIRIS sensor (ftp://ftp.enc.purdue.edu/biehl/MultiSpec/92AV3C). The image was captured over a rural area in the northern part of Indiana (Figure 2a). Most of the scene represents agricultural land and wooded areas. The image also includes some buildings and road infrastructure. A field survey map is available and has been used as a ground truth reference in previous experimental studies. The survey information is high-level (polygon rather than pixel based) and incomplete – it covers only about 50% of the

Figure 2. Hyperspectral image: (a) original data using bands 35, 130, and 194; (b) O-Cluster results.

(a)

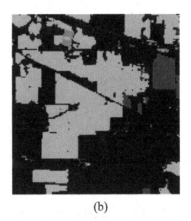

(b)

image and omits some prominent features. The current set of experiments uses this reference information for evaluation of the unsupervised method results and as a source of labels for the supervised classification method.

Typically, hyperspectral data can be subjected to a number of pre-processing steps (e.g., contrast enhancement, texture processing). In addition, to avoid the "curse of dimensionality" associated with mining high-dimensional spaces, dimensionality reduction techniques are often employed. Dimensionality reduction can also be used to reduce storage requirements. Such transformations and preprocessing are either built-in or can be easily integrated within an RDBMS framework (e.g., Oracle's PL/SQL procedures, table functions). Among the built-in dimensionality reduction techniques, the most noteworthy one is Singular Value Decomposition that is available through Oracle's UTL_NLA package. Here, however, the raw hyperspectral sensor data is used directly. The data mining algorithms presented here are well suited for high-dimensional spaces.

Clustering

Orthogonal partitioning clustering (O-Cluster) is a density-based method that was developed to handle large high-dimensional databases (Mile-

nova & Campos, 2002). It is a fast (one-pass) algorithm that is well suited for handling large volumes of continuous sensor data. O-Cluster creates a binary tree hierarchy. The topology of the hierarchy, along with its splitting predicates, can be used to gain insights into the clustering solution. The number of leaf clusters is determined automatically.

Figure 2b shows an annotated map based on the results produced by O-Cluster. The algorithm identified six distinct clusters. Each of these clusters is assigned a different color in the map. The individual clusters capture distinctive features in the image. Further insight into the results and their quality can be gained by exploring the model. Figure 3 depicts the hierarchical clustering tree.

Every branching node contains split predicate information (band number and the split condition). The selected bands were found to be most discriminative in the identification of dense regions with low overlap. O-Cluster's model transparency can be used to gain insight into the underlying structure of the data and can assist feature selection. The six leaf nodes in Figure 3 map to corresponding areas in Figure 2b (the same color coding is used in both figures). To assess the purity of the clusters and the quality of the results, each leaf node was labeled using ground truth information. Only pixels that were part of the survey map (~50%

Figure 3. O-Cluster model

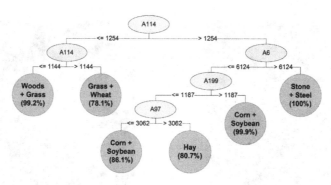

of the image) were used. The percentage value within each leaf indicates the fraction of survey map pixels within a given cluster that belong to the chosen label. Even though the survey information is incomplete and does not fully reflect all relevant features, the resulting clusters have reasonably high purity. The transparent nature of the hierarchy also allows the extraction of simple descriptive rules, such as:

If Band114 > 1254 and Band6 > 6124 then the pixel belongs to a stone/steel structure.

The maps derived from the clustering model are created in a fully unsupervised manner. Such high-level maps can be used to identify the most salient image features. They represent a good starting point that can be further improved upon by human experts or supervised data mining algorithms.

Classification

In the present example, a linear SVM model was built to produce an annotated map of the area. Table 1 shows the confusion matrix on the test dataset. The rows represent actual ground truth values while the columns represent the predictions made by the model. The overall accuracy is 85.7%. The main source of error is confusion between corn and soybean fields. There are also some misclassifications between the woods and building/grass/tree/drives categories.

Figure 4a shows a map based on SVM's predictions. While the overall map quality is reasonable and the different types of crops are easily identifiable, the noise due to the corn-soybean and woods-buildings/grass/ tree/drives errors is evident. Various approaches can smooth the predictions based on neighborhood information – one possibility would be to provide the classifier with spatial input (e.g., coordinate values or predictor averages). Here we chose to post-process the classifier predictions using Oracle spatial functionality.

The predictions for each class can be treated as thematic layers – each layer represents a 2D array of binary values where each bit corresponds to a pixel in the image. If a bit is set, a positive prediction within this thematic layer will be made. A simple smoothing strategy would be to set or unset a bit based on the values of its immediate neighbors. Here, a bit is set only if the SVM classifier made positive predictions for 2/3 of the neighborhood. Otherwise the bit remains unset.

Figure 4b illustrates the results of such smoothing. The amount of noise is greatly reduced and the dominating features of the scene are well delineated – the 'smoothed' predictions have 90% accuracy. However, due to the stringent 2/3 smoothing criterion some of the transition areas result in no predictions and some localized

Table 1. SVM confusion matrix; target classes: corn (C), grass (G), hay (H), soybeans (S), wheat (W), woods (D), buildings/grass/tree/drives (B), stone/steel (SS).

	C	G	H	S	W	D	B	SS
C	538	2	0	260	0	0	0	0
G	10	393	7	7	0	2	0	0
H	0	0	170	0	0	0	0	0
S	96	8	0	1237	0	0	0	1
W	0	0	0	0	61	0	0	0
D	2	12	0	0	0	414	4	0
B	0	23	0	2	3	39	52	0
SS	2	0	0	1	0	0	0	26

features (e.g., correct predictions on small man-made structures) are smoothed out. The level of detail is usually application specific and can be adjusted accordingly. The smoothed results were computed using spatial queries. A spatial index was built on the table column containing coordinate data. This index can improve the performance of nearest-neighbor queries. The k-nearest neighbors of a point can be retrieved using the Oracle SDO_NN operator. Alternatively, SDO_WITHIN_DISTANCE retrieves all points within a given radius. The second approach is preferable here as it handles the edge effects.

Spatial Queries

To further illustrate the flexibility and expressive power of such an integrated approach, we include a sample SQL query that combines data mining and spatial analytic features to perform the following task: Within the left upper quadrant of the image, find the soybean fields that are no further than 150m from stone-steel structures.

```
1 WITH
2   quadrant AS(
3     SELECT *
```

Figure 4. SVM predictions: (a) raw classifier output; (b) smoothed classifier output

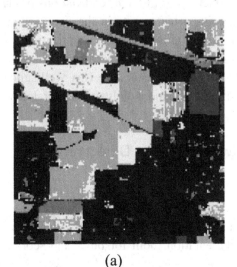

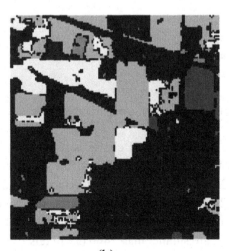

(a)   (b)

```
4     FROM hyperspectral_data a
5     WHERE a.pixel.sdo_point.x < 1460
6     AND a.pixel.sdo_point.y < 1460),
7  soybean AS (
8     SELECT a.pixel.sdo_point.x x,
9     a.pixel.sdo_point.y y
10    FROM quadrant a
11    WHERE PREDICTION(
12       SVM_model
13       using *) = 'soybean'
14    AND PREDICTION_PROBABILITY(
15       SVM_model,
16       'soybean'
17       using *) > 0.5),
18 stone_steel AS (
19    SELECT pixel FROM quadrant
20    WHERE PREDICTION(
21       SVM_model
22       using *) = 'stone-steel'
23    AND PREDICTION_PROBABILITY(
24       SVM_model,
25       'stone-steel'
26       using *) > 0.5),
27 stone_steel_150_radius AS(
28    SELECT DISTINCT
29       a.pixel.sdo_point.x x,
30       a.pixel.sdo_point.y y
31    FROM quadrant a, stone_steel b
32    WHERE
33       SDO_WITHIN_DISTANCE(
34       a.pixel, b.pixel,
35       'distance=150')='TRUE')
36 SELECT a.x, a.y
37 FROM soybean a,
38       stone_steel_150_radius b
39 WHERE a.x=b.x AND a.y=b.y;
```

The optional WITH clause is used here for clarity and improved readability of the subqueries (lines 2-35). The first subquery, named quadrant (lines 2-6), restricts the search to the left upper quadrant. The spatial coordinates in the left upper quadrant are smaller than 1460 (73 pixels x 20m raster grid). All subsequent subqueries run against this restricted set of points. The second subquery, soybean (lines 7-17), identifies the coordinates of the pixels that were classified by SVM as soybean with probability greater than 0.5. The query uses the PREDICTION and PREDICTION_PROBABILITY SQL operators. Predictions are made using the SVM model described earlier. The third subquery, stone_steel (lines 18-26), selects the pixels that were classified as stone-steel structures with probability greater than 0.5. Unlike the previous subquery that retrieved the pixel coordinates, here we return pixel spatial objects. These objects will be used in the final subquery, stone_steel_radius_150 (lines 27-35). This subquery retrieves the coordinates of all pixels that fall within 150m of a stone-steel structure. It makes use of the SDO_WITHIN_DISTANCE operator. This operator leverages the spatial index on the pixel column to efficiently retrieve all objects within the specified radius. Since the operation is equivalent to pair-wise comparisons between the two groups of pixels, the DISTINCT clause limits the output to unique coordinate pairs. The main query (lines 36-39) returns the coordinates of the pixels from the upper left quadrant that are soybean fields and lie within the 150m zone.

This example highlights the ease of combining individual ADW components and the wealth of expression that can be achieved via SQL queries. The terseness of the SQL code and its modularity are important assets in the development of complex and mission critical applications. Even though the example here performs a batch operation, the SQL operators are very well suited for real-time applications and can be integrated within the ETL process.

Network Intrusion Detection

The second case study involves intrusion detection in computer networks. In this domain, it is essential to have a highly scalable mining algorithm implementation, both for build and scoring. In our prototype, we simulated a network environment by streaming previously collected network activity data. This dataset was originally created by DARPA and later used in the KDD'99 Cup (http://www-cse.ucsd.edu/users/elkan/clresults.html). The intrusion detection dataset includes examples of normal behavior and four high-level groups of attacks - probing, denial of service (dos), unauthorized access to local superuser/root (u2r), and unauthorized access from a remote machine (r2l). These four groups summarize 22 subclasses of attacks. The test dataset includes 37 subclasses of attacks under the same four generic categories. The test data is used to simulate the performance in detection mode.

An SVM model was used to classify the network activity as normal or as belonging to one of the four types of attack. The overall accuracy of the system was 92.1%. Additionally, a one-class SVM model was used to identify the network activity as normal or anomalous. Since anomaly detection does not rely on instances of previous attacks, the one-class model was built on the subset of normal cases in the DARPA dataset. On the test dataset, the model had excellent discrimination with an ROC area of 0.989. Sliding the probability decision threshold allows a trade-off between the rate of true positives and rate of false alarms – for example, in this model a true positive rate of 96% corresponded to a false alarm rate of 5%.

One benefit from having in-database analytics is that no data movement to an external analytic server is required. An example of efficient implementation of analytics in the database is Oracle's SVM (Milenova et al., 2005). Figure 5a depicts the build scalability of a linear SVM classification model with increasing number of records. The datasets of smaller size represent random samples of the original intrusion detection data. The fast training times allow for frequent model rebuilds. Figure 5b illustrates the PREDICTION operator's scalability. The scalability results were generated using a linear SVM model built on 500,000 connection records. The same hardware was used as in the build timing tests. Tests were run on a machine with the following hardware and software specifications: single 3GHz i86 processor, 2GB RAM memory, and Red Hat enterprise Linux OS 3.0.

In our application prototype periodic model updates are scheduled as new data is accumulated. A model rebuild is also triggered when the performance accuracy falls below a predefined level. Also, as part of the intrusion detection prototype, an OBIEE-based dashboard was created. The Intrusion Detection Center (IDC) dashboard monitors the state of the network and displays relevant information. One of the tasks of the IDC dashboard is to monitor the alert queue and dis-

Figure 5. Build and scoring scalability: (a) build; (b) prediction.

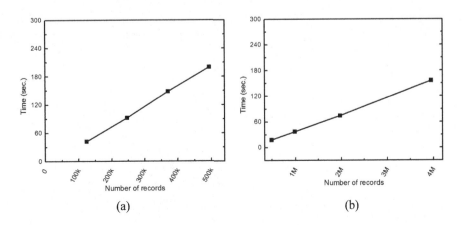

(a)　　　　　(b)

play alert notifications. Figure 6 shows a screen shot of the IDC dashboard. The alert notification information includes the type of attack (based on the model prediction) and some of the important connection details.

The dashboard leverages the tools available in OBIEE to instrument a network activity reporting and analysis mechanism. On the main page (Figure 6), users can monitor:

- Number of intrusion attempts during the past 24 hours (top left panel);
- Breakdown of the network activity into normal and specific types of attack (bottom left panel);
- Log of recent alerts (center panel);
- Detector error rate over the last 7 days (top panel).

On the details page (Figure 7), users can review historic data. When a date is selected in the top left panel, the graphs are updated with the corresponding network activity information. The bottom left panel displays the breakdown of network activity into normal and specific types of attack for the selected date. The middle and right panels show the distribution of connection activity over different types of protocol and service.

The dials indicate how compromised a service or protocol is. Clicking on any of the rows in these panels produces a breakdown by attack type of the respective protocol or service.

FUTURE TRENDS

Contemporary RDBMSs offer a wide variety of useful functionality in the context of sensor network applications. Every new RDBMS release has also increased the number of integrated features for data reduction and analytics. However, these data reduction and analytics features can be further enhanced by the addition of dedicated analytic stream processing functionality. Enabling incremental mining model builds against streaming of data would simplify the model maintenance aspects of a deployed application. Another important trend is the addition of features for managing large number of data mining models in RDBMSs. These include: scheduling of model building and deployment, model versioning, and auditing.

Semantic Web Technologies have growing relevance to sensor network analysis. Usage includes data and content integration across heterogeneous sources, probabilistic reasoning, and inference. Commercial databases have already

Figure 6. Intrusion Detection Center dashboard.

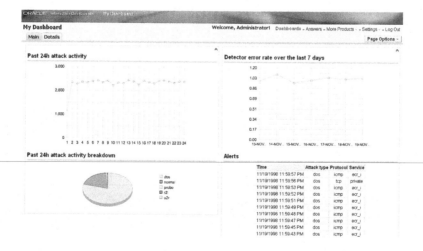

Figure 7. Intrusion Detection Center dashboard detail statistics.

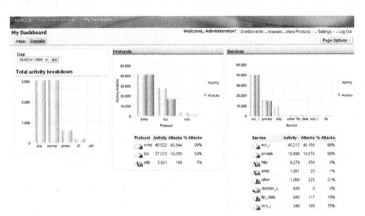

provided some support for these technologies. For example, the Oracle RDBMS has introduced native RDF/RDFS/OWL support, enabling application developers to benefit from a scalable and integrated platform (Lopez & Das, 2007). Based on a graph data model, RDF triples are persisted, indexed, and queried, similar to other object-relational data types.

Another trend that can have a large impact on sensor network analysis is the increasing support of spatio-temporal data in commercial RDBMS systems. Recent work by Kothiri et al. (2008) demonstrates how existing database technology can be leveraged to support spatio-temporal data processing. It uses the Oracle extensibility framework to implement R-tree indexes for spatio-temporal data.

CONCLUSION

A database-centric platform for building sensor data applications offers many advantages. Integrated database analytics enable effective data integration, in-depth data analysis, and real-time online monitoring capabilities. Additionally, the RDBMS framework offers applications inherent security, scalability, and high availability. Current trends in RDBMSs are moving towards providing all key components for delivering comprehensive state-of-the-art analytic applications supporting streaming sensor data. Major database vendors already incorporate key functionalities, including strong data mining, OLAP, and spatial analysis features. As illustrated above, these features provide great flexibility and analytic power. By leveraging an existing RDBMS-based technology stack, a full-fledged sensor data application can be developed in a reasonably short time and at low development cost.

REFERENCES

Aggarwal, C. C., Han, J., Wang, J., & Yu, P. S. (2004). On demand classification of data streams. In *Proceedings of the Tenth ACM SIGKDD international Conference on Knowledge Discovery and Data Mining* (pp. 503-508).

Babcock, B., Babu, S., Datar, M., Motwani, R., & Widom, J. (2002). Models and issues in data stream systems. In *Proceedings of the Twenty-First ACM SIGMOD-SIGACT-SIGART Symposium on Principles of Database Systems*.

Babu, S., & Widom, J. (2001). Continuous queries over data streams. *SIGMOD Record*, *30*(3), 109–120. doi:10.1145/603867.603884

Brand, M. (2006). Fast low-rank modifications of the thin singular value decomposition. *Linear Algebra and Its Applications, 415*(1), 20–30. doi:10.1016/j.laa.2005.07.021

Campos, M. M., & Milenova, B. L. (2005). Creation and deployment of data mining-based intrusion detection systems in Oracle Database l0g. *In Proceedings of the 2005 International Conference on Machine Learning and Applications* (pp. 105-112).

Gaber, M. M., Zaslavsky, A., & Krishnaswamy, S. (2005). Mining data streams: A review. *SIGMOD Record, 34*(2), 18–26. doi:10.1145/1083784.1083789

Garofalakis, M., Gehrke, J., & Rastogi, R. (2002). Querying and mining data streams: You only get one look a tutorial. In *Proceedings of the 2002 ACM SIGMOD International Conference on Management of Data.*

Golab, L., & Özsu, M. T. (2003). Issues in data stream management. *SIGMOD Record, 32*(2), 5–14. doi:10.1145/776985.776986

Guha, S., Meyerson, A., Mishra, N., Motwani, R., & O'Callaghan, L. (2003). Clustering data streams: Theory and practice. *IEEE Transactions on Knowledge and Data Engineering, 15*(3), 515–528. doi:10.1109/TKDE.2003.1198387

Hulten, G., Spencer, L., & Domingos, P. (2001). Mining time-changing data streams. In *Proceedings of the Seventh ACM SIGKDD international Conference on Knowledge Discovery and Data Mining* (pp. 97-106).

Kothuri, R., Hanckel, R., & Yalamanchi, A. (2008). Using Oracle Extensibility Framework for Supporting Temporal and Spatio-Temporal Applications. In *Proceedings of the fifteenth International Symposium on Temporal Representation and Reasoning* (pp. 15-18).

Lopez, X., & Das, S. (2007). *Semantic data integration for the enterprise.* Retrieved from http://www.oracle.com/technology/tech/semantic_technologies/pdf/semantic11g_dataint_twp.pdf

Margaritis, D., Faloutsos, C., & Thrun, S. (2001). NetCube: A scalable tool for fast data mining and compression. In *Proceedings of the 27th International Conference on Very Large Data Bases* (pp. 311-320).

Milenova, B. L., & Campos, M. M. (2002). O-Cluster: Scalable clustering of large high-dimensional data sets. In *Proceedings of the IEEE International Conference on Data Mining* (pp. 290-297).

Milenova, B. L., & Campos, M. M. (2005). Mining high-dimensional data for information fusion: A database-centric approach. In *Proceedings of the 2005 International Conference on Information Fusion.*

Milenova, B. L., Yarmus, J., & Campos, M. M. (2005). SVM in Oracle Database 10g: Removing the barriers to widespread adoption of support vector machines. In *Proceedings of the 31st International Conference on Very Large Data Bases* (pp. 1152-1163).

Smith, P., & Hobbs, L. (2008). *Comparing materialized views and analytic workspaces in Oracle Database 11g.* Retrieved from http://www.oracle.com/technology/products/bi/db/11g/pdf/comparision_aw_mv_11g_twp.pdf

Stackowiak, R., Rayman, J., & Greenwald, R. (2007). *Oracle data warehousing and business intelligence solutions.* Indianapolis, IN: Wiley Publishing Inc.

Witkowski, A., Bellamkonda, S., Li, H., Liang, V., Sheng, L., Smith, W., et al. (2007). Continuous queries in Oracle. In *Proceedings of the 33rd International Conference on Very Large Data Bases* (pp. 1173-1184).

Chapter 2

Improving OLAP Analysis of Multidimensional Data Streams via Efficient Compression Techniques

Alfredo Cuzzocrea
ICAR-CNR, Italy and University of Calabria, Italy

Filippo Furfaro
University of Calabria, Italy

Elio Masciari
ICAR-CNR, Italy

Domenico Saccà
University of Calabria, Italy

ABSTRACT

Sensor networks represent a leading case of data stream sources coming from real-life application scenarios. Sensors are non-reactive elements which are used to monitor real-life phenomena, such as live weather conditions, network traffic etc. They are usually organized into networks where their readings are transmitted using low level protocols. A relevant problem in dealing with data streams consists in the fact that they are intrinsically multi-level and multidimensional in nature, so that they require to be analyzed by means of a multi-level and a multi-resolution (analysis) model accordingly, like OLAP, beyond traditional solutions provided by primitive SQL-based DBMS interfaces. Despite this, a significant issue in dealing with OLAP is represented by the so-called curse of dimensionality problem, which consists in the fact that, when the number of dimensions of the target data cube increases, multidimensional data cannot be accessed and queried efficiently, due to their enormous size. Starting from this practical evidence, several data cube compression techniques have been proposed during the last years, with alternate fortune. Briefly, the main idea of these techniques consists in computing compressed representations of input data cubes in order to evaluate time-consuming OLAP queries against them, thus

DOI: 10.4018/978-1-60566-328-9.ch002

obtaining approximate answers. Similarly to static data, approximate query answering techniques can be applied to streaming data, in order to improve OLAP analysis of such kind of data. Unfortunately, the data cube compression computational paradigm gets worse when OLAP aggregations are computed on top of a continuously flooding multidimensional data stream. In order to efficiently deal with the curse of dimensionality problem and achieve high efficiency in processing and querying multidimensional data streams, thus efficiently supporting OLAP analysis of such kind of data, in this chapter we propose novel compression techniques over data stream readings that are materialized for OLAP purposes. This allows us to tame the unbounded nature of streaming data, thus dealing with bounded memory issues exposed by conventional DBMS tools. Overall, in this chapter we introduce an innovative, complex technique for efficiently supporting OLAP analysis of multidimensional data streams.

INTRODUCTION

Data Stream Management Systems (DSMS) have captured the attention of large communities of both academic and industrial researchers. *Data streams* pose novel and previously-unrecognized research challenges due to the fact that traditional DBMS (Henzinger, Raghavan & Rajagopalan, 1998, Cortes, Fisher, Pregibon, Rogers & Smith, 2000), which are based on an exact and detailed representation of information, are not suitable in this context, as the whole information carried by streaming data cannot be stored within a bounded storage space (Babcock, Babu, Datar, Motwani & Widom, 2002). From this practical evidence, a plethora of recent research initiatives have been focused on the problem of efficiently representing, querying and mining data streams (Babu & Widom, 2001,Yao & Gehrke, 2003, Acharya, Gibbons, Poosala, & Ramaswamy, 1999, Avnur & Hellerstein, 2000).

Sensor networks (Bonnet, Gehrke & Seshadri, 2000, Bonnet, Gehrke & Seshadri, 2001) represent a leading case of data stream sources coming from real-life application scenarios. Sensors are non-reactive elements which are used to monitor real-life phenomena, such as live weather conditions, network traffic etc. They are usually organized into networks where their *readings* are transmitted using low level protocols (Gehrke & Madden, 2004, Madden & Franklin, 2002, Madden, Franklin, &

Hellerstein, 2002, Madden, Szewczyk, Franklin & Culler, 2002). Under a broader vision, sensor networks represent a non-traditional source of information, as readings generated by sensors flow continuously, leading to an infinite, memory-unbounded stream of data.

A relevant problem in dealing with data streams consists in the fact that *they are intrinsically multi-level and multidimensional in nature* (Cai, Clutterx, Papex, Han, Welgex & Auvilx, 2004; Han, Chen, Dong, Pei, Wah, Wang & Cai, 2005), hence they require to be analyzed by means of a multi-level and a multi-resolution (analysis) model accordingly. Furthermore, it is a matter of fact to note that enormous data flows generated by a collection of stream sources like sensors *naturally* require to be processed by means of advanced analysis/mining models, beyond traditional solutions provided by primitive SQL-based DBMS interfaces. Consider, for instance, the application scenario drawn by a *Supply Chain Management System* (SCMS) (Gonzalez, Han, Li & Klabjan, 2006), which can be intended as a sort of sensor network distributed over a wide geographical area. Here, due to the characteristics of the particular application domain, data embedded in streams generated by supply providers (i.e., the sensors, in this case) are *intrinsically* multidimensional, and, in addition to this, correlated in nature. In more detail, multidimensionality of data is dictated by the fact that, in a typical supply chain scenario, the domain model is captured by several attributes like store region, warehouse region, location, product category, and so forth. Here, *hierarchies of data* naturally arise, as real-life data produced and processed by knowledge management processes are typically organized into weak or strong hierarchical relationships (e.g., *StoreCountry → StoreRegion → Store*). Correlation of data is instead due to the fact that, for instance, stock quotations strictly depend on the actual market trend, and market prices strictly depend on the actual capability of suppliers in delivering products timely. The same happens with the monitoring of environmental parameters, in the context of environmental sensor networks. Here, geographical coordinates naturally define a multidimensional space, and, consequentially, a multidimensional data model, very often enriched by additional metadata attributes, like in *Geographical Information Systems* (GIS). For what regards correlation of data, it is a matter of fact to note that temperature, pressure, and humidity of a given geographical area are very often correlated, even highly correlated.

Conventional analysis/mining tools (e.g., DBMS-inspired) cannot carefully take into consideration these kinds of multidimensionality and correlation of real-life data, as stated in (Cai, Clutterx, Papex, Han, Welgex & Auvilx, 2004; Han, Chen, Dong, Pei, Wah, Wang & Cai, 2005), so that, if one tries to process multidimensional and correlated data streams by means of such tools, rough errors are obtained in practice, thus seriously affecting the quality of decision making processes that found on analytical results mined from streaming data.

Contrary to conventional tools, *multidimensional analysis* provided by *OnLine Analytical Processing* (OLAP) technology (Gray, Chaudhuri, Bosworth, Layman, Reichart, Venkatrao, Pellow & Pirahesh, 1997; Chaudhuri & Dayal, 1997), which has already reached a high-level of maturity, allows us to efficiently exploit data multidimensionality and correlation, in order

to improve the quality of both analysis/mining tasks and decision making. OLAP allows us to aggregate data according to (*i*) a fixed logical schema (Vassiliadis & Sellis, 1999) that can be a *star-* or a *snowflake-schema* (Colliat, 1996; Han & Kamber, 2000), and (*ii*) a given SQL aggregate operator, such as SUM, COUNT, AVG etc. The resulting data structures, called *data cubes*, which are usually materialized within multidimensional arrays (Agarwal, Agrawal, Deshpande, Gupta, Naughton, Ramakrishnan & Sarawagi, 1996; Zhao, Deshpande & Naughton, 1997), allow us to meaningfully take advantages from the amenity of querying and mining data according to a multidimensional and a multi-resolution vision of the target data domain, and from the rich availability of a wide set of OLAP operators (Han & Kamber, 2000), such as roll-up, drill-down, slice-&-dice, pivot etc, and OLAP queries, such as range- (Ho, Agrawal, Megiddo & Srikant, 1997), top-*k* (Xin, Han, Cheng & Li, 2006), and iceberg (Fang, Shivakumar, Garcia-Molina, Motwani & Ullman, 1998) queries.

Technique Overview

On the basis of these considerations, the idea of analyzing massive data streams by means of OLAP technology makes sense perfectly, and puts the foundations for novel models and computational paradigms that can be used to efficiently extract summarized, OLAP-like knowledge from data streams, thus overcoming limitations of conventional DBMS-inherited analysis/mining tools. By meaningfully designing the underlying OLAP (logical) model in dependence on the specific application domain and analysis goals, multidimensional models can efficiently provide support to intelligent tools for a wide set of real-life data-stream-based application scenarios such as weather monitoring systems, environment monitoring systems, systems for controlling telecommunication networks, network traffic monitoring systems, alerting/alarming systems in time-critical applications (Alert System, 2007), sensor network data analysis tools etc. In all such scenarios, multidimensional analysis can represent the critical "add-in" value to improve the quality of knowledge extraction processes, as most data streams are multi-level and multidimensional in nature. This specific characteristic of data streams puts the basis for an extremely variegated collection of stream data mining tools with powerful capabilities, even beyond those of conventional data mining tools running on transactional data, such as clustering (e.g., (Guha, Meyerson, Mishra, Motwani & O'Callaghan, 2003)), correlation (e.g., (Ananthakrishna, Das, Gehrke, Korn, Muthukrishnan & Srivastava, 2003)), classification (e.g, (Domingos & Hulten, 2000)), frequent item-set mining (e.g., (Manku & Motwani, 2002)), XML stream processing (Ives, Levy & Weld, 2000), and so forth.

The resulting representation/analysis model constitutes what we call *OLAP stream model*, which can be reasonably intended as a novel model for processing multidimensional data streams, and supporting multidimensional and multi-resolution analysis over data streams. (Han, Chen, Dong, Pei, Wah, Wang & Cai, 2005) asserts similar motivations we provide in our research, and proposes the so-called *stream cube* model, a multidimensional approach to analyze data streams. According to the OLAP stream model, data stream readings are collected and stored by means of the so-called *application-oriented OLAP based acquisition model*. This model allows us to capture and tame the multidimensionality of streaming data, and efficiently support OLAP over data streams. *Acquisition models and tools* (Madden & Franklin, 2002) are critical components of DSMS, as they allow us to deal with the distributed and delocalized nature of data streams, which initially are not suitable to be analyzed and mined by means of conventional tools (e.g., DBMS-inspired). Acquisition tools define the way data streams are collected and organized according to a pre-fixed

scheme, e.g. *model-driven* (Madden & Franklin, 2002), for processing and analysis purposes.

In our framework, collected data stream readings are, in turn, materialized in a *multidimensional fashion* via meaningfully exploiting the *same* OLAP logical model used during the acquisition phase. As a consequence, the latter is again the model used to query and mine the continuously flowing (multidimensional) data stream. Storing repositories of materialized data stream readings allows us to efficiently support OLAP analysis/ mining tasks over data streams. This approach is well-founded and well-motivated by the evidence stating that it is particularly hard to OLAPing and mining data streams on the fly, as clearly highlighted by recent studies (Gonzalez, Han, Li & Klabjan, 2006). From this breaking evidence, it follows that accessing and querying summarized data stream readings stored off-line by means of *aggregate queries*, also called *range queries* (Ho, Agrawal, Megiddo & Srikant, 1997), which are baseline operations of complex OLAP analysis/ mining tasks, makes sense perfectly in order to gain efficiency and performance during the execution of OLAP tasks over data streams.

A significant issue in dealing with OLAP is represented by the so-called *curse of dimensionality problem* (e.g., (Berchtold, Böhm & Kriegel, 1998; Li, Han & Gonzalez, 2004; Intanagonwiwat, Estrin, Govindan & Heidemann, 2002; Zhang, Gunopulos, Tsotras & Seeger, 2002)), which, briefly, consists in the fact that when the number of dimensions of the target data cube increases, multidimensional data cannot be accessed and queried efficiently, due to their enormous size. Starting from this practical evidence, several *data cube compression techniques* have been proposed during the last years (e.g., (Gonzalez, Han, Li & Klabjan, 2006; Deligiannakis, Kotidis & Roussopoulos, 2004, Deligiannakis, Kotidis & Roussopoulos, 2003)), with alternate fortune. Briefly, the main idea of these techniques consists in computing compressed representations (Qiao, Agrawal &

El Abbadi, 2002; Gilbert, Kotidis, Muthukrishnan & Strauss, 2001) of input data cubes in order to evaluate time-consuming OLAP queries against them, thus obtaining *approximate answers* (Dobra, Garofalakis, Gehrke & Rastogi, 2002; Ganti, Li Lee & Ramakrishnan, 2000). Despite compression introduces some approximation in the retrieved answers, it has been demonstrated (e.g., see (Cuzzocrea, 2005)) that fast and approximate answers are perfectly suitable to OLAP analysis goals, whereas exact and time-consuming answers introduce excessive computational overheads that, in general, are very often incompatible with the requirements posed by an online computation for decision making, as a very large number of tuples/data-items must be accessed in order to retrieve the desired exact answers. The above-described computational paradigm gets worse when OLAP aggregations are computed on top of a continuously flooding multidimensional data stream, as traditional DBMS-inspired query methodologies are basically transaction-oriented, i.e. their main goal is to guarantee data consistency, and they do not pay particular attention to query efficiency, neither are suitable to deal with an unbounded data stream.

In order to efficiently deal with the curse of dimensionality problem, we propose an innovative solution consisting in the so-called *OLAP dimension flattening process* (see Figure 1), which is a fundamental component of our OLAP stream model. Basically, this process consists in flattening a multidimensional data cube model onto a two-dimensional OLAP view model, whose dimensions, called *flattening dimensions*, are selected from the original dimension set in dependence on specific application requirements. The OLAP dimension flattening process is finally performed via systematically merging original hierarchies defined on dimensions of the multidimensional data stream model. The final flattening dimensions are thus equipped with *specialized hierarchies* generated in dependence on application-oriented requirements. Due to the particular application

scenario considered (e.g., data streams generated by sensor networks), one of the flattening dimensions is *always* the temporal dimension, which allows us to carefully represent and capture how data evolve over time. It is a matter of fact to note that time is always aggregated according to the natural temporal hierarchy (e.g., *Year → Quarter → Month → Day*), which is chosen in dependence on application requirements, thus representing a sort of "invariant" of our OLAP stream model. However, without loss of generality, our innovative OLAP dimension flattening process is general enough to deal with non-temporal dimensions selected as flattening dimensions as well. The OLAP dimension flattening process permits us to finally introduce a *two-dimensional OLAP view based acquisition model* for the multidimensional data stream (as a specialized instance of the above-discussed OLAP based acquisition models), and, in consequence of this, an effective *transformation* of the multidimensional data stream into a *flattened two-dimensional data stream* (see Figure 1). The latter data stream is the one used to finally populate the summarized repositories of data stream readings.

In more detail, the continuously flooding two-dimensional data stream is collected and materialized in a *collection of two-dimensional OLAP views* that evolve over time. The "flooding nature" of the so-materialized OLAP views suggest us to adopt a very efficient solution for their in-memory-representation, which indeed represents a critical issue for any DSMS. This solution consists in representing the data stream by means of a *list of quad-tree windows*, which are able to dynamically represent and store new data stream readings as time passes. Briefly, a quad-tree window is a summarized representation of a bulk of collected data stream readings related to a given time window. A quad-tree window is implemented as a two-dimensional array such that (*i*) the first dimension represents data stream sources organized according to the hierarchy given by the above-mentioned OLAP dimension

Figure 1. Technique overview

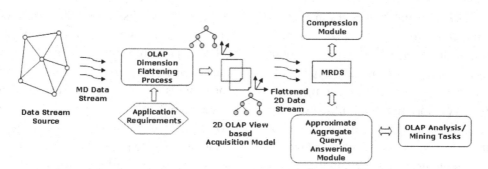

flattening process, and (*ii*) the second dimension represents the time organized according to the natural temporal hierarchy. Also, the overall list of quad-tree windows, which represents a "snapshot" of the actual multidimensional data stream, is embedded into a highly-efficient data structure called *Multi-Resolution Data Stream Summary* (MRDS) (see Figure 1). Briefly, the MRDS is a hierarchical summarization of the data stream embedded into a flexible indexing structure, which permits us to both access and update compressed data efficiently, and support an efficient evaluation of aggregate queries. Such compressed representation of data is updated continuously, as new data stream readings arrive.

Finally, in order to achieve high efficiency in processing and querying multidimensional data streams, we propose *novel compression techniques over materialized data stream readings*. This allows us to tame the unbounded nature of streaming data, thus dealing with bounded memory issues exposed by conventional DBMS tools. According to our compression scheme, the MRDS is *progressively compressed* as time passes and the available storage space is not enough to store new readings. It should be noted that while the OLAP dimension flattening process, which is indeed a transformation of the multidimensional data stream into a two-dimensional data stream, realizes the first-stage compression of the multidimensional data stream, called *semantic-oriented compression task*, the compression of materialized

data stream readings realizes the second-stage compression of the multidimensional data stream, called *data-oriented compression task*. The meaningful combination of both tasks finally allows us to obtain and effective and efficient compression of the multidimensional data stream.

In the data-oriented compression task, the general idea consists in obtaining free space when needed via compressing the "oldest" stored data so that recent information, which is usually the most relevant one to retrieve, can be represented with more detail than old information. This allows us to obtain a very efficient solution to represent a multidimensional data stream in a bounded amount of memory, and achieve high compression ratios accordingly. Our approach for compressing the MRDS is innovative with respect to the state-of-the-art in what follows. We propose a two-step (data) compression scheme. In the first step, the two MRDS dimensions are submitted to a *progressive coalescing process*, as time passes. This process aims at progressively decreasing the granularities of both the dimensions of the oldest quad-tree windows of the MRDS, thus obtaining an initial compression of them by means of progressively aggregating data at the coarser granularities. In the second step of the compression scheme, when a given application-driven threshold is reached and it is no longer possible to further decrease the granularities of dimensions of the oldest quad-tree windows in order to obtain free space, the MRDS is further compressed by meaningfully erasing the oldest quad-tree windows, and maintaining aggregate information about them only. It should be noted that both the first and the second compression step cause the loss of details of oldest information in favor of newer information, which is reasonably retained more significant that oldest one for the goal of extracting summarized, OLAP-like knowledge from data streams.

Querying the compressed MRDS is another relevant aspect of our work. In fact, it should be noted that efficiently answering baseline range queries, e.g. range-SUM and range-COUNT queries, over multidimensional data streams, also called *window queries*, puts the basis for efficiently supporting even complex OLAP analysis/mining models, which clearly found on the range queries above (see Figure 1). Without loss of generality, the issue of defining new query evaluation paradigms to provide fast answers to aggregate queries is very relevant in the context of data streams. In fact, the amount of data produced by data stream sources like sensors is very large and grows continuously, and queries need to be evaluated very quickly, in order to make performing a timely "reaction to the world" possible. Moreover, in order to make the information produced by data stream sources useful, it should be possible to retrieve an up-to-date snapshot of the monitored world continuously, as time passes and new readings are collected. For instance, a climate disaster prevention system would benefit from the availability of continuous information on atmospheric conditions during the last hour. Similarly, a network congestion detection system would be able to prevent network failures exploiting the knowledge of network traffic during the last minutes. If the answer to these queries, called *continuous queries*, is not fast enough, we could observe an increasing delay between the query answer and the arrival of new data, and thus a non-timely reaction to the world. In order to evaluate approximate answers to both window and continuous queries, we introduce efficient query algorithms able to handle the compressed MRDS,

and retrieve highly-accurate approximate answers that are perfectly suitable to OLAP analysis goals (e.g., see (Cuzzocrea, 2005)).

Overall, in this chapter we introduce an innovative, complex technique for efficiently supporting OLAP analysis of multidimensional data streams. We highlight since here that our proposed representation and analysis models are indeed general enough to deal with data streams generated by any source of intermittent data, regardless from the particular application scenario considered in this chapter and represented by data streams generated by sensor networks. Figure 1 provides a comprehensive overview of our technique.

THE MULTIDIMENSIONAL DATA STREAM MODEL

Consider a set S of N data stream sources (e.g., sensors) denoted by $S = \{s_0, s_1, ..., s_{N-1}\}$. Let $M_S = \langle D(M_S), H(M_S), M(M_S) \rangle$ be the N-dimensional (OLAP) model of S, such that: (*i*) $D(M_S) = \{d_0, d_1, ..., d_{N-1}\}$ denotes the set of N dimensions of M_S; (*ii*) $H(M_S) = \{h_0, h_1, ..., h_{N-1}\}$ denotes the set of N hierarchies of M_S, where h_k in $H(M_S)$ denotes the hierarchy associated to the dimension d_k in $D(M_S)$; (*iii*) $M(M_S)$ denotes the set of measures of M_S. For the sake of simplicity, in the following we will assume to deal with single-measure OLAP models, i.e. $M(M_S) = \{m\}$. However, models and algorithms presented in this chapter can be straightforwardly extended to the more challenging case in which multiple-measure OLAP models (i.e., $|M(M_S)| > 1$) are considered.

For the sake of simplicity, the stream source name $s_i \in S$ will also denote the data stream generated by the source itself. Each stream source $s_i \in S$ produces a multidimensional stream of data composed by an unbounded sequence of (data stream) readings of kind: $r_{i,j}$, i.e. $s_i = \langle r_{i,0}, r_{i,1}, r_{i,2}... \rangle$ with $|s_i| \to \infty$. In more detail, $r_{i,j}$ denotes the *j*-th reading of the data stream s_i, and it is defined as a tuple $r_{i,j} = \langle id_i, v_{i,j}, ts_{i,j}, a_{i,j,k_0}, a_{i,j,k_1}, ..., a_{i,j,k_{P-1}} \rangle$, where:

1. $id_i \in \{0,..,N-1\}$ is the stream source (absolute) identifier;
2. $v_{i,j}$ is a non-negative integer value representing the measure produced by the stream source s_i identified by id_i, i.e. the reading value;
3. $ts_{i,j}$ is a *timestamp* that indicates the time when the reading $r_{i,j}$ was produced by the stream source s_i identified by id_i, i.e. the reading timestamp;
4. a_{i,j,k_p} is the value associated to the dimensional attribute A_{k_p} of the P-dimensional model of the stream source s_i identified by id_i, denoted by $M_{s_i} = \langle D(M_{s_i}), H(M_{s_i}), M(M_{s_i}) \rangle$, being $D(M_{s_i})$, $H(M_{s_i})$ and $M(M_{s_i})$ the set of dimensions, the set of hierarchies and the set of measures of M_{s_i}, respectively.

The definition above adheres to the so-called *multidimensional data stream model*, which is a fundamental component of the OLAP stream model introduced in the first Section. According to the multidimensional data stream model, each reading $r_{i,j}$ embeds a *dimensionality*, which is used to meaningfully handle the overall multidimensional stream. This dimensionality is captured by the set of values $\{ a_{i,j,k_0}, a_{i,j,k_1}, ..., a_{i,j,k_{P-1}} \}$ associated to the dimensional attributes $\{ A_{i,k_0}, A_{i,k_1}, ..., A_{i,k_{P-1}} \}$ of M_{s_i}. Also, dimensional attribute values in $r_{i,j}$ are logically organized in an (OLAP) hierarchy, denoted by $h_{i,j}$.

As a demonstrative example, consider the case of a 3-dimensional stream source s_i having identifier $id_i = 4556113$ and whose multidimensional model is the following: $M_{s_i} = \langle D(M_{s_i}), H(M_{s_i}), M(M_{s_i}) \rangle = \langle \{Country, State, City\}, \{h_{Country}, h_{State}, h_{City}\}, \{Temperature\} \rangle$, such that $h_{Country}$ is the hier-

archy associated to the dimension *Country*, h_{State} is the hierarchy associated to the dimension *State*, and h_{City} is the hierarchy associated to the dimension *City*, respectively. Possible readings produced by s_i are the following: (i) $r_{i,h} = \langle 4556113, 77, 8992, USA, California, LosAngeles \rangle$, which records the temperature reading $v_{i,h} = 77$ F monitored by s_i at the timestamp $ts_{i,h} = 8992$ in Los Angeles that is located in California, USA; (ii) $r_{i,k} = \langle 4556113, 84, 9114, USA, California, SanFrancisco \rangle$, which records the temperature reading $v_{i,k} = 84$ F monitored by s_i at the timestamp $t_{i,k} = 9114$ in San Francisco that is located in California, USA. Note that, in this case, the OLAP hierarchies associated to readings of s_i model the geographical hierarchy of the state of California.

For the sake of simplicity, in the following we will refer the set S of stream sources as the "stream source" itself. To give insights, S could identify a sensor network that, in the vest of collection of sensors, is a stream source itself. Another important assertion states that the OLAP stream model assumes that the multidimensional model of S, M_S, is a-priori known, as happens in several real-life scenarios such as sensor networks monitoring environmental parameters.

Given a stream source S and its multidimensional model, M_S, the multidimensional model of each stream source $s_i \in S$, M_{s_i}, can either be totally or partially mapped onto the multidimensional model of S, M_S. The total or partial mapping relationship only depends on the mutual correspondence between dimensions of the multidimensional models, as the (single) measure is always the same, thus playing the role of invariant for both models. In the first case (i.e., total mapping), M_{s_i} and M_S are equivalent, i.e. $M_{s_i} \equiv M_S$. In the second case (i.e., partial mapping), M_{s_i} is a *multidimensional (proper) sub-model* of M_S, i.e. $M_{s_i} \subset M_S$. Basically, this defines a *containment relationship* between models, and, consequentially, a containment relationship between the multidimensional data models. It should be noted that mapping relationships above are able to capture even complex scenarios occurring in real-life data stream applications and systems. Also, in our research, mapping relationships define the way readings are aggregated during the acquisition phase (see the fifth Section).

To give an example, consider a sensor network S composed by a set of sensors monitoring environmental parameters. Assume that the multidimensional model of S, M_S, is the following: $M_S = \langle D(M_S), H(M_S), M(M_S) \rangle = \langle \{Longitude, Latitude, Azimuth, ElevationAngle, DistanceFromLocalPoint, DistanceAboveSeaLevel\}, \{h_{Long}, h_{Lat}, h_{Azi}, h_{ElA}, h_{DFP}, h_{DSL}\}, \{Temperature\} \rangle$, such that h_{Long} is the hierarchy associated to the dimension *Longitude*, h_{Lat} is the hierarchy associated to the dimension *Latitude*, and so forth. This leads to the definition of a 6-dimensional model. Consider a data stream source $s_i \in S$, which is characterized by the following multidimensional model: $M_{s_i} = \langle D(M_{s_i}), H(M_{s_i}), M(M_{s_i}) \rangle = \langle \{Longitude, Latitude, ElevationAngle, DistanceFromLocalPoint, \{h_{Long}, h_{Lat}, h_{Azi}, h_{ElA}, h_{DFP}\}, \{Temperature\} \rangle$. This leads to the definition of a 4-dimensional model, and the partial containment relationship is determined accordingly, i.e. $M_{s_i} \subset M_S$. Consider instead a data stream source $s_j \in S$, which is characterized by the following multidimensional model: $M_{s_j} = \langle D(M_{s_j}), H(M_{s_j}), M(M_{s_j}) \rangle = \langle \{Longitude, Latitude, ElevationAngle, DistanceFromLocalPoint\}, \{h_{Long}, h_{Lat}, h_{Azi}, h_{ElA}, h_{DFP}, h_{DSL}\}, \{Temperature\} \rangle$. This leads to the definition of a 6-dimensional model. Since $M_{s_j} \equiv M_S$, the total containment relationship is determined accordingly.

Flattening OLAP Dimensions

The OLAP dimension flattening process allows us to obtain a transformation of the multidimensional data stream into a flattened two-dimensional data stream, which is indeed a semantics-based compression of the original stream (see the first

Section). This process is driven by application requirements, according to which two flattening dimensions are selected from the multidimensional model of the stream source S, M_S, and used to generate a two-dimensional OLAP view based acquisition model which, in turn, is exploited to populate the summarized repositories of data stream readings (see the first Section).

The two flattening dimensions, denoted by d_{f_0} and d_{f_1}, respectively, are selected from the set $D(M_S)$ and then equipped with specialized hierarchies, denoted by h_{f_0} and h_{f_1}, respectively, such that each hierarchy h_{f_i}, with $i \in \{0, 1\}$, is built by meaningfully merging the "original" hierarchy of d_{f_i} with hierarchies of other dimensions in $D(M_S)$, according to application requirements driven by specific OLAP analysis goals over the target multidimensional data stream. To theoretical consistency purposes, here we assume that $h_{f_0} \in H(M_S)$ and $h_{f_1} \in H(H_S)$, respectively.

The final shape of each hierarchy h_{f_i}, with $i \in \{0, 1\}$, depends on the so-called *ordered definition set Def*(d_{f_i}), with $i \in \{0, 1\}$, which constitutes an input parameter for the OLAP dimension flattening process. This set is composed by tuples of kind: $\langle L_j, d_{j+1}, P_{j+1} \rangle$, such that, given two consecutive tuples $\langle L_j, d_{j+1}, P_{j+1} \rangle$ and $\langle L_{j+1}, d_{j+2}, P_{j+2} \rangle$ in *Def*(h_{d_i}), the sub-tree of h_{j+2} (i.e., the hierarchy of d_{j+2}) rooted at the root node of h_{j+2} and having depth equal to P_{j+2}, said $T_{P_{j+2}}(h_{j+2})$, is merged to h_{j+1} (i.e., the hierarchy of d_{j+1}) via appending a *clone* of $T_{P_{j+2}}(h_{j+2})$ to each member $\sigma_{i,L_{j+1}}$ of h_{j+1} at level L_{j+1}, and erasing the original sub-tree rooted at $\sigma_{i,Lj+1}$. This process is iterated for each tuple $\langle L_j, d_{j+1}, P_{j+1} \rangle$ in *Def*(d_{f_i}) until the final hierarchy h_{f_i} is obtained.

From the described approach, it follows that: (*i*) the ordering of tuples in the definition set *Def*(d_{f_i}) determines itself the way of building h_{f_i} thus the final shape of h_{f_i}; (*ii*) during the creation of the hierarchy h_{f_i} the first hierarchy to be processed is just the hierarchy of the corresponding dimen-

Figure 2. Merging OLAP hierarchies

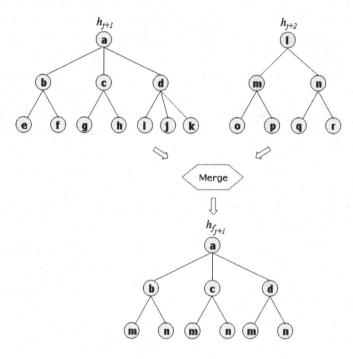

sion d_{f_i} chosen as flattening dimension, and then sub-trees extracted from the other hierarchies are progressively appended to such hierarchy. Therefore: (*i*) the first tuple of *Def*(d_{f_i})is of kind: $\langle NULL, d_{f_i}, NULL \rangle$; (*ii*) the second tuple of *Def*(d_{f_i})is of kind: $\langle L_i, d_j, P_j \rangle$, with $j \neq 0$ and $j \neq 1$, such that L_i is the level of h_{f_i} to which sub-trees of h_j are appended; (*iii*) starting from the third tuple of *Def*(d_{f_i}), $\langle L_j, d_{j+1}, P_{j+1} \rangle$, with $j \neq 0$ and $j \neq 1$, L_j is set equal to the *current* depth of h_{f_i}.

To give an example, consider Figure 2, where the two hierarchies h_{j+1} and h_{j+2} are merged in order to obtain the new hierarchy $h_{f_{j+1}}$ that can also be intended as a "modified" version of the original hierarchy h_{j+1}. Specifically, $h_{f_{j+1}}$ is obtained via setting both L_{j+1} and P_{j+2} equal to 1.

In our framework, algorithm MergeOLAPHierarchies implements the merging of two OLAP hierarchies. It takes as arguments the two OLAP hierarchies to be merged, h_{j+1} and h_{j+2}, and the parameters needed to perform the merging task (i.e.,

the level of h_{j+1}, L_{j+1}, to which clones of the sub-tree of h_{j+2} rooted at the root node of h_{j+2} have to be appended, and the depth P_{j+2} of such sub-tree), and returns the modified hierarchy. Furthermore, algorithm FlattenMultidimensionalModel implements the overall OLAP dimension flattening process, and makes use of algorithm MergeOLAPHierarchies. It takes as arguments the multidimensional model of the stream source S, M_S, the definition sets of the two flattening dimensions, *Def*(d_{f_0}) and *Def*(d_{f_1}), respectively, and returns the two flattening dimensions d_{f_0} and d_{f_1} with modified hierarchies, h_{f_0} and h_{f_1}, respectively.

THE TWO-DIMENSIONAL OLAP VIEW BASED ACQUISITION MODEL

Consider a stream source S and its multidimensional model M_S. Given the two-dimensional flattened data stream $s_{2D} \in S$, which is generated by the OLAP dimension flattening process (see the fourth Section), the two-dimensional OLAP view based acquisition model deals with the problem of populating the summarized repositories of data stream readings by means of s_{2D} and according to the underlying OLAP stream model. From the first Section, recall that in our proposed research the OLAP view based acquisition model and the OLAP aggregation model of the summarized data stream readings (i.e., the storage model) are coincident. Also, recall that each summarized repository of data stream readings is finally represented as a quad-tree window, i.e. a two-dimensional array where the first dimension represents the stream sources, and the second dimension represents the time.

As highlighted in the first Section, while our OLAP stream model is general enough to handle any kind of dimension, we will consider the specialized case in which the temporal dimension is always one of the two flattening dimensions, as this allows us to meaningfully capture how data streams evolve over time. Therefore, for the sake of notation, we hereafter denote as d_{f_N} the *normal* flattening dimension, whose hierarchy h_{f_N} is obtained according to the OLAP dimension flattening process described in the fourth Section, and as d_{f_T} the *temporal* flattening dimension, whose hierarchy h_{f_T} follows the natural temporal hierarchy (e.g., *Year* → *Quarter* → *Month* → *Day*), respectively.

The main idea of the two-dimensional OLAP view based acquisition model consists in determining how readings in s_{2D} participate to the aggregations defined by the OLAP storage model of the summarized repositories of data stream readings, which, in turn, finally determines the way of populating these repositories. To this end, given a reading $r_{2D,j}$ embedded in s_{2D} of S, on the basis of the conventional OLAP aggregation scheme (e.g., (Chaudhuri & Dayal, 1997)), the measure $v_{2D,j}$ of $r_{2D,j}$ have to be aggregated along all the dimensions of the multidimensional model $M_{s_{2D}}$. In our proposed OLAP stream model, this means that the

measure $v_{2D,j}$ contributes to a certain (array) cell of the target summarized repository (and updates its value) based on the *membership* of dimensional attribute values a_{i,j,k_0}, a_{i,j,k_1} ,..., $a_{i,j,k_{P-1}}$ and the timestamp $ts_{i,j}$ of the reading $r_{i,j}$ with respect to the normal and temporal hierarchy associated to the dimensions of the repository, respectively. This way, we obtain a *specialized* aggregation scheme for our proposed OLAP stream model able of (*i*) taming the curse of dimensionality problem arising when multidimensional data streams are handled (see the first Section), and (*ii*) effectively supporting the simultaneous multidimensional aggregation of data stream readings.

It should be noted that the OLAP dimension flattening process plays a role in the final way readings are aggregated during the acquisition phase. We briefly describe this dependency in the following. Focus the attention on the normal flattening dimension d_{f_N} and the associated hierarchy h_{f_N}. Assume that $D_N(M_S) = \{ d_{k_0}, d_{k_1} ,..., d_{k_{F-1}} \}$ is the sub-set of $D(M_S)$ used to generate d_{f_N} ($D_N(M_S) \subset D(M_S)$). Let us now focus on the collection of stream sources $\{s_0, s_1 ,..., s_{N-1}\}$ of S. Although each stream source $s_i \in S$ could define a total (i.e., $M_{s_i} \equiv M_S$) or partial (i.e., $M_{s_i} \subset M_S$) containment relationship with respect to the multidimensional model of S, the OLAP dimension flattening process essentially combines dimensions in $D_N(M_S)$ and, as a consequence, the final multidimensional model of the flattened two-dimensional data stream s_{2D}, $M_{s_{2D}}$, results to be a "combination" of the multidimensional models of data stream sources in S. Intuitively enough, it is easy to observe that, if the multidimensional models $M_{s_{2D}}$ and M_S are coincident (i.e., $M_{s_{2D}} \equiv M_S$), then readings embedded in s_{2D} are simultaneously aggregated along *all* the dimensions in M_S to obtain the final aggregate value in the corresponding repository cell. Otherwise, if the multidimensional models $M_{s_{2D}}$ and M_S define a proper containment relationship (i.e., $M_{s_{2D}} \subset M_S$), then readings embedded in s_{2D} are simultaneously aggregated along a *partition* of the dimensions in M_S to obtain the final aggregate value in the corresponding repository cell.

Formally, given a reading $r_{2D,j} = \langle id_{2D}, v_{2D,j}, ts_{2D,j}, a_{2D,j,k_0}, a_{2D,j,k_1}, ..., a_{2D,j,k_{P-1}} \rangle$ embedded in s_{2D}, on the basis of a top-down approach, starting from the dimensional attribute value at the highest aggregation level of $h_{2D,j}$, a_{2D,j,k_0}, we first search the hierarchy of the normal flattening dimension d_{f_N}, h_{f_N}, starting from the member at the highest aggregation level, denoted by $\sigma_{0,0}^N$, and by means of a *breadth-first tree visiting strategy*, and we check whether a_{2D,j,k_0} belongs to the class defined by the current member of h_{f_N}, σ_{i,L_j}^N (when $i = 0$, then $\sigma_{i,L_j}^N \equiv \sigma_{0,0}^N$). When a member of h_{f_N} such that a_{2D,j,k_0} belongs to the class it defines, denoted by $\sigma_{i^*,L_j^*}^N$, is found, then (i) the breadth-first search is contextualized to the sub-tree of h_{f_N} rooted at $\sigma_{i^*,L_j^*}^N$, denoted by $T^*(h_{f_N})$, and (ii) the current search dimensional attribute value becomes the value that immediately follows a_{2D,j,k_0} in the hierarchy $h_{2D,j}$, i.e. a_{2D,j,k_1}. After that, the whole search is repeated again, and it ends when a leaf node of h_{f_N} is reached, denoted by $\sigma_{i^*,Depth(h_{f_N})}^N$, such that $Depth(h_{f_N})$ denotes the depth of h_{f_N}. Note that the search should end when the last dimensional attribute value $a_{2D,j,k_{P-1}}$ is processed accordingly, but, due to the OLAP dimension flattening process and the possible presence of imprecise or incomplete data it could be the case that the search ends before that. For the sake of simplicity, hereafter we assume to deal with hierarchies and readings adhering to the simplest case in which the search ends by reaching a leaf node of h_{f_N} while the last dimensional attribute value $a_{2D,j,k_{P-1}}$ is processed. The described search task allows us to determine an indexer on the first dimension of the array-based repository, i.e. the dimension representing the stream sources. Let us denote as I_S^* this indexer. The second indexer on the temporal dimension of the array-based repository, denoted by I_T^*, is determined by means of the same approach exploited for the case of the sensor dimension, with the difference that, in this case, the search term is fixed and represented by the reading timestamp $ts_{2D,j}$.

When both indexers I_S^* and I_T^* are determined, a repository cell is univocally located, and the reading measure $v_{2D,j}$ is used to finally update the value of this cell.

Let us now focus on a running example showing how our proposed OLAP aggregation scheme for multidimensional data stream readings works in practice. Figure 3 shows the hierarchy h_{f_N} associated to the normal flattening dimension d_{f_N} of the running example, whereas Figure 4 shows instead the hierarchy h_{f_T} associated to the temporal flattening dimension d_{f_T}. As suggested by Figure 3 and Figure 4, the multidimensional data stream model of the running example describes an application scenario focused on sales of electrics and personal computer parts sold in Europe, Asia and America during 2008. The hierarchy h_{f_N} derives from the OLAP dimension flattening process, whereas the hierarchy h_{f_T} follows the natural temporal hierarchy organized by months and groups of months. Readings are produced by different locations distributed in Europe, Asia and America, thus defining a proper network of data stream sources. In more detail, the described one is a typical application scenario of modern *Radio Frequency IDentifiers* (RFID) (Gonzalez, Han, Li & Klabjan, 2006) based applications and systems. Figure 5 shows the array-based repository that represents summarized information on readings produced by the sources, equipped with the normal and temporal hierarchies. In particular, each (array) cell of the repository stores a SUM-based OLAP aggregation of readings according to both the normal and temporal dimension, simultaneously.

Now consider the reading $r_{2D,k} = \langle id_{2D}, 5, 6/15/08, India, Delta-Power, Power2500 \rangle$ embed-

Figure 3. The hierarchy associated to the normal flattening dimension of the running example

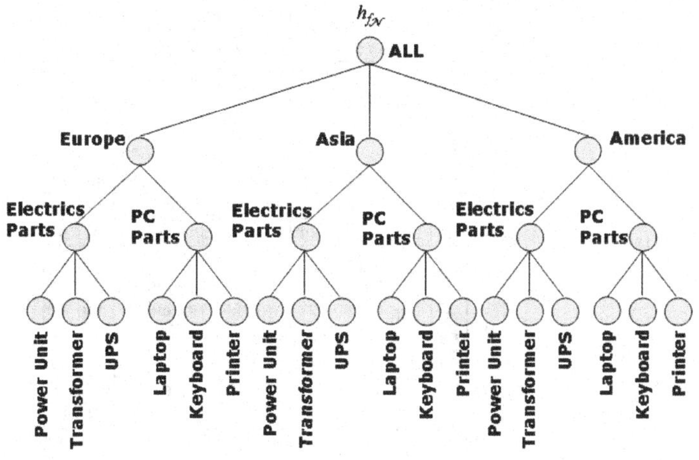

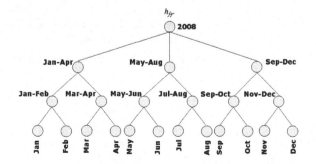

ded in s_{2D}, which records the sale of a *Power2500* transformer, produced by the company *Delta-Power* at the price of 5 \$, sold in India on June 15, 2008. Focus the attention on how the value of $r_{2D,k}$ is aggregated in cells of the summarized repository. On the basis of our proposed OLAP aggregation scheme, the final repository cell to be updated is finally located by means on two distinct paths on the respective hierarchies h_{f_N} and h_{f_T} determined by the simultaneous membership of dimensional attribute values and timestamp of the reading $r_{2D,k}$ to classes defined by members of these hierarchies, in a top-down manner. Figure 6 shows the configuration of the summarized repository after the update. Note that the value

of the target cell has been updated to the new value $69 + 5 = 74$.

Finally, algorithm UpdateRepository implements the proposed OLAP aggregation scheme that allows us to populate the target array-based repository of summarized data stream readings by means of the two-dimensional flattened stream s_{2D}. It takes as arguments the repository R and the input reading $r_{2D,j}$ of s_{2D}, and updates R by the measure value embedded in $r_{2D,j}$, $v_{2D,j}$, according to the simultaneous membership-based multidimensional aggregation approach described above.

Figure 4. The hierarchy associated to the temporal flattening dimension of the running example

Figure 5. The array-based repository of summarized data stream readings of the running example

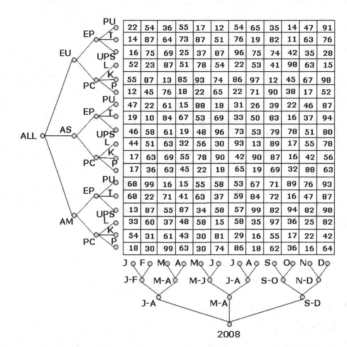

Figure 6. The array-based repository of Figure 5 after the update

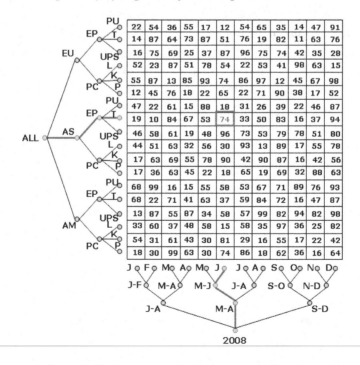

TWO-DIMENSIONAL SUMMARIZED REPRESENTATION OF DATA STREAM READINGS

As highlighted in the first Section, evaluating aggregate queries over summarized data stream readings represents a critical aspect in order to gain efficiency and performance in OLAP analysis/ mining tasks over multidimensional data streams. To this end, aggregating the values produced by a subset of sources within a time interval is an important issue in managing multidimensional data streams. This means answering range queries on the overall stream of data generated by the source S. Formally, a range query Q is a pair $Q = \langle s_i...s_j, [t_{start}.t_{end}]\rangle$, such that (*i*) $s_i \in S$, (*ii*) $s_j \in S$, (*iii*) $i < j$, (*iv*) $t_{start} < t_{end}$, whose answer is the evaluation of an aggregate operator (such as SUM, COUNT, AVG etc) on the values produced by the sources $s_i, s_{i+1}, ..., s_j$ within the time interval $[t_{start}.t_{end}]$. Although a wide class of range queries for OLAP exists (e.g., see (Ho, Agrawal, Megiddo & Srikant, 1997)), in this chapter we focus on the issue of efficiently answering range-SUM queries over data streams, as SUM-based aggregations are a very popular way to materialize data cubes and extract summarized knowledge from data cubes (Gray, Chaudhuri, Bosworth, Layman, Reichart, Venkatrao, Pellow & Pirahesh, 1997; Chaudhuri & Dayal, 1997). However, models and algorithms presented in this chapter are indeed general enough to be straightforwardly extended as to deal with other aggregations different from SUM (e.g., COUNT, AVG etc).

In order to efficiently evaluate range queries over data streams, we propose to represent the data stream by means of a two-dimensional array, where the first dimension corresponds to the set of sources organized according to the hierarchy given by the OLAP dimension flattening process, and the other one corresponds to time organized to a given natural temporal hierarchy. In particular, the time is divided into intervals Δt_j of the same size. Each element $\langle s_i, \Delta t_j \rangle$ of the array stores the sum of all the values generated by the source s_i whose timestamp is within the time interval Δt_j. Obviously, the use of a time granularity generates a loss of information, as readings of a stream source belonging to the same time interval are aggregated. Indeed, if a time granularity which is appropriate for the particular context monitored by stream sources is chosen, the loss of information will be negligible.

Using this representation, an estimate of the answer to a range-SUM query $Q = \langle s_i...s_j, [t_{start}.t_{end}]\rangle$ over the summarized data stream readings can be obtained by summing two contributions. The first one is given by the sum of those elements which are completely contained inside the range of the query, i.e. the elements $\langle s_k, \Delta t_l \rangle$ such that $i \le k \le j$ and Δt_l is completely contained into $[t_{start}.t_{end}]$. The second one is given by those elements which partially overlap the range of the query, i.e. the elements $\langle s_k, \Delta t_l \rangle$ such that $i \le k \le j$ and $t_{start} \in \Delta t_l$ or $t_{end} \in \Delta t_l$. The first of these two contributions does not introduce any approximation, whereas the second one is generally approximate, as the use of the time granularity makes it unfeasible to retrieve the exact distribution of values generated by each sensor within the same interval Δt_l. The latter contribution can be evaluated by performing *linear interpolation*, i.e. assuming that the data distribution inside each interval Δt_i is uniform (*Continuous Values Assumption* – CVA (Colliat, 1996)). For instance, the contribution of the element $\langle s_2, \Delta_3 \rangle$ to the range-SUM query represented in Figure 7 is given by $((6 - 5) / 2) \times 4 = 2$.

As the stream of readings produced by every source is potentially "infinite", detailed information on the stream (i.e., the exact sequence of values generated by every source) cannot be stored, so that exact answers to every possible range query cannot be provided. However, exact answers to aggregate queries are often not necessary, as approximate answers usually suffice to get useful reports on the content of data streams, and to provide a meaningful description of the world monitored by stream sources. This evidence makes more sense

Figure 7. Two-dimensional representation of data stream readings

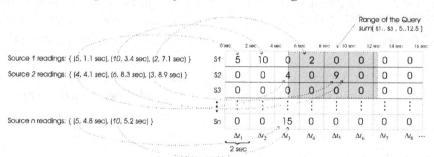

and relevance in the context of OLAP over data streams (e.g., see (Cuzzocrea, 2005; Han, Chen, Dong, Pei, Wah, Wang & Cai, 2005)).

A solution for providing approximate answers to aggregate queries is to store a compressed representation of the overall data stream, and then to run queries on the compressed data. The use of a time granularity introduces a form of compression, but it does not suffice to represent the whole stream of data, as the stream length is possibly infinite. An effective structure for storing the information carried by the data stream should have the following characteristics:

1. it should be efficient to update, in order to catch the continuous stream of data coming from the sources;

2. it should provide an up-to-date representation of the data stream readings, where recent information is possibly represented more accurately than old one;

3. it should permit us to answer range queries efficiently.

As we demonstrate throughout the chapter, thanks to high-efficient data compression techniques, our framework for efficiently supporting OLAP over multidimensional data streams is indeed able to achieve the above-listed requirements. For the sake of simplicity, in the remaining part of the chapter we will refer to two-dimensional array-based representations of data stream readings

directly, by omitting to detail OLAP hierarchies associated to the dimensions.

REPRESENTING TIME WINDOWS

Preliminary Definitions

Consider given a two-dimensional $n_1 \times n_2$ array A. Without loss of generality, array indices are assumed to range respectively in $1..n_1$ and $1..n_2$. A *block r* (of the array) is a two-dimensional interval $[l_1..u_1, l_2..u_2]$ such that $1 \le l_1 \le u_1 \le n_1$ and $1 \le l_2 \le u_2 \le n_2$. Informally, a block represents a "rectangular" region of the array. We denote by $size(r)$ the size of the block r, i.e. the value $(u_1 - l_1 + 1) \times (u_2 - l_2 + 1)$. Given a pair $\langle v_1, v_2 \rangle$ we say that $\langle v_1, v_2 \rangle$ is inside r if $v_1 \in [l_1..u_1]$ and $v_2 \in [l_2..u_2]$. We denote by $sum(r)$ the sum of the array elements occurring in r, i.e. $sum(r) = \Sigma_{\langle i,j \rangle \text{inside} r} A[i, j]$. If r is a block corresponding to the whole array (i.e., $r = [1..n_1, 1..n_2]$), $sum(r)$ is also denoted by $sum(A)$. A block r such that $sum(r) = 0$ is called a *null block*. Given a block $r = [l_1..u_1, l_2..u_2]$ in A, we denote by r_i the i-th quadrant of r, i.e. $r_1 = [l_1..m_1, l_2..m_2]$, $r_2 = [m_1 + 1..u_1, l_2..m_2]$, $r_3 = [l_1..m_1, m_2 + 1..u_2]$, and $r_4 = [m_1 + 1..u_1, m_2 + 1..u_2]$, where $m_1 = \lfloor (l_1 + u_1)/2 \rfloor$ and $m_2 = \lfloor (l_2 + u_2)/2 \rfloor$.

Given a time interval $t = [t_{start}..t_{end}]$ we denote by $size(t)$ the size of the time interval t, i.e. $size(t) = t_{end} - t_{start}$. Furthermore we denote by $t_{i/4}$ the i-th quarter of t. That is $t_{i/4} = [t_{is}..t_{ie}]$ with $t_{is} = t_{start} +$

Figure 8. A time window and the corresponding quad-tree partition

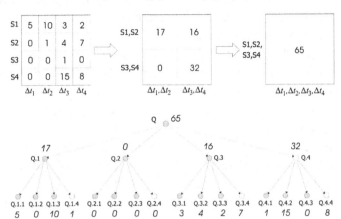

$(i - 1) \times size(t) / 4$ and $t_{ie} = t_{start} + i \times size(t) / 4$, $i = 1,..,4$.

Given a 4-ary tree T, we denote by $Root(T)$ the root node of T and, if p is a non leaf node, we denote the i-th child node of p by $Child(p, i)$.

Given a triplet $x = \langle id_s, v, t_s \rangle$, representing a value generated by a source, id_s is denoted by $id_s(x)$, v by $value(x)$ and t_s by $t_s(x)$.

The Quad-Tree Window

In order to represent data occurring in a time window, we do not store directly the corresponding two-dimensional array, indeed we choose a hierarchical data structure, called quad-tree window, which offers some advantages:

- it makes answering (portions of) range queries internal to the time window more efficient to perform (w.r.t. a "flat" array representation),
- it stores data in a straight compressible format. That is, data is organized according to a scheme that can be directly exploited to perform compression.

This hierarchical data organization consists of storing multiple aggregations performed over the time window array according to a quad-tree

partition. This means that we store the sum of the values contained in the whole array, as well as the sum of the values contained in each quarter of the array, in each eighth of the array and so on, until the single elements of the array are stored. Figure 8 shows an example of quad-tree partition, where each node of the quad-tree is associated to the sum of the values contained in the corresponding portion of the array.

The quad-tree structure is very effective for answering (sum) range queries inside a time window efficiently, as we can generally use the pre-aggregated sum values in the quad-tree nodes for evaluating the answer. Moreover, the space needed for storing the quad-tree representation of a time window is about the same as the space needed for a flat representation, as we will explain later.

Furthermore, the quad-tree structure is particularly prone to progressive compressions. In fact, the information represented in each node is summarized in its ancestor nodes. For instance, the node Q of the quad-tree in Figure 8 contains the sum of its children $Q.1, Q.2, Q.3, Q.4$; analogously, $Q.1$ is associated to the sum of $Q.1.1$, $Q.1.2, Q.1.3, Q.1.4$, and so on. Therefore, if we prune some nodes from the quad-tree, we do not loose every information about the corresponding portions of the time window array, but we represent them with less accuracy. For instance, if we removed the nodes *Q.1.1, Q.1.2, Q.1.3, Q.1.4*, then the detailed values of the readings produced by the sensors S_1 and S_2 during the time intervals Δt_1 and Δt_2 would be lost, but it would be kept summarized in the node *Q.1*. The compression paradigm that we use for quad-tree windows will be better explained in the next Section.

We will next describe the quad-tree based data representation of a time window formally. Denoting by *u* the time granularity (i.e., the width of each interval Δt_j), let $T = n \times u$ be the time window width (where *n* is the number of sources). We refer to a *Time Window* starting at time *t* as a two-dimensional array *W* of size $n \times n$ such that $W[i, j]$ represents the sum of the values generated by a source s_i within the *j*-th unitary time interval of *W*. That is $W[i,j] = \Sigma_{x:ids(x)=i \wedge ts(x) \in \Delta tj}$ *value(x)*, where Δt_j is the time interval $[t + (j - 1) \times u..t + j \times u]$.

The whole data stream consists of an infinite sequence $W_1, W_2, ...$ of time windows such that the *i*-th one starts at $t_i = (i - 1) \times T$ and ends at $t_{i+1} = i \times T$.

In the following, for the sake of presentation, we assume that the number of sources is a power of 2 (i.e., $n = 2^k$, where $k > 1$).

A Quad-Tree Window on the time window *W*, called *QTW(W)*, is a full 4-ary tree whose nodes are pairs $\langle r, sum(r) \rangle$ (where *r* is a block of *W*) such that:

1. $Root(QTW(W)) = \langle [1..n, 1..n], sum([1..n, 1..n]) \rangle$;
2. each non leaf node $q = \langle r, sum(r) \rangle$ of *QTW(W)* has four children representing the four quadrants of *r*; that is, $Child(q, i) = \langle r_i, sum(r_i) \rangle$ for $i = 1, ..., 4$;
3. the depth of *QTW(W)* is $log_2 n + 1$.

Property 3 implies that each leaf node of *QTW(W)* corresponds to a single element of the time window array *W*.

Given a node $q = \langle r, sum(r) \rangle$ of *QTW(W)*, *r* is referred to as *q.range* and *sum(r)* as *q.sum*.

Compact Physical Representation of Quad-Tree Windows

The space needed for storing all the nodes of a quad-tree window *QTW(W)* is larger than the one needed for a flat representation of *W*. In fact, it can be easily shown that the number of nodes of *QTW(W)* is $(4 \times n^2 - 1) / 3$, whereas the number of elements in *W* is n^2. Indeed, *QTW(W)* can be represented compactly, as it is not necessary to store the sum values of all the nodes of the quad-tree. That is, if we have the sum values associated to a node and to three of its children, we can easily compute the sum value of its fourth child. This value can be obtained by subtracting the sum of the three children from the sum of the parent node. We say that the fourth child is a *derivable* node.

For instance, the node *Q.4* of the quad-tree window in Figure 8 is derivable, as its sum is given by $Q.sum - (Q.1.sum + Q.2.sum + Q.3.sum)$. Derivable nodes of the quad-tree window in Figure 8 are all colored in white.

Using this storing strategy, the number of nodes that are not derivable (i.e., nodes whose sum must be necessarily stored) is n^2, that is the same as the size of *W*.

This compact representation of *QTW(W)* can be further refined to manage occurrences of null values efficiently. If a node of the quad-tree is null, all of its descendants will be null. Therefore, we can avoid to store the sum associated to every descendant of a null node, as its value is implied. For instance, the sums of the nodes *Q.2.1, Q.2.2, Q.2.3, Q.2.4* need not be stored: their value (i.e., the value 0) can be retrieved by accessing their parent.

We point out that the physically represented quad-tree describing a time window is generally not full. Indeed, null nodes having a non null parent are treated as leaves, as none of their children is

Figure 9. A quad-tree window and its physical representation

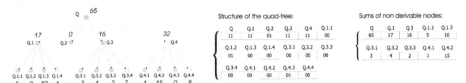

physically stored. We will next focus our attention on the physical compact representation of a quad-tree window.

A quad-tree window can be stored representing separately the tree structure and the content of the nodes. The tree structure can be represented by a string of bits: two bits per node of the tree indicate whether the node is a leaf or not, and whether it is associated with a null block or not. Obviously, in this physical representation, an internal node cannot be null.

In more detail, the encoding pairs are: (1) $\langle 0,0 \rangle$ meaning non null leaf node, (2) $\langle 0,1 \rangle$ meaning null leaf node, (3) $\langle 1,1 \rangle$ meaning non leaf node. It remains one available configuration (i.e., $\langle 1,0 \rangle$) which will be used when compressing quad-tree windows. The mapping between the stored pair of bits and the corresponding nodes of the quad-tree is obtained storing the string of bits according to a predetermined linear ordering of the quad-tree nodes. In Figure 9, the physically represented QTW corresponding to the QTW of Figure 8 is shown. The children of *Q.2* are not explicitly stored, as they are inferable. The string of bits describing the structure of the QTW corresponds to a breadth-first visit of the quad-tree.

Note that, since the blocks in the quad-tree nodes are obtained by consecutive splits into four equally sized quadrants, the above string of bits stores enough information to reconstruct the boundaries of each of these blocks. This means that the boundaries of the blocks corresponding to the nodes do not need to be represented explicitly, as they can be retrieved by visiting the quad-tree structure. It follows that the content of the quad-tree can be represented by an array containing just the sums occurring in the nodes. Some storage space can be further saved observing that:

- we can avoid to store the sums of the null blocks, since the structure bits give enough information to identify them;
- we can avoid to store the sums contained in the *derivable* nodes of the quad-tree window, i.e. the nodes *p* such that $p = Child(q,4)$, for some other node *q*. As explained above, the sum of *p* can be derived as $p.sum = q.sum - \Sigma_{i=1..3}Child(q,i).sum$.

Altogether, the quad-tree window content can be represented by an array storing the set { *p.sum* | *p* is a non-derivable quad-tree node and *p.sum* > 0}.

The above sums are stored according to the same ordering criterion used for storing the structure, in order to associate the sum values to the nodes consistently. For instance, the string of sums reported on the right-hand side of Figure 9 corresponds to the breadth-first visit which has been performed to generate the string of bits on the center of the same figure. The sums of the nodes *Q.2*, *Q.1.2* and *Q.4.3* are not represented in the string of sums as they are null, whereas the sums of the nodes *Q.4*, *Q.1.4*, *Q.3.4* and *Q.4.4* are not stored, as these nodes are derivable.

It can be shown that, if we use 32 bits for representing a sum, the largest storage space needed for a quad-tree window is $S_{QTW}^{\max} = (32 + 8 / 3) \times n^2 - 2 / 3$ bits (assuming that the window does not contain any null value).

Figure 10. Populating a quad-tree window

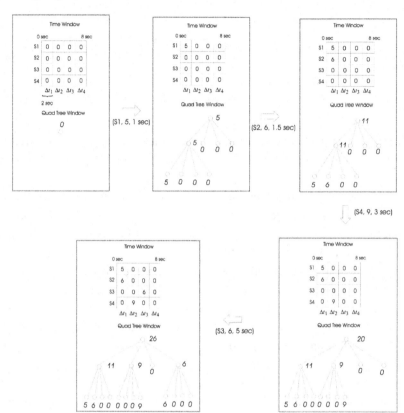

Populating Quad-Tree Windows

In this Section we describe how a quad-tree window is populated as new data arrive. Let W_k be the time window associated to a given time interval $[(k-1) \times T .. k \times T]$, and $QTW(W_k)$ the corresponding quad-tree window. Let $x = \langle id_s, v, t_s \rangle$ be a new sensor reading such that t_s is in $[(k-1) \times T .. k \times T]$. We next describe how $QTW(W_k)$ is updated on the fly, to represent the change of the content of W_k.

Let $QTW(W_k)_{old}$ be the quad-tree windows representing the content of W_k before the arrival of x. If x is the first received reading whose time-stamp belongs to the time interval of W_k, $QTW(W_k)_{old}$ consists of a unique null node (the root). The following algorithm takes as arguments x and $QTW(W_k)_{old}$, and returns the up-to-date quad-tree window Q_{new} on W_k.

First, the old quad-tree window $QTW(W_k)_{old}$ is assigned to Q_{new}. Then, the algorithm determines the coordinates $\langle id_s, j \rangle$ of the element of W_k which must be updated according to the arrival of x, and visits Q_{new} starting from its root. At each step of the visit, the algorithm processes a node of Q_{new} corresponding to a block of W_k which contains $\langle id_s, j \rangle$. The sum associated to the node is updated by adding $value(x)$ to it (see Figure 10). If the visited node was null (before the updating), it is split into four new null children. After updating the current node (and possibly splitting it), the visit goes on processing the child of the current node which contains $\langle id_s, j \rangle$. Algorithm ends after updating the node of Q_{new} corresponding to the single element $\langle id_s, j \rangle$.

Figure 11. Space-time regions

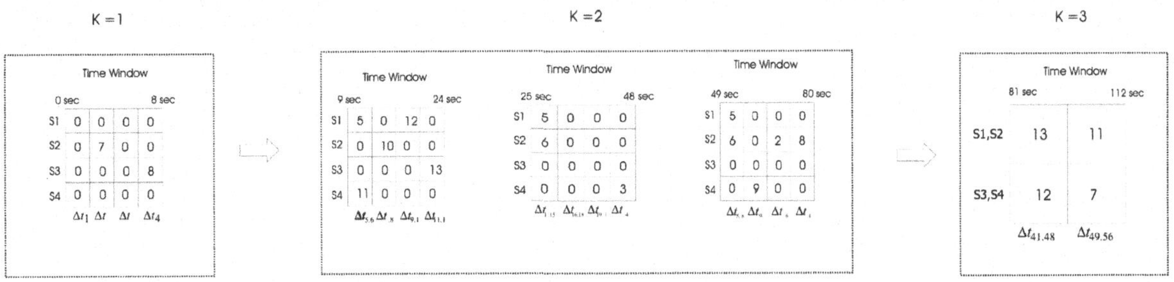

K = 1

Time Window

	0 sec			8 sec
S1	0	0	0	0
S2	0	7	0	0
S3	0	0	0	8
S4	0	0	0	0
	Δt_1	Δt	Δt	Δt_4

K = 2

Time Window

	9 sec			24 sec
S1	5	0	12	0
S2	0	10	0	0
S3	0	0	0	13
S4	11	0	0	0
	$\Delta t_{5,6}$	Δt_n	$\Delta t_{6,1}$	$\Delta t_{1,1}$

Time Window

	25 sec			48 sec
S1	5	0	0	0
S2	6	0	0	0
S3	0	0	0	0
S4	0	0	0	3
	$\Delta t_{i,2}$	$\Delta t_{n,}$	Δt_{j}	$\Delta t_{,}$

Time Window

	49 sec			80 sec
S1	5	0	0	0
S2	6	0	2	8
S3	0	0	0	0
S4	0	9	0	0
	$\Delta t_{k,}$	Δt_o	$\Delta t_{,}$	$\Delta t_{,}$

K = 3

Time Window

	81 sec	112 sec
S1,S2	13	11
S3,S4	12	7
	$\Delta t_{41,48}$	$\Delta t_{49,56}$

THE MULTI-RESOLUTION DATA STREAM SUMMARY

A quad-tree window represents the readings generated within a time interval of size T. The whole sensor data stream can be represented by a sequence of quad-tree windows $QTW(W_1)$, $QTW(W_2)$, ... When a new sensor reading x arrives, it is inserted in the corresponding quad-tree window $QTW(W_k)$, where $ts(x) \in [(k-1) \times T.. k \times T]$. A quad-tree window $QTW(W_k)$ is physically created when the first reading belonging to $[(k-1) \times T.. k \times T]$ arrives.

In this Section we define a structure for representing exploiting different resolution level the values carried by the stream. This structure is called MRDS and pursues two aims:

1. making range queries involving more than one time window efficient to evaluate;
2. making the stored data easy to compress.

We give first an intuition of the overall strategy then we go deep in details of the overall technique. In order to obtain a multi-resolution structure we divide the available storage space in n regions. The first $n - 1$ regions contain time windows whose time granularity is bounded by a maximum level (e.g. a region whose maximum granularity is $4 \times \Delta t$ may also contain QTWs whose granularity is $2 \times \Delta t$ and $3 \times \Delta t$). The n-th region stores QTWs that have been compressed also w.r.t. the spatial source dimension. As new data arrive the algorithm tries to store them in the "youngest" region, two cases may occur: 1) there exists enough storage space so data can be stored in existing QTWs or in newly created ones; 2) the available storage space does not suffice to store all data. In the latter case we need to release some storage space. We proceed as follows, we first compute the minimum number of QTW that has to be dropped away from the current region and try to merge them with QTWs in the nearest region having lower granularity. If the region does not have enough available storage space the merge operation continues backward eventually till the region containing data aggregated both on time and source domain. If we need further storage space we discard "oldest" information since they are assumed to be less relevant.

Managing Input Data Streams

Incoming data stream is stored in a flexible data structure called *Space-Time Multiresolution structure* (STM). In order to store readings we divide the available storage space in region using the following strategy. Depending of the monitored context we fix a maximum time granularity we want to store, say it k, and divide the storage space in $k + 1$ regions as shown in Figure 11.

First k regions (in our example regions 1 and 2) store information pertaining to all the sensors being monitored we refer to them as *Time Compression* regions, while the $(k + 1)$-th region (region 3 in our example) stores information that have been

Figure 12. Merging two quad-tree windows

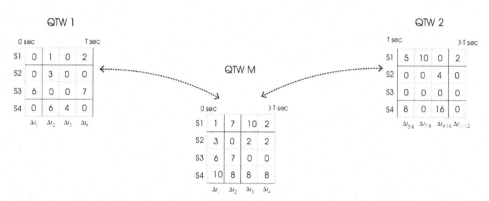

Merging Quad-Tree Windows

compressed both on time and sensor domain and is referred as a *Space-Time Compression* region. Time Compression region k (except the most recent one having $k = 1$ whose maximum granularity is Δt) stores data whose time granularity is bounded by $2^k \times \Delta t$, this imply that inside a given region we store information that can have different time granularity, however as we will show this is not a limitation. As new readings arrive the available storage space in the region having $k = 1$ (i.e., the one that stores most recent data) could be not sufficient to store new data (as a stream is assumed to be possibly infinite) so we need to release some storage space. In order to get necessary storage space we move some QTWs from their original region to the nearest region in the STM. To perform this operation while preserving data quality we merge the $QTW(Q_o)$ to be removed from the original region with the closest $QTW(Q_d)$ in the destination region, i.e. the closest region in the STM, obviously this imply that the resulting QTW will have a coarser granularity than the original ones. If after this merging the available storage space is not enough the merge operation can be performed backward possibly till the last region where the spatio-temporal compression is performed and oldest data could be possibly discarded.

In this Section we describe how the merging operation of two QTWs is performed in order to release some storage space. First of all we need to compute the overall granularity of the resulting QTW. Given two quad-tree windows QTW_1 and QTW_2 whose granularity is respectively $k_1 \times \Delta t$ and $k_2 \times \Delta t$ the quad-tree window QTW_M obtained by merging QTW_1 and QTW_2 exhibit a time granularity $(k_1 + k_2) \times \Delta t$. In order to properly populate QTW_M we have to take into account the fact that k_1 may be different than k_2 thus it may happen that the readings falling in the new unitary time interval should be obtained by splitting the original ones as shown in Figure 12. Once computed the time granularity of QTW_M (say it k_3) we populate it by using the following strategy. We first visit the "youngest" QTW (suppose for the sake of simplicity QTW_1) and merge k_3 / k_1 time interval for each source. Two cases may occur, either the number of time interval contained in QTW_1 is a multiple of k_3 or not. In the former case after visiting QTW_1 completely we populate a portion of QTW_M that does not need further computation. The latter case is more interesting, since in order to compute the readings belonging to a given time interval in QTW_M we need to sum up the contribution of both QTW_1 and QTW_2. In order to better understand this case consider the QTW in Figure 12. QTW_1 has granularity Δt and

QTW_2 has granularity $2 \times \Delta t$ thus the granularity of QTW_M is $3 \times \Delta t$.

As can be easily observed time interval Δt_4 in QTW_1 has to be merged with time interval Δt_{5-6} in order to populate time interval Δt_2 in QTW_M. This example is particularly useful to understand what could happen when time interval in QTW_M does not include a whole time interval in the original QTW. In fact, consider in previous example the interval in QTW_M that includes Δt_{7-8} and only half of Δt_{9-10}. In this case we use CVA and assign half of the value of Δt_{9-10} to both the time interval in QTW_M involving it.

Spatio-Temporal Compression

In this Section we describe the spatio-temporal compression performed in the last region of our data structure. When the merging operation described above does not suffice to get enough storage space to store new readings, we need to adopt a different compression strategy and eventually discard some of the oldest readings. In order to get the desired amount of space we proceed as follows. Suppose that we need an amount of space S, and the last region contains n quad-trees. We try to get the desired amount (S) by releasing an amount of space from each quad-tree that is proportional to its age, i.e. the oldest is the quad-tree the highest is the portion that will be discarded. More formally, given a QTW, say it QTW_i the amount of storage space that will be released from it is the (i / n) % of the storage space actually used by it. Two cases may occur: (1) the released storage space R is greater than S, or (2) we need further storage space to be released. In the former case we proceed with the compression strategy described in next Section. In the latter case we delete some QTWs starting from the oldest one. In order to release some storage space from a given QTW we progressively delete the leaves till the released storage space is greater than the required storage space. Note that the deletion of a level could release more storage space than needed but this is not a limitation since as explained above removing storage space from oldest QTWs is a desirable feature of our technique. The following algorithm perform the compression of an input quad-tree.

COMPRESSION OF THE MULTI-RESOLUTION DATA STREAM SUMMARY

Due to the bounded storage space which is available to store the information carried by the sensor data stream, the MRDS (which consists of a list of indexed clusters of quad-tree windows) cannot be physically represented, as the stream is potentially infinite.

As new sensor readings arrive, the available storage space decreases till no other reading can be stored. Indeed, we can assume that recent information is more relevant than older one for answering user queries, which usually investigate the recent evolution of the monitored world. Therefore, older information can be reasonably represented with less detail than recent data. This suggests us the following approach: as new readings arrive, if there is not enough storage space to represent them, the needed storage space is obtained by discarding some detailed information about "old" data.

We next describe our approach in detail. Let x be the new sensor reading to be inserted, and let $4TI(C_1)$, $4TI(C_2)$, ..., $4TI(C_k)$ be the list of *4-ary Tree Indices* (4TI) representing all the sensor readings preceding x. This means that x must be inserted into $4TI(C_k)$. The insertion of x is done by performing the following steps:

1. the storage space $Space(x)$ needed to represent x into $4TI(C_k)$ is computed by evaluating how the insertion of x modifies the structure and the content of $4TI(C_k)$. $Space(x)$ can be easily computed using the same visiting strategy used for insertion;

2. if $Space(x)$ is larger than the left amount $Space_a$ of available storage space, then the storage space $Space(x) - Space_a$ is obtained by compressing (using a lossy technique) the oldest 4-ary tree indices, starting from $4TI(C_1)$ towards $4TI(C_k)$, till enough space is released;

3. x is inserted into $4TI(C_k)$.

We next describe in detail how the needed storage space is released from the list $4TI(C_1)$, $4TI(C_2)$, ..., $4TI(C_k)$. First, the oldest 4-ary tree index is compressed (using a technique that will be described later) trying to release the needed storage space. If the released amount of storage space is not enough, then the oldest 4-ary tree index is removed from the list, and the same compression step is executed on the new list $4TI(C_2)$, $4TI(C_3)$, ..., $4TI(C_k)$. The compression process ends when enough storage space has been released from the list of 4-ary tree indices.

The compression strategy adopted exploits the hierarchical structure of the 4-ary tree indices: each internal node of a 4TI contains the sum of its child nodes, and the leaf nodes contain the sum of all the reading values contained in the referred quad-tree windows. This means that the information stored in a node of a 4TI is replicated with a coarser "resolution" in its ancestor nodes. Therefore, if we delete four sibling nodes from a 4-ary tree index, we do not loose every information carried by these nodes: the sum of their values is kept in their ancestor nodes.

Analogously, if we delete a quad-tree window QTW_k, we do not loose every information about the values of the readings belonging to the time interval $[(k-1) \times T. \; k \times T]$, as their sum is kept in a leaf node of the 4TI.

The compression of a 4TI consists of removing its nodes progressively, so that the detailed information carried by the removed nodes is kept summarized in their ancestors. This summarized data will be exploited to estimate the original information represented in the removed QTWs

underlying the 4TI. The depth of a 4TI (or, equivalently, the number of QTWs in the corresponding cluster) determines the maximum degree of aggregation which is reached in the MRDS. This parameter depends on the application context. That is, the particular dynamics of the monitored world determines the average size of the time intervals which need to be investigated in order to retrieve useful information. Data summarizing time intervals which are too large w.r.t. this average size are ineffective to exploit in order to estimate relevant information. For instance, the root of a 4TI whose depth is 100 contains the sum of the readings produced within 4^{99} consecutive time windows. Therefore, the value associated to the root cannot be profitably used to estimate the sum of the readings in a single time window effectively (unless additional information about the particular data distribution carried by the stream is available). This issue will be clearer as the estimation process on a compressed MRDS will be explained.

Compressing Quad-Tree Windows

The strategy used for compressing 4-ary tree indices could be adapted for compressing quad-tree windows, as quad-trees can be also viewed as 4-ary trees. For instance, we could compress a quad-tree window incrementally (i.e., as new data arrive) by searching for the left-most node N having 4 child leaf nodes, and then deleting these children.

Indeed, we refine this compression strategy in order to delay the loss of detailed information inside a QTW. Instead of simply deleting a group of nodes, we try to release the needed storage space by replacing their representation with a less accurate one, obtained by using a lower numeric resolution for storing the values of the sums.

To this end, we use a compact structure (called *n Level Tree index - nLT*) for representing approximately a portion of the QTW.

Figure 13. A 3LT index associated to a portion of a quad-tree window

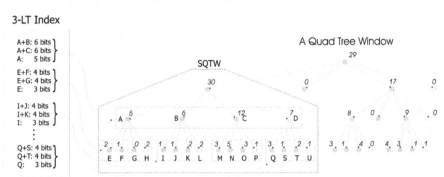

nLT indices have been first proposed in (Buccafurri, Furfaro, Saccà & Sirangelo, 2003) where they are shown to be very effective for the compression of two-dimensional data. A nLT index occupies 64 bits and describes approximately both the structure and the content of a sub-tree with depth at most n of the QTW.

An example of nLT index (called "*3 Level Tree index*" - 3LT) is shown in Figure 13. The left-most sub-tree *SQTW* of the quad-tree of this figure consists of 21 nodes, which occupy 2 × 21 + 32 × 16 = 554 bits (2 × 21 bits are used to represent their structure, whereas 32 × 16 bits to represent the sums of all non derivable nodes). The 64 bits of the nLT index used for *SQTW* are organized as follows: the first 17 bits are used to represent the second level of *SQTW*, the second 44 bits for the third level, and the remainder 3 bits for some structural information about the index. That is, the four nodes in the second level of *SQTW* occupy 3 × 32 + 4 × 2 = 104 bits in the exact representation, whereas they consume only 17 bits in the index. Analogously, the 16 nodes of the third level of *SQTW* occupy 4 × (3 × 32 + 4 × 2) = 416 bits, and only 44 bits in the index. In Figure 13 the first 17 bits of the 3LT index are described in more detail.

Two strings of 6 bits are used for storing *A.sum* + *B.sum* and *A.sum* + *C.sum*, respectively, and further 5 bits are used to store *A.sum*. These string of bits do not represent the exact value of

the corresponding sums, but they represent the sums as fractions of the sum of the parent node. For instance, if *R.sum* is 100 and *A.sum* = 25, *B.sum* = 30, the 6 bit string representing *A.sum* + *B.sum* stores the value: $L_{A+B} = round((A.sum + B.sum) / R.sum \times (2^6 - 1)) = 35$, whereas the 5 bit string representing *A.sum* stores the value: $L_A = round(A.sum / (A.sum + B.sum) \times (2^5 - 1)) = 14$. An estimate of the sums of *A, B, C, D* can be evaluated from the stored string of bits. For instance, an estimate of *A.sum* + *B.sum* is given by: $A.sum + B.sum = L_{A+B} / (2^6 - 1) \times R.sum = 55.6$, whereas an estimate of *B.sum* is computed by subtracting the estimate of *A.sum* (obtained by using L_A) from the latter value.

The 44 bits representing the third level of *SQTW* are organized in a similar way. For instance, two strings of 4 bits are used to represent *E.sum* + *F.sum* and *E.sum* + *G.sum*, respectively, and a string of 3 bits is used for *E.sum*. The other nodes at the third level are represented analogously.

We point out that saving one bit for storing the sum of *A* w.r.t. *A* + *B* can be justified by considering that, on average, the value of the sum of the elements inside *A* is an half of the sum corresponding to *A* + *B*, since the size of *A* is an half of the size of *A* + *B*. Thus, on the average, the accuracy of representing *A* + *B* using 6 bits is the same as the accuracy of representing *A* using 5 bits.

The family of nLT indices includes several types of index other than the 3LT one. Each of these indices reflects a different quad-tree structure: 3LT describes a balanced quad-tree with 3 levels, 4LT (*4 Level Tree*) an unbalanced quad-tree with at most 4 levels, and so on.

However, the exact description of nLT indices is beyond the aim of this chapter. The detailed description of these indices can be found in (Buccafurri, Furfaro, Saccà & Sirangelo, 2003).

The same portion of a quad-tree window could be represented approximately by any of the proposed nLT indices. In (Buccafurri, Furfaro, Saccà & Sirangelo, 2003) a metric for choosing the most "*suitable*" nLT index to approximate a portion of a quad-tree is provided: that is, the index which permits us to re-construct the original data distribution most accurately. As it will be clear next, this metric is adopted in our compression technique: the oldest "portions" of the quad-tree window are not deleted, but they are replaced with the most suitable nLT index.

The QTW to be compressed is visited in order to reach the left-most node N (i.e., the oldest node) having one of the following properties:

1. N is an internal node of the QTW such that $size(N.range) = 16$;
2. The node N has 4 child leaf nodes, and each child is either null or equipped with an index.

Once the node with one of these properties is found, it is equipped with the most suitable nLT index, and all its descending nodes are deleted. In particular, in case 1 (i.e., N is at the last but two level of the uncompressed QTW) N is equipped with a 3LT index.

In case 2 the following steps are performed:

1. All the children of N which are equipped with an index are "expanded": that is, the quad-trees represented by the indices are approximately re-constructed;
2. The most suitable nLT index I for the quad-tree rooted in N is chosen, using the above

cited metric in (Buccafurri, Furfaro, Saccà & Sirangelo, 2003);

3. N is equipped with I and all the nodes descending from N are deleted.

The compressed QTW obtained as described above is not, in general, a full 4-ary tree, as nodes can be deleted during the compression process. Furthermore leaf nodes can be possibly equipped with an nLT index. Thus, the compact physical representation of a QTW, has to be modified in order to represent a compressed QTW. In particular:

*   The pairs of bits which encode the tree structure are redefined as follows: (1) $\langle 0,0 \rangle$ means non null leaf node equipped with nLT index, (2) $\langle 0,1 \rangle$ means null leaf node, (3) $\langle 1,0 \rangle$ means non null leaf node not equipped with nLT index, (4) $\langle 1,1 \rangle$ means non leaf node.
*   The array of sums representing the content of the tree is augmented with the nLT indices associated to the leaves of the compressed QTW.

ESTIMATING RANGE QUERIES ON A MULTI-RESOLUTION DATA STREAM SUMMARY

A range-SUM query $Q = \langle s_i..s_j, [t_{start}..t_{end}] \rangle$ can be evaluated by summing the contributions of every QTW corresponding to a time window overlapping $[t_{start}..t_{end}]$. The QTWs underlying the list of 4TIs are represented by means of a linked list in time ascending order. Therefore the sub-list of QTWs giving some contribution to the query result can be extracted by locating the first (i.e., the oldest) and the last (i.e., the most recent) QTW involved in the query (denoted, respectively, as QTW_{start} and QTW_{end}). This can be done efficiently by accessing the list of 4TIs indexing the QTWs, and locating the first and the last 4TI involved in the query. That is, the

Figure 14. A range query on the MRDS

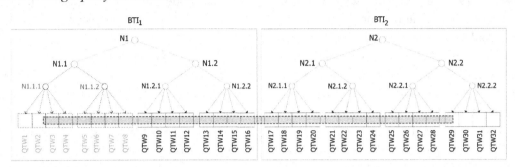

4-ary tree indices $4TI_{start}$ and $4TI_{end}$ which contain a reference to QTW_{start} and QTW_{end}, respectively. $4TI_{start}$ and $4TI_{end}$ can be located efficiently, by performing a binary search on the list of 4TIs. Then, QTW_{start} and QTW_{end} are identified by visiting $4TI_{start}$ and $4TI_{end}$. The answer to the query consists of the sum of the contributions of every QTW between QTW_{start} and QTW_{end}. The evaluation of each of these contributions is explained in detail in the next Section.

Indeed, as the Sensor Data Stream Summary is progressively compressed, it can happen that QTW_{start} has been removed, and the information it contained is only represented in the overlying 4TI with less detail. Therefore, the query can be evaluated as follows:

1. The contribution of all the removed QTWs is estimated by accessing the content of the nodes of the 4TIs where these QTWs are summarized;
2. The contribution of the QTWs which have not been removed is evaluated after locating the oldest QTW involved in the query which is still stored. This QTW will be denoted as QTW'_{start}.

Indeed, it can happen that QTW_{end} has been removed either. This means that all the QTWs involved in the query have been removed by the compression process to release some space, as the QTWs are removed in time ascending order. In this case, the query is evaluated by estimating the contribution of each involved QTW by accessing only the nodes of the overlying 4TIs.

For instance, consider the MRDS consisting of two 4TIs shown in Figure 14. The QTWs whose perimeter is dashed (i.e., QTW_1, QTW_2, ..., QTW_8) have been removed by the compression process. The query represented with a grey box is evaluated by summing the contributions of the $4TI_1$ nodes *N1.1* and *N1.2* with the contribution of each QTW belonging to the sequence QTW_9, QTW_{10}, ..., QTW_{29}.

The query estimation strategy algorithm uses a function *4TIBinarySearch* which takes as arguments a MRDS and the time boundaries of the range query, and returns the first and the last of the summary involved in the query. Moreover, it uses the function *EstimateAndLocate*. This function is first invoked on $4TI_{start}$ and performs two tasks: (1) it evaluates the contribution of the 4TI nodes involved in the query where the information of the removed QTWs is summarized, and (2) it locates (if possible) QTW'_{start}, i.e. the first QTW involved in the query which has not been removed. If QTW'_{start} is not referred by $4TI_{start}$, *EstimateAndLocate* is iteratively invoked on the subsequent 4TIs, till either QTW'_{start} is found or all the 4TIs involved in the query have been visited. The contribution of the 4TI leaf nodes to the query estimate is evaluated by performing linear interpolation. This is a simple estimation technique which is widely used on summarized data, such as *histograms*, in the context of selectivity estimation (Acharya, Gibbons, Poosala &

Ramaswamy, 1999), and compressed datacubes, in the context of OLAP applications. The use of linear interpolation on a leaf node N of a 4TI is based on the assumption that data are uniformly distributed inside the two-dimensional range $N.range$ (CVA). If we denote the two dimensional range corresponding to the intersection between $N.range$ and the range of the query Q as $N \cap Q$, and the size of the whole two dimensional range delimited by the node N as $size(N)$, the contribution of N to the query estimate is given by: $size(N \cap Q) / size(N) \times N.sum)$.

Estimating a Range-SUM Query Inside a QTW

The contribution of a QTW to a query Q is evaluated as follows. The quad-tree underlying the QTW is visited starting from its root (which corresponds to the whole time window).

When a node N is being visited, three cases may occur:

1. *The range corresponding to the node is external to the range of Q*: the node gives no contribution to the estimate;
2. *The range corresponding to the node is entirely contained into the range of Q*: the contribution of the node is given by the value of its sum;
3. *The range corresponding to the node partially overlaps the range of Q*: if N is a leaf and is not equipped with any index, linear interpolation is performed for evaluating which portion of the sum associated to the node lies onto the range of the query. If N has an index, the index is "expanded" (i.e., an approximate quad-tree rooted in N is reconstructed using the information contained in the index). Then the new quad-tree is visited with the same strategy as the QTW to evaluate the contribution of its nodes. Finally, if the node N is internal, the contribution of

the node is the sum of the contributions of its children, which are recursively evaluated.

The pre-aggregations stored in the nodes of quad-tree windows make the estimation inside a QTW very efficient. In fact, if a QTW node whose range is completely contained in the query range is visited during the estimation process, its sum contributes to the query result exactly, so that none of its descending nodes must be visited. This means that, generally, not all the leaf nodes involved in the query need to be accessed when evaluating the query estimate. The overall estimation process turns out to be efficient thanks to the hierarchical organization of data in the QTWs, as well as the use of the overlying 4TIs which permits us to locate the quad-tree windows efficiently. We point out that the 4TIs involved in the query can be located efficiently too, i.e. by performing a binary search on the ordered list of 4TIs stored in the MRDS. The cost of this operation is logarithmic with respect to the list length, which is, in turn, proportional to the number of readings represented in the MRDS.

Answering Continuous (Range) Queries

The range query evaluation paradigm on the data summary can be easily extended to deal with *continuous range queries*. A *continuous query* is a triplet $Q = \langle s_i..s_j, \Delta T_{start}, \Delta T_{end} \rangle$ (where $\Delta T_{start} > \Delta T_{end}$) whose answer, at the current time t, is the evaluation of an aggregate operator (such as *sum, count, avg*, etc.) on the values produced by the sources $s_i, s_{i+1}, ..., s_j$ within the time interval $[t - \Delta T_{start}..t - \Delta T_{end}]$. In other words, a continuous query can be viewed as range query whose time interval "moves" continuously, as time goes on. The output of a continuous query is a stream of (simple) range query answers which are evaluated with a given frequency. That is, the answer to a continuous query $Q = \langle s_i..s_j, \Delta T_{start}, \Delta T_{end} \rangle$

issued at time t_0 with frequency Δt is the stream consisting of the answers of the queries $Q_0 = \langle s_i. s_j, t_0 - \Delta T_{start}, t_0 - \Delta T_{end} \rangle$, $Q_1 = \langle s_i..s_j, t_0 - \Delta T_{start} + \Delta t, t_0 - \Delta T_{end} + \Delta t \rangle$, $Q_2 = \langle s_i..s_j, t_0 - \Delta T_{start} + 2 \times \Delta t, t_0 - \Delta T_{end} + 2 \times \Delta t \rangle$, ... The i-th term of this stream can be evaluated efficiently if we exploit the knowledge of the $(i-1)$-th value of the stream. In this case the ranges of two consecutive queries Q_{i-1} and Q_i are overlapping, and Q_i can be evaluated by answering two range queries whose size is much less than the size of Q_i. These two range queries are $Q' = \langle s_i..s_j, t_0 - \Delta T_{start} + (i-1) \times \Delta t, t_0 - \Delta T_{start} + i \times \Delta t \rangle$, and $Q'' = \langle s_i..s_j, t_0 - \Delta T_{end} + (i-1) \times \Delta t, t_0 - \Delta T_{end} + i \times \Delta t \rangle$. Thus, we have: $Q_i = Q_{i-1} - Q' + Q''$.

CONCLUSION

In this chapter, a comprehensive framework for efficiently supporting OLAP analysis over multidimensional data streams has been presented. Thanks to intelligent multidimensional data processing and compression paradigms, the proposed framework allows us to gain effectiveness and efficiency during data stream analysis in comparison with capabilities of conventional DBMS-inspired tools.

Future work is oriented towards the definition of new models and algorithms for supporting the evaluation of OLAP analysis tasks over multidimensional data streams based on complex aggregate queries (e.g., (Dobra, Garofalakis, Gehrke & Rastogi, 2002)) rather than simple range queries considered in this research effort.

REFERENCES

Acharya, S., Gibbons, P. B., Poosala, V., & Ramaswamy, S. (1999). Join synopses for approximate query answering. In *Proceedings of 19th ACM Symposium on Principles of Database Systems* (pp. 275-286).

Agarwal, S., Agrawal, R., Deshpande, P. M., Gupta, A., Naughton, J. F., Ramakrishnan, R., & Sarawagi, S. (1996). On the computation of multidimensional aggregates. In *Proceedings of the 22th International Conference on Very Large Data Bases* (pp. 506-521).

Agrawal, R., & Srikant, R. (1994). Fast algorithms for mining association rules. In *Proceedings of the 20th International Conference on Very Large Data Bases* (pp. 487-499).

Alert System. (2007). Retrieved from http://www.alertsystems.org

Ananthakrishna, R., Das, A., Gehrke, J., Korn, F., Muthukrishnan, S., & Srivastava, D. (2003). Efficient approximation of correlated sums on data streams. *IEEE Transactions on Knowledge and Data Engineering, 15*(3), 569–572. doi:10.1109/TKDE.2003.1198391

Avnur, R., & Hellerstein, J. M. (2000). Eddies: Continuously adaptive query processing. In *Proceedings of the 2000 ACM International Conference on Management of Data* (pp. 261-272).

Babcock, B., Babu, S., Datar, M., Motwani, R., & Widom, J. (2002). Models, issues in data stream systems. In *Proceedings of the 21st ACM Symposium on Principles of Database Systems* (pp. 1-16).

Babu, S., & Widom, J. (2001). Continuous queries over data streams. *SIGMOD Record, 30*(3), 109–120. doi:10.1145/603867.603884

Barbarà, D., Du Mouchel, W., Faloutsos, C., Haas, P. J., Hellerstein, J. M., & Ioannidis, Y. E. (1997). The New Jersey data reduction report. *A Quarterly Bulletin of the Computer Society of the IEEE Technical Committee on Data Engineering, 20*(4), 3–45.

Berchtold, S., Böhm, C., & Kriegel, H.-P. (1998). The pyramid-technique: Towards breaking the curse of dimensionality. In *Proceedings of the 1998 ACM International Conference on Management of Data* (pp. 142-153).

Beyer, K., & Ramakrishnan, R. (1999). Bottom-up computation of sparse, iceberg cubes. In *Proceedings of the 1999 ACM International Conference on Management of Data* (pp. 359-370).

Bonnet, P., Gehrke, J., & Seshadri, P. (2000). Querying the physical world. *IEEE Personal Communications*, *7*(5), 10–15. doi:10.1109/98.878531

Bonnet, P., Gehrke, J., & Seshadri, P. (2001). Towards sensor database systems. In *Proceedings of 2nd International Conference on Mobile Data Management* (pp. 3-14).

Buccafurri, F., Furfaro, F., Saccà, D., & Sirangelo, C. (2003). A quad-tree based multiresolution approach for two-dimensional summary data. In *Proceedings of the 15th IEEE International Conference on Scientific, Statistical Database Management* (pp. 127-137).

Cai, Y. D., Clutterx, D., Papex, G., Han, J., Welgex, M., & Auvilx, L. (2004). MAIDS: Mining alarming incidents from data streams. In *Proceedings of the 2004 ACM International Conference on Management of Data* (pp. 919-920).

Chaudhuri, S., & Dayal, U. (1997). An overview of data warehousing and OLAP technology. *SIGMOD Record*, *26*(1), 65–74. doi:10.1145/248603.248616

Chen, Y., Dong, G., Han, J., Wah, B. W., & Wang, J. (2002). Multi-dimensional regression analysis of time-series data streams. In *Proceedings of the 28th International Conference on Very Large Data Bases* (pp. 323-334).

Colliat, G. (1996). OLAP, relational, and multidimensional database systems. *SIGMOD Record*, *25*(3), 64–69. doi:10.1145/234889.234901

Cortes, C., Fisher, K., Pregibon, D., Rogers, A., & Smith, F. (2000). Hancock: A language for extracting signatures from data streams. In *Proceedings of the 6th ACM International Conference on Knowledge Discovery and Data Mining* (pp. 9-17).

Cuzzocrea, A. (2005). Overcoming limitations of approximate query answering in OLAP. In *Proceedings of the 9th IEEE International Database Engineering, Applications Symposium* (pp. 200-209).

Deligiannakis, A., Kotidis, Y., & Roussopoulos, N. (2003). *Data reduction techniques for sensor networks*. (Technical Report CS-TR-4512). UM Computer Science Department.

Deligiannakis, A., Kotidis, Y., & Roussopoulos, N. (2004). Compressing historical information in sensor networks. In *Proceedings of the ACM International Conference on Management of Data* (pp. 527-538).

Deligiannakis, A., Kotidis, Y., & Roussopoulos, N. (2004). Hierarchical In-network data aggregation with quality guarantees. In *Proceedings of 9th International Conference on Extending Database Technology* (pp. 658-675).

Dobra, A., Garofalakis, M., Gehrke, J., & Rastogi, R. (2002). Processing Complex aggregate queries over data streams. In *Proceedings of the 2002 ACM International Conference on Management of Data* (pp. 61-72).

Domingos, P., & Hulten, G. (2000). Mining high-speed data streams. In *Proceedings of the 6th ACM International Conference on Knowledge Discovery, Data Mining* (pp. 71-80).

Fang, M., Shivakumar, N., Garcia-Molina, H., Motwani, R., & Ullman, J. D. (1998). Computing iceberg queries efficiently. In *Proceedings of the 24th International Conference on Very Large Data Bases* (pp. 299-310).

Ganti, V., Li Lee, M., & Ramakrishnan, R. (2000). ICICLES: Self-tuning samples for approximate query answering. In *Proceedings of 26th International Conference on Very Large Data Bases* (pp. 176-187).

Gehrke, J., & Madden, S. (2004). Query processing in sensor networks. *IEEE Pervasive Computing / IEEE Computer Society [and] IEEE Communications Society, 3*(1), 46–55. doi:10.1109/MPRV.2004.1269131

Gilbert, A. C., Kotidis, Y., Muthukrishnan, S., & Strauss, M. J. (2001). Surfing wavelets on streams: One-pass summaries for approximate aggregate queries. In *Proceedings of 27th International Conference on Very Large Data Bases* (pp. 79-88).

Gonzalez, H., Han, J., Li, X., & Klabjan, D. (2006). Warehousing and analyzing massive RFID data sets. In *Proceedings of the 22nd IEEE International Conference on Data Engineering* (pp. 83-93).

Gray, J., Chaudhuri, S., Bosworth, A., Layman, A., Reichart, D., & Venkatrao, M. (1997). Data cube: A relational aggregation operator generalizing group-by, cross-tab and sub-totals. *Data Mining and Knowledge Discovery, 1*(1), 29–54. doi:10.1023/A:1009726021843

Guha, S., Meyerson, A., Mishra, N., Motwani, R., & O'Callaghan, L. (2003). Clustering data streams: Theory, practice. *IEEE Transactions on Knowledge and Data Engineering, 15*(3), 515–528. doi:10.1109/TKDE.2003.1198387

Han, J., Chen, Y., Dong, G., Pei, J., Wah, B. W., Wang, J., & Cai, Y. D. (2005). Stream cube: An architecture for multi-dimensional analysis of data streams. *Distributed and Parallel Databases, 18*(2), 173–197. doi:10.1007/s10619-005-3296-1

Han, J., & Kamber, M. (2000). *Data mining: Concepts and techniques*. Morgan Kaufmann.

Han, J., Pei, J., Dong, G., & Wang, K. (2001). Efficient computation of iceberg cubes with complex measures. In *Proceedings of the 2001 ACM International Conference on Management of Data* (pp. 1-12).

Henzinger, M. R., Raghavan, P., & Rajagopalan, S. (1998). Computing on data streams (Technical Report 1998-011). Digital Systems Research Center.

Ho, C.-T., Agrawal, R., Megiddo, N., & Srikant, R. (1997). Range queries in OLAP data cubes. In *Proceedings of the 1997 ACM International Conference on Management of Data* (pp. 73-88).

Intanagonwiwat, C., Estrin, D., Govindan, R., & Heidemann, J. S. (2002). Impact of network density on data aggregation in wireless sensor networks. In *Proceedings of the 22nd IEEE International Conference on Distributed Computing Systems* (pp. 457-458).

Ives, Z. G., Levy, A. Y., & Weld, D. S. (2000). Efficient evaluation of regular path expressions on streaming XML data (Technical Report UW-CSE-2000-05-02). University of Washington.

Li, X., Han, J., & Gonzalez, H. (2004). High-dimensional OLAP: A minimal cubing approach. In *Proceedings of the 30th International Conference on Very Large Data Bases* (pp. 528-539).

Madden, S., & Franklin, M. J. (2002). Fjording the stream: An architecture for queries over streaming sensor data. In *Proceedings of the 18th IEEE International Conference on Data Engineering* (pp. 555-566).

Madden, S., Franklin, M. J., & Hellerstein, J. M. (2002). TAG: A Tiny AGgregation service for ad-hoc sensor networks. *ACM SIGOPS Operating Systems Review, 36*, 131–146. doi:10.1145/844128.844142

Madden, S., & Hellerstein, J. M. (2002). Distributing queries over low-power wireless sensor networks. In *Proceedings of the 2002 ACM International Conference on Management of Data* (pp. 622).

Madden, S., Szewczyk, R., Franklin, M. J., & Culler, M. J. (2002). Supporting aggregate queries over ad-hoc wireless sensor networks. In *Proceedings of the 4th IEEE Workshop on Mobile Computing and Systems, Applications* (pp. 49-58).

Manku, G. S., & Motwani, R. (2002). Approximate frequency counts over data streams. In *Proceedings of the 28th International Conference on Very Large Data Bases* (pp. 346-357).

Qiao, L., Agrawal, D., & El Abbadi, A. (2002). RHist: Adaptive summarization over continuous data streams. In *Proceedings of the 11th ACM International Conference on Information and Knowledge Management* (pp. 469-476).

Vassiliadis, P., & Sellis, T. K. (1999). A survey of logical models for OLAP databases. *SIGMOD Record, 28*(4), 64–69. doi:10.1145/344816.344869

Xin, D., Han, J., Cheng, H., & Li, X. (2006). Answering top-k queries with multi-dimensional selections: The ranking cube approach. In *Proceedings of the 32th International Conference on Very Large Data Bases* (pp. 463-475).

Yao, Y., & Gehrke, J. (2003). Query Processing in Sensor Networks. In *Proceedings of the 1st International Conference on Innovative Data Systems Research*. Retireved from http://www-db.cs.wisc.edu/cidr/cidr2003/program/p21.pdf

Zhang, D., Gunopulos, D., Tsotras, V. J., & Seeger, B. (2002). Temporal aggregation over data streams using multiple granularities. In *Proceedings of 8th International Conference on Extending Database Technology* (pp. 646-663).

Zhao, Y., Deshande, P. M., & Naughton, J. F. (1997). An array-based algorithm for simultaneous multidimensional aggregates. In *Proceedings of the 1997 ACM International Conference on Management of Data* (pp. 159-170).

KEY TERMS AND DEFINITIONS

Approximate Query Answer: Answer to a query evaluated over a given compressed data structure for query efficiency purposes. The main goal of approximate query answering techniques is represented by the minimization of the query error.

Compression: The re-encoding of data into a form that uses fewer bits of information than the original data. Compression is often used to minimize memory resources needed to store and manage such data.

Continuous Value Assumption: It is an a-priori assumption on the nature of data according to which data are uniformly distributed over the interval being considered for analysis.

Data Stream: A sequence of data that flow continuously from a stream source. It could be assumed as infinite, thus leading to different computation needs. A data stream can be generally defined as a pair $\langle T_{id}, t_s \rangle$ where T_{id} denotes a tuple and t_s a timestamp.

Data Warehouse: A large store of data supporting multidimensional analysis. Organizations make use of data warehouses to help them analyze historic transactional data in order to detect useful patterns and trends. Data from operational source are transferred into the data warehouse by means of a process called ETL (Extracting, Transforming and Loading). Then, data are organized and stored in the data warehouse in ways that optimize them for high-performance analysis.

Dimension Flattening: A process meant to reduce a multidimensional data stream into a flattened two-dimensional data stream, which is a semantics-based compression of the original stream.

OLAP: On-Line Analytical Processing (OLAP) is a category of software technology that enables analysts, managers and executives to gain insight into data through fast, consistent, interactive access to a wide variety of possible views of information that has been transformed from raw data in order to reflect the real dimensionality of the enterprise as understood by the user.

Quad-Tree: A tree-based data structure whose key feature is that each internal node has up to four children. Quad trees are used to partition a two dimensional space by recursively subdividing it into four quadrants or regions.

Sensor Networks: Sensors are non-reactive elements used to monitor real life phenomena, such as live weather conditions, network traffic, and so forth. Sensor networks can be defined as large collections of linked sensors.

Chapter 3
Warehousing RFID and Location–Based Sensor Data

Hector Gonzalez
University of Illinois at Urbana-Champaign, USA

Jiawei Han
University of Illinois at Urbana-Champaign, USA

Hong Cheng
University of Illinois at Urbana-Champaign, USA

Tianyi Wu
University of Illinois at Urbana-Champaign, USA

ABSTRACT

Massive Radio Frequency Identification (RFID) datasets are expected to become commonplace in supply-chain management systems. Warehousing and mining this data is an essential problem with great potential benefits for inventory management, object tracking, and product procurement processes. Since RFID tags can be used to identify each individual item, enormous amounts of location-tracking data are generated. Furthermore, RFID tags can record sensor information such as temperature or humidity. With such data, object movements can be modeled by movement graphs, where nodes correspond to locations, and edges record the history of item transitions between locations and sensor readings recorded during the transition. This chapter shows the benefits of the movement graph model in terms of compact representation, complete recording of spatio-temporal and item level information, and its role in facilitating multidimensional analysis. Compression power and efficiency in query processing are gained by organizing the model around the concept of gateway nodes, which serve as bridges connecting different regions of graph, and provide a natural partition of item trajectories. Multi-dimensional analysis is provided by a graph-based object movement data cube that is constructed by merging and collapsing nodes and edges according to an application-oriented topological structure.

DOI: 10.4018/978-1-60566-328-9.ch003

1 INTRODUCTION

The increasingly wide adoption of RFID technology by retailers to track containers, pallets, and even individual items as they move through the global supply chain, from factories in producer countries, through transportation ports, and finally to stores in consumer countries, creates enormous datasets containing rich multi-dimensional information on the movement patterns associated with objects along with massive amounts of important sensor information collected for each object. However, this information is usually hidden in terabytes of low-level RFID readings, making it difficult for data analysts to gain insight into the set of interesting patterns influencing the operation and efficiency of the procurement process. For example, we may discover a pattern that relates humidity and temperature during transportation to return rates for dairy products. In order to realize the full benefits of detailed object tracking and sensing information, we need a compact and efficient RFID cube model that provides OLAP-style operators useful to navigate through the movement data at different levels of abstraction of both spatio-temporal, sensor, and item information dimensions. This is a challenging problem that cannot be efficiently solved by traditional data cube operators, as RFID datasets require the aggregation of high-dimensional graphs representing object movements, not just that of entries in a flat fact table.

The problem of constructing a warehouse for RFID datasets has been studied in (Gonzalez, Han, Li, & Klabjan, 2006; Gonzalez, Han, & Li, 2006; Gonzalez, Han, Li, & Klabjan, 2006) introduced the concept of the RFID-cuboid, which compresses and summarizes an RFID dataset by recording information on items that stay together at a location with *stay* records, and linking such records through the use of a *map* table that connects groups of items that move and stay together through several locations. This view carries an implicit notion of the graph structure of RFID

datasets but it fails to explicitly recognize the concept of *movement graph* as a natural model for item movements, and thus neglects the study of the topological characteristics of such a graph and its implications for query processing, cube computation, and data mining. In this chapter, we approach the RFID data warehouse from a *movement graph-centric* perspective, which makes the warehouse conceptually clear, better organized, and obtaining significantly deeper compression and an order of magnitude performance gain over (Gonzalez, Han, Li, & Klabjan, 2006) in the processing of path queries.

The importance of the *movement graph* approach to RFID data warehousing can be illustrated with an example.

Example 1. Consider a large retailer with a global supplier and distribution network that spans several countries, and that tracks objects with RFID tags placed at the item level. Such a retailer sells millions of items per day through thousands of stores around the world, and for each such item it records the complete set of movements between locations starting at factories in producing countries, going through the transportation network, and finally arriving at a particular store where it is purchased by a customer. The complete path traversed by each item can be quite long as readers are placed at very specific locations within factories, ships, or stores (*e.g.*, a production lane, a particular truck, or an individual shelf inside a store). Further, for each object movement, we can record sensor readings, such as weight loss, humidity, or temperature. These lead not only to a tremendous amount of "scattered" data but also to a rather complicated picture.

The questions become "*how can we present a clean and well organized picture about RFID objects and their movements?*" and "*whether such a picture may facilitate data compression, data cleaning, query processing, multi-level, multi-dimensional OLAPing, and data mining?*"

The movement-graph approach provides a nice and clean picture for modeling RFID tagged objects at multiple levels of abstraction. Further, it facilitates data compression, data cleaning, and answering rather sophisticated queries, such as

- Q_1 (High-level aggregate/OLAP query): *What is the average shipping cost of transporting electronic goods from factories in Shanghai, to stores in San Francisco in 2007? And then click to drill-down to month and see the trend.*

- Q_2 (Sensor query): *Print the transportation paths for the meat products from Argentina sold at L.A. on April 5, that were exposed to over C heat for over 5 hours on the route.*

- Q_3 (Data mining query): *Why did the 20 packages of Dairyland milk at this Walmart store go bad today? Is it more related to farm, or store, or transportation?*

In this chapter we propose a *movement graph-based* model, which leads to concise and clean modeling of massive RFID datasets and facilitates RFID data compression, query answering, cubing, and data mining. The *movement graph* is a graph that contains a node for every *distinct* (or more exactly, *interesting*) location in the system, and edges between locations record the history of shipments (groups of items that travel together) between locations. For each shipment we record a set of interesting measures such as, travel time, transportation cost, or sensor readings like temperature or humidity. We show that this graph can be partitioned and materialized according to its topology to speedup a large number of queries, and that it can be aggregated into cuboids at different abstraction levels according to location, time, and item dimensions, to provide multi-dimensional and multi-level summaries of item movements. The core technical aspects that we will cover in the chapter are as follows:

1. **Gateway-based partitioning of the movement graph.** The *movement graph* can be divided into disjoint partitions that are connected through special gateway nodes. Most paths with locations in more than one partition include the gateway nodes. For example, most items travel from China to the United States by going through major shipping ports in both countries. Gateways can be given by a user or be discovered by analyzing traffic patterns. Further, materialization can be performed for indirect edges between individual locations and their corresponding gateways and between gateways. Such materialization facilitates computing measures on the paths connecting locations in different partitions. An efficient graph partitioning algorithm, is developed, that uses the gateways to split the graph into clusters in a single scan of the path database.

2. **Partition-based group movement compression.** (Gonzalez, Han, & Li, 2006a) Introduces the idea of compressing RFID datasets by mapping the *EPCs* of multiple individual objects traveling together into a single generalized identifier *gid*. This compression is based on the assumption that items move in large groups near factories, and are then split into smaller groups as they approach stores. However, in global supply-chain applications, one may observe a *"merge-split"* process, *e.g.*, shipments grow in size as they approach the major ports, and then after a long distance bulky shipping, they gradually split (or even recombined) when approaching stores. Thus the mappings based on the one-way splitting model (Gonzalez, Han, & Li, 2006a) provide limited compression. One can use instead a partitioned map table that creates a separate mapping for each partition, rooted at the gateways.

3. **Movement graph aggregation.** The *movement graph* can be aggregated to different levels of abstraction according to the location concept hierarchy that determines which subset of locations are interesting for

Table 1. An example path database

Tag	Path
t1	$(A,1,2)(D,4,5)(G_1,6,8)(G_2,9,10)(F,11,12)(J,14,17)$
t2	$(A,2,3)(D,4,5)(G_1,6,8)(G_2,9,10)(F,11,12)(K,16,18)$
t3	$(A,2,3)(D,4,6)$
t4	$(B,1,2)(D,3,5)(G_1,6,7)(G_2,9,10)(I,11,13)(K,14,15)$
t5	$(C,1,3)(E,4,5)(G_1,6,7)(G_2,9,10)(I,11,14)$
t6	$(A,3,3)(B,4,5)(I,10,11)$

analysis, and at which level. For example, the *movement graph* may only have locations inside Massachusetts, and it may aggregate every individual location inside factories, warehouses, and stores to a single node. This aggregation mechanism is very different from the one present in traditional data cubes as nodes and edges in the graph can be merged or collapsed, and the shipments along edges and their measures need to be recomputed using different semantics for the cases of node merging and node collapsing. A second view of *movement graph* aggregation is the process of merging shipment entries according to time and item dimensions.

4. **Partitioned-based cubing.** One can take advantage of the partitioned graph to conduct partition level cubing. The idea is, that for each partition, one can transforms the original path database into a compressed path prefix tree representation, that is used to compute the map table, and perform simultaneous aggregation of paths and items to every interesting level of abstraction.

2 GATEWAY-BASED MOVEMENT GRAPH MODEL

In this section we provide a brief introduction to RFID data, present a gateway-based movement graph model for RFID warehouse, and outline the overall system architecture.

2.1 RFID Data

An RFID object tracking system is composed of a collection of RFID readers scanning for tags at periodic intervals. Each such reader is associated with a location, and generates a stream of time-ordered tuples of the form (*EPC, location, time*) where, *EPC*[1] is a unique 96 bit electronic product code associated with a particular item, *location* is the place where item was detected, and *time* is the time when the detection took place. Significant data compression is possible by merging all the readings for an item that stays at a location for a period of time, into a tuple of the form (*EPC, location*,, where, is the time when the item identified by *EPC* entered *location*, and is the time when it left.

By sorting tag readings on *EPC* one can generate a *path database*, where the sequence of locations traversed by each item is stored. Entries in the path database are of the form: (*EPC*, ... Table 1 presents an example path database for six items identified with tags *t*1 to *t*6, traveling through locations *A, B, C, D, G*$_1$, G$_2$, *F, I*, and *J*.

Sensor Data

In many cases we can also have extra information describing sensor readings collected during item shipments, this data has the form (*from, to, t*$_1$*, t*$_2$: where, *from* and *to* are the initial and final locations of the shipment, t_1 and t_2 are the starting and ending time of the shipment, is the set of

Figure 1. An example movement graph

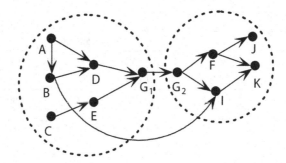

items transported, and describes properties such as temperature, humidity, or weight.

Path Independent Data

In addition to the location information collected by RFID readers, an RFID application has detailed information on the characteristics of each item in the system, such information can be represented with tuples of the form (EPC, d_1, d_2, ..., d_m), where each d_i is a particular value for dimension D_i, typical dimensions could be *product type, manufacturer*, or *price*.

2.2 Gateway-Based Movement Graph

Among many possible models for RFID data warehouses, we believe the gate-way movement graph model not only provides a concise and clear view over the movement data, but also facilitates data compression, querying, and analysis of massive RFID datasets (which will be clear later).

Definition 1. *A movement graph G(V,E) is a directed graph representing object movements; V is the set of locations, E is the set of transitions between locations. An edge e(i,j) indicates that objects moved from location v_i to location v_j. Each edge is annotated with the history of object movements along the edge, each entry in the history is a tuple of the form,,.:, where all the objects in took the transition together, starting at time t_{start} and ending at time, and records properties of the shipment.*

Figure 1 presents the *movement graph* for the path database in Table 1.

A large RFID dataset may involve many locations and movements. It is important to put such movements in an organized picture. In a global supply chain it is possible to identify important locations that serve to connect remote sets of locations in the transportation network. Gateways generally aggregate relatively small shipments from many distinct, regional locations into large shipments destined for a few well-known remote locations; or they distribute large shipments from remote locations into smaller shipments destined to local regional locations. These special nodes are usually associated with shipping ports, *e.g.*, the port in Shanghai aggregates traffic from multiple factories, and makes large shipments to ports in the United States, which in turn split the shipments into smaller units destined for individual stores. The concept of gateways is important because it allows us to naturally partition the *movement graph* to improve query processing efficiency and reduce the cost of cube computation.

Gateways can be categorized into three classes: *Out-Gateways, In-Gateways*, and In-Out-Gateways, as shown below.

Out-Gateways: In the supply chain it is common to observe locations, such as ports, that receive relatively low volume shipments from a multitude of locations and send large volume shipments to a few remote locations. For example, a port in Shanghai may receive products from a multitude

Figure 2. Three types of gateways

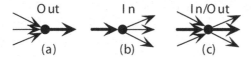

of factories and logistics centers throughout China to later send the products through ship to a port in San Francisco. We call this type of node an *Out-Gateway*, and it is characterized by having multiple incoming edges with relatively low average shipment sizes and a few outgoing edges with much larger average shipment sizes and traffic. One important characteristic of these locations is that products usually can only reach remote locations in the graph by first going through an *Out-Gateway*. Figure 2-a presents an *Out-gateway*. For our running example, Figure 1, location is an *Out-gateway*.

In-Gateways: *In-gateways* are the symmetric complement of *Out-Gateways*, they are characterized by a few incoming edges with very large average shipment sizes and traffic, and a multitude of relatively low average shipment size outgoing edges. An example of an *In-Gateway* may be sea port in New York where a large volume of imported goods arrive at the United States and are redirected to a multitude of distribution centers throughout the country before reaching individual stores. As with *Out-Gateways* these nodes dominate a large portion of the flow in the *movement graph*; most products entering a partition of the graph will do so through an *In-Gateway*. Figure 2-b presents an example *Out-gateway*. For our running example, Figure 1, location G_2 is an *In-gateway*.

In-Out-Gateways: *In-Out-gateways* are the locations that serve as both *In-gateways* and *Out-gateways*. This is the case of many ports that may for example serve as an *In-gateway* for raw materials being imported, and an *Out-gateway* for manufactured goods being exported. Figure 2-c presents such an example. It is possible to split an *In-Out-gateway* into separate In- and Out-

gateways by matching incoming and outgoing edges carrying the same subset of items into the corresponding single direction traffic gateways.

Notice that gateways may naturally form hierarchies. For example, one may see a hierarchy of gateways, *e.g.*, country level sea ports ® region level distribution centers ® state level hubs.

2.3 Overall System Architecture

Based on the gateway-based movement graph model, Figure 3 presents the RFID warehouse architecture. The system receives as input a stream of RFID readings in the form of (*EPC*, *location*, *time*), these readings are stored in a path database containing the complete path for each *EPC* in the system. From this path database, gateway information is extracted, which serves a natural marker to partition long paths, *e.g.*, a long path from factories in China to retail stores at the United States can be partitioned into segments defined by the ports it went through in both countries. Such partitioned movement graph guides us to group and compress bulky objects movements at the partition level. Moreover, materialization can be explored by precomputing node-to-gateway, gateway-to-gateway, and gateway-to-node movements, which provides significant query speedups. Cubing can be done at the partition level in a single scan, which can be in turn used in query processing, OLAP, and data mining. Each component described in this overall architecture will be detailed in the subsequent sections.

Figure 3. System Architecture

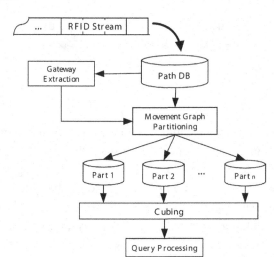

3 DATA PARTITIONING

In this section we discuss the methods for identifying gateways, partitioning based on the *movement graph*, and associating partitions to gateways.

3.1 Gateway Identification

In many applications it is possible for data analysts to provide the system with the complete list of gateways, this is realistic in a typical supply-chain application where the set of transportation ports is well known in advance, *e.g.*, Walmart knows all the major ports connecting its suppliers in Asia to the entry ports in the United States. In some other cases, we need to discover gateways automatically. One can use existing graph partitioning techniques such as *balanced minimum cut* or *average minimum cut* (Chung, 1997), to find a small set of edges that can be removed from the graph so that the graph is split into two disconnected components; such edges will typically be associated with the strong traffic edges of in- or out- gateways. Gateways could also be identified by using the concept of betweenness and centrality in social network analysis as they will correspond to nodes with high betweenness

as defined in (Freeman, 1977) and we can use an efficient algorithm such as (Brandes, 2001) to find them.

We propose a simple but effective approach to discover gateways that works well for typical supply-chain operations where gateways have strong characteristics that are easy to identify. We can take a *movement graph*, and rank nodes as potential gateways based on the following observations: (i) a large volume of traffic goes through gateway nodes, (ii) gateways carry *unbalanced traffic*, *i.e.*, incoming and outgoing edges carrying the same tags but having very different average shipment sizes, and (iii) gateways split paths into largely disjoint sets of nodes that only communicate through the gateway. The algorithm can find gateways by eliminating first low traffic nodes, and then the nodes with balanced traffic, *i.e.*, checking the number of incoming and outgoing edges, and the ratio of the average incoming (outgoing) shipment sizes vs. the average of the outgoing (incoming) shipment sizes. Finally, for the edges that pass the above two filters, check which locations split the paths going through the location into two largely disjoint sets. That is, the locations in paths involving the gateway can be split into two subsets, locations occurring in the path before the gateway and those occurring in the path after the gateway.

3.2 Movement Graph Partitioning

The movement graph partitioning problem can be framed as a traditional graph clustering problem and we could use techniques such as spectral clustering (Harinarayan, Rajaraman, & Ullman, 1996; Chung, 1997). But for the specific problem of partitioning supply-chain *movement graphs* we can design a less costly algorithm that takes advantage of the topology of the graph to associate locations to those gateways to which they are more strongly connected.

The key idea behind the partitioning algorithm is that in the movement graph for a typical supply chain application locations only connect directly (without going through another gateway) to a few gateway nodes. That is, very few items in Europe reach the major ports in the United States without first having gone through Europe's main shipping ports. Using this idea, we can associate each location to the set of gateways that it directly reaches (we use a frequency threshold to filter out gateways that are reached only rarely), when two locations l_i and l_j have a gateway in common we merge their groups into a single partition containing the two locations and all their associated gateways. We repeat this process until no additional merge is possible. At the end we do a postprocessing step where we associate very small partitions to the larger partition to which it most frequently directly connects to.

3.3 Handling Sporadic Movements: Virtual Gateways

An important property of gateways is that all the traffic leaving or entering a partition goes through them. However, in practice it is still possible to have small sets of sporadic item movements between partitions that bypass gateways. Such movements reduce the effectiveness of gateway-based materialization because path queries involving multiple partitions will need to examine some path segments of the graph unrelated to gateways. This problem can be easily solved by adding a special *virtual gateway* to each partition for all outgoing and incoming traffic from and to other partitions that does not go through a gateway. Virtual gateways guarantee that inter-gateway path queries can be resolved by looking at gateway-related traffic only.

For our running example, Figure 1, we can partition the *movement graph* along the dotted circles, and associate gateway G_1 with the first partition, and gateway G_2 with the second one. In this case we need to create a virtual gateway G_x to send outgoing traffic from the first partition (*i.e.* traffic from B) that skips G_1, and another virtual gateway G_y to receive incoming traffic into the second partition (*i.e.* traffic to I), that skips G_2.

4 STORAGE MODEL

With the tremendous amounts of RFID data, it is crucial to study the storage model. We propose to use three data structures to store both compressed and generalized data: (1) an edge table, storing the list of edges, (2) a map table, linking groups of items moving together, and (3) an information table, registering path-independent information related to the items in the graph.

4.1 Edge Table

This table registers information on the edges of the *movement graph*, the format is ⟨*from,to,history*⟩, where *from* is the originating node, *to* is the destination node, and *history* is a set of tuples recording the history of item shipments along the edge; each history tuple is of the form,, *direct*,, where is the time when the items departed the location *from*, is the time when the items arrived at the location *to*, *direct* is a boolean value that is true if the items moved directly between *from* and *to* and false if intermediate nodes were visited, is the list of items that traveled together, and is a set of aggregate functions computed on the items in while they took the transition, *e.g.*, it can be count, average temperature, average movement cost, *etc.*. We will elaborate more on concept of *gids* in the section describing the map table.

An alternative to recording edges in the graph is to record nodes (Gonzalez, Han, Li, & Klabjan, 2006), and the history of items staying at each node. In some applications this may be useful if most queries inquire about the properties of items as they stay at a particular location more than the properties of items as they move between locations. Our model can accommodate this view by just switching locations and edges in the *movement graph*, *i.e.*, in this view edges represent locations, and nodes represent transitions. We can also adapt the model for the case when we record item properties for both locations and transitions, by modifying the *movement graph* to associate both transitions and locations with edges. Which model to choose will depend on applications. For example, if the truck temperature during the transportation is essential, take the edge table. If storage room situation is critical, take the stay table. If both are important, use both tables. In the rest of this chapter we will assume, without loss of generality, that edges in the *movement graph* are associated with transitions, but it should be clear that all the techniques and algorithms would work under alternative edge interpretations.

4.2 Map Table

Bulky movement means a large number of items move together. A generalized identifier *gid* can be assigned to every group of items that moves together, which will substantially reduce the size of the at each edge, similar to that proposed in (Gonzalez, Han, Li, & Klabjan, 2006). When groups of items split into smaller groups, *gid* (original group) can be split into a set of children *gids*, representing these smaller groups. The map

table contains entries of the form ⟨*partition*,*gid*,,, where *partition* is the subgraph of the *movement graph* where this map is applicable, is the list of all *gids* with a list of items that is a superset of the items *gid*, is the list *gids* with item lists that are a subset of *gid*, or a list of individual tags if *gid* did not split into smaller groups.

The proposed design of the map table differs from the one presented in (Gonzalez, Han, Li, & Klabjan, 2006). (Gonzalez, Han, Li, & Klabjan, 2006) adopts a split-only model: It starts with the most generalized *gids* near factories and splits into smaller sets in distributions. Since the real case may involve both merge and split, the split-only model cannot have much sharing and compression. Here we adopt a "merge-split" model, where objects can be merged, shuffled, and split in many different combinations during transportation. Our mapping table takes a *gateway-centered model*, where mapping is centered around gateways, *i.e.*, the largest merged and collective moving sets at the gateways become the root *gids*, and their children *gids* can be spread in both directions along the gateways. This will lead to the maximal *gid* sharing and *gid*_list compression.

4.3 Information Table

The information table records other attributes that describe properties of the items traveling through the edges of the *movement graph*. The format of the tuples in the information table is, $D_1, \ldots$,, where is the list of items that share the same values on the dimensions D_1 to D_n, and each dimension D_i describes a property of the items in . An example of attributes that may appear in the information table could be *product*, *manufacturer*, or *price*. Each dimension of the information table can have an associated concept hierarchy, *e.g.*, the *product* dimension may have a hierarchy such as *EPC* ® *SKU* ® *product* ® *category*.

5 MATERIALIZATION STRATEGY

Materialization of path segments in the *movement graph* may speedup a large number of path-related queries. Since there are an exponential number of possible path segments that can be pre-computed in a large *movement graph*, it is only realistic to partially materialize only those path segments that provide the highest expected benefit at a reasonable cost. We will develop such a strategy here.

5.1 Path Queries

A path query requires the computation of a measure over all the items with a path that matches a given *path pattern*. It is of the form: *info,, measure›*, where *info* is a selection on the information table that retrieves the relevant items for analysis; is a sequence of stage conditions on location and time that should appear in every path, in the given order but possibly with gaps; and *measure* is a function to be computed on the matching paths. An example path query may be *info* = {product = meat, sale_date = 2006}, Argentina farm A, San Mateo store S}, and *measure=average temperature*, which asks for the average temperature recorded for each meat package, traveling from a certain farm in Argentina to a particular store in San Mateo.

There may be many strategies to answer a path query, but in general we will need to retrieve the appropriate tag lists and measures for the edges along the paths involving the locations in the *path expression*; retrieve the tag list for the items matching the info selection; intersect the lists to get the set of relevant tags; and finally, if needed, retrieve the relevant paths to compute the measure.

5.2 Path-Segment Materialization

We can model path segments as indirect edges in the *movement graph*. For example, if we want to pre-compute the list of items moving from location l_i to location l_j through any possible path, we can materialize an edge from l_i to l_j that records a history of all tag movements between the nodes, including movements that involve an arbitrary number of intermediate locations. Indirect edges are stored in the same format as direct ones, but with the flag *direct* set to *false*.

The benefit of materializing a given indirect edge in the *movement graph* is proportional to the number of queries for which this edge reduces the total processing cost. Indirect edges, involved in a path query, reduce the number of edges that need to be analyzed, and provide shorter tag lists that are faster to retrieve and intersect. In order for an indirect edge to help a large number of queries, it should have three properties: (i) carry a large volume of traffic, (ii) be part of a large portion of all the paths going from nodes in one partition of the graph to nodes in any other partition, and (iii) be involved directly or indirectly in a large number of path queries. The set of edges that best match these characteristics are the following.

Node-to-gateway. In supply-chain implementations it is common to find a few well defined *Out-gateways* that carry most of the traffic leaving a partition of the graph where items are produced, before reaching a partition of the graph where items are consumed. For example, products manufactured in China destined for exports to the United States leave the country through a set of ports. We propose to *materialize the (virtual) edges from every node to the Out-gateways that it first reaches*. Such materialization, for example, would allow us to quickly determine the properties of shipments originating at any location inside China and leaving the country.

Gateway-to-node. Another set of important nodes for indirect edge materialization are *In-gateways*, as most of the item traffic entering a region of the graph where items are consumed has to go through an *In-gateway*. For example, imported products coming into the United States all arrive through a set of major ports. When we need to determine which items sold in the U.S. have paths that involve locations in foreign countries, we can easily get this information by pre-computing the list of items that arrived at the location from each of the *In-gateways*. We propose to *materialize all the (virtual) edges from an* In-gateway *to the nodes that it reaches without passing through any other gateway.*

Gateway-to-gateway. Another interesting set of indirect edges to materialize are *the ones carrying inter-gateway traffic*. For example, we want to pre-compute which items leaving the Shanghai port finally arrive at the New York port. The benefit of such indirect edge is twofold: First, it aggregates a large number possible paths between two gateways and pre-computes important measures on the shipments; and second, it allows us to quickly determine which items travel between partitions.

In general, when we need to answer a path query involving all paths between two nodes, we need to retrieve all edges between the nodes and aggregate their measures. This can be very expensive if the locations are connected by a large number of distinct paths, which is usually the case when nodes are in different partitions of the graph. By using gateway materialization we reduce this cost significantly as remote nodes can always be connected through a few edges to, from, and between gateways.

6 GRAPH CUBE

So far we have examined the *movement graph* at a single level of abstraction. Since items, locations (as nodes in the graph), and the history of shipments along each edge all have associated concept hierarchies; aggregations can be performed at different levels of abstraction. We propose a data cube model for such *movement graphs*. The main difference between the traditional cube model and the *movement graph* cube is that the former aggregates on simple dimensions and levels but the latter needs to aggregate on path-dimensions as well, which may involve path collapsing as a new

form of generalization. In this section we develop a model for *movement graph* cubes and introduce an efficient algorithm to compute them.

6.1 Movement Graph Aggregation

With a concept hierarchy associated with locations, a path can be aggregated to an abstract location by aggregating each location to a generalized location, collapsing the corresponding movements, and rebuilding the *movement graph* according to the new path database.

Location aggregation. We can use the location concept hierarchy to aggregate particular locations inside a store to a single store location, or particular stores in a region to a single region location. We can also completely disregard certain locations not interesting for analysis, *e.g.*, a store manager may want to eliminate all factory-related locations from the *movement graph* in order to see a more concise representation of the data.

Figure 4 presents a *movement graph* and some aggregations. All the locations are initially at the lowest abstraction level. By generalization, the transportation-related locations are collapsed into a single node *Transportation*, and the store-related locations into a single node *Store* (shown as dotted circles). Then the original single path: *Factory ® Dock ® Hub ® Backroom ® shelf* is collapsed to the path *Factory ® Transportation ® Store* in the aggregated graph. If we completely remove transportation locations, we will get the path *Factory ® Store*.

Edge aggregation semantics. From the point of view of the edge table, graph aggregation corresponds to merging of edge entries, but it is different from regular grouping of fact table entries in a data cube because collapsing paths will create edges that did not exist before, and some edges can be completely removed if they are not important for analysis. In a traditional data cube, fact table entries are never created or removed, they are just aggregated into larger or smaller groups. For example, in Figure 4, if we remove

Figure 4. Graph Aggregation

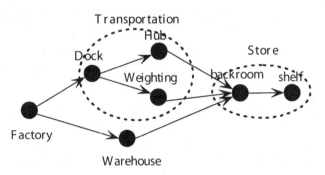

all transportation-related locations, a new edge (*Factory*, *Store*) will be created, and all edges to and from transportation locations removed.

Graph aggregation involves different operations over the at each edge, when we remove nodes we need to intersect to determine the items traveling through the new edge, but when we simply aggregate locations to higher levels of abstraction (without removing them) we need instead to compute the union of the of several edges. For example, looking at Figure 4 in order to determine the for the edge (*Factory*, *Store*) we need to intersect the of all outgoing edges from the node *Factory* with the incoming edges to the node *Store*; on the other hand if we aggregate transportation locations to a single node, in order to determine the for the edge (*Transportation*, *Store*), we need to union the of the edges (*Hub*,*Store*) and (*Weighting*,*Store*).

Item aggregation. Each item traveling in the *movement graph* has an associated set of dimensions that describe its properties. These dimensions can also be aggregated, *e.g.*, an analyst may ask for the *movement graph* of *laptops* in *March 2006*, while another analyst may be interested in looking at more general movements for *electronic goods* in *2006*. Aggregation along info dimensions is the same type of aggregation that we see in traditional data cubes, we group items according to the values they share at a given abstraction level of the info dimensions.

There are thus, two distinct but related views on *movement graph* aggregation. The first view, which we call *path aggregation* corresponds to the merging of nodes in the *movement graph* according to the location concept hierarchy. The second view, corresponds to aggregation of shipments and items traveling through the edges of the *movement graph*, according concept hierarchies for time and info dimensions.

6.2 Cube Structure

Fact table. The fact table contains information on the *movement graph*, and the items aggregated to the minimum level of abstraction that is interesting for analysis. Each entry in the fact table is a tuple of the form: ⟨*from, to,,, d_1, d_2, ..., d_k*, where is list of *gids* that contains all the items that took the transition between *from* and *to* locations, starting at time and ending at time, and all share the dimension values d_1, ..., d_k for dimensions D_1, ..., D_k in the info table, contains a set of measures computed on the *gids* in .

Measure. For each entry in the fact table we register the corresponding to the tags that match the dimension values in the entry. We can also record for each *gid* in the list a set of measures recorded during shipping, such as average temperature, total weight, or count. We can use the to quickly retrieve those paths that match a given *slice*, *dice*, or *path selection* query at any level of

Table 2. Cuboid example

From	To	t1	t2	Product	tag_list
⊢	A	*	*	*	t1,t2,t3,t6
⊢	B	*	*	*	t4
⊢	C	*	*	*	t5
A	G_1	*	*	*	t1,t2
A	G_x	*	*	*	t6
A	⊣	*	*	*	t3
B	G_1	*	*	*	t4
B	G_x	*	*	*	t6
C	G_1	*	*	*	t5

abstraction. When a query is issued for aggregate measure that are already pre-computed in the cube, we do not need to access the path database, and all query processing can be done directly on the aggregated *movement graph*. For example, if we record *count* as a measure, any query asking for counts of items moving between locations can be answered directly by retrieving the appropriate cells in the cube. When a query asks for a measure that has not been pre-computed in the cube, we can still use the aggregated cells to quickly determine the list of relevant *gids* and retrieve the corresponding paths to compute the measure on the fly.

Cuboids. A cuboid in the *movement graph* resides at a level of abstraction of the location concept hierarchy, a level of abstraction of the time dimension, and a level of abstraction of the *info dimensions*. *Path aggregation* is used to collapse uninteresting locations, and *item aggregation* is used to group-related items. *Cells* in a *movement graph* cuboid group both items, and edges that share the same values at the cuboid abstraction level.

It is possible for two separate cuboids to share a large number of common cells, namely, all those corresponding to portions of the *movement graph* that are common to both cuboids, and that share the same *item aggregation* level. A natural optimization is to compute cells only once. The size

of the full *movement graph* cube is thus the total number of distinct cells in all the cuboids. When materializing the cube or a subset of cuboids we compute all relevant cells to those cuboids without duplicating shared cells between cuboids.

Table 2 presents an example cuboid computed on the first partition of the *movement graph* of Figure 1. In this cuboid locations *C* and *D* are uninteresting, and both time and item dimensions have been aggregated to *all*. Each entry is a cell, and the measure is a tag_list[2].

6.3 Querying, OLAP, and Data Mining

By compressing and aggregating RFID data in an organized way, the movement graph-based data warehouse can be readily used for efficient query processing, OLAP and data mining.

In Example 1, we presented three queries with different flavor, from high-level aggregate/OLAP query (Q_1) to low-level path query (Q_2) and data mining query (Q_3). Q_1 can be answered directly using a precomputed graph cube. Q_2 can be answered by first extracting set of transportation paths from the cube using a set of query constants, and then selecting the transportation route for those paths with temperature greater than C and sum of the durations of these hot paths is over 5 hours. For Q_3, a classification model can be constructed by taking the set of rotten milk as positive examples and other package of milk as negatives, and perform induction at multi-level abstraction on the cube to find the essential features that distinguish the two sets. With the organized data warehouse and pre-computed cube, such a process can be done efficiently.

7 PERFORMANCE STUDY

In this section we present a comprehensive evaluation of the *movement graph* model and algorithms. All the experiments were conducted on a Pentium 4 3.0 GHz, with 1.5Gb RAM, running Win XP; the code was written in C++ and compiled with Microsoft Visual Studio 2003.

7.1 Data Synthesis

The path databases used for performance evaluation were generated using a synthetic path generator. We first generate a random *movement graph* with 5 partitions and 100 locations, each partition has a random number of gateways. Lo-

cations inside a partition are arranged according to a *producer configuration*, where we simulate factories connecting to intermediate locations that aggregate traffic, which in turn connect to *Out-gateways*; or a *consumer configuration*, where we simulate products moving from *In-gateways*, to intermediate locations such as distribution centers, and finally to stores. We generate paths by simulating groups of items moving inside a partition, or between partitions, and going usually through gateways, but sometimes, also "jumping" directly between non-gateway nodes; we increase shipment sizes close to gateways. We control the number of items moving together by a *shipment size* parameter, which indicates the smallest number of items moving together in a partition. Each item in the system has an entry in the path database, and an associated set of *item dimensions*. We characterize a dataset by N the number of paths, and S the minimum shipment size.

In most of the experiments in this section we compare three competing models. The first model is *part gw mat*, it represents a partitioned *movement graph* where we have performed materialization

Figure 5. Fact table size vs. Path db size (S=300)

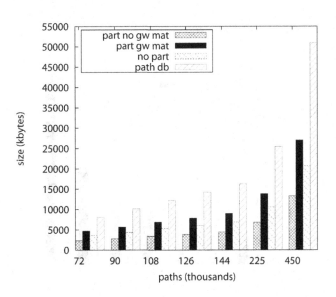

of path segments to, from, and between gateways. The second model is *part gw no mat*, it represents a model where we partition the *movement graph* but we do not perform gateway materialization. The third model is *no part*, which represents a *movement graph* that has not been partitioned and corresponds to the model introduced in (Gonzalez, Han, Li, & Klabjan, 2006).

7.2 Model Size

In these experiments we compare the sizes of three models of *movement graph* materialization, and the size of the original path database *path db*. For all the experiments, we materialize the graph at the same level of abstraction as the one in the path database and thus is a lossless representation of the data.

Figure 5 presents the size of the four models on path databases with a varying number of paths. For this experiment we can clearly see that the partitioned graph without gateway materialization *part no gw mat* is always significantly smaller than the non partition model *non part*, this comparison is fair as both models materialize only direct edges. When we perform gateway materialization the size of the model increases, but the increase

is still linear on the size of the *movement graph* (much better than full materialization of edges between every pair of nodes which is quadratic in the size of the *movement graph*), and close to the size of the model in (Gonzalez, Han, Li, & Klabjan, 2006).

Figure 6 presents the size of the map table for the partitioned *part gw mat* and non-partitioned *no part* models. The difference in size is almost a full order of magnitude. The reason is that our partition level maps capture the semantics of collective object movements much better than (Gonzalez, Han, Li, & Klabjan, 2006). This has very important implications in compression power, and more importantly, in query processing efficiency. Figures 7 and 8 present the same analysis for a dataset where we vary the minimum shipment size. As expected, our model works better when shipment sizes are relatively large.

7.3 Query Processing

In these experiments we generate 100 random path queries that, ask for a measure on the path segments, for items matching an *item dimension* condition, that go from a single initial location to a single ending location, and that occurred within

Figure 6. Map table size vs. Path db size (S=300)

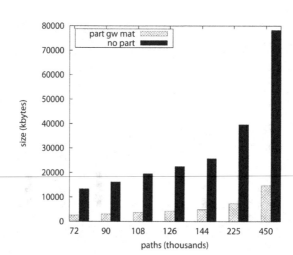

Figure 7. Fact table size vs. Shipment size (N=108,000)

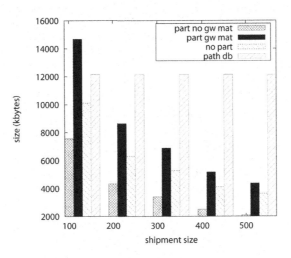

Figure 8. Map table size vs. Shipment size (N=108,000)

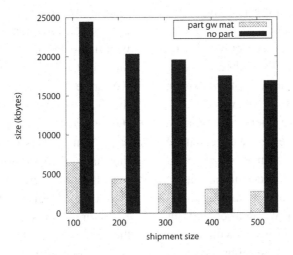

a certain time interval. We compare the partition *movement graph* with gateway materialization *part gw mat*, against the partitioned graph *part no gw mat* without gateway materialization, and the non-partitioned graph *no part*. All the queries were answering a *movement graph* at the same abstraction level as the original path database. For lack of space we restrict the analysis queries with starting and ending locations in different partitions as those are in general more challenging to answer. Based on our experiments on single

partition queries, our method has a big advantage given by their compact map tables.

For the case of non-partitioned graph, we use the same query processing algorithm presented in (Gonzalez, Han, Li, & Klabjan, 2006). For the case of partitioned graph without gateway materialization, we retrieve all the relevant edges from the initial node to the gateways in its partition, edges between gateways, and edges from the gateways in the ending location's partition and the location. In this case we do not perform inter-gateway join

Figure 9. Query IO vs. Path db size (S=300)

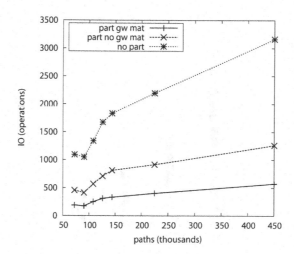

of the *gid* lists, but the overhead of such join can be small if we keep an inter-gateway join table, or if our measure does not require matching of the relevant edges in both partitions. For the gateway materialization case, we retrieve only relevant gateway-related edges. For this method we compute the cost using tag lists instead of *gid* lists on materialized edges to and from gateways.

In Figure 9 we analyze query performance with respect to path database size. We see that the gateway-based materialization method is the clear winner, its cost is almost an order of magnitude smaller than the method proposed in (Gonzalez, Han, Li, & Klabjan, 2006). We also see that our method has the lowest growth in cost with respect to database size. Figure 10 presents the same analysis but for a path database with different minimum shipment sizes. Our model is the clear winner in all cases, and as expected performance improves with larger shipment sizes.

7.4 Cubing

For the cubing experiments we compute a set of 5 random cuboids, with significant shared dimensions among them, *i.e.*, the cuboids share a large number of interesting locations, and *item dimensions*. We are interested in the study of

such cuboids because it captures the gains in efficiency that we would obtain if we used our algorithm to compute a full *movement graph* cube, as ancestor/descendant cuboids in the cube lattice benefit most from shared computation. Figure 11 presents the total runtime to compute 5 cuboids, we can see that *shared* significantly outperforms the level by level cubing algorithm presented in (Gonzalez, Han, Li, & Klabjan, 2006). For the case when cuboids are very far apart in the lattice the shared computation has a smaller effect and our algorithm performs similarly to (Gonzalez, Han, Li, & Klabjan, 2006).

Figure 12 presents the total size of the cells in the 5 cuboids for the case of a partitioned graph without gateway materialization, and a non-partitioned graph. The compression advantage of our method increases for larger database sizes. This advantage becomes even more important as more cuboids are materialized. We can thus use our model to create compact *movement graphs* at different levels of abstraction, and furthermore, use them to answer queries significantly more efficiently than competing models. If we want even better query processing speed, we can sacrifice some compression and perform gateway-based materialization.

Figure 10. Query IO vs. Shipment size (N=108,000)

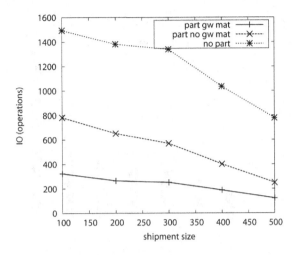

Figure 11. Cubing time vs. Path db size (S=300 N=108,000)

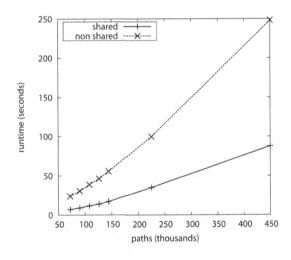

7.5 Partitioning

The final experiment evaluates the scalability of our *movement graph* partitioning algorithm. For this experiment we assume that the set of gateway nodes is given, which is realistic in many applications as this set is small and well known (*e.g.*, major shipping ports). Figure 13 shows that the algorithm scales linearly with the size of the path database. And for our test cases, which are generated following the operation of a typi-

cal global supply chain operation, the algorithm always finds the correct partitions. This algorithm can partition very large datasets quickly, and at a fraction of the cost of standard graph clustering algorithms. It is important to note that for the case when we are dealing with a more general *movement graph*, that does not represent a supply chain, more expensive algorithms are likely required to find good partitions.

Figure 12. Cube size vs. Path db size (S=300 N=108,000)

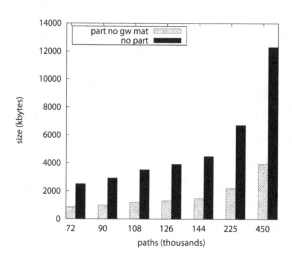

Figure 13. Partitioning time vs. Path db size

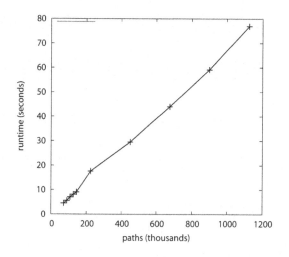

8 RELATED WORK

Management of RFID systems is an active area of research that has received extensive attention both from the hardware and software communities. Hardware research has focused on the development of communications mechanisms and protocols aimed at fast, secure, and accurate detection of tags (Finkenzeller, 2003). The software community studies the problem of online processing of RFID datasets (Sarma, Brock, & Ashton, 2000), cleaning of noisy readings (Jeffery, Alonso,

Franklin, Hong, & Widom, 2006; Gonzalez, Han, & Shen, 2007; Jeffery, Garofalakis, & Franklin, 2006; Rao, Doraiswamy, Thakar, & Colby, 2006), and warehousing, mining, and cubing of large RFID datasets (Gonzalez, Han, Li, & Klabjan, 2006; Gonzalez, Han, & Li, 2006a; Gonzalez, Han, & Li, 2006b; Chawathe, Krishnamurthy, Ramachandran, & Sarma, 2004).

(Gonzalez, Han, Li, & Klabjan, 2006) introduces the concept of the RFID-cuboid, a model that compresses and summarizes RFID datasets by recording information on items that stay and move together through the use of stay and map tables. Our work builds on these ideas, but we make the key observation that materialization, query processing, and cubing should be guided by the topology of the *movement graph*, a concept absent in (Gonzalez, Han, Li, & Klabjan, 2006). Our analysis and experiments show that our *movement graph* centric model provides better compression and significantly more efficient query processing and cubing than (Gonzalez, Han, Li, & Klabjan, 2006). (Gonzalez, Han, & Li, 2006a) develops the idea of the *FlowCube*, which is a data cube where cells contain a probabilistic *FlowGraph* (Gonzalez, Han, & Li, 2006b), which can be used to compute path probabilities and identify correlations. The *FlowGraph* is a lossy model that sacrifices accuracy for compactness. Our cube model is a lossless and compact representation of the data, and can be used to obtain precise answers to any query on item or path behavior, something that the *FlowCube* cannot provide. We can position the *movement graph* cube as a building block on top of which a multitude of mining problems, including the *FlowCube*, can be implemented efficiently.

Our work benefits from the RFID cleaning line of research (Jeffery, Garofalakis, & Franklin, 2006; Rao, Doraiswamy, Thakar, & Colby, 2006; Gonzalez, Han, & Shen, 2007), but it can also help the cleaning process. For example, we can use the RFID warehouse to improve the accuracy of tag and transition detections presented in (Jeffery, Garofalakis, & Franklin, 2006; Gonzalez, Han, & Shen, 2007) by cross checking missed tag readings with nearby locations to see if a transition has occurred. We can also improve the performance of the deferred cleansing techniques proposed in (Rao, Doraiswamy, Thakar, & Colby, 2006) by providing a more efficient data model than that obtained from standard relational databases.

The problem of gateway identification is similar to the problem of measuring the centrality notion of node betweenness in social networks (Freeman, 1979) and we can benefit from studies

on fast betweenness computation algorithms such as (Brandes, 2001). In our work we propose an efficient gateway-based partitioning algorithm that performs well for supply-chain *movement graphs*. In the more general case of warehousing arbitrary moving objects, we can make use of the extensive literature on minimum balanced cut clustering through spectral techniques (Chung, 1997) to partition such *movement graphs*. Our cube model shares many of the general cubing ideas introduced in (Agarwal et al., 1996), but we differ in that the unit of cubing is a graph, not a relational table. The decision on which cuboids to compute can use the extensive research on partial cube materialization (Harinarayan, Rajaraman, & Ullman, 1996). Our cubing algorithm builds on the ideas of shared computation of cuboids presented in (Zhao, Deshpande, & Naughton, 1997).

9 CONCLUSION

We have presented the *gateway-based movement graph model* for warehousing massive, transportation-based RFID datasets. This model captures the essential semantics of supply-chain application as well as many other RFID applications that explore object movements of similar nature. It provides a clean and concise representation of large RFID datasets. Moreover, it sets up a solid foundation for modeling RFID data and facilitates efficient and effective RFID data compression, data cleaning, multi-dimensional data aggregation, query processing, and data mining.

A set of efficient methods has been presented for movement graph construction, gateway identification, gateway-based graph partitioning, efficient storage structuring, multi-dimensional aggregation, graph cube computation, and cube-based query processing. This weaves an organized picture for systematic modeling and implementation of such an RFID data warehouse. The implementation and performance study shows that the methods proposed here are more efficient in both storage cost, cube computation, and query processing comparing with a previous study (Gonzalez, Han, Li, & Klabjan, 2006) that uses a global map table without gateway-based movement graph modeling and partitioning.

The gateway-based movement graph model captures the semantics of bulky, sophisticated, but collective object movements, including merging, shuffling, and splitting processes. Its applications are not confined to RFID data sets but also extensible to other bulky object movement data. However, further study is needed to model and warehouse objects with scattered movements, such as traffic on highways where each vehicle moves differently from others. We are currently studying such modeling techniques and will report our progress in the future.

REFERENCES

Agarwal, S., Agrawal, R., Deshpande, P. M., Gupta, A., Naughton, J. F., Ramakrishnan, R., & Sarawagi, S. (1996). On the computation of multidimensional aggregates. In *VLDB'96* (pp. 506-521).

Brandes, U. (2001). A faster algorithm for betweenness centrality. *The Journal of Mathematical Sociology*, *25*, 163–177.

Chawathe, S., Krishnamurthy, V., Ramachandran, S., & Sarma, S. (2004). Managing RFID data. In *VLDB'04*.

Chung, R. (1997). *Spectral Graph Theory* (Vol. 92). CBMS Regional Conference Series in Mathematics.

Finkenzeller, K. (2003). *RFID Handbook: Fundamentals and Applications in Contactless Smart Cards and Identification*. John Wiley and Sons.

Freeman, L. (1977). A set of measures of centrality based on betweenness. *Sociometry*, *40*, 35–41. doi:10.2307/3033543

Freeman, L. (1979). Centrality in social networks: conceptual clarifications. *Social Networks*, *1*, 215–239. doi:10.1016/0378-8733(78)90021-7

Gonzalez, H., Han, J., & Li, X. (2006a). Flowcube: Constructuing RFID flowcubes for multidimensional analysis of commodity flows. In *VLDB'06*, Seoul, Korea.

Gonzalez, H., Han, J., & Li, X. (2006b,). Mining compressed commodity workflows from massive RFID data sets. In *CIKM'06*, Virginia.

Gonzalez, H., Han, J., Li, X., & Klabjan, D. (2006). Warehousing and analysis of massive RFID data sets. In *ICDE'06*, Atlanta, Georgia.

Gonzalez, H., Han, J., & Shen, X. (2007). Cost-Conscious cleaning of massive RFID data sets. In *ICDE'07*, Istanbul, Turkey.

Harinarayan, V., Rajaraman, A., & Ullman, J. D. (1996). Implementing data cubes efficiently. In *SIGMOD'96*.

Jeffery, S. R., Garofalakis, M., & Franklin, M. J. (2006). Adaptive cleaning for RFID data streams. In *VLDB'06*, Seoul, Korea.

Jeffrey, S. R., Alonso, G., Franklin, M., Hong, W., & Widom, J. (2006). A pipelined framework for online cleaning of sensor data streams. In *ICDE'06*, Atlanta, Georgia.

Rao, J., Doraiswamy, S., Thakar, H., & Colby, L. (2006). A deferred cleansing method for RFID data analytics. In *VLDB'06*, Seoul, Korea.

Sarma, S., Brock, D. L., & Ashton, K. (2000). *The networked physical world* [White paper]. MIT Auto-ID Center.

Zhao, Y., Deshpande, P. M., & Naughton, J. F. (1997). An array-based algorithm for simultaneous multidimensional aggregates. In *SIGMOD '97*.

ENDNOTES

[1] We use EPC and tag interchangeably in the chapter

[2] In reality this list would be a gid_list, but we use tag_list in order to make the example easier to understand

Chapter 4
Warehousing and Mining Streams of Mobile Object Observations

S. Orlando
Università Ca' Foscari di Venezia, Italy

A. Raffaetà
Università Ca' Foscari di Venezia, Italy

A. Roncato
Università Ca' Foscari di Venezia, Italy

C. Silvestri
Università Ca' Foscari di Venezia, Italy

ABSTRACT

In this chapter, the authors discuss how data warehousing technology can be used to store aggregate information about trajectories of mobile objects, and to perform OLAP operations over them. To this end, the authors define a data cube with spatial and temporal dimensions, discretized according to a hierarchy of regular grids. This chapter analyses some measures of interest related to trajectories, such as the number of distinct trajectories in a cell or starting from a cell, the distance covered by the trajectories in a cell, the average and maximum speed and the average acceleration of the trajectories in the cell, and the frequent patterns obtained by a data mining process on trajectories. The authors focus on some specialised algorithms to transform data, and load the measures in the base cells. Such stored values are used, along with suitable aggregate functions, to compute the roll-up operations. The main issues derive, in this case, from the characteristics of input data (i.e., trajectory observations of mobile objects), which are usually produced at different rates, and arrive in streams in an unpredictable and unbounded way. Finally, the authors also discuss some use cases that would benefit from such a framework, in particular in the domain of supervision systems to monitor road traffic (or movements of individuals) in a given geographical area.

DOI: 10.4018/978-1-60566-328-9.ch004

INTRODUCTION

The widespread diffusion of modern technologies such as low-cost sensors, wireless, ubiquitous and location-aware mobile devices, allows for collecting overwhelming amounts of data about trajectories of moving objects. Such data are usually produced at different rates, and arrive in streams in an unpredictable and unbounded way. This opens new opportunities for monitoring and decision making applications in a variety of domains, such as traffic control management and location-based services. However, for these applications to become reality, new technical advances in spatial information management are needed. Typically analytical and reasoning processes for a large set of data require, as a starting point, their organisation in repositories, or data warehouses (DWs), where they can be extracted with powerful operators and further elaborated by means of sophisticated algorithms.

In this chapter we define a Trajectory DW (TDW) model for storing aggregates about trajectories, implementable using off-the-shelf DW systems. More specifically, it is a data cube with spatial and temporal dimensions, discretized according to a hierarchy of regular grids. The model abstracts from the identifiers of the objects in favour of aggregate information concerning global properties of a set of moving objects, such as the distance travelled by these objects inside an area, or their average speed or acceleration, or spatial patterns co-visited by many trajectories.

There are good reasons for storing only this aggregate information: in some cases personal data should not be stored due to legal or privacy issues; individual data may be irrelevant or unavailable; and individual data may be highly volatile and involve huge space requirements. In addition, current spatio-temporal applications are much more interested in aggregates, rather than information about individual objects (Tao & Papadias, 2005). For example, traffic supervision systems usually monitor the number of cars in an area of interest rather than their ids. Also mobile phone companies can exploit the number of phone-calls per cell in order to identify trends and prevent potential network congestion.

Note that a different solution, alternative to our TDW, could be based on the exploitation of Moving Object Databases (MODs) (Güting & Schneider, 2005), which extend database technologies for modelling, indexing and query processing raw trajectories. One of the main drawback of this MOD-based solution is space complexity and privacy: we need to store and maintain huge raw data and individual information. The other drawback is time complexity to compute a spatio-temporal window aggregate query (which usually specifies a spatial rectangle, a time interval, and an aggregate function to compute): we first need to perform an expensive step to extract from the MOD all the relevant trajectory segments, and then compute on them the requested aggregate function.

Concerning the TDW measures about trajectory data, in this chapter we go beyond numerical ones (Orlando, Orsini, Raffaetà, Roncato & Silvestri, 2007). We are interested also in aggregate properties, obtained through a knowledge discovery process from the raw data. In particular, we focus on the knowledge extracted by a Spatio-Temporal Frequent Pattern Mining (ST-FPM) tool. Apart from the transformation phase of ST raw data, the above problem can be reduced to the well-known Frequent Itemset Mining (FIM) (Agrawal & Srikant, 1994). This implies that the trajectory properties we store in and retrieve from the TDW are sets of spatial regions which are *frequently* visited, in any order, by a large number of trajectories (beyond a certain threshold). The extraction of frequent patterns is a time consuming task. Similarly to the way we process and load the other measures, as soon as data arrive, we transform data, extract patterns, and load the base cells of our TDW with the mined patterns. Such partial aggregations stored in the base cells can be aggregated in order to answer roll-up queries about patterns occurring in larger ST cells.

The knowledge stored in our TDW can be exploited in many ways. A mobility manager can look at spatio-temporal aggregates about either the presence or the average speed of vehicles for decision making, for example for rerouting traffic and reducing congestion in the road network. To this end, s/he can also explore data aggregates using visual analytic tools, like the one mentioned in Section 9. Other uses of the TDW may concern the citizens. Consider for example a tourist, who, in order to plan a city sightseeing, may want to look at the common sets of spatial regions visited by a multitude of people. From them it is possible to derive association rules, used to suggest other regions to visit, given a set of regions already visited by the tourist. In addition, an analyst, interested in understanding social behaviour, can exploit this revealed knowledge for his/her studies.

The main research challenges for the implementation of our TDW are its (extraction & transformation & loading) ETL process, and the ST aggregation functions used to answer roll-up queries. Starting from the received stream of raw sampled observations, we first reconstruct trajectories and compute measures to load the TDW base cells. In order to avoid the use of unbounded amounts of memory buffer for storing complete trajectories, we suggest methods for feeding the TDW before the complete reception of trajectories. In addition, some approximations must be introduced in aggregate computation, in order to give a feasible solution to the problem posed by the *holistic* nature of the functions used to aggregate some of our TDW measures.

The rest of the chapter is organised as follows. Section 2 discusses some related work. Section 3 describes the spatio-temporal data that are subject of our analysis and their representations. In Section 4 we illustrate the data warehouse model that is used to store measures concerning the trajectories, while some specific measures of interest related to

trajectories are introduced in Section 5. Section 6 deals with transformation and loading issues of the various measures previously discussed. Section 7 describes algorithms to compute the measures to store in the base cells of our data cube. Section 8 details the devised functions used to aggregate measures and to perform roll-ups, focusing on the methods proposed to approximate holistic aggregate functions for some of the measures. In Section 9 we discuss some possible use cases of the trajectory data warehouse, in particular the post-processing of the ST aggregate measures, and the graphic visualisation of OLAP reports. Finally, Section 10 draws some conclusion and gives possible directions of our future work.

RELATED WORK

The tools traditionally used to manage geographical data are spatial databases, GISs (Rigaux & Scholl, 2001; Shekhar & Chawla, 2003; Worboys & Duckham, 2004) and spatial DWs (Han, Stefanovic & Kopersky, 1998; Marchant, Briseboi, Bédard & Edwards, 2004; Rivest, Bédard & Marchand, 2001; Shekhar, Lu, Tan, Chawla & Vatsavai, 2001). However, spatial data warehousing is in its infancy. The pioneering work of Han et al. (Han, Stefanovic & Kopersky, 1998) introduces a spatial data cube model which consists of both spatial and non-spatial dimensions and measures. Additionally, it analyses how OLAP operations can be performed in such a spatial data cube. Recently, several authors have proposed data models for spatial DWs at a conceptual level (e.g., (Damiani & Spaccapietra, 2006; Pedersen & Tryfona, 2001)). Unfortunately, none of these approaches deal with objects moving continuously in time.

A related research issue, which is gaining an increasing interest in the last years and is relevant to the development of DWs for spatio-temporal data, concerns the specification and efficient implementation of the operators for spatio-temporal aggregation. A first comprehensive classification and formalisation of spatio-temporal aggregate functions is presented in (Lopez, Snodgrass & Moon, 2005) whereas in (Papadias, Tao, Kalnis & Zhang, 2002; Tao & Papadias, 2005) techniques for the computation of aggregate queries are developed based on the combined use of specialised indexes and materialisation of aggregate measures.

The problem of generating global patterns from sub-aggregates, such as the ones stored in our TDW base cells, has some affinity with the Frequent Pattern Mining problem in a distributed (Kargupta & Sivakumar, 2004; Silvestri & Orlando, 2005) or stream (Giannella, Han, Pei, Yan & Yu, 2003; Silvestri & Orlando, 2007) settings. In both cases, even if for different reasons, the database is partitioned and each partition/chunk is mined separately. The referred items are the same in each partition. The models extracted, though locally coherent and accurate, pose several problems and complexities when combined/aggregated to infer a global model. The main issue is the correct computation of the support $f(p)$ of a pattern p, since it may be infrequent in a partition/chunk, and be frequent in another.

In addition, the maintenance of a DW of frequent itemsets has been already presented in (Monteiro, Zimbrao, Schwarz, Mitschang & Souza, 2005). Nonetheless, in our case data and patterns refer to trajectories instead of generic categorical data, and this poses several peculiar research challenges. Moreover, our TDW has both spatial and temporal dimensions, which requires distinct aggregation functions, but also more specific methods that leverage the peculiarities of trajectories. In (Gonzalez, Han, Li & Klabjan, 2006) a non traditional DW model is presented, designed and tailored for logistic applications, where RFID technologies are used for goods tracing. Unlike our TDW, this DW has to preserve object transition information, while allowing multi-dimensional analysis of path-dependent aggregates.

TRAJECTORY REPRESENTATION

In real-world applications the movements of a spatio-temporal object, i.e. its *trajectory*, is often given by means of a finite set of *observations*. This is a finite subset of points, called *sampling*, taken from the actual continuous trajectory. An observation is a tuple (id,t,x,y), meaning that the object *id* is at location (x,y) at time t. Figure 1 (a) shows a trajectory of a moving object in a 2D space and a possible sampling, where each point is annotated with the corresponding time-stamp. It is reasonable to expect that observations are taken at irregular rates for each object, and that there is no temporal alignment between the observations of different objects. Moreover, they arrive in streaming, temporally ordered, and the observations of the different trajectories can interleave.

In many situations, e.g., when one is interested in computing the cumulative number of trajectories in a given area, an (approximate) reconstruction of each trajectory from its sampling is needed. Among the several possible solutions, we will focus on *local interpolation*. According to this method, although there is not a global function describing the whole trajectory, objects are assumed to move between the observed points following some rule. For instance, a *linear* interpolation function models a straight movement with constant speed, while other polynomial interpolations can represent smooth changes of direction. The linear (local) interpolation, in particular, seems to be a quite standard approach to the problem (see, for example, (Pfoser, Jensen & Theodoridis, 2000)), and yields a good trade-off between flexibility and simplicity. Hence in this chapter we will adopt this kind of interpolation. Figure 1 (b) illustrates the reconstructed trajectory of the moving object in Figure 1 (a). However, it is straightforward to use a different interpolation, based, for example, on additional information concerning the environment traversed by the moving objects.

Figure 1. (a) The sampling of 2D trajectory, (b) the reconstructed trajectory by local linear interpolation

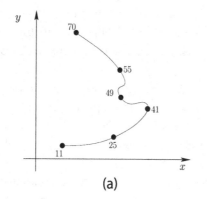

(a)

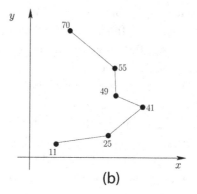

(b)

Region-Based Trajectories

This point-based representation is not adequate when we want to extract meaningful patterns from trajectories. In fact it is hard to find trajectories that traverse the same points. As a consequence, we introduce the concept of *region of interest*, which is a natural way to highlight meaningful spatial areas in our domain and correspondly, to associate spatial points with region labels. Such regions can be indicated by users, or extracted by a tool from the trajectories themselves on the basis of the popularity of some areas, e.g., a clustering tool could discover that a given region turns out to be of interest since it corresponds to a very dense area in which many trajectories observations have been recorded. We denote with $\mathcal{R} = \{r_1, \ldots, r_k\}$ the set of all spatial regions of interest occurring in our spatial domain.

Given the set of regions of interest $\mathcal{R}$ we have to transform the trajectories from sequences of *point-based* observations into sequences of *region-based* observations. This data transformation makes it possible to mine more valuable patterns, i.e., patterns that turn out to occur more frequently in the input trajectories, since their spatial references are looser than the original observations. First of all, we notice that a trajectory can traverse a region of interest r_i even if no observation falls into r_i. For instance, consider Figure 2 (a): the

moving object will likely cross region r_4, but no observation is present in r_4. By reconstructing the trajectory by local linear interpolation, we build a region-based observation (t, r_i), where t is the (interpolated) time in which the trajectory presumably entered the region r_i (see Figure 2 (b)). Once all the interpolated region-based observations have been added, the original observations can be transformed as follows. Given a trajectory observation (t, x, y), if point (x, y) falls into a region $r_i \in \mathcal{R}$ it is transformed into (t', r_i) (see below for the choice of the timestamp), otherwise it is removed from the trajectory.

In order to choose the timestamp t' of the region-based observation (t', r_i), we use the following criteria (Giannotti, Nanni, Pedreschi & Pinelli, 2007):

- if the *starting observation* of a given trajectory is (t, x, y), and point (x, y) falls into region r_i, the transformation simply yields the region-based observation (t, r_i);
- in all other cases, take the (interpolated) entering times t' of the trajectory in each region of interest, and associate it with the region identifier.

If *consecutive* trajectory observations $o_1, \ldots, o_j$ fall into the same region r_i, this criteria can generate duplicates. In this case, in the transformed

Figure 2. (a) The point-based trajectory traversing some regions of interest, and (b) the corresponding region-based trajectory

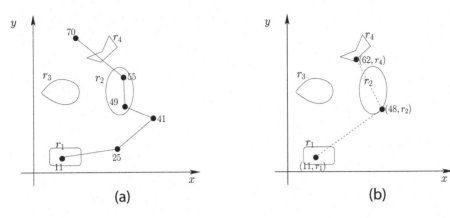

(a)   (b)

trajectory, we will keep only the region-based observation corresponding to o_1.

Example 1. In Figure 2 (a) we show the regions of interest in our spatial domain and which ones our trajectory traverses, whereas Figure 2 (b) illustrates the transformed region-based trajectory. Notice that the point-based observations timestamped with 25, 41, and 70 have been removed. In addition, the two observations with timestamps 49 and 55 have been collapsed into a single region-based observation. The timestamps associated with the region-based observations are 48 and 62 for regions r_2 and r_4, corresponding to the entering time of the trajectory into these regions, while the timestamp associated with r_1 is the original starting time of the trajectory. The resulting region-based trajectory is:

$$T_R = \{ID, \langle (11, r_1), (48, r_2), (62, r_4) \rangle\}.$$

Finally, we say that a trajectory $T_R = \{ID, \langle (t_1, r_1), ..., (t_n, r_n) \rangle\}$ *traverses* a set of regions of interest R if $R \subseteq \bigcup_{i=1}^{n} \{r_i\}$. For instance, the trajectory in Figure 2(b) traverses $\{r_1, r_4\}$.

These two kinds of representation for a trajectory will be useful in the loading phase in order to compute the measures for the base cells.

FACTS, DIMENSIONS AND AGGREGATE FUNCTIONS

The data cube of our TDW has temporal and spatial dimensions, discretized according to a hierarchy of regular grids. The model abstracts from the identifiers of the objects in favour of aggregate information concerning global properties of a set of moving objects. We can model our TDW by means of a star schema, as simple and generic as possible. The *facts* of the TDW are the set of trajectories which intersect each cell of the grid. Some typical properties we want to describe, i.e., the *measures*, are the number of trajectories, their average, maximum/minimum speed, the covered distance, the frequent patterns, etc. The spatial dimensions of analysis are X and Y, ranging over spatial intervals, while the temporal one T ranges over temporal intervals. We thus assume a regular three-dimensional grid obtained by discretizing the corresponding values of the dimensions, and associate with them a set-grouping hierarchy. A partial order can thus be defined among groups of values, as illustrated in Figure 3 (a) for a temporal or a spatial dimension. Note that in the TDW of Figure 3 (b), the basic information we represent concerns the set of trajectories intersecting the spatio-temporal cell having for each X, Y, or T dimensions as

Figure 3. (a) A concept (set-grouping) hierarchy and (b) a TDW example

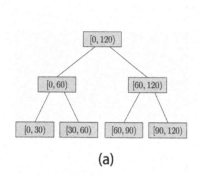

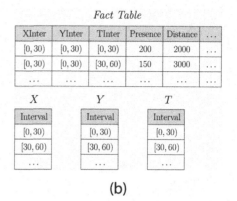

(a)

(b)

minimum granularity 30 units, measured in the corresponding unit of measure.

We can build the spatial data cube (Gray et al., 1997) as the lattice of cuboids, where the lowest one (*base cuboid*) references all the dimensions at the primitive abstraction level, while the others are obtained by summarising on different subsets of the dimensions, and at different abstraction levels along the concept hierarchy. In order to denote a component of the base cuboid we will use the term *base cell*, while we will simply use *cell* for a component of a generic cuboid. We use two different notations to indicate a cell:

- $C_{x,y,t}$, to specify the three independent dimensions;
- (Z,I), to focus on the extent of the 2D spatial area and the temporal interval composing the cell.

In the rest of the chapter, the 2D spatial area of a cell will be referred to as *zone* and denoted by Z (possibly with subscripts).

In order to summarise the information contained in the base cells, Gray et al. (Gray et al.., 1997) categorise the aggregate functions into three classes based on the space complexity needed for computing a super-aggregate starting from a set of sub-aggregates already provided, e.g., the

sub-aggregates associated with the base cells of the DW. The classes are the following:

1. *Distributive*. The super-aggregates can be computed from the sub-aggregates.
2. *Algebraic*. The super-aggregates can be computed from the sub-aggregates together with a *finite* set of auxiliary measures.
3. *Holistic*. The super-aggregates cannot be computed from sub-aggregates, not even using any finite number of auxiliary measures.

In Section 8 we will classify the aggregate functions, used to roll-up the various measures stored in our TDW cube, according to the above classes.

TRAJECTORY MEASURES

In this section we present the measures of our TDW. The complexity and the memory space to load and aggregate them starting from a stream of observations, can vary from simple computation with no buffer requirements to very expensive and approximate functions.

We studied the following measures to store in each cell of our TDW:

m1: the number of observations in the cell, *numObs*;

m2: the number of trajectories starting in the cell, *trajInit*;

m3: the number of distinct trajectories in the cell, *presence*;

m4: the total distance covered by trajectories in the cell, *distance*;

m5: the average speed of trajectories in the cell, *speed*;

m6: the maximum speed of trajectories in the cell, *maxSpeed*;

m7: the average acceleration of trajectories in the cell, *acc*;

m8: the frequent sets of spatial regions, *freqRegs*.

The measures from *m1* to *m7* are numerical, and can be computed starting from the raw observations by introducing, in some cases, some approximation. For example, consider measures *m4*: the actual distance covered by a moving object cannot be exactly determined by simply using the trajectory observations, and the reconstruction carried out through local linear interpolation (see Figure 1 (b)) only approximates the actual one. Among these various measures, the most complex one is *m3*, due to the *distinctness* property of the count we need to compute (Orlando, Orsini, Raffaetà, Roncato & Silvestri, 2007; Tao, Kollios, Considine, Li & Papadias, 2004).

The last measure *m8* is non numerical, and models significant *behavioural patterns* obtained through a knowledge discovery process starting from the trajectories intersecting the base cells of our TDW. In order to extract patterns we exploit a Frequent Itemset Mining (FIM) tool. The FIM problem (Agrawal & Srikant, 1994), introduced in the context of Market Basket Analysis (MBA), deals with the extraction of all the frequent itemsets from a database $\mathcal{D}$ of transactions. A set of items $\mathcal{I} = \{a_1,...,a_M\}$ is fixed and each transaction $t \in \mathcal{D}$ is associated with a transaction

identifier *TID*, and contains a subset of items in $\mathcal{I}$. Given a *k-itemset*, i.e., a subset $i \subseteq \mathcal{I}$, with $|i| = k$, let $f(i)$ denote its *support*, defined as $f(i) = |\{t \in \mathcal{D} \mid i \subseteq t\}|$. Mining all the frequent itemsets from $\mathcal{D}$ requires to discover all the *k-itemsets i* (for *k*=1,2,...) having a support greater than or equal to $\sigma_{min} * |\mathcal{D}|$, i.e. $f(i) \geq \sigma_{min} * |\mathcal{D}|$ where σ_{min} is a fixed threshold stating which is the minimal fraction of the transactions in $\mathcal{D}$ that must support a given itemset *i*.

In the context of trajectories, interesting itemsets to be extracted can be *frequent sets of regions of interest*, i.e., collections of regions of interest which are visited together, even if in different orders, by a large number of trajectories. Formally we define:

Definition 1 (Frequent Set of Spatial Regions). Let $p = \{r_1,...,r_n\}$ be a pattern, i.e, a subset of regions of interest ($p \subseteq \mathcal{R}$), let σ_{min} be the minimum support threshold and let T be the set of trajectories. Let T_p be the set of all trajectories which traverse p and let $s(p)$ be the set of trajectory identifiers in T_p.

Then the frequency of *p* is $f(p) = |T_p|$ (or, equivalently, the cardinality of $s(p)$). We say that the pattern *p* is *frequent* w.r.t. σ_{min} iff $f(p) \geq \sigma_{min} * |T|$.

ETL: EXTRACTION, TRANSFORMATION, AND LOADING

So far, we have presented our TDW, based on a cube model composed of spatio-temporal cells, and the list of measures associated with a cell which model properties of the set of trajectories intersecting such a cell. Since each trajectory usually spans over different cells, in order to correctly compute the measures for the base cells, we have to define adequate operations that *restrict* our trajectories to these cells.

In order to discuss restriction and loading of trajectories, we first introduce an original classification of our TDW measures with respect to the complexity of the associated ETL process. For example, some of them require little pre-computation, and can be updated in the TDW as soon as single observations of the various trajectories arrive. Measures are ordered according to the increasing amount of pre-calculation effort required:

a. **No pre-computation:** The measure can be updated in the data warehouse by directly using each single observation;

b. **Per trajectory local pre-computation:** The measure can be updated by exploiting a simple pre-computation, which only involves a few and close observations of the same trajectory;

c. *Per trajectory global pre-computation:* The measure update requires a pre-computation which considers all the observations of a single trajectory;

d. **Global pre-computation:** The measure requires a pre-computation which considers all the observations of all the trajectories.

Measure *m1* is of type a), *m2* of type a) or b), *m3 - m8* are of type b). Notice that the loading of *m2* can be simplified, and becomes of type a), only if the first observation of each trajectory is suitably marked. As far as measure *m8*, in Section 7 we discuss a way to prepare data for loading that only requires a few and temporally consecutive observations, and can thus be classified of type b). Finally, as an example of a measure of type c) and d) consider respectively the number of trajectories longer than a given value *d* and the number of trajectories that intersect another trajectory only in the cell.

The amount of pre-computation associated with each type of measure impacts upon the amount of memory required to buffer incoming trajectory observations. For example, measures of type c) need all the observations of each trajectory to be received, before starting to load/update a measure in the data warehouse. Measures of type a) are the

less expensive in terms of space and time, since it is enough to consider observations one at a time, without buffering anything. Therefore a measure can be updated as soon as each single observation $Obs_i^j = (id_i, t_i^j, x_i^j, y_i^j)$ of the various trajectories arrives. Conversely, for type b) the measure must be computed starting from a finite set of neighbours of each observation $Obs_i^j = (id_i, t_i^j, x_i^j, y_i^j)$. In general, this could require a *k*-window of observations $Obs_i^{j-k-1}, \ldots, Obs_i^j$ to be considered and stored in a buffer (see the end of this section for details).

Restriction of Trajectories to Base Cells

Restriction of Point-Based Trajectories

To cope with measures from *m3* to *m7*, we have to reconstruct a trajectory from its sampling.

In fact, the set of streaming observations are not enough to correctly compute such measures, since a trajectory can cross a cell but no observation falls into it (see the dark square in Figure 4 (a)). This is why we have to generate new points through linear interpolation. We propose to consider as additional interpolated points for each cell traversed by a trajectory, those points corresponding to the intersections between the trajectory and the border of the spatio-temporal cell. Figure 4(b) shows the resulting interpolated points as white and gray circles. The white points, with temporal labels 30 and 60 correspond to cross points of a temporal border of some 3D cell. The gray points, labelled with 16, 33, 47, 57, and 70, correspond to the cross points of the spatial borders of some 3D cell, or, equivalently, the cross points of the spatial 2D squares shown in Figure 4 (b). All these additional points do not cause a space explosion, since they are generated, considered on-the-fly and finally thrown away during the stream processing.

Thanks to these additional interpolated points, we have a greater number of 3D base cells in which

Figure 4. (a) Linear interpolation of the 2D trajectory with the grid, and (b) the interpolated trajectory with additional points matching the spatial and temporal minimum granularity

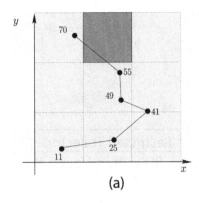
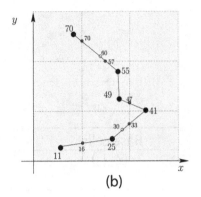

(a)   (b)

we can store significant measures. In particular, by construction, the new points subdivide the interpolated trajectory into small segments, each one completely included in some 3D base cell. Thus, the restriction of a trajectory to a base cell can be obtained as the set of segments contained in that cell, and these segments can be used to update its measures.

Restriction of Region-Based Trajectories

In order to compute the measure *freqRegs* we have to consider the region-based representation of our trajectories. Even in this case, adequate operations are required to partition such trajectories among the different base cells *(Z,I)* where *Z* refers to the spatial zone and *I* to the temporal interval composing the cell.

Without loss of generality, we can assume that, for each region of interest $r \in \mathcal{R}$ there exists only one spatial zone *Z* that *covers r*, denoted by $Z \triangleright r$. We can eventually split the region into sub-regions in order to satisfy such a constraint.

Let us now define the restrictions of a trajectory to a temporal interval and to a spatial zone.

Definition 2 (Restriction to a temporal interval). Let $T_R = \{ID, \langle (t_1, r_1), ..., (t_n, r_n) \rangle\}$ be a trajectory and let *I* be a temporal interval. The trajectory T_R

restricted to *I*, denoted by $T_R|_I$, is the maximal sequence of region-based observations in T_R:

$$TR\,|_I = \{ID, \langle (t_h, r_h), (t_{h+1}, r_{h+1}), ..., (t_{h+m}, r_{h+m}) \rangle\},$$

such that $[t_h, t_{h+m}] \subseteq I$.

Notice that $T_R|_I$ is a subsequence of *consecutive* region-based observations of the trajectory T_R.

Definition 3 (Restriction to a spatial zone). Let $T_R = \{ID, \langle (t_1, r_1), ..., (t_n, r_n) \rangle\}$ be a trajectory and let *Z* be a spatial zone.

The trajectory restricted to *Z*, denoted by $T_R|_Z$, is the maximal sequence of region-based observations in T_R:

$$TR\,|_Z = \{ID, \langle (t_{i_1}, r_{i_1}), ..., (t_{i_m}, r_{i_m}) \rangle\},$$

such that $1 \le i_1 \le ... \le i_m \le n$ and $Z \triangleright r_{i_j}$ for all $j = 1, ..., m$.

It is worth remarking that the restriction to a spatial zone *Z* can consist of several trajectory subsequences of T_R. This is due to the fact that an object can exit from *Z*, visit other zones, and then go back to *Z*, thus entering the zone several times.

Definition 4 (ST Restriction to base cells). Let $T_R = \{ID, \langle (t_1, r_1), ..., (t_n, r_n) \rangle\}$ be a trajectory, let *Z* be a spatial region and let *I* be a temporal

Figure 5. (a) Two trajectories, and (b) their restriction to the TDW base cells

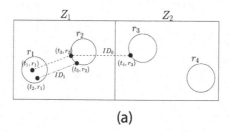

(a)

Base cells	Restricted Trajectories
(Z_1, I_1)	$\{ID_0, \langle (t_1, r_1), (t_3, r_2) \rangle\}$ $\{ID_1, \langle (t2, r1) \rangle\}$
(Z_1, I_2)	$\{ID_1, \langle (t_5, r_2) \rangle\}$
(Z_2, I_2)	$\{ID_0, \langle (t_4, r_3) \rangle\}$

(b)

interval. The restriction of T_R to the base cell (Z, I) is obtained by restricting the trajectory to Z, and to I in any order. We denote the result as:

$$T_R|_{Z,I}$$

if T is the set of trajectories then $T|_{Z,I} = \{T_R|_{Z,I} | T_R \in T\}$.

Observe that the order is not relevant, i.e., $(T_R|_Z)|_I = (T_R|_I)|_Z$ by definition of temporal and spatial restriction. Thus the definition above is well given.

Example 2. Figure 5 (a) illustrates two trajectories ID_0 and ID_1. The regions of interest are $\{r_1, ..., r_4\}$ and ID_0 traverses the regions of interest r_1, r_2 and r_3 while ID_1 traverses r_1 and r_2.
(b)

We consider two spatial zones Z_1 and Z_2 and two time intervals I_1 and I_2. Thus $Z_1 \triangleright r_1, r_2$ and $Z_2 \triangleright r_3, r_4$. Concerning the timestamps, t_1, t_2, and t_3 range over the time interval I_1, whereas t_4 and t_5 over I_2.

Given the two spatial zones and the two temporal intervals, we obtain four base cells: (Z_1, I_1), (Z_1, I_2), (Z_2, I_1), (Z_2, I_2). The ST restriction

of trajectories ID_0 and ID_1 to the TDW base cells partitions ID_0 and ID_1 into the sub-trajectories shown in Figure 5 (b).

Loading Methods

As mentioned above, different levels of pre-computation may be involved in feeding the TDW, depending on the type of measures and the required accuracy. We recall that trajectory observations arrive in streams at different rates, in an unpredictable and unbounded way. In order to limit the amount of *buffer memory* needed, it is essential to store information only about *active*, i.e., not ended trajectories. In our simple model of trajectory sampling, since we do not have an end-mark associated with the last observation of a given trajectory, the system module that is responsible for feeding data could decide to consider a trajectory as *ended* when for a long time interval no further observation for the object has been received.

In the following, we limit our discussion to the feeding methods for measures of type a) and b) only, which cover the needs of our measures ($m1 - m8$). In particular, for measures of type b), for which only a few and close observations of the same trajectory can be involved, we further distinguish between measures for which a single pair of observation is enough, and other measures that need to buffer a given number of previous positions per each trajectory.

- **Single observations (SO).** We load the TDW on the basis of single observations only. This method is not suitable to update measures that can only be computed on the basis of the knowledge of more than one observation, such as the covered distance or the average speed. Moreover, it cannot be used when we need to first reconstruct the trajectory, by exploiting, for example, local linear interpolation.

- **Pairs of observations (PO).** We consider a pair of observations consisting of the currently received observation Obs_i^j of trajectory T_i, along with the previous buffered one, Obs_i^{j-1}. Using this pair of points, for example, we can linearly interpolate the trajectory, compute the distance covered between the two points.

- **Observation window (OW).** We consider a window of k observations. In particular, let $Obs_i^j = (id_i, x_i^j, y_i^j, t_i^j)$ be the currently received observation of trajectory T_i. The k window thus includes $Obs_i^{j-k-1}, Obs_i^{j-k-2}, \dots, Obs_i^{j-1}, Obs_i^j$. The window size k is dynamically adjusted according to the following constraints: (1) All $t_{j-k-1}, \dots, t_j$ must fall within the same temporal interval $[l, u)$, characterising a base cell of our cuboid, i.e., $l \leq t_{j-k-1} < t_{j-k-2} < \dots < t_j < u$. (2) Obs_i^{j-k} is not included in the window because it is associated with a timestamp $t_{j-k} < l$.

Buffering all these points (and some related information) not only guarantees the linear interpolation of the associated trajectory, but also permits us to consider the recent history of a trajectory inside the base cells. For example it is possible to see if a trajectory crossed multiple times a base cell.

It is straightforward to show that if we encounter a new observation Obs_i^{j+1}, with $t_{j+1} \geq u$, we can forget (un-buffer) all the points of the window, since we are surely going to update a new cell, associated with a different and successive temporal interval.

ALGORITHMS FOR LOADING THE FACT TABLE

In this section we present the algorithms used to feed the base cells of our TDW. We distinguish between numerical measures and measure *m8* which requires a knowledge discovery process. We thus assume that a cell C is associated with a tuple consisting of the following components: *C.numObs, C.trajInit, C.vmax, C.distance, C.time, C.speed, C.Δv, C.acc, C.presence, C.crossX, C.crossY, C.crossT, C.freqRegs*. It is worth noticing that besides the numerical measures already introduced, the tuple contains the total time spent by the trajectories in the cell (*C.time*), the sum of the variations of speed of the trajectories in the cell (*C.Δv*), the number of trajectories crossing the X border, the Y border and the T border of the cell (*C.crossX, C.crossY, C.crossT*), which are auxiliary measures required for the roll-up operation, as we will discuss in Section 8.

Numerical Measures

In order to compute measures *m1* and *m2*, we do not need interpolation. In fact measure *m1* requires only the observations and no auxiliary storage, hence we can use the SO method for loading. Instead, in the case of measure *m2*, in order to understand whether an incoming observation starts a new trajectory, we store in the buffer, for each active trajectory, the identifier and the time-stamp of the last processed point of the trajectory, thus it is necessary the PO method for loading. As discussed in the previous section, using the time-stamp, we can decide if a trajectory must be considered as ended. In this case the corresponding data in the buffer can be removed, thus freeing the space for storing new incoming observations. This procedure is not detailed in the pseudo-code for the sake of readability.

Algorithm 1 consists of an infinite loop which processes the observations arriving in streaming. At each step, the function *getNext* gets the next observation *obs* in the stream and we determine the base cell C_{cur} which *obs* belongs to (function *findCell*). Then we check whether the base cell C_{cur} is in the Fact Table *FT*. In this case we correspondingly update the measures, otherwise we insert the cell in the Fact Table and we initialise the corresponding tuple.

Algorithm 1: Algorithm for Measures Computed Without Interpolation

```
INPUT: Stream ST of observations (obs) in
the form (id, x, y, t).
OUTPUT: Fact Table FT.
1: FT ← Æ
2: buffer ← Æ
3: loop
4:      obs ← getNext(ST)
5:      C_cur ← findCell (obs.x, obs.y,
obs.t)
6:      if (C_cur Ï FT) then
7:          C_cur.numObs ← 1
8:          if (obs.id Ï buffer) then
9:              C_cur.trajinit ← 1
10:         end if
11:     else
12:         C_cur.numObs ← C_cur.numObs + 1
13:         if (obs.id Ï buffer) then
14:             C_cur.trajinit ← Cell.
trajinit + 1
```

```
15:         end if
16:     end if
17:     buffer[obs.id] ← obs.t
18: end loop
```

Algorithm 2 is used to compute measures *m3* to *m7*. The buffer contains the identifiers of the active trajectories, their last processed point, the cell such point belongs to, and the speed in the segment ending at such a point (for the first point of a trajectory we assume that the speed is 0).

As in the previous case, the algorithm consists of an infinite loop which processes the observations arriving in streaming. For any observation *obs*, after initialising the measures and the buffer, we consider the cell C_{cur} which *obs* belongs to. If C_{cur} is different from the cell C_{pred} associated with the last stored observation $point_{pred}$ of the trajectory, we linearly interpolate the two points (*obs.x*, *obs.y*, *obs.t*) and $point_{pred}$ and consider all the intersections with borders of spatio-temporal cells (function *interpNewPoints*) as described in Section 6. The points corresponding to such intersections are inserted in a queue (ordered with respect to the time) and then processed in order to update the measures of the cells which these points belong to. We remark that as a result of the addition of these new points, the trajectory is divided in segments, each one completely included in a base cell, a fact which allows us to update the measures *m4–m7* of the base cells in a correct way. Since we used the local linear interpolation, we chose the PO method for loading which is the most memory-efficient one. In case a more complex trajectory reconstruction is needed, for example if moving objects are subject to network constraints, the use of the observation window (OW) method could be a better choice.

Algorithm 2: Algorithm for Measures Requiring Interpolation

```
INPUT: Stream ST of observations (obs) in
the form (id, x, y, t).
```

OUTPUT: Fact Table FT.

```
1:  FT ← Æ
2:  buffer ← Æ
3:  loop
4:        obs ← getNext(ST)
5:        C_cur ← findCell(obs.x, obs.y,
    obs.t)
6:        if (C_cur Ï FT) then
7:              insert(C_cur, FT)
8:              initMeasures(C_cur)
9:        end if
10:       if (obs.id Ï buffer) then
11:             insert(obs.id, buffer)
12:             initBuffer(obs, C_cur, buffer)
13:             C_cur.P ← 1
14:       else
15:             point_pred ←
    extractPoint(buffer[obs.id])
16:             C_pred ← extractCell
    (buffer[obs.id])
17:             if (C_pred ≠ C_cur) then
18:                   Q ←
    interpNewPoints(point_pred, (obs.x, obs.y,
    obs.t))
19:                   repeat
20:                         p ← extract(Q)
21:                         C_p ← findNextCell
    (p.x, p.y, p.t, C_pred)
22:                         if(C_p Ï FT)then
23:                               insert(C_p,
    FT)
24:
    initMeasures(C_p)
25:                         end if
26:
    updateMeas&Buf(C_pred, point_pred, C_p, p, obs.
    id, buffer)
27:
    updatePresence(C_pred, C_p, p)
28:                         C_pred ← C_p
29:                         point_pred ← p
30:                   untilQ = Æ
31:             end if
32:             updateMeas&Buf(C_pred,
```

```
    point_pred, (obs.x, obs.y, obs.t), obs.id,
    buffer)
33:       end if
34: end loop
```

Auxiliary Procedures for Algorithm 2

```
1:  procedure initMeasures(Cell)
2:        Cell.v_max ← 0
3:        Cell.distance ← 0
4:        Cell.time ← 0
5:        Cell.speed ← 0
6:        Cell.D v ← 0
7:        Cell.acc ← 0
8:        Cell.P ← 0
9:        Cell.CrossX ← 0
10:       Cell.CrossY ← 0
11:       Cell.CrossT ← 0
12: end procedure
13: procedure initBuffer(obs, Cell, buf-
    fer)
14:       buffer[obs.id] ←
    ⟨(obs.x, obs.y, obs.t), Cell, 0⟩
15: end procedure
16: procedure updateMeas&Buf(Cell_pred,
    point_pred, Cell_curr, point_cur, id, buffer)
17:       Cell_pred.distance ← Cell_pred.dis-
    tance + dist(point_pred, point_cur)
18:       Cell_pred.time ← Cell_pred.time +
    (point_cur.t - point_pred.t)
19:       Cell_pred.speed ← Cell_pred.distance/
    Cell_pred.time
20:       v_cur ← (dist(point_pred, point_cur))/
    (point_cur.t - point_pred.t)
21:       if (v_cur > Cell_pred.v_max) then
22:             Cell_pred.v_max ← v_cur
23:       end if
24:       v_init ← extractSpeed(buffer[id])
25:       Cell_pred.D v ← Cell_pred.Dv + (v_cur
    - v_init)
26:       Cell_pred.acc ← Cell_pred.Dv /
    Cell_pred.time
27:       buffer[id] ← ⟨point_cur, Cell_pred, v_cur⟩
```

```
28: end procedure
29: procedure updatePresence (Cell_pred,
Cell_p, p)
30:        Cell_p.P ← Cell_p.P + 1
31:        if (p ∈ Cell_pred.X) then
32:                Cell_pred.CrossX ← Cell_pred.
CrossX + 1
33:        else if (p ∈ Cell_p.X) then
34:                Cell_p.CrossX ← Cell_p.CrossX
+ 1
35:        else if (p ∈ Cell_pred.Y) then
36:                Cell_pred.CrossY ← Cell_pred.
CrossY + 1
37:        else if (p ∈ Cell_p.Y) then
38:                Cell_p.CrossY ← Cell_p.CrossY
+ 1
39:        else if (p ∈ Cell_pred.T) then
40:                Cell_pred.CrossT ← Cell_pred.
CrossT + 1
41:        else
42:                Cell_p.CrossT ← Cell_p.CrossT
+ 1
43:        end if
44: end procedure
```

At line 31, the symbol *Cell.X* denotes the face of the cell *Cell* whose points have the same value for x coordinate and $Cell.X \subseteq Cell$ (remember that cells are close on one face, and open on the opposite one). The same applies for *Cell.Y* and *Cell.T*.

As far as the measure presence is concerned, we implemented the PO method which reduces the number of duplicates simply by avoiding to update the measure when the new point falls in the same base cell of the previous one. Nonetheless, only the duplicates due to consecutive observations in the same base cells can be avoided using the PO method. The OW method, instead, allows for the exact computation of the presence measure. To this end, it is enough to record the cells whose presence measures have been updated on the basis of the window points, and avoid following updates to the same cells. We recall that observations arrive in a temporarily ordered stream. Thus, whenever a new observation falls in a time interval that follows the current one, i.e. the ones in which fall all the buffered observations, we can forget them, and remove them from the buffer along with the current set of already visited cells.

Frequent Sets of Spatial Regions

The base cell load operation for Frequent Region Sets is more complex. The loading phase is split into three parts: first we transform the point-based trajectories into region-based trajectories. Then for each base cell $C_i^j = (Z_i, I_j)$, we restrict the set of trajectories T to each C_i^j, thus obtaining the set $T|_{Z_i, I_j}$. We drop from these region-based trajectories the timestamps and we collect the traversed regions of interest. As an example the table below shows the result of the elimination of the timestamps from the restricted trajectories in Table 1

Once obtained the identifiers of the trajectory associated with the sets of unordered spatial regions, for each base cell we apply on these data an FIM algorithm, with a given threshold σ_{min}. The frequent sets of spatial regions extracted represent interesting patterns occurred inside a given base cell, and are thus stored in the corresponding TDW base cell.

The FIM phase needs to use the complete sets of regions traversed by each trajectory inside a base cell. Thus, it is necessary to use the observation window (OW) loading method.

SPATIO-TEMPORAL AGGREGATE FUNCTIONS

Once loaded the base cells with the measures, we want to define aggregate functions to answer roll-up queries. We will see that for all the numerical measures except for the presence, we are able to define either distributive or algebraic aggregate

Table 1.

Base cells	Input for FIM
(Z_1, I_1)	$(ID_0, \{r_1, r_2\})$
	$(ID_1, \{r_1\})$
(Z_1, I_2)	$(ID_1, \{r_2\})$
(Z_2, I_2)	$(ID_0, \{r_3\})$

functions, which can compute the exact value associated with the aggregate cell. On the other hand, for both presence and frequent sets of spatial regions, we have to introduce approximate aggregate functions, due to the holistic nature of such aggregations.

Numerical Measures

The super-aggregates for *m1, m2, m4* and *m6* are simple to compute because the corresponding aggregate functions are *distributive*. In fact once the base cells have been loaded with the exact measure, for *m1, m2* and *m4* we can accumulate such measures by using the function *sum* whereas for *m6* we can apply the function *max*. The super-aggregate for *m5* and *m7* are algebraic: we need some auxiliary measures in order to compute the aggregate functions. For the average speed of trajectories in a cell, a pair $\langle distance, time \rangle$ where *distance* is the measure *m4* and *time* is the total time spent by trajectories in the cell must be considered. For a cell C arising as the union of adjacent cells, the cumulative function performs a component-wise addition, thus producing a pair $\langle distance_f, time_f \rangle$. Then the average speed in C is given by $distance_f / time_f$. In a similar way, to compute the average acceleration, the sum of the variations of the speed (Δv) in the cell and the total time (*time*) spent by the trajectories in the cell are required as auxiliary measures.

The aggregate function for *m3* is holistic since it is a sort of COUNT_DISTINCT() aggregate. It needs the base data to compute the result in all levels of dimensions. Such a kind of function represents a big issue for DW technology, and, in particular, in our context, where the amount of data is huge and unbounded. A common solution consists of computing holistic functions in an *approximate* way.

We propose two alternative and *non-holistic* aggregate functions that *approximate* the exact value of the presence. These functions only need a small and constant memory size to maintain the information to be associated with each base cell of our DW, from which we can start computing a super-aggregate.

The first aggregate function is distributive, i.e., the super-aggregate can be computed from the sub-aggregate, and it is called $Presence_{Distributive}$. We assume that the only measure associated with each base cell is the exact (or approximate) count of all the *distinct* trajectories intersecting the cell. Therefore, the super-aggregate corresponding to a roll-up operation is simply obtained by summing up all the measures associated with the cells. This is a common approach (exploited, e.g., in (Papadias, Tao, Kalnis & Zhang, 2002) to aggregate spatio-temporal data. However, our experiments (Orlando, Orsini, Raffaetà, Roncato & Silvestri, 2007) have shown that this aggregate function may produce a very inexact approximation of the effective *presence*, because the same trajectory might be counted multiple times. This is due to the fact that in the base cell we do not have enough information to perform a *distinct count* when rolling-up.

The second aggregate function is algebraic, i.e., the super-aggregate can be computed from the sub-aggregate together with a *finite* set of auxiliary measures, and it is called *Presence*$_{Algebraic}$. In this case each base cell stores a tuple of measures. Besides the exact (or approximate) count of all the *distinct* trajectories intersecting the cell, the tuple includes other measures which are used when we compute the super-aggregate. These are helpful to correct the errors, caused by the duplicates, introduced by the function *Presence*$_{Distributive}$.

More formally, let $C_{x,y,t}$ be a base cell of our cuboid, where x, y, and t identify intervals of the form $[l,u)$, in which the spatial and temporal dimensions are partitioned. The tuple associated with the cell consists of $C_{x,y,t}.presence$, $C_{x,y,t}.crossX$, $C_{x,y,t}.crossY$, and $C_{x,y,t}.crossT$.

- $C_{x,y,t}.presence$ is the number of *distinct* trajectories intersecting the cell.
- $C_{x,y,t}.crossX$ is the number of *distinct* trajectories crossing the *spatial* border between $C_{x-1,y,t}$ and $C_{x,y,t}$.
- $C_{x,y,t}.crossY$ is the number of *distinct* trajectories crossing the *spatial* border between $C_{x,y-1,t}$ and $C_{x,y,t}$.
- $C_{x,y,t}.crossT$ is the number of *distinct* trajectories crossing the *temporal* border between $C_{x,y,t-1}$ and $C_{x,y,t}$.

Let $C_{x',y',t'}$ be a cell consisting of the union of two adjacent cells with respect to a given dimension, namely $C_{x',y',t'} = C_{x,y,t} \cup C_{x+1,y,t}$. In order to compute the super-aggregate corresponding to $C_{x',y',t'}$, we proceed as follows:

$$C_{x',y',t'}.presence \approx Presence_{Algebraic}(C_{x,y,t} \cup C_{x+1,y,t}) =$$

$$= C_{x,y,t}.presence + C_{x+1,y,t}.presence - C_{x+1,y,t}.crossX$$

$$(1)$$

The other measures associated with $C_{x',y',t'}$ can be computed in this way:

$$C_{x',y',t'}.crossX = C_{x,y,t}.crossX$$

$$C_{x',y',t'}.crossY = C_{x,y,t}.crossY + C_{x+1,y,t}.crossY$$

$$C_{x',y',t'}.crossT = C_{x,y,t}.crossT + C_{x+1,y,t}.crossT$$

Equation (1) can be thought of as an application of the well known Inclusion/Exclusion principle: $|A \cup B| = |A| + |B| - |A \cap B|$ for all sets A, B. Suppose that the elements included in the sets A and B are just the distinct trajectories intersecting the cells $C_{x,y,t}$ and $C_{x+1,y,t}$, respectively. Hence, their cardinalities $|A|$ and $|B|$ exactly correspond to $C_{x,y,t}.presence$ and $C_{x+1,y,t}.presence$. Then $C_{x+1,y,t}.crossX$ is intended to approximate $|A \cap B|$, but, notice that, unfortunately, in some cases $C_{x+1,y,t}.crossX$ is not equal to $|A \cap B|$, and this may introduce errors in the values returned by *Presence*$_{Algebraic}$. Figure 6 (a) shows a trajectory that will be correctly counted, since it crosses the border between the two cells to be rolled-up. Conversely, Figure 6 (b) shows a more agile trajectory, which will be counted twice during the roll-up, since it is not accounted in $C_{x+1,y,t}.crossX$, even if it should appear in $|A \cap B|$. In fact, the trajectory intersects both $C_{x,y,t}$ and $C_{x+1,y,t}$, but does not cross the border between the two cells.

We compared our approximate algebraic function with the method proposed in (Tao, Kollios, Considine, Li & Papadias, 2004), based on sketches, and we showed, with the help of various experiments, that in our case the error in the computation of the distinct number of trajectories is in general much smaller (see (Orlando, Orsini, Raffaetà, Roncato & Silvestri, 2007) for details).

Frequent Sets of Spatial Regions

We want to define aggregate functions to answer roll-up queries to determine which are the frequent sets of spatial regions occurring in a specific spatial zone and during a given temporal interval, where either the zone or the interval can be larger than the granularity of base cells of our data cube. When we aggregate adjacent cells of the TDW,

Figure 6. A trajectory (a) that is correctly counted, and (b) that entails duplicates during the roll-up

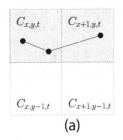
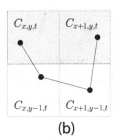

(a)   (b)

spatially and/or *temporally*, we have to exploit only the patterns stored in each of these cells in order to approximate the exact answer, without resorting to the raw data T.

Temporal Aggregation

Given $v+1$ consecutive temporal intervals, $I_i, I_{i+1}, \ldots, I_{i+v}$, and a zone Z, we want to compute the frequent sets of spatial regions (frequent patterns) in the ST aggregate cell (Z,I), where $I = I_i \cup I_{i+1} \cup \ldots \cup I_{i+v}$, by using the frequent patterns contained in each base cell (Z,I_j). Notice that the possible regions of interest are the same in all the cells, since we are referring to the same spatial zone Z.

As discussed in Section 2, this problem has apparently some similarity with the problem of evaluating the frequency patterns in either a stream environments (Giannella, Han, Pei, Yan & Yu, 2003; Silvestri & Orlando, 2007) or a distributed setting (Kargupta & Sivakumar, 2004). In fact a pattern may be infrequent in a partition/chunk (in our case, a cell (Z,I_j)), and be frequent in another (in our case, a cell (Z,I_h)). If p is infrequent in a partition/chunk, i.e. its frequency count is below a given threshold, nothing is stored about p. Hence, to avoid generating too large errors during aggregation, in accordance with (Giannella, Han, Pei, Yan & Yu, 2003), we also store the count of some infrequent patterns in each cell, i.e., not only frequent ones but also *sub-frequent* ones.

Definition 5. Let p be a set of regions of interest, let σ_{min} be the minimum support threshold, let ε be the maximum support error, let Z be a spatial zone and I be a temporal interval, and let W be the number of trajectories crossing $C=(Z,I)$. Let $f_{(Z,I)}(p)$ be the frequency of p in the cell C, i.e. the number of (restricted) trajectories in C traversing p. We say that

- p is *frequent* in C if $f_C(p) \geq \sigma_{min} * W$.
- p is *sub-frequent* in C if $(\sigma_{min}-\varepsilon) * W \leq f_C(p) < \sigma_{min} * W$.
- p is *infrequent* in C if $f_C(p) < (\sigma_{min}-\varepsilon) * W$.

In each base cell of our TDW we store patterns that are either *sub-frequent* or *frequent*. Notice that we cannot always compute the exact aggregate frequency for a pattern over a compound time period $I = I_i \cup I_{i+1} \cup \ldots \cup I_{i+v}$ since when the pattern is infrequent in a base cell we ignore it. Hence we introduce the following approximate support count $\hat{f}$ for a pattern p.

Definition 6. Let p be a pattern, Z a spatial zone, and $I = I_i \cup I_{i+1} \ldots \cup I_{i+v}$ an aggregate time interval. We assume that $f_{(Z,I_j)}(p)$ is the known support count of p in the base cell (Z,I_j). Notice that $f_{(Z,I_j)}(p) \neq 0$ only if p is either frequent or sub-frequent. Then, the approximate frequency $\hat{f}_{(Z,I)}(p)$ over I is defined as

$$\hat{f}_{(Z,I)}(p) = \sum_{j=i}^{i+v} f_{(Z,I_j)}(p)$$

In Algorithm 4 we present the pseudo code for the temporal aggregation over the interval $I = I_i \cup I_{i+1} \ldots \cup I_{i+v}$. For the sake of conciseness, $W_{(Z,I)}$ stands for *(Z,I).presence*, i.e. the value of measure *m3* in the cell *(Z,I)*. Our final goal is to mine all patterns occurring in the ST-cell $(Z,I) = (Z, I_i \cup \ldots \cup I_{i+v})$, whose supports are not less than $\sigma_{min} * W_{(Z,I)}$. Note that, due to the possible split of trajectories among the various base cells involved in the aggregation, in general we have that $W_{(Z,I)} \leq W' = \sum_{j=i}^{i+v} W_{(Z,I_j)}$.

Algorithm 4 proceeds by joining the input sets of patterns one by one (line 3), producing the set P. It extracts from P, patterns of increasing length k (line 6 – 11) and it exploits the anti-monotonicity property to select only those patterns which can be candidate to be frequent (line 9). Among the candidates, we compute their approximate frequency (lines 7 and 10). However, to avoid to remove patterns that can become frequent at a subsequent iteration, we add to the frequency of a certain pattern p, $\sum_{c=i+j+1}^{i+v} W_{(Z,I_c)}$, i.e., the number of trajectories in each cell following the current one (Z,I_{i+j}). Thus, if a pattern p has a frequency less than this value, it can never become frequent during the time interval I, hence it can be safely removed. Finally, the final set of frequent sets is returned (line 12).

Algorithm 4: Algorithm for Temporal Aggregation of Frequent Sets of Spatial Regions

INPUT: P_i, ..., P_{i+v}, with $0 < i < i + v$ $\leq n$, where each P_j is the collection of frequent and sub-frequent sets of regions extracted from $(Z,I_j), I = I_i \cup I_{i+1} \ldots \cup I_{i+v}$

σ_{min}: minimum support. e: maximum support error. $W_{(Z,I)} \leq W' = \sum_{j=i}^{i+v} W_{(Z,I_j)}$.

OUTPUT: F, the collection of frequent sets of spatial regions in (Z, I).

```
1:  P ← P_i
2:  T ← I_i
3:  for j = 1 to v do
4:       P ← P ∪ P_{i+j}
5:       T ← T ∪ I_{i+j}
6:       P^1 ← { x ∈ P | |x| = 1 }
7:       F^1 ← { x ∈ P^1 |
```

$$\hat{f}_{(Z,T)}(x) + \sum_{c=i+j+1}^{i+v} W_{(Z,I_c)} \geq (\sigma_{min} - \epsilon) * W_{(Z,I)} \}$$

```
8:       for k = 2 to s (where s is the
    size of the largest frequent set) do
9:            P^k ← {x ∈ P |
```
$|x| = k \wedge \forall y, y \subseteq x, |y| = k-1 \Rightarrow y \in F^{k-1}$ }
```
10:           F^k ← {x ∈ P^k |
```
$$\hat{f}_{(Z,T)}(x) + \sum_{c=i+j+1}^{i+v} W_{(Z,I_c)} \geq (\sigma_{min} - \epsilon) * W_{(Z,I)}$$ }
```
11:      end for
12:           F ← ⋃_{k=1}^{s} F^k
13:      P ← F
14: end for
15: return F
```

This algorithm was inspired by the work on stream mining (Giannella, Han, Pei, Yan & Yu, 2003). Unfortunately the presence of the spatio-temporal grid and the peculiarities of trajectories make two problems arise:

- some patterns spanning over several base cells cannot be extracted;
- the frequency of some patterns can be overestimated.

Both problems are related to the fact that our trajectories may be split and assigned to more than one base cell (Z,I_j). Consider the trajectory ID_1 in Figure 5 (a) that traverses regions r_1 and

r_2 both covered by zone Z_1. Since r_1 is visited at timestamp $t_2 \in I_1$ and r_2 is visited at timestamp $t_5 \in I_2$ the pattern $\{r_1, r_2\}$ correctly does not appear to be supported by trajectory ID_1 either in (Z_1, I_1) or in (Z_1, I_2). However, when we join I_1 and I_2 and we want to extract the patterns in $(Z_1, I_1 \cup I_2)$, Algorithm 4 cannot recover the information that $\{r_1, r_2\}$ is supported by ID_1.

The overestimate of the frequency is due to the well-known *distinct count problem* (Orlando, Orsini, Raffaetà, Roncato & Silvestri, 2007; Tao, Kollios, Considine, Li & Papadias, 2004). Consider the moving object ID_0 shown in Figure 7 (a) exiting and coming back in the same zone Z_1 during different temporal intervals I_1 and I_2. When we compute the frequency of the pattern $\{r_1, r_2\}$ in the larger cell $(Z_1, I_1 \cup I_2)$ Algorithm 4 simply sums the frequency of the pattern in both cells (Z_1, I_1) and (Z_2, I_2), thus counting ID_0 twice.

A possible solution to these issues is to store along with the extracted patterns the set of the trajectory identifiers supporting them, called *IDSets*. Due to space limitations this improvement is not described in depth here. The reader can grasp the idea in the next section, where we cope with this problem for spatial aggregation.

Spatial Aggregation

Given m spatially adjacent zones, $Z_1, \ldots, Z_m$, and a temporal interval I, we want to compute the frequent sets of spatial regions in the ST aggregate cell (Z, I), where $Z = Z_1 \cup \ldots \cup Z_m$, by using the patterns contained in each base cell (Z_i, I). In the following, without loss of generality, we will focus on the aggregation of a pair of adjacent zones, namely Z_j and Z_h. The algorithm for pairs can be naturally extended to a set of adjacent zones.

The proposed algorithm aggregates local results, which in turn have been obtained by mining projected trajectories in each (Z_j, I) and (Z_h, I). It is worth noticing that the regions of interest in the patterns occurring in different base cells are distinct. Hence in order to find the patterns for the larger cell $(Z_j \cup Z_h, I)$ we exploit the set of trajectory identifiers, supporting the patterns.

Proposition 1. Let Z_j and Z_h be two adjacent zones and let I be a temporal interval. Let p_j and p_h two patterns respectively occurring in (Z_j, I) and (Z_h, I) and let $s_{(Z_j, I)}(p_j)$ and $s_{(Z_h, I)}(p_h)$ be the sets of trajectory identifiers supporting respectively p_j and p_h in (Z_j, I) and in (Z_h, I). Then the support of the aggregated pattern $p_j \cup p_h$ in $(Z_j \cup Z_h, I)$ can be computed as follows:

$$f_{(Z_j \cup Z_h, I)}(p_j \cup p_h) = |\, s_{(Z_j, I)}(p_j) \cap s_{(Z_h, I)}(p_h)\, |$$

Algorithm 5 starts by removing from P_h the patterns which are no longer frequent in the larger cell $(Z_j \cup Z_h, I)$ whereas the remaining ones initialise the output F (lines 2-7). For each pattern p_j in (Z_j, I) we check if it is frequent in the larger cell (lines 8-10). Then for each pattern $p_h \in P_h$ we try to join it with p_j. The new pattern $p_h \cup p_j$ is frequent iff the cardinality of the set obtained as the intersection of the *IDSets* associated with p_j and p_h is greater or equal than $\sigma_{min} * W_{(Z_j \cup Z_h, I)}$ (lines 11-13).

Algorithm 5: Algorithm for Spatial Aggregation of Frequent Sets of Spatial Regions

INPUT: P_j and P_h, the collections of frequent sets of spatial regions extracted from (Z_j, I) and (Z_h, I), two adjacent cells with respect to the spatial dimension. A minimum support threshold σ_{min}. $W_{(Z_j \cup Z_h, I)} \leq W' = W_{(Z_j, I)} + W_{(Z_h, I)}$.

OUTPUT: F, the collection of frequent sets of spatial regions in $(Z_j \cup Z_h, I)$.

```
1: F ← Æ
2: for all p_h ∈ P_h do
3:   if ( f_(Z_h,I)(p_h) ≥ σ_min * W_(Z_j∪Z_h,I) ) then
4:     F ← F ∪ {p_h}
5:   end if
```

Figure 7. (a) A trajectory, and (b) the extracted patterns with their frequency

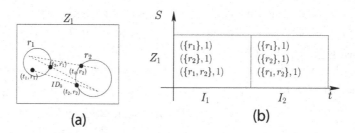

(a)   (b)

```
6:  end for
7:  P_h ← F
8:  for all p_j ∈ P_j do
9:  if ( f_(Z_j,I)(p_j) ≥ σ_min * W_(Z_j∪Z_h,I) ) then
10: F ← F ∪ {p_j}
11: for all p_h ∈ P_h do
12: if (| s_(Z_j,I)(p_j) ∩ s_(Z_h,I)(p_h) |≥ σ_min * W_(Z_j∪Z_h,I) )
then
13: F ← F ∪ { p_h ∪ p_j }
14: end if
15: end for
16: end if
17: end for
18: return F
```

EXPLOITING COLLECTED DATA

Measures extracted from our TDW could be used to discover properties of the real world, either directly by means of standard OLAP operators, or using them as an input for subsequent analyses, e.g., for data mining purposes.

As a possible use case of our system, consider the data collected automatically by mobile phone companies about the movements of individuals, and stored in our TDW. In this context, observations correspond to the information recorded by the base stations about mobile phones. The measures stored in the TDW could be used to discover traffic phenomena, like traffic jams. Unfortunately, it seems difficult to detect them by only considering single measures. As an example, consider that a traffic jam in a cell implies

a low average speed of trajectories crossing the cell itself. In many cases this simple argument is not reliable, when observations not only refer to car drivers but also to pedestrians, bikers, etc. Thus a cell crossed by a lot of pedestrians could be associated with a low average speed, even if no traffic jam actually occurred. In addition, a velocity of 50 km/h is normal for a urban road, but it surely corresponds to a traffic jam for a highway. These considerations seem to exclude that a single measure, along with a trigger level, can be sufficient to detect a complex event, like a traffic jam, with enough accuracy.

In order to discover interesting phenomena, analysts could take advantage from visual tools providing a visual representation of the extracted measures. For example, they could visualise the time series of the measures associated with a given zone in order to understand specific temporal trends in that area. Moreover, if a combination of measures is recognised as predictive for some phenomena, such as a traffic jam or a very crowded area, their values can be used to colour chromatic maps.

Figure 8 shows an example of these two types of visualisations coexisting in the same user interface. The chromatic map on the left shows the discretized value of the measure *presence* in a given temporal interval (the 90th time slice) for different zones of the centre of Milan. Only zones traversed by at least one trajectory are coloured. In particular, the various colours (appearing as gray levels in this printout) indicate different concentrations of trajectories. The user can modify the map by selecting the granularity, the time slice and the measures to display, and, also, the discretization criteria to be associated with different colours. Finally, if the user selects a specific zone using the mouse, a thick border appears around the zone itself, while a set of charts is shown on the right side. Each chart represents the temporal evolution of a measure in the selected zone.

The data extracted from the TDW, referenced in both space and time, can also be used as input for a subsequent *data mining* analysis (Han & Kamber, 2006). Even if this can be considered a well known and common activity, in our case some additional complexity comes from the spatial and temporal dimensions of the data (see, e.g., (Giannotti, Nanni, Pedreschi & Pinelli, 2007; Malerba, Appice & Ceci, 2004)) extracted from the TDW, which make it challenging to exploit mining tools that were not thought for analysing such data.

Typical data extracted from our TDW can simply be the *records of measures* associated with the various cells (base or aggregate ones),

possibly integrated with other information/measures related to town/road planning, demography, economics, etc., which can come from external sources like a GIS.

According to (Tan et al.., 2001), in order to extract knowledge we can mine either *intra-zone* or *inter-zone* data, where a zone is identified by its spatial coordinates. An intra-zone analysis is conducted on data records referring to the same spatial zone observed at different times. An inter-zone analysis is instead conducted on records referring to various zones – e.g., zones that are geographically close or that have some common features – observed during different time intervals.

In addition, if we know that a given phenomenon (e.g, a traffic jam) actually occurred in specific spatio-temporal cells, we can exploit this supervised knowledge to label our extracted data. For example, such labelled records can then be used to build a *classification model*, or to understand which combination of inter- or intra-

Figure 8. Visualisation interface

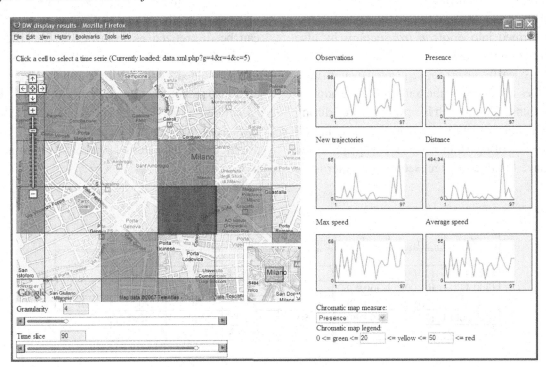

zone measures implies, with high probability, the occurrence of the given phenomenon.

In other terms we can associate with each record, in turn corresponding to a given spatio-temporal cell referring to a specific zone, a *class label*, e.g., either YES_{jam} or NO_{jam}. Such records can then be used to build a *classification model*, which can also be employed to understand which combination of stored measures implies, with high probability, the occurrence of a jam.

In addition, a descriptive data mining method that can be applied to our intra- or inter-zone data is *clustering*. For example, inter-zone clustering could be used to directly discover similar/dissimilar cells, independently of the spatial/temporal dimensions associated with the cells themselves. Conversely, intra-zone clustering could be used to determine similar/dissimilar cells referring to the same zone. The analyst could discover unsuspected similarities between cells, and use this knowledge along with other information of interest coming from other sources.

Finally, also classical *frequent (not-sequential) pattern* or *frequent sequential pattern* mining can be exploited as well. Such analyses can be applied to the inter- or intra-zone records discussed above, transformed into *Boolean transactions* by also discretizing the various measures.

Classical frequent (not-sequential) pattern mining does not consider the temporal dimension of each transaction, which is simply ignored during knowledge extraction. Conversely, the temporal dimension of each transaction is considered when we have to mine frequent sequential patterns.

An example of inter-zone analysis is discussed in the following. First, suppose that we are interested in analysing some zones of interest Z_i, each corresponding to the same spatial intervals X_i and Y_i. For each Z_i and each time interval T_j – i.e., for each base or aggregate cell of our TDW – we can ask the TDW for a record of stored/aggregate measures. For example, this record can contain the number of observations (*numObs*), the numbers of starting trajectories (*trajInit*), the average speed of the trajectories (*speed*), the total distance covered by the trajectories crossing the cell (*distance*), etc.

A snapshot of such extracted records could be the following, where each triple (X_i, Y_i, T_j) identifies a zone (X_i, Y_i) during a given time interval T_j, (See Figure 9).

Note that each record can easily be transformed into a Boolean transaction, where the triple (X_i, Y_i, T_j) univocally identifies the transaction (*Transaction ID*). For example, in order to apply an algorithm like Apriori (Agrawal & Srikant, 1994) to extract classical frequent (non-sequential) patterns, and make these results more meaningful, the attributes involved may need to be discretized. For example, the domain of attribute *numObs* can be discretized as follows: [0,10), [10,20), …, [190,200), [200,210). So, the two distinct *items* obtained for the attribute *numObs* from the above snapshot are "*numObs* = [200,210)" and "*numObs* = [150,160)". As previously discussed, further information can be added to each record/transaction using external sources of information like a GIS.

Figure 9.

X_i	Y_i	T_j	numObs	trajInit	speed	distance	...
...	...	...	...	...	...	...	...
[300,400)	[0,100)	[30,60)	200	20	50	2000	...
[300,400)	[0,100)	[60,90)	150	15	80	3000	...
[400,500)	[0,100)	[30,60)	200	20	50	2000	...
[400,500)	[0,100)	[60,90)	150	15	80	3000	...
...	...	...	...	...	...	...	...

From the above analysis we can discover which combinations of measures appear to be correlated. If such frequent patterns describe some real phenomenon, then the analyst will need to understand/identify it in order to arrive at a true knowledge. For example, looking at the transactions supporting the most significant frequent patterns, we can for example discover which are the spatio-temporal cells that share such common characteristics.

In addition, if we are interested in employing a temporal analysis like the extraction of frequent sequential patterns (Agrawal & Srikant, 1995), we need to explicitly consider the temporal attribute T_j. For example, in the above snapshot we have two temporal sequences, the first corresponding to the ID $(X_i, Y_i) = [300,400), [0,100)$, and the second corresponding to the ID $(X_i, Y_i) = [400,500), [0,100)$.

CONCLUSION AND FUTURE WORK

In this chapter we have discussed relevant issues arising in the design of a data warehouse that maintain aggregate measures about trajectory of mobile objects. The scope of this chapter covers different aspects of the design of the data warehouse, from the underlying data model to the ETL and aggregation algorithms.

In our study we consider several measures of interest related to trajectories. These measure range from relatively simple ones, like travelling time or average speed of trajectories inside a cell, to more complex measures, like the number of distinct trajectory intersecting a cell and the newly introduced frequent region patterns. This last measure consists of a set of patterns, where each pattern concerns a collection of spatial regions of interest that are visited, in any order, by large numbers of moving objects.

Finally, we hinted at the possibility of using the aggregate information stored in the DW in order to discover properties of real world entities,

either by visualisation techniques or by using such data as input for subsequent data mining analysis (supervised and unsupervised data mining operations, such as clustering and classification, applied either in an intra-cell or inter-cell context).

Our TDW has been implemented and tested by using the DW tools of the Oracle DBMS suite. For performance reasons, the most complex transformations of the ETL process, like interpolation, have been realised by using a Java application. This Java application is also in charge of extracting the observations from the incoming stream (see (Orlando, Orsini, Raffaetà, Roncato & Silvestri, 2007) for details).

In the near future, we plan to experimentally assess the accuracy and the performance of the algorithms proposed for the OLAP operations concerning the *frequent region patterns* measure, following the same methodology of our previous studies on the *presence* measure (distinct trajectory count) (Orlando, Orsini, Raffaetà, Roncato & Silvestri, 2007).

An important issue we have to face in the design of our TDW, which will be a subject of future work, concerns the data structures used to store, in a compressed form, patterns about trajectories. This compression can largely reduce the needed memory, when similar patterns occur in close cells, like for example a sequence of cells along the temporal dimension. In fact, such cells refer to the same spatial zone, and thus contain patterns in turn referring to regions of interest within that zone. We may adopt a solution based on an FP-tree, like the one discussed in (Giannella, Han, Pei, Yan & Yu, 2003), where a window table is maintained for each frequent pattern of the FP-tree that spans multiple time intervals.

Moreover, while in this work we adopted a simple multi-dimensional cube model, where the spatio-temporal dimensions are discretized according to a regular grid, we aim at considering more complex concept hierarchies, to be associated with pure spatial (i.e., coordinate roadway, district, cell, city, province, country) and temporal (i.e., second, minute, hour, day, month, year) dimensions, corresponding to the spatio-temporal framework wherein objects actually travel. This should also influence the way in which we reconstruct trajectories from the set of samplings, thus supporting different kinds of interpolations.

Finally, note that if we need to store and compute measures concerning specific classes of objects having common characteristics, our simple cube model can be easily extended by adding a fourth dimension. In this way, a base cell becomes a place where to store information concerning a given object class.

ACKNOWLEDGMENT

This work has been partially funded by the European Commission project IST-6FP-014915 "GeoPKDD: Geographic Privacy-aware Knowledge Discovery and Delivery (GeoPKDD)" (web site: http://www.geopkdd.eu).

REFERENCES

Agrawal, R., & Srikant, R. (1994). Fast algorithms for mining association rules. In *VLDB'94* (pp. 487-499).

Agrawal, R., & Srikant, R. (1995). Mining sequential patterns. In *ICDE'95* (pp. 3-14).

Damiani, M. L., & Spaccapietra, S. (2006). Spatial data warehouse modelling. In *Processing and managing complex data for decision support* (pp. 21-27).

Giannella, C., Han, J., Pei, J., Yan, X., & Yu, P. (2003). Mining frequent patterns in data streams at multiple time granularities. In *NSF Workshop on Next Generation Data Mining*.

Giannotti, F., Nanni, M., Pedreschi, D., & Pinelli, F. (2007). Trajectory pattern mining. In *KDD'07* (pp. 330-339). ACM.

Gonzalez, H., Han, J., Li, X., & Klabjan, D. (2006). Warehousing and analyzing massive RFID data sets. In *ICDE'06* (p. 83).

Gray, J., Chaudhuri, S., Bosworth, A., Layman, A., Reichart, D., & Venkatrao, M. (1997). Data cube: A relational aggregation operator generalizing group-by, cross-tab and sub-totals. *Data Mining and Knowledge Discovery*, *1*(1), 29–54. doi:10.1023/A:1009726021843

Güting, R. H., & Schneider, M. (2005). *Moving object databases*. Morgan Kaufman.

Han, J., & Kamber, M. (2006). *Data mining: Concepts and techniques* (2nd ed.). Morgan Kaufmann.

Han, J., Stefanovic, N., & Kopersky, K. (1998). Selective materialization: An efficient method for spatial data cube construction. In *PAKDD'98* (pp. 144-158).

Kargupta, H., & Sivakumar, K. (2004). Existential pleasures of distributed data mining. In *Data mining: Next generation challenges and future directions*. AAAI/MIT Press.

Lopez, I., Snodgrass, R., & Moon, B. (2005). Spatiotemporal aggregate computation: A survey. *IEEE TKDE*, *17*(2), 271–286.

Malerba, D., Appice, A., & Ceci, M. (2004). A data mining query language for knowledge discovery in a geographical information system. In *Database Support for Data Mining Applications* (pp. 95-116).

Marchant, P., Briseboi, A., Bédard, Y., & Edwards, G. (2004). Implementation and evaluation of a hypercube-based method for spatiotemporal exploration and analysis. *ISPRS Journal of Photogrammetry and Remote Sensing*, *59*, 6–20. doi:10.1016/j.isprsjprs.2003.12.002

Monteiro, R. S., Zimbrao, G., Schwarz, H., Mitschang, B., & de Souza, J. M. (2005). Building the data warehouse of frequent itemsets in the dwfist approach. In *ISMIS* (pp. 294-303).

Orlando, S., Orsini, R., Raffaetà, A., Roncato, A., & Silvestri, C. (2007). Trajectory data warehouses: Design and implementation issues. *Journal of Computing Science and Engineering, 1*(2), 240–261.

Papadias, D., Tao, Y., Kalnis, P., & Zhang, J. (2002). Indexing spatio-temporal data warehouses. In *ICDE'02* (p. 166-175).

Pedersen, T., & Tryfona, N. (2001). Pre-aggregation in spatial data warehouses. In *SSTD'01* (Vol. 2121, pp. 460-480).

Pfoser, D., Jensen, C. S., & Theodoridis, Y. (2000). Novel approaches in query processing for moving object trajectories. In *VLDB'00* (pp. 395-406).

Rigaux, P., & Scholl, M. A. V. (2001). *Spatial database: With application to GIS*. Morgan Kaufmann.

Rivest, S., Bédard, Y., & Marchand, P. (2001). Towards better support for spatial decision making: Defining the characteristics of spatial on-line analytical processing (SOLAP). *Geomatica, 55*(4), 539–555.

Shekhar, S., & Chawla, S. (2003). *Spatial databases: A tour*. Prentice Hall.

Shekhar, S., Lu, C., Tan, X., Chawla, S., & Vatsavai, R. (2001). Map cube: A visualization tool for spatial data warehouse. In H. J. Miller & J. Han (Eds.), *Geographic data mining and knowledge discovery*. Taylor and Francis.

Silvestri, C., & Orlando, S. (2005). Distributed approximate mining of frequent patterns. In *SAC'05* (p. 529-536). ACM.

Silvestri, C., & Orlando, S. (2007). Approximate mining of frequent patterns on streams. *Int. Journal of Intelligent Data Analysis, 11*(1), 49–73.

Tan, P.-N., Steinbach, M., Kumar, V., Potter, C., Klooster, S., & Torregrosa, A. (2001). Finding spatio-temporal patterns in earth science data. In *KDD Workshop on Temporal Data Mining*.

Tao, Y., Kollios, G., Considine, J., Li, F., & Papadias, D. (2004). Spatio-temporal aggregation using sketches. In *ICDE'04* (p. 214-225). IEEE.

Tao, Y., & Papadias, D. (2005). Historical spatio-temporal aggregation. *ACM TOIS, 23*, 61–102. doi:10.1145/1055709.1055713

Worboys, M., & Duckham, M. (2004). *GIS: A computing perspective* (2nd ed.). CRC Press.

Section 2
Mining Sensor Network Data

Chapter 5
Anomaly Detection in Streaming Sensor Data

Alec Pawling
University of Notre Dame, USA

Ping Yan
University of Notre Dame, USA

Julián Candia
Northeastern University, USA

Tim Schoenharl
University of Notre Dame, USA

Greg Madey
University of Notre Dame, USA

ABSTRACT

This chapter considers a cell phone network as a set of automatically deployed sensors that records movement and interaction patterns of the population. The authors discuss methods for detecting anomalies in the streaming data produced by the cell phone network. The authors motivate this discussion by describing the Wireless Phone Based Emergency Response (WIPER) system, a proof-of-concept decision support system for emergency response managers. This chapter also discusses some of the scientific work enabled by this type of sensor data and the related privacy issues. The authors describe scientific studies that use the cell phone data set and steps we have taken to ensure the security of the data. The authors also describe the overall decision support system and discuss three methods of anomaly detection that they have applied to the data.

INTRODUCTION

The Wireless Phone-Based Emergency Response System (WIPER) is a laboratory proof-of-concept, Dynamic Data Driven Application System (DD-DAS) prototype that uses cell phone network data to identify potential emergency situations and monitor aggregated population movement and calling activity. The system is designed to complement existing emergency response management tools by provid-

DOI: 10.4018/978-1-60566-328-9.ch005

ing a high level view of human activity during a crisis situation using real-time data from the cell phone network in conjunction with geographical information systems (GIS). Using cell phones as sensors has the advantages of automatic deployment and sensor maintenance; however, the data available from the network is limited. Currently only service usage data and coarse location data, approximated by a Voronoi lattice defined by the cell towers, are available, although cell-tower triangulation and GPS could greatly improve the location data (Madey, Szabó, & Barabási, 2006, Madey et al., 2007, Pawling, Schoenharl, Yan, & Madey, 2008, Schoenharl, Bravo, & Madey, 2006, Schoenharl, Madey, Szabó, & Barabási, 2006, Schoenharl & Madey, 2008).

The viability of using cell phones as a sensor network has been established through the use of phone location data for traffic management (Associated Press, 2005). WIPER applies this finding to fill a need in emergency response management for a high level view of an emergency situation that is updated in near real-time. Tatomir and Rothkrantz (2005) and Thomas, Andoh-Baidoo, and George (2005) describe systems for gathering on-site information about emergency situations directly from response worker on the ground via ad-hoc networks of PDAs. While these systems can provide detailed information about some aspects of the situation, such as the location of victims and environmental conditions, the information is limited to what can be observed and reported by the responders. This provides valuable but local information, though there may be observations from different, geographically dispersed locations. In contrast, WIPER provides less detail, but instead gives an overall view of population movements that may be valuable in refining the response plans or directing response workers to gather more detailed information at a particular location.

Dynamic data driven applications systems (DDDAS) provide a framework in which running simulations incorporate data from a sensor network to improve accuracy. To achieve this, the simulations dynamically steer the measurement process to obtain the most useful data. The development of DDDAS applications is motivated by the limited ability to predict phenomena such as weather and wildfire via simulation. Such phenomena are quite complex, and the correct simulation parameterization is extremely difficult. The goal of DDDAS is to provide robustness to such simulations by allowing them to combine sub-optimal initial parameterizations with newly available, real-world data to improve performance without the expense of rerunning the simulations from the beginning (Douglas & Deshmukh, 2000).

In this chapter, we focus on one component of WIPER: the detection and alert system. This module monitors streaming data from the cell phone network for anomalous activity. Detected anomalies are used to initiate an ensemble of predictive simulations with the goal of aiding emergency response managers in taking effective steps to mitigate crisis events. We discuss methods for anomaly detection on two aspects of the call data: the call activity (the number of calls made in a fixed time interval) and the spatial distribution of network usage.

The remainder of the chapter is organized as follows: we discuss background literature related to mining data from a cell phone network. We start with a discussion of methods for detecting outliers in our data, with a focus on using data clustering to model normality in data. Those clusters of outliers in the streaming data could be indicators of a problem, disaster or emergency in a geographical area (e.g., an industrial explosion, a civil disturbance, progress of a mandated evacuation prior to a hurricane, a terrorist bombing). We then give an overview of the data set and the WIPER system, followed by descriptions of algorithms used in the detection and alert system. Finally, we discuss some of the privacy issues related to this work and our plans for future work in the spatial, graph and temporal analysis of the underlying social network.

BACKGROUND

In this section, we discuss background literature on outlier detection and clustering especially that relevant to our application of detecting anomalies in streaming cell phone sensor data: both (1) location and movement data and (2) calling patterns of the population carrying the cell phones.

Outlier Detection

An outlier is an item in a data set that does not appear to be consistent with the rest of the set (Barnett & Lewis, 1994). There is a great deal of literature on the problem of outlier detection as well as a number of applications, including fraud detection, intrusion detection, and time series monitoring (Hodge & Austin, 2004).

There are three fundamental approaches for outlier detection:

- *Model both normality and abnormality*: this approach assumes that a training set representative of both normal and abnormal data exists.
- *Model either normality or abnormality*: this approach typically models normality and is well suited for dynamic data.
- *Assume no a priori knowledge of the data*: this approach is well suited for static distributions and assumes that outliers are, in some sense, far from normal data. (Hodge & Austin, 2004).

Additionally, there are four statistical techniques for outlier detection: parametric, semi-parametric, non-parametric, and proximity based methods. Parametric outlier detection techniques assume that the data follows a particular probability distribution. These techniques tend to be fast but inflexible. They depend on a correct assumption of the underlying data distribution and are not suitable for dynamic data. Semi-parametric models use mixture models or kernel density estimators rather than a single global model. Both mixture models and kernel models estimate a probability distribution as the combination of multiple probability distributions. Non-parametric techniques make no assumptions about the underlying distribution of the data and tend to be computationally expensive. Proximity based techniques define outliers in terms of their distance from other points in the data set and, like non-parametric techniques, make no assumptions about the data distribution (Hodge & Austin, 2004).

In this chapter, we approach the outlier detection problem by modeling normal behavior. One technique for modeling normality in multidimensional space is data clustering, which enables outlier detection using proximity based techniques.

Data Clustering

The goal of data clustering is to group similar data items together. Often, similarity is defined in terms of distance: the distance between similar items is small. Data items that do not belong to any cluster, or data items that belong to very small clusters, may be viewed as outliers, depending on the clustering algorithm and application.

The clustering problem is defined as follows: let a data set D consist of a set of data items $\{\vec{d}_1, \vec{d}_2, ...\}$ such that each data item is a vector of measurements, $\vec{d}_i = \langle d_{i,1}, d_{i,2}, ..., d_{i,n} \rangle$. Clustering provides a convenient way for finding anomalous data items: anomalies are the data items that are far from all other data items. These may be data items that belong to no cluster, or they may be the data items that belong to small clusters. If we take the view that anomalies belong to no cluster, we can use a clustering of the data to model normal behavior. If we view each cluster as a single point, we can greatly reduce the cost of proximity based anomaly detection, assuming the number of clusters is small relative to the total number of data items and that we can cluster the data quickly.

Traditional clustering algorithms can be divided into two types: partitional and hierarchical. Partitional algorithms, such as *k*-means and expectation maximization, divide the data into some number, often a predefined number, of disjoint subsets. These algorithms often start with a random set of clusters and iterate until some stopping condition is met. As a result, these algorithms have a tendency to converge on local minima. Hierarchical algorithms divide the data into a nested set of partitions and are useful for discovering taxonomies in data. They may either take a top-down approach, in which an initial data cluster containing all of the data items is iteratively split until each data item is in its own cluster, or a bottom-up approach, in which clusters initially consisting of only a single element are iteratively merged until all of the data items belong to a single cluster. Often, hierarchical algorithms must compute the distance between each pair of data items in the data set, and, therefore, tend to be computationally expensive, though there are techniques for making this process more efficient (Jain, Murty, & Flynn, 1999).

Partitional and hierarchical clustering algorithms may also incrementally incorporate new data into the cluster model (Jain et al, 1999). The leader algorithm incrementally partitions the data into cluster using a distance threshold to determine if a new data item should be added to an existing cluster or placed in a new cluster (Hartigan, 1975). Fisher (1987) describes the COBWEB algorithm, an incremental clustering algorithm that identifies a conceptual hierarchy. The algorithm uses category utility, a function that provides a measure of similarity of items in the same cluster and dissimilarity of items in different cluster, to determine whether a new object should be classified using an existing concept in the hierarchy or whether a new concept should be added. The COBWEB algorithm also combines and splits classes as necessary, based on category utility. Charikar et al (1997) describe several incremental agglomerative clustering algorithms for information retrieval applications. In these algorithms, when a new data item does not meet the criteria for inclusion in one of the existing clusters, a new cluster is created and two other clusters are merged so that *k* clusters exist at all times. The algorithms differ in their approach to determining the two clusters to be merged.

Stream clustering algorithms are similar to incremental algorithms. In addition to processing each item only once, stream algorithms typically use no more that order polylogarithmic memory with respect to the number of data items. Guha, Meyerson, Mishra, Motwani, and O'Callaghan (2003) present a method based on *k*-medoids— an algorithm similar to *k*-means. The clusters are computed periodically as the stream arrives, using a combination of the streaming data and cluster centers from previous iterations to keep memory usage low. Aggarwal, Han, Wang, and Yu (2003) present a method that takes into account the evolution of streaming data, giving more importance to more recent data items rather than letting the clustering results be dominated by a significant amount of outdated data. The algorithm computes *micro-clusters*, which are statistical summaries of the data, periodically throughout the stream. These micro-clusters serve as the data points for a modified *k*-means clustering algorithm.

Hybrid techniques combine two clustering algorithms. Cheu, Keongg, and Zhou (2004) examine the use of iterated, partitional algorithms, such as *k*-means, as a method of reducing a data set before applying hierarchical algorithms. Chipman and Tibshirani (2006) combine agglomerative algorithms, which tend to effectively discover small clusters, with divisive methods, which tend to effectively discover large clusters. Surdeanu, Turmo, and Ageno (2005) propose a hybrid clustering algorithm that uses hierarchical clustering to determine initial parameters for expectation maximization.

Percolation Theory

We use concepts from percolation theory to discover spatial anomalies. Percolation theory studies the emergence of connected components, or clusters, in a d-dimensional lattice as the probability of an edge existing between a pair of neighbors in the lattice approaches 1. At some critical probability a percolating cluster, a connected component containing most of the vertices in the lattice, appears. Percolation theory is typically interested in three quantities: the fraction of the lattice in the percolating cluster, the average cluster size, and the cluster size distribution (Stauffer & Aharony, 1992, Albert & Barabási, 2002).

THE DATASET

We use a large primary dataset consisting of actual call record data from a cellular service provider to support the development of WIPER. The data set contains records describing phone calls and SMS messages that traverse the network. Each record contains the date, time, the ID of the individual making the call, the ID of the individual receiving the call, the tower the caller's phone is communicating with, and the type of transaction: voice call or SMS. The IDs of the individuals making and receiving the calls are generated by the service provider using an undisclosed hash function. Currently, we have approximately two years of call record data, taking up 1.25 TB of disk space after being compressed with gzip. One month of data contains approximately five hundred million records. Roughly one quarter of these are SMS transactions, and the remaining three-quarters are voice phone calls. An additional dataset provides the latitude and longitude of the cell towers. In addition to the primary dataset (actual call record data) from the service provider, we generate a secondary data set (synthetic call record data) using validated simulations from a component of the WIPER system. Although there are many anomalies in the primary dataset of actual call data, in most cases we have not been able to determine the cause (although highly newsworthy events have been correlated with anomalies in the data streams). The synthetic data helps us test our anomaly detection algorithms and other components of the WIPER system in a more controlled manner.

Since the call record includes the location and the time of the transactions (voice and SMS), when aggregated, it forms a times series data stream with normal patterns varying with day of the week, time of the day, and cell-tower location. Abnormal patterns in the data, what we call anomalies, could be indications of a disaster, e.g., an airplane crash, a political riot, or a terrorist attack. Of course the anomaly could be caused by a more benign event such as traffic after a championship football game, a public holiday, or a news event outside the area under watch. Such anomalous patterns could reflect many individuals making phone calls or sending SMS text messages because of a traffic jam, a public holiday, or a nearby disaster. In all cases the level of calling would increase above a baseline and be visible in the time series data streams as an anomaly.

The primary data has been used in a number of other studies. Onnela et al. (2007a) analyze a wide range of characteristics of the call graph, including degree distribution, tie strength distribution, topological and weighted assortativity, clustering coefficient, and percolation properties.

In a second study, Onnela et al. (2007b) explore the role of tie strength in information diffusion. A graph is built from 18 months of data, using the total call duration between pairs of users as the edge weights. The analysis of the graph shows a positive correlation between the strength of an edge and the number of common neighbors shared by the two vertices, indicating that strong ties tend to form within community structures and that weak ties form the connections between communities. To measure the importance of strong and weak ties, two experiments are performed: the edges are removed in increasing order of strength and the edges are removed in decreasing order of strength. Removing weak ties causes the graph to collapse into many small components very quickly, where removing the strong ties causes the graph to shrink slowly.

The usefulness of this data goes beyond social network analysis and the development of emergency response tools. González, Hidalgo, & Barabási (2008) study human movement patterns over a six month period. Information from this type of study can be used for a number of applications, including design of public transportation systems, traffic engineering, and prediction and control of disease outbreak. Candia et al (2008) and González & Barabási (2007) discuss the privacy implications of working with this type of data in the context of scientific research.

WIPER: CELL PHONES AS SENSORS FOR SITUATIONAL AWARENESS

One goal of the WIPER project is to develop a laboratory proof-of-concept to evaluate the potential of using cell phones as sensors to increase situational awareness of emergency response managers during an ongoing crisis. The WIPER system is designed to accept streams of near real-time aggregated data about calling activity and location data of the cell phones in a geographical area. This data could be monitored for anomalies that could serve as alerts of potential problems or emergencies (the primary focus of this chapter), but could also be displayed on maps to provide emergency mangers a view of where the citizens are, their movements, and potential "hot spots" indicated by above normal calling activity. The system runs simulations that attempt to infer the nature of the anomalous event and to predict future behavior of the cell phone network and, hence, the population affected by the crisis. New data is used as it becomes available from the cell phone

network to validate and steer running simulations in order to improve their predictive utility.

The WIPER system consists of five components, each of which is described briefly below.

- The *Decision Support System* (DSS) is a web-based front end through which emergency response managers interact with the WIPER system.
- The *Detection and Alert System* (DAS) monitors streaming network data for anomalous activity. There are various aspects of the cell phone network data that may be of interest, including overall usage levels, spatial distribution of users, and the underlying social network.
- The *Simulation and Prediction System* (SPS) receives anomaly alerts from the DAS, produces hypotheses that describe the anomaly, and uses simulations in conjunction with streaming activity data to validate or reject the hypotheses. We also use the simulations resident in the SPS to generate our synthetic datasets described earlier.
- The *Historical Data Source* (HIS) is a repository of cellular network data that resides in secondary storage. This data is used to determine the base-line behavior of the network against which anomalies are detected and to periodically calibrate and update the DAS.
- The *Real-Time Data Source* (RTDS) is a real-time system that will receive transaction data directly from a cellular service provider. The RTDS is responsible for handling requests for streaming data from the DAS, SPS, and DDS and streaming incoming data to these components in a timely manner.

Figure 1 shows an architectural overview of the WIPER system. The RTDS and HIS will provide the bridge from the service provider

Figure 1. WIPER system architecture

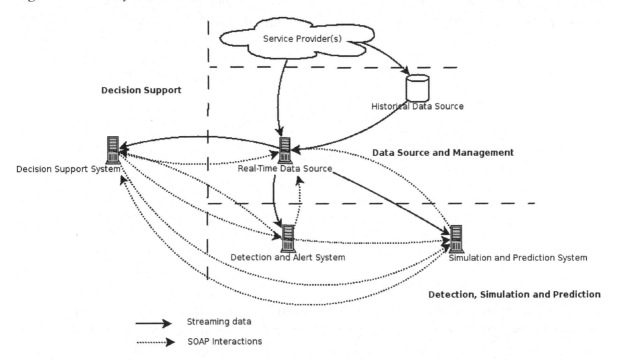

and the WIPER system. The figure shows the flow of streaming data from the service provider through the RTDS, possibly by way of the HIS for development and training, and to the remaining components. Requests for streaming data from the RTDS occur via SOAP messages. SOAP messages are also used by the Detection and Alert System to inform the Simulation and Prediction system of potential anomalies in the streaming data.

The Detection and Alert System The detection and alert system is designed to examine the streaming data from the cellular service provider for anomalous activity on two axes: call activity (the number of calls made in a fixed time interval), and spatial distribution (location and movement) of the cell phones based on calls made using them. Three data mining techniques have been implemented and evaluated for use in the Detection and Alert System of WIPER: (1) a model that uses a Markov modulated Poisson process technique, (2) a method for spatial analysis based on percolation theory, and 3) a method for spatial analysis using online hybrid clustering. These techniques

and their adaptation to data mining of cell phone data for anomalies within the WIPER system are described below.

Call Activity Analysis using Markov Modulated Poisson Processes

The most basic indicator of anomalous behavior in a cell phone network is an increase or a decrease in cell phone call activity within a given geographical area. This type of anomaly can be detected by monitoring a time series consisting of the number of calls made in disjoint time intervals of a fixed size, *e.g.* the number of calls made every 10 minutes. The Markov modulated Poisson process, which uses a Poisson process in conjunction with a hidden Markov model to identify anomalies in the data, is described by Ihler, Hutchins, and Smyth (2006, 2007) and is summarized below.

A Poisson process, which models the number of random events that occur during a sequence of time intervals, can be used to model the baseline

Figure 2. The overall average rate (λ_0), day effect combined with the overall average ($\lambda_0 \delta_{d(t)}$), and time of day effect combined with the overall average and the day effect ($\lambda_0 \delta_{d(t)} \eta_{d(t),h(t)}$).

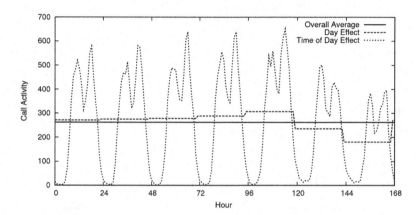

behavior of such a time series: the number of events per time interval follows a Poisson distribution with an expected value of λ, the rate parameter for the process. In this model, the probability of N events occurring in a time step is:

$$P(N; \lambda) = \frac{e^{-\lambda} \lambda^N}{N!} \tag{5}$$

for $N = 0, 1, \ldots$ (Mitzenmacher & Upfal, 2005).

The standard Poisson process is not sufficient for modeling many real-world phenomena since the rate of many natural processes varies over time. In the case of human activities, there are often daily and weekly cycles, so the rate becomes a function of the day of the week and time of day. The overall average, λ_0, is the average rate over all time intervals and establishes the baseline rate of the process. The day effect, $\delta_{d(t)}$, $d(t) \in \{1 \ldots 7\}$, is the average rate over all time intervals for each day of the week, normalized such that the average day effect is 1, *i.e.* $\sum \delta_{d(t)} = 7$. The day effect expresses the call activity of the day relative to the overall average. The time of day effect, $\eta_{d(t),h(t)}$, $h(t) \in \{1 \ldots D\}$ is the average rate for each time interval for each day of the week. The time of

day effect for each of the 7 days of the week is normalized such that average time of day effect for each day is 1, *i.e.* $\forall d(t), \sum \eta_{d(t),h(t)} = D$. The time of day effect expresses the call activity of the interval relative to the product of the overall average and the day effect.

The rate function for a Markov modulated Poisson process is

$$\lambda(t) = \lambda_0 \delta_{d(t)} \eta_{d(t),h(t)} \tag{6}$$

To illustrate the components of the rate function, we compute the overall average rate, the day effect, and the time of day effect from two weeks of real cell phone data. Figure 2 shows each component of the rate function.

The Poisson process described above is used in conjunction with a hidden Markov model to identify anomalies in the call data. The hidden Markov model has two states for describing the current call activity: normal and anomalous. The transition probability matrix for the hidden Markov model is

$$M = \begin{bmatrix} 1 - A_0 & A_1 \\ A_0 & 1 - A_1 \end{bmatrix} \tag{7}$$

where each entry $m_{ij} \in M$ corresponds to the probability of moving from state i to state j. Intuitively, $1 / A_0$ is the expected time between anomalies and $1 / A_1$ is the expected duration of an anomaly. Initially, we guess that $A_0 = 0.01$ and $A_1 = 0.25$. These guesses are updated based on the state sequence generated in each iteration.

Anomalies are identified using the Markov Chain Monte Carlo method. For each iteration of the method, the forward-backward algorithm is used to determine a sample state sequence of the hidden Markov model. For each interval in the forward recursion $t : 1, 2, \ldots T$, the probability of each hidden Markov state is computed by

$$p(A(t) \mid N(t)) = \pi_0 \sum M \cdot p(A(t-1) \mid N(t-1)) p(N(t) \mid A(t))$$

$$(8)$$

where π_0 is the initial distribution of the Markov chain. If the hidden Markov model is in the normal state, the likelihood function, $p(N(t) \mid A(t))$, is simply the probability $N(t)$ is generated by the Poisson process at time t. If the hidden Markov model is in the anomalous state, the likelihood function takes into account the range of possible number of observations, $i \in \{0, 1, \ldots N(t)\}$, beyond the expected number. The probability that i of the $N(t)$ observations are normal is computed using a negative binomial distribution. Let $\mathrm{NBIN}(N, n, p)$ be the probability of N observations given a negative binomial distribution with parameters n, p, and let this negative binomial distribution model the number of anomalous observations, $N(t) - i$, in an interval. The likelihood function is

$$p(N(t) \mid A(t)) = \begin{cases} P(N(t); \lambda(t)) & A(t) = 0 \\ \sum_{i=0}^{N(t)} P(i, \lambda(t)) \mathrm{NBIN}(N(t) - i; a^E, \frac{b^E}{1 - b^E}) & A(t) = 1 \end{cases}$$

$$(9)$$

where $a^E = 5$ and $b^E = 1/3$ are empirically determined parameters of the negative binomial distribution.

For each interval in the backward recursion, $t : T, T-1, \ldots 1$, samples are drawn from the conditional distribution $M' \cdot p(A(t) \mid N(t+1))$, where M' is the inverse of the transition probability matrix, to refine the probability of the current state t.

Once the forward-backward algorithm has generated a sample hidden state sequence, the values of the transition probability matrix are updated using the empirical transition probabilities from the sample state sequence, and the process is repeated.

We apply this approach to two weeks of call activity data taken from our primary data set (i.e., actual call data), using 50 iterations of the Markov Chain Monte Carlo simulations described above to determine the probability of anomalous behavior for each 10 minute interval. Figure 3 shows the actual call activity and the call activity modeled by the Markov modulated Poisson process for two weeks of for a small town with 4 cell towers. Visual inspection of the graph indicates that the Markov modulated Poisson process models the real call activity well. We do not have information about any emergency events that may be present in this dataset; therefore, this figure shows the posterior probability of an anomaly at each time step in the lower frame based on the hidden Markov model. Note that on the last day of observation, the Markov modulated Poisson process identifies an anomaly corresponding to an observed call activity that is significantly higher than expected. Additionally, an anomaly is detected on the second Tuesday; however, we cannot see a major deviation from the expected call activity raising the possibility that this is a false positive. For each remaining interval, the posterior probability of an anomaly is no greater than 0.5. This analysis indicates that outliers in the call activity time series data can be identified

Figure 3. This figure shows the result of using a Markov modulated Poisson process to detect anomalies in 2 weeks of call activity. The top frame shows the expected and observed number of calls for each time interval, and the bottom frame shows the probability that the observed behavior is anomalous at each time step.

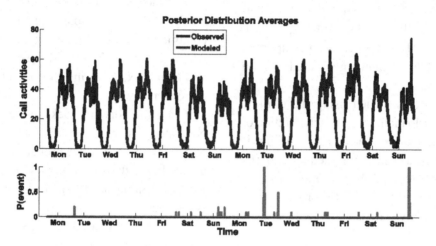

using a Markov modulated Poisson process and could be useful as an alerting method to indicate possible anomalies and emergency events. Such a system would need a second stage of analysis to determine if the outlier is a true positive for an emergency event. These detected anomalies trigger an alert that is sent to the Decision Support System and the Simulation and Prediction System of the WIPER system. Yan, Schoenharl, Pawling, and Madey (2007) describe in greater detail this application of a Markov modulated Poisson process to the problem of detecting outliers and anomalies in call activity data.

Spatial Analysis using Percolation Theory

We have determined that models based on percolation theory can be used to detect spatial anomalies in the cell phone data. The geographical area covered by the data set is divided into a two dimensional lattice, and the call activities through the towers within each cell of the lattice are aggregated. The normal activity for each cell is defined by the mean and standard deviation of

the call activity, and a cell is in an anomalous state when its current observed call activity deviates from the mean by some factor, l, of the standard deviation. In the percolation theory model, neighboring anomalous sites are connected with an edge. When an anomaly occurs in the cell phone network, the number of clusters and the distribution of cluster sizes are statistically different from those that arise due to a random configuration of connected neighbors. In contrast, when the cell phone network is behaving normally, the number of clusters and distribution of cluster sizes match what is expected. Candia et al. (2008) provide a more detailed discussion of percolation theory and how the spatial anomalies of the cell phone data can be detected.

Spatial Analysis using Online Hybrid Clustering

We have evaluated a hybrid clustering algorithm for online anomaly detection for the WIPER system. This hybrid algorithm is motivated by the fact that streaming algorithms for clustering, such as those described by Guha et al (2003) and Aggarwal et al (2003), require *a priori* knowledge of the number of clusters. Due to the dynamic nature of the data stream, we believe that an algorithm that dynamically creates new clusters as needed, such as the leader algorithm, is more appropriate for this application. However, we also believe that the leader algorithm is too inflexible since it produces clusters of a constant size.

The hybrid algorithm combines a variant of the leader algorithm with *k*-means clustering to overcome these issues. The basic idea behind the algorithm is to use *k*-means to establish a set of clusters and the leader algorithm in conjunction with statistical process control to update the clusters as new data arrives. For detecting anomalies in the spatial distribution of call activity, the feature vectors consist of the call activities for each cell tower in the area of interest.

Statistical process control aims to distinguish between "assignable" and "random" variation. Assignable variations are assumed to have low probability and indicate some anomaly in the underlying process. Random variations, in contrast, are assumed to be quite common and to have little effect on the measurable qualities of the process. These two types of variation may be distinguished based on the difference in some measure on the process output from the mean, μ, of that measure. The threshold is typically some multiple, l, of the standard deviation, σ. Therefore, if the measured output falls in the range $\mu \pm l\sigma$, the variance is considered random; otherwise, it is assignable (Bicking & Gryna, Jr., 1979).

The algorithm represents the data using two structures: the cluster set and the outlier set. To save space, the cluster set does not store the examples that make up each cluster. Instead, we use the summarization approached described by Zhang, Ramakrishnan & Livny (1996), where each cluster is summarized by the sum and sum squared values of its feature vectors along with the number of items in the cluster. The outlier set consists of the examples that do not belong to any cluster. The means and the standard deviations

describe the location and size of the clusters, so clusters are only accepted when they contain some minimum number of examples, *m*, such that these values are meaningful. The algorithm periodically clusters the examples in the outlier set using *k*-means. Clusters that contain at least *m* items are reduced to the summary described above and added to the cluster set. If a new data point is within the threshold, $l\sigma$, of the closest cluster center, it is added to the cluster and the summary values are updated. Otherwise, it is placed in the outlier set.

By using mean values as the components of the cluster center and updating the centers whenever a new example is added to a cluster, the algorithm can handle a certain amount of concept drift. At the same time, the use of statistical process control to filter out anomalous data prevents the cluster centers from being affected by outlying points. This algorithm does not require *a priori* knowledge of the number of clusters, since new clusters will form as necessary.

This approach does have some drawbacks. There are cases in which the *k*-means clustering component will fail to produce any clusters of sufficient size; however, we have successfully used this algorithm on data vectors containing usage counts of 5 services provided by a cellular communication company at one minute intervals and simulated spatial data. This hybrid clustering algorithm used for online anomaly detection is described in more detail in Pawling, Chawla, and Madey (2007).

DISCUSSION

Results and Limitations

WIPER is a proof-of-concept prototype that illustrates the feasibility of dynamic data driven application systems. It has been shown that anomalies in real world data can be detected using Markov modulated Poisson processes (Yan et al, 2007) and percolation theory (Candia et al, 2008). The hybrid clustering algorithm has been evaluated using synthetic spatial data generated from simulations based on real-world data with promising results.

The detection and alert system assumes that emergency events are accompanied by a change in underlying call activity. In cases where this does not hold, the system will fail to identify the emergency. Additionally, in cases where the underlying call activity changes very gradually, the system may fail to detect the situation.

In its current state, WIPER can only identify that an anomaly has occurred, it cannot make any determination of its cause. Therefore, the system cannot distinguish between elevated call activity due to an emergency, such as a fire, from a benign event such as a football game. The WIPER system is a laboratory prototype with no immediate plans for deployment. Laboratory tests have demonstrated that the individual components perform as desired and that the multiple modules can work in a distributed manner using SOAP messaging.

Data Mining and Privacy

As the fields of database systems and data mining advance, concerns arise regarding their effects on privacy. Moor (1997) discusses a theory of privacy in the context of "greased" data, data that is easily moved, shared, and accessed due to advances in electronic storage and information retrieval. Moor argues that as societies become large and highly interactive, privacy becomes necessary for security.

"Greased" data is difficult to anonymize because it can be linked with other databases, and there have been cases where data has been "de-identified" but not "anonymized". That is, all identifying fields, such as name and phone number, have been removed or replaced but at least one person's identity can be determined by linking the records to other databases. In these cases, the remaining fields uniquely identify one or more individuals (Sweeney, 1997). With the development of new technologies, data sets thought to be anonymized when collected can become de-anonymized as additional data sets become available in the future. Thus anonymizing "greased" data is extremely difficult. (National Research Council, 2007).

Geographic Information Systems (GIS) provide additional data against which records can be linked. For safety reasons, some governments require that telecommunication companies be able to locate cell phones with some specified accuracy so that people calling for emergency services can be quickly located. Emergency responders can easily find a phone by plotting the location on maps using GIS technology. This method of locating phones can also be used to provide subscribers with location-based services, or it can be used to track an individual's movements (Armstrong, 2002).

A significant issue that arises in the discussion of data mining and privacy is the difficulty of precisely defining privacy. Solove (2002) surveys the ways in which privacy has been conceptualized throughout the history of the U.S. legal system, and points out serious shortcomings of each. Complicating the issue further is the fact that ideas of privacy are determined by culture and are constantly evolving, driven in part by advances in technology (Armstrong & Ruggles, 2005).

Clifton, Kantarcioglu, and Vaidya (2002) describe a framework of privacy for data mining. This paper looks at two types of privacy: individual privacy, which governs information about specific people, and corporate privacy, which governs information about groups of people. In general, individual privacy is maintained, from a legal standpoint, if information cannot be tied to a single individual. Corporate privacy aims to protect a data set, which includes protecting the results of analysis of the data. In a follow-up paper, Kantarcioglu, Jin, and Clifton (2004) propose framework for measuring the privacy preserving properties of data mining results. This framework assumes that the data includes fields that are public, sensitive, and unknown but not sensitive. The framework provides measures of how well the sensitive fields are protected against various attacks using the classifier, such as attempting to infer the values of sensitive fields using public fields.

In response to privacy concerns relating to data mining, researchers are developing data mining methods that preserve privacy. Agrawal and Srikant (2000) propose an approach to classification that achieves privacy by modifying values such that a reliable model may be built without knowing the true data values for an individual. Two methods are used for modifying attribute values: (1) value-class membership is essentially a discretization method that aggregates values into intervals, each of which has a single associated class, and (2) value distortion in which random noise is added to the real value. In the case of value distortion, the data distribution is recovered based on the result of the distortion and the distribution of the distorting values, but the actual attribute values remain hidden.

Lindell and Pinkas (2002) describe a privacy preserving data mining protocol that allows two parties with confidential databases to build a data mining model on the union of the databases without revealing any information. This approach utilizes homomorphic encryption functions. Homomorphic encryption functions allow computations on encrypted values without revealing the actual values. Benaloh (1994) and Paillier (1999) describe additively homomorphic public key encryption functions. Let E be an encryption function and x and y be plaintext messages. If E is additively homomorphic, $E(x)$ and $E(y)$ can be used to compute $E(x+y)$ without revealing x or y. This classification method assumes "semi-honest" parties that correctly follow the protocol but try to obtain further information from the messages passed during the computation.

Friedman, Schuster, and Wolff (2006) describe a decision tree algorithm that produces k-anony-

mous results with the goal of preventing linking attacks that use public information and a classifier to infer private information about an individual. They describe a method for inducing a decision tree in which any result from the decision tree can be linked to no fewer than k individuals.

The nature of the phone data set raises some concerns about privacy issues in relation to our work. Data stored by service providers allows fairly detailed tracking of individuals based on the triangulation of radio signals received by cell towers from phones, as well as the capability to identify an individual's current location. A major concern is the potential for abuse of this technology by the government and law enforcement, especially considering that there is no consensus on what level of evidence is required to gain this information from cellular service providers. Some judges require law enforcement to show probable cause before allowing this data to be accessed, while others view this information as public, since cell phone users choose to keep their device powered on (Nakashima, 2007).

Compounding this concern is the fact that following the terrorist attacks on September 11, 2001 in the U.S., a number of U.S. airlines provided the U.S. government with their passenger records, in direct violation of their own privacy policies. The courts did not accept arguments that this was a breach of contract since no evidence was provided that this breach of contract caused any harm. Solove (2007) argues that the harm here is a loss of trust in companies and the rise of an imbalance in power, since, apparently once a company has information about an individual, the individual loses control over that information completely. In a similar, and more widely known case, U.S. telecommunication companies provided the U.S. government with call records for their subscribers, violating a long held tradition of only releasing customer information when ordered to do so by a court (Cauley, 2006).

In the European Union, privacy is viewed as a Human Right. As a result, the privacy laws are much more comprehensive and are extensive in their coverage of both private and public institutions. In 1968, the Council of Europe discussed the impact of scientific and technological advances on personal privacy, with a focus on bugging devices and large-scale computerized storage of personal information. This discussion led to an evaluation of the adequacy of privacy protection provided by the national laws of member states given recent advances in technology, and preliminary reports indicated that improvement was needed. In 1973 Sweden passed the Data Protection Act requiring governmental approval and oversight of any "personal data register". This was followed by similar legislation in Germany, France, Denmark, Norway, Austria, and Luxembourg by 1979 (Evans, 1981) and the European Data Privacy Directive in 1995 (European Parliament and Council of the European Union, 1995).

The European Data Privacy Directive requires "adequate" data privacy protections be in place before personal data of European Union citizens can be exported to a country outside the Union (European Parliament and Council of the European Union, 1995). In general, the United States does not provide an "adequate" level of protection; however, the U.S. Department of Commerce developed the "Safe Harbor" program that allows American businesses to continue receiving data from Europe by certifying that their data protection policies meet the requirements of the European Union (Murray, 2001).

"Safe Harbor" requires that companies notify customers of how their personal data is used, provide customers with ways in which to make inquiries and lodge complaints relating to their personal information held by the company, and provide customers with information about data sharing policies along with avenues for allowing the customer to limit the use and sharing of their personal data. In cases where personal data is shared with third parties or used for a new purpose, users must be given an opportunity to "opt out", and in cases where this data is particularly

sensitive, *e.g.* medical or health data, religious affiliation, or political views, the customer must "opt in" before the data can be shared (Murray, 2001).

Issues of data security, integrity, and access are also addressed by "Safe Harbor". Companies in possession of personal data are required to take "reasonable precautions" to prevent security compromises, including unauthorized access, disclosure, and alteration of the data. Data integrity refers to the relevance and reliability of the data. Companies must have a specific use for each item of personal information in order to obtain it and may not use that data for any other purpose without consent of the individual described by the data. Finally, users are required to have access to their personal data possessed by the company and the company must provide mechanisms that allow individuals to correct any inaccuracies in the data or request its deletion (Murray, 2001).

FUTURE DIRECTIONS

Several tasks remain to be completed on this project: incorporation of link mining and social network analysis into the stream mining component of the WIPER system, the development of a better understanding of the relationship between outliers, anomalies, and emergencies in our data, and finally the field testing of the system, both with emergency managers within an emergency operations center and with a live stream from a cellular carrier.

Much of the previous work in identifying anomalies in graphs is based on subgraph matching; however, these approaches tend to be computationally expensive. Another possibility is clustering graphs based on some vector of metrics. Like the call activity, graph properties such as assortativity and clustering coefficient exhibit daily and weekly periodic behavior. It may be possible to identify outliers and classify emergency situations using vectors of graph metrics computed on graphs built from a sliding window of call transactions.

There are still important issues that must be resolved. It is not clear what graph properties should be used, and the appropriate window size must be determined. Unsupervised feature selection methods (Dy & Brodley, 2004, Mitra, Murthy, & Pal, 2002) from data mining may be used to identify the best set of graph properties from those that can be computed quickly.

SUMMARY

In this chapter, we have described the detection and alert component of the Wireless Phone-based Emergency Response System, a proof of concept dynamic data-driven application system. This system draws from research in data mining and percolation theory to analyze data from a cell phone network on multiple axes of analysis to support dynamic data-driven simulations.

ACKNOWLEDGMENT

This material is based upon work supported in part by the National Science Foundation, DDDAS program, under Grant No. CNS-0540348, ITR program (DMR-0426737), and IIS-0513650 program, the James S. McDonnell Foundation 21st Century Initiative in Studying Complex Systems, the U.S. Office of Naval Research Award N00014-07-C, the NAP Project sponsored by the National Office for Research and Technology (CKCHA005). Data analysis was performed on the Notre Dame Biocomplexity Cluster supported in part by NSF MRI Grant No DBI-0420980.

REFERENCES

Aggarwal, C. C., Han, J., Wang, J., & Yu, P. S. (2003). A framework for clustering evolving data streams. In *Proceedings of the 29th Conference on Very Large Data Bases*. Berlin, Germany: VLDB Endowment.

Agrawal, R., & Srikant, R. (2000). Privacy-preserving data mining. In *Proceedings of the 2000 ACM SIGMOD Conference on Management of Data*. New York: ACM.

Albert, R., & Barabási, A.-L. (2002). Statistical mechanics of complex networks. *Reviews of Modern Physics, 74*, 47–97. doi:10.1103/RevModPhys.74.47

Albert, R., Jeong, H., & Barabási, A.-L. (1999). Diameter of the world-wide web. *Nature, 401*, 130. doi:10.1038/43601

Albert, R., Jeong, H., & Barabási, A.-L. (2000). Error and attack tolerance of complex networks. *Nature, 406*, 378–382. doi:10.1038/35019019

Armstrong, M. P. (2002). Geographic information technologies and their potentially erosive effects on personal privacy. *Studies in the Social Sciences, 27*, 19–28. doi:10.1016/S0165-4896(01)00085-3

Armstrong, M. P., & Ruggles, A. J. (2005). Geographic information technologies and personal privacy. *Cartographica, 40*, 4.

Associated Press. (2005). Tracking cell phones for real-time traffic data. Retrieved from http://www.wired.com/news/wireless/0,1382,69227,00.html

Barabási, A. L. (2002). *Linked: The new science of networks*. New York: Penguin.

Barabási, A.-L., & Albert, R. (1999). Emergence of scaling in random networks. *Science, 286*, 509–512. doi:10.1126/science.286.5439.509

Barnett, V., & Lewis, T. (1994). *Outliers in statistical data* (3rd ed.). New York: John Wiley & Sons.

Barrat, A., & Weigt, W. (2000). On the Properties of Small-World Networks. *The European Physical Journal B, 13*, 547–560. doi:10.1007/s100510050067

Benaloh, J. (1994). Dense probabilistic encryption. In *Proceedings of the Workshop on Selected Areas of Cryptography* (pp. 120-128).

Bicking, C., & Gryna, F. M., Jr. (1979). Quality control handbook. In J. M. Juran, F. M. Gryna, Jr., & R. Bingham, Jr. (Eds.), (pp. 23-1-23-35). New York: McGraw Hill.

Boccaletti, S., Latora, V., Moreno, Y., Chavez, M., & Hwang, D.-U. (2006). Complex networks: Structure and dynamics. *Physics Reports, 424*, 175–308. doi:10.1016/j.physrep.2005.10.009

Brin, S., & Page, L. (1998). The anatomy of a large-scale hypertextual Web search engine. *Computer Networks and ISDN Systems, 30*(1-7), 107-117.

Candia, J., González, M. C., Wang, P., Schoenharl, T., Madey, G., & Barabási, A.-L. (2008, June). Uncovering individual and collective human dynamics from mobile phone records. *Journal of Physics A . Mathematical and Theoretical, 41*, 224015. doi:10.1088/1751-8113/41/22/224015

Cauley, L. (2006). *NSA has massive database of Americans' phone calls.* USA Today.

Charikar, M., Chekuri, C., Feder, T., & Motwani, R. (1997) Incremental clustering and dynamic information retrieval. In *Proceedings of the 29th Annual ACM Symposium on Theory of Computing.*

Cheu, E. Y., Keongg, C., & Zhou, Z. (2004). On the two-level hybrid clustering algorithm. In *International Conference on Artificial Intelligence in Science and Technology* (pp. 138-142).

Chipman, H., & Tibshirani, R. (2006). Hybrid hierarchical clustering with applications to microarray data. *Biostatistics (Oxford, England), 7*(2), 286–301. doi:10.1093/biostatistics/kxj007

Clifton, C., Kantarcioglu, M., & Vaidya, J. (2002). Defining privacy for data mining. In *Proceedings of the National Science Foundation Workshop on Next Generation Data Mining.*

Coble, J., Cook, D. J., & Holder, L. B. (2006). Structure discovery in sequentially-connected data streams. *International Journal of Artificial Intelligence Tools, 15*(6), 917–944. doi:10.1142/S0218213006003041

Cook, D. J., & Holder, L. B. (1994). Substructure discovery using minimum description length and background knowledge. *Journal of Artificial Intelligence Research, 1*, 231–255.

Cortes, C., Pregibon, D., & Volinsky, C. (2003). Computational methods for dynamic graphs. *Journal of Computational and Graphical Statistics, 12*, 950–970. doi:10.1198/1061860032742

Dy, J. G., & Brodley, C. E. (2004). Feature selection for unsupervised learning. *Journal of Machine Learning Research, 5*, 845–889.

European Parliament and Council of the European Union. (1995). Directive 95/64/EC of the European Parliament and of the Council of 24 October 1995. *Official Journal of the European Communities, 281*, 30–51.

Evans, A. C. (1981). European data protection law. *The American Journal of Comparative Law, 29*, 571–582. doi:10.2307/839754

Fisher, D. H. (1987). Knowledge acquisition via incremental conceptual clustering. *Machine Learning, 2*, 139–172.

Friedman, A., Schuster, A., & Wolff, R. (2006). *k*-Anonymous decision tree induction. In J. Fürnkranz, T. Scheffer, & M. Spiliopoulou (Eds.), *Proceedings of the 10th European Conference on Principles and Practice of Knowledge Discovery in Databases.*

Getoor, L., & Diehl, C. P. (2005). Link mining: A survey. *SIGKDD Explorations Newsletter*, *7*(2), 3–12. doi:10.1145/1117454.1117456

González, M. C., & Barabási, A.-L. (2007). Complex networks: From data to models. *Nature Physics*, *3*, 224–225. doi:10.1038/nphys581

González, M. C., Hidalgo, C. A., & Barabási, A.-L. (2008). Understanding individual mobility patterns. *Nature*, *453*, 479–482. doi:10.1038/nature06958

Guha, S., Meyerson, A., Mishra, N., Motwani, R., & O'Callaghan, L. (2003, May/June). Clustering data streams: Theory and practice. *IEEE Transactions on Knowledge and Data Engineering*, *3*, 515–528. doi:10.1109/TKDE.2003.1198387

Hartigan, J. A. (1975). *Clustering algorithms.* New York: Wiley.

Hodge, V. J., & Austin, J. (2004). A survey of outlier detection methodologies. *Artificial Intelligence Review*, *22*, 85–126.

Ihler, A., Hutchins, J., & Smyth, P. (2006). Adaptive event detection with time-varying poisson processes. In *Proceedings of the 12th ACM SIGKDD International Conference on Knowledge Discovery and Data Mining.* New York: ACM.

Ihler, A., Hutchins, J., & Smyth, P. (2007). Learning to detect events with markov-modulated poisson processes. *ACM Transactions on Knowledge Discovery from Data, 1*(3).

Jain, A. K., Murty, M. N., & Flynn, P. J. (1999, September). Data clustering: A review. *ACM Computing Surveys*, *31*(3), 264–323. doi:10.1145/331499.331504

Kantarcioglu, M., Jin, J., & Clifton, C. (2004). When do data mining results violate privacy? In *Proceedings of the 10th ACM SIGKDD International Conference on Knowledge Discovery and Data Mining* (pp. 599-604). New York: ACM Press.

Lindell, Y., & Pinkas, B. (2002). Privacy preserving data mining. *Journal of Cryptology*, *15*(3), 177–206. doi:10.1007/s00145-001-0019-2

Madey, G. R., Barabási, A.-L., Chawla, N. V., Gonzalez, M., Hachen, D., Lantz, B., et al. (2007). Enhanced situational awareness: Application of DDDAS concepts to emergency and disaster management. In Y. Shi, G. D. van Albada, J. Dongarra, & P. M. A. Sloot (Eds.), *Proceedings of the International Conference on Computational Science* (Vol. 4487, pp. 1090-1097). Berlin, Germany: Springer.

Madey, G. R., Szabó, G., & Barabási, A.-L. (2006). WIPER: The integrated wireless phone based emergency response system. In V. N. Alexandrov, G. D. val Albada, P. M. A. Sloot, & J. Dongarra (Eds.), *Proceedings of the International Conference on Computational Science* (Vol. 3993, pp. 417-424). Berlin, Germany: Springer-Verlag.

Mitra, P., Murthy, C., & Pal, S. K. (2002). Unsupervised feature selection using feature similarity. *IEEE Transactions on Pattern Analysis and Machine Intelligence*, *24*(3), 301–312. doi:10.1109/34.990133

Mitzenmacher, M., & Upfal, E. (2005). *Probability and computing: Randomized algorithms and probabilistic analysis.* Cambridge, UK: Cambridge University Press.

Moor, J. H. (1997). Towards a theory of privacy in the information age. *ACM SIGCAS Computers and Society*, *27*(3), 27–32. doi:10.1145/270858.270866

Murray, S. D. (Ed.). (2001). U.S.-EU "Safe Harbor" Data Privacy Arrangement. *The American Journal of International Law, 91,* 169–169.

Nakashima, E. (2007). *Cellphone tracking powers on request.* Washington Post.

National Research Council. (2007). *Putting people on the map: Protecting confidentiality with linked social-spatial data* (M. P. Gutmann & P. C. Stern, Eds.). Washington, D.C.: The National Academies Press.

Newman, M. (2003). The structure and function of complex networks. *SIAM Review, 45*(2), 167–256. doi:10.1137/S003614450342480

Newman, M., Barabási, A.-L., & Watts, D. J. (Eds.). (2006). *The structure and dynamics of networks.* Princeton, NJ: Princeton University Press.

Newman, M. E. J. (2001). Clustering and preferential attachment in growing networks. *Physical Review E: Statistical, Nonlinear, and Soft Matter Physics, 64*(025102).

Newman, M. E. J. (2002). Assortative mixing in networks. *Physical Review Letters, 89*(208701).

Newman, M. E. J. (2004). Detecting community structure in networks. *The European Physical Journal B, 38,* 321–330. doi:10.1140/epjb/e2004-00124-y

Noble, C. C., & Cook, D. (2003). Graph-based anomaly detection. In *Proceedings of the 9th ACM SIGKDD International Conference on Knowledge Discovery and Data Mining.*

Onnela, J.-P., Saramäki, J., Hyvönen, J., Szabó, G., de Menezes, M. A., & Kaski, K. (2007a). Analysis of a large-scale weighted network of one-to-one human communication. *New Journal of Physics, 9,* 179. doi:10.1088/1367-2630/9/6/179

Onnela, J.-P., Saramäki, J., Hyvövnen, J., Szabó, G., Lazer, D., & Kaski, K. (2007b). Structure and tie strengths in mobile communication networks. *Proceedings of the National Academy of Sciences of the United States of America, 104*(18), 7332–7336. doi:10.1073/pnas.0610245104

Paillier, P. (1999). Public-key cryptosystems based on composite degree residuosity classes. In *Advances in Cryptology -Eeurocrypt '99 Proceedings* (LNCS 1592, pp. 223-238).

Pawling, A., Chawla, N. V., & Madey, G. (2007). Anomaly detection in a mobile communication network. *Computational & Mathematical Organization Theory, 13*(4), 407–422. doi:10.1007/s10588-007-9018-7

Pawling, A., Schoenharl, T., Yan, P., & Madey, G. (2008). WIPER: An emergency response system. In *Proceedings of the 5th International Information Systems for Crisis Response and Management Conference.*

Schoenharl, T., Bravo, R., & Madey, G. (2006). WIPER: Leveraging the cell phone network for emergency response. *International Journal of Intelligent Control and Systems, 11*(4), 209–216.

Schoenharl, T., & Madey, G. (2008). Evaluation of measurement techniques for the validation of agent-based simulations agains streaming data. In M. Bubak, G.D. van Albada, J. Dongarra, & P.M.A. Sloot (Eds.), Proceedings of the *International Conference on Computational Science* (Vol. 5103, pp. 6-15). Heidelberg, Germany: Springer.

Schoenharl, T., Madey, G., Szabó, G., & Barabási, A.-L. (2006). WIPER: A multi-agent system for emergency response. In B. van de Walle & M. Turoff (Eds.), *Proceedings of the 3rd International Information Systems for Crisis Response and Management Conference.*

Solove, D. J. (2002). Conceptualizing privacy. *California Law Review, 90,* 1088–1155. doi:10.2307/3481326

Solove, D. J. (2007). "I've got nothing to hide" and other misunderstandings of privacy. *The San Diego Law Review*, 44.

Stauffer, D., & Aharony, A. (1992). *Introduction to Percolation Theory*. (2nd ed.) London: Taylor and Francis.

Stinson, D. R. (2006). *Cryptography: Theory and practice*. Boca Raton, FL: Chipman & Hall.

Surdeanu, M., Turmo, J., & Ageno, A. (2005). A hybrid unsupervised approach for document clustering. In *Proceedings of the 11th ACM SIG-KDD International Conference on Knowledge Discovery and Data Mining*.

Sweeney, L. (1997). Weaving technology and policy together to maintain confidentiality. *The Journal of Law, Medicine & Ethics*, 25, 98–110. doi:10.1111/j.1748-720X.1997.tb01885.x

Taomir, B., & Rothkrantz, L. (2005). Crisis management using mobile ad-hoc wireless networks. In *Proceedings of the 2nd International Information Systems for Crisis Response Management Conference*.

Thomas, M., Andoh-Baidoo, F., & George, S. (2005). Evresponse–moving beyond traditional emergency response notification. In *Proceedings of the 11th Americas Conference on Information Systems*.

Watts, D. J. (1999). *Small worlds*. Princeton, NJ: Princeton University Press.

Watts, D. J., & Strogatz, S. H. (1998). Collective dynamics of 'small-world' networks. *Nature, 393*, 440–442. doi:10.1038/30918

Yan, P., Schoenharl, T., Pawling, A., & Madey, G. (2007). *Anomaly detection in the WIPER system using a Markov modulated Poisson process*. Working Paper. http://www.nd.edu/~dddas/Papers/MMPP.pdf

Zhang, T., Ramakrishnan, R., & Livny, M. (1996). BIRCH: An effective data clustering method for very large databases. In *Proceedings of the ACM SIGMOD International Conference on Management of Data*. New York: ACM.

Chapter 6
Knowledge Discovery for Sensor Network Comprehension

Pedro Pereira Rodrigues
LIAAD - INESC Porto L.A. & University of Porto, Portugal

João Gama
LIAAD - INESC Porto L.A. & University of Porto, Portugal

Luís Lopes
CRACS - INESC Porto L.A. & University of Porto, Portugal

1 INTRODUCTION

Knowledge discovery is a wide area of research where machine learning, data mining and data warehousing techniques converge to the common goal of describing and understanding the world. Nowadays applications produce infinite streams of data distributed across wide sensor networks. This ubiquitous scenario raises several obstacles to the usual knowledge discovery work flow, enforcing the need to develop new techniques, with different conceptualizations and adaptive decision making. The current setting of having a web of sensory devices, some of them enclosing processing ability, represents now a new knowledge discovery environment, possibly not completely observable,

that is much less controlled by both the human user and a common centralized control process. This ubiquitous and fast-changing scenario is nowadays subject to the same interactions required by previous static and centralized applications. Hence the need to inspect how different knowledge discovery techniques adapt to ubiquitous scenarios such as wired/wireless sensor networks.

In this chapter we explore different characteristics of sensor networks which define new requirements for knowledge discovery, with the common goal of extracting some kind of comprehension about sensor data and sensor networks, focusing on clustering techniques which provide useful information about sensor networks as it represents the interactions between sensors. This network comprehension ability is related with sensor data clustering and clustering of the data streams produced by the sensors. A wide

DOI: 10.4018/978-1-60566-328-9.ch006

range of techniques already exists to assess these interactions in centralized scenarios, but the seizable processing abilities of sensors in distributed algorithms present several benefits that shall be considered in future designs. Also, sensors produce data at high rate. Often, human experts need to inspect these data streams visually in order to decide on some corrective or proactive operations (Rodrigues & Gama, 2008). Visualization of data streams, and of data mining results, is therefore extremely relevant to sensor data management, and can enhance sensor network comprehension, and should be addressed in future works.

1.1 Sensor Network Data Streams

Sensors are usually small, low-cost devices capable of sensing some attribute of a physical phenomenon. In terms of hardware development, the state-of-the-art is well represented by a class of multi-purpose sensor nodes called *motes* (Culler & Mulder, 2004). In most of the current applications sensor nodes are controlled by module-based operating systems such as TinyOS (TinyOS, 2000) and are programmed using arguably somewhat *ad-hoc* languages such as nesC (Gay et al., 2003). Sensor networks are composed of a variable number of sensors (depending on the application), which have several features that put them in an entirely new class when compared to other wireless networks, namely: (a) the number of nodes is potentially very large and thus scalability is a problem, (b) the individual sensors are prone to failure given the often challenging conditions they experiment in the field, (c) the network topology changes dynamically, (d) broadcast protocols are used to route messages in the network, (e) limited power, computational, and memory capacity, and (f) lack of global identifiers (Akyildiz et al., 2002).

Sensor network applications are, for the most part, data-centric in that they focus on gathering data about some attribute of a physical phenomenon. The data is usually returned in the form of streams of simple data types without any local

processing. In some cases more complex data patterns or processing is possible. *Data aggregation* is used to solve routing problems (e.g. *implosion, overlap*) in data-centric networks (Akyildiz et al., 2002). In this approach, the data gathered from a neighborhood of sensor nodes is combined in a receiving node along the path to the sink. Data aggregation uses the limited processing power and memory of the sensing devices to process data online.

Sensor data is usually produced at high rate, in a stream. A data stream is an ordered sequence of instances that can be read only once or a small number of times using limited computing and storage capabilities (Gama & Rodrigues, 2007a). The data elements in the stream arrive online, being potentially unbounded in size. Once an element from a data stream has been processed it is discarded or archived. It cannot be retrieved easily unless it is explicitly stored in memory, which is small relative to the size of the data streams. These sources of data are characterized by being open-ended, flowing at high-speed, and generated by non stationary distributions.

1.2 Knowledge Discovery in Streaming Scenarios

In streaming scenarios, data flows at huge rates, reducing the ability to store and analyze it, even though predictive procedures may also be required to be applied to it. Predictions are usually followed by the real label value in a short future (e.g., prediction of next value of a time series). Nevertheless, there are also scenarios where the real label value is only available after a long term, such as predicting one week ahead electrical power consumption (Gama & Rodrigues, 2007b). Learning techniques which operate through fixed training sets and generate static models are obsolete in these contexts. Faster answers are usually required, keeping an anytime model of the data and enabling better decisions, possibly forgetting older information. The sequences of data points are not independent, and are not generated by stationary distributions. We need dynamic models that evolve over time and are able to adapt to changes in the distribution generating examples. If the process is not strictly stationary (as most of real-world applications), the target concept may gradually change over time. Hence data stream mining is an incremental task that requires incremental learning algorithms that take drift into account (Gama et al., 2004). Hulten et al. (Hulten et al., 2001) presented some desirable properties for data stream learning systems. Overall, they should process examples at the rate they arrive, use a single scan of data and fixed memory, maintain a decision model at any time and be able to adapt the model to the most recent data. Successful data stream learning systems were proposed for prediction: decision rules (Ferrer et al., 2005), decision trees (Gama et al., 2006), neural networks (Gama & Rodrigues, 2007b); example clustering: centralized (Aggarwal et al., 2003) and distributed (Cormode et al., 2007); and clustering of time series (Rodrigues et al., 2008b). All of them share the aim to discover knowledge, by producing reliable predictions, unsupervised clustering structures or data relations.

1.3 Sensor Network Comprehension

Sensor networks can include a variable number of small sensors, with dynamic network topologies and evolvable concepts producing data. In real-world applications, data flows at huge rates, with information being usually forwarded throughout the network into a common sink node, being afterwards available for analysis. However, common applications usually inspect behaviors of single sensors, looking for threshold-breaking values or failures. To increase the ability to understand the inner dynamics of the entire network, deeper knowledge should be extracted.

Sensor network comprehension tries to extract information about global interaction between sensors by looking at the data they produce. When no other information is available, usual knowledge discovery approaches are based on unsupervised techniques. Clustering is probably the most frequently used data mining algorithm (Halkidi et al., 2001), used in exploratory data analysis. It consists on the process of partitioning data into groups, where elements in the same group are more similar than elements in different groups. There are two different approaches how sensor networks could be clustered: by examples or by sensors. Clustering examples searches for dense regions of the sensor data space, identifying *hot-spots* where sensors tend to produce data. For example, usually, sensor 1 is at high values when sensor 2 is in mid-range values, and this happens more often than any other combination. Clustering sensors finds groups of sensors that behave similarly through time. For example, sensors 1 and 2 are highly correlated in the sense that when one's values are increasing the other's are also increasing.

From the previous two procedures additional knowledge can be exploited. Consider mobile sensor networks where each sensor produces a stream with its current GPS location. Clustering the examples would give an indication of usual dispersion patterns, while clustering the sensors could give indication of physical binding between sensors, forcing them to move with similar paths. Another application could rise from temperature/pressure sensors placed around geographical sites such as volcanoes or seismic faults. Furthermore, the evolution of these clustering definitions is also relevant. If each sensor's stream consists of IDs of the sensors for which this sensor is forwarding messages, changes in the clustering structure would indicate changes in the physical topology of the network, as dynamic routing strategies are commonly encountered in current sensor network applications. Overall, the main goal of sensor network comprehension is to apply automatic unsupervised procedures in order to discover interactions between sensors, trying to exploit dynamism and robustness of the network being deployed in the objective site.

1.4 Chapter Overview

In this chapter we try to cover some different characteristics of sensor networks which define new requirements for global sensor network comprehension, mainly clustering techniques for sensor networks' data. Next sections present both state-of-the-art and future trends in these topics. We explore the concept of sensor network comprehension in two dimensions: sensor data clustering and clustering of sensor data streams. In the following, we address scenarios where each sensor produces one stream of data, being afterwards combined in such way with the remaining network streams to achieve a global clustering definition. First, we shall inspect clustering definitions for the horizontal examples (values of all sensors at a given point in time) in Section 2. This process tries to extract knowledge in order to define dense regions of the sensor data space. Then, in Section 3, clustering of the streams produced by sensors is addressed. This process tries to extract knowledge about the similarity between data series produced by different sensors. Sections 2 and 3 present both relevant issues and recent approaches and proposals do deal with them, ending by discussing how the proposals presented in this chapter augment the ability to extract comprehension about sensor networks. Finally, in Section 4, we point out some future directions for sensor network comprehension using knowledge discovery.

2 NETWORK COMPREHENSION BY SENSOR DATA CLUSTERING

A recent and useful information about sensor networks is the interaction between sensors. This network comprehension is related with sensor data clustering. Clustering streaming examples is widely spread in the data mining community as a technique used to discover structures in data over time (Barbará, 2002; Guha et al., 2003). This task

also requires high-speed processing of examples and compact representation of clusters, yielding adaptivity issues. A wide range of techniques already exists to assess this characteristic in centralized scenarios, but distributed algorithms seem more adaptable and reliable as data and processing is distributed across sensors in the network. In this section, we shall inspect clustering definitions for the horizontal examples (values of all sensors at a given point in time), trying to extract knowledge in order to define dense regions of the sensor data space. Nowadays applications produce infinite streams of data distributed across wide sensor networks. In this topic we study the problem of continuously maintain a cluster structure over the data points generated by the entire network. Usual techniques operate by forwarding and concentrating the entire data in a central server, processing it as a multivariate stream. The seizable processing abilities of sensors present several benefits that must be considered in the design of clustering algorithms.

2.1 Centralized Clustering of Multivariate Streaming Examples

It is known that solving a clustering problem is the equivalent to finding the global optimal solution of a non-linear optimization problem, hence NP-hard, suggesting the application of optimization heuristics (Bern & Eppstein, 1996). The main problem in applying clustering to data streams is that systems should consider data evolution, being able to compress old information and adapt to new concepts. The range of clustering algorithms that operate online over data streams is wide, including *partitional* or *hierarchical*, *density-based* and *grid-based* methods. A common connecting feature is the definition of unit cells or representative points, from which clustering can be obtained with less computational costs. Since the *density* concept can be applied to either points (e.g. using weights) or grid cells (e.g. using counts), we can inspect different methods in terms of data processing and storage, mainly differentiating between *point-based* and *grid-based* approaches.

2.1.1 Point-Based Clustering

Several algorithms operate over summaries or samples of the original stream. Bradley et al. (Bradley et al., 1998) proposed the *Single Pass K-Means*, increasing the capabilities of *K-Means* for large datasets, by using a buffer where points of the dataset are kept in a compressed way. The *STREAM* (O'Callaghan et al., 2002) system can be seen as an extension of (Bradley et al., 1998) which keeps the same goal but has as restriction the use of available memory. After filling the buffer, STREAM clusters the buffer into *k* clusters, retaining only the *k* centroids weighted by the number of examples in each cluster. The process is iteratively repeated with new points. The BIRCH hierarchical method (Zhang et al., 1996), uses *Clustering Features* to keep sufficient statistics for each cluster at the nodes of a balanced tree, the *CF-tree*. Given its hierarchical structure, each nonleaf node in the tree aggregates the information gathered in the descendant nodes. This algorithm tries to find the best groups with respect to the available memory, while minimizing the amount of input and output. Another use of the *CF-tree* appears in *CluStream*, where an online component produces summary statistics of the data, while an offline component which computes the cluster definition based on the summaries (Aggarwal et al., 2003). A different strategy is used in another hierarchical method, the *CURE* system (Guha et al., 1998), where each cluster is represented by a constant number of points well distributed within the cluster, which capture the extension and shape of the cluster. This process allows the identification of clusters with arbitrary shapes on a random sample of the dataset, using Chernoff bounds in order to obtain the minimum number of required examples. The same principle of error-bounded results was recently used in *VFKM* to apply con-

secutive runs of K-Means, with increasing number of examples, until the error bounds were satisfied (Domingos & Hulten, 2001). This strategy supports itself on the idea of guaranteeing that the clustering definition does not differ significantly from the one gather with infinite data. Hence, it does not consider data evolution. Evolutionary clustering tries to optimize this issue (Chakrabarti et al., 2006). In evolving streaming scenarios, summary statistics tend to include a higher burden in the process, as incremental computation is extremely easy but decremental computation is extremely hard (Gama & Rodrigues, 2007a). The use of representative points throughout the streaming process presents possibly more robust and self-contained representations of the evolving clusters.

2.1.2 Grid-Based Clustering

The main focus of *grid-based* algorithms is the so called *spatial data*, which model the geometric structure of objects in space. These algorithms divide the data space in small units, defining a grid, and assigning each object to one of those units, proceeding to divisive and aggregate operations hierarchically. These features make this type of methods similar to hierarchical algorithms, with the main difference of applying operations based on a parameter rather than the dissimilarities between objects. A sophisticated example of this type of algorithms is *STING* (Wang et al., 1997), where the space area is divided in cells with different levels of resolution, creating a layered structure. The main features and advantages of this algorithm include being incremental and able of parallel execution. Also, the idea of dense units, usually present in *density-based* methods (Ester et al., 1996), has been successfully introduced in *grid-based* systems. The *CLIQUE* algorithm tries to identify sub-spaces of a large dimensional space which can allow a better clustering of the original data (Agrawal et al., 1998). It divides each dimension on the same number of equally ranged intervals, resulting in exclusive units. One unit is accepted as dense if the fraction of the total number of points within the unit is higher than a parameter value. A cluster is the largest set of contiguous dense units within a subspace. This technique's main advantage is the fact that it automatically finds subspaces of maximum dimensionality in a way that high density clusters exist in those subspaces. The *Statistical Grid-based Clustering* system (Park & Lee, 2004) was especially designed for data stream applications, where clusters are constituted by adjacent dense cells. It works by applying three different divisive methods, based on the statistics of objects belonging to each cell: *μ-partition*, which divides one cluster in two setting the border at the mean of the parent group; *σ-partition*, which divides the group in two, one with 68% of the objects belonging to $[\mu-\sigma, \mu+\sigma]$ (assuming a normal distribution of objects) and another with the remainder tail objects; and a third method which includes the efficient features of the previous, *hybrid-partition*. Another recent application of the concept of dense grid cells to evolvable data streams is *D-Stream* (Chen & Tu, 2007). The inclusion of the notion of dense units in simpler *grid-based* methods presents several benefits. However, in distributed systems, the increase in communication given the need to keep sufficient statistics may be prejudicial. Nevertheless, keeping counts of hits in grid cells is lighter than keeping summary statistics, hence the usability of *grid-based* clustering in distributed environments.

2.2 Recent Approaches in Distributed Clustering of Sensor Data

Since current applications generate many pervasive distributed computing environments, data mining systems must nowadays be designed to work not as a monolithic centralized application but as a distributed collaborative process. The centralization of information yields problems not only with resources such as communication and memory, but also with privacy of sensitive information.

Methods that aim to cluster sensor network data must consider combinations of local and central processing in order to achieve good results without centralizing the whole data. Given the extent of common sensor networks, the old client-server model is essentially useless to help the process of clustering data streams produced on sensors. Instead of centralizing relevant data in a single server and afterwards perform the data mining operations, the entire process should be distributed and, therefore, paralleled throughout the entire network of processing units.

Recent research developments are directed towards distributed algorithms for continuous clustering of examples over distributed data streams. For example, Kargupta et al. presented a collective principal component analysis (PCA), and its application to distributed example cluster analysis (Kargupta et al., 2001). However, this technique still considers a centralized process to define the clusters, which could become overloaded if sensors were required to react to the definition of clusters, forcing the server to communicate with all sensors. Klusch et al. proposed a kernel density based clustering method over homogeneous distributed data (Klusch et al., 2003), which, in fact, does not find a single clustering definition for all data set. It defines local clustering for each node, based on a global kernel density function, approximated at each node using sampling from signal processing theory. Nevertheless, these techniques present a good feature as they perform only two rounds of data transmission through the network. Other approaches using the K-Means algorithm have been developed for peer-to-peer environments and sensor networks settings. In (Datta et al., 2006) the authors present a distributed majority vote algorithm which can be seen as a primitive to monitor a K-Means clustering over peer-to-peer networks. The K-Means monitoring algorithm has two major parts: monitoring the data distribution in order to trigger a new run of K-Means algorithm and computing the centroids actually using the K-Means algorithm. The monitoring part is carried out by an exact local algorithm, while the centroid computation is carried out by a centralization approach. The local algorithm raises an alert if the centroids need to be updated. At this point data is centralized, a new run of K-Means is executed, and the new centroids are shipped back to all peers. Cormode et al. (Cormode et al., 2007) proposed different strategies to achieve the same goal, with local and global computations, in order to balance the communication costs. They considered techniques based on the *furthest point* algorithm (Gonzalez, 1985), which gives a approximation for the radius and diameter of clusters with guaranteed cost of two times the cost of the optimal clustering. They also present the *parallel guessing* strategy, which gives a slightly worse approximation but requires only a single pass over the data. They conclude that, in actual distributed settings, it is frequently preferable to track each site locally and combine the results at the coordinator site. Recent developments have concentrated on these scenarios. *CluDistream* is an *Expectation-Maximization* (Dempster et al., 1977) (EM) variant which addresses evolvable distributed data streams using a model-based approach (Zhou et al., 2007). The idea is that each site keeps a pool of previously known clustering models and tests if current chunk of data is generated by any of those models, triggering the coordinator's merging procedure if a new model fits the current data. However, the need to detect and track changes in clusters is not enough, and it is also often required to provide some information about the nature of changes (Spiliopoulou et al., 2006). Future systems should take this issue into account.

2.3 Grid-Based Clustering of Distributed Sensor Data

If data streams are being produced separately (each variable in each sensor) in distributed sites, and although each site should process its own univariate stream locally before any clustering procedure, a coordinator site must execute some kind of processing (actually it should execute the clustering procedure) on the whole gathered data, possibly feeding the remote sites with the current clustering model. A recent data-driven approach combines both *grid-based* approximations with *point-based* clustering.

DGClust (Rodrigues et al., 2008a) is a distributed grid clustering system for sensor data streams, where each local site continuously receives a data stream from a a given sensor, being incrementally discretized into a univariate adaptive grid. Each new data point triggers a cell in this grid, reflecting the current state of the data stream at the local site. Whenever the triggered cell changes, and only then, the new state is communicated to a central site, which keeps the global state of the entire network where each local site's state is the cell number of each local site's grid. Given the huge number of sensors possibly included in the network, an exponential number of cell combinations should be monitored by the central site. However, it is expected that only a small number of this combinations are frequently triggered by the whole network, so, parallel to the aggregation, the central site keeps a small list of counters of the most frequent global states. Finally, the current clustering structure is defined and maintained by a simple adaptive partitional clustering algorithm applied on the frequent states central points. The advantages proposed by this method are mainly two: the reduction of communication and the reduction of the clustering process dimensionality. The reduction on communication of data points to the coordinator site is achieved by communicating only the triggered cell number, and only when this changes which, since real-world sensor data tend to be a highly autocorrelated time series (Gama & Rodrigues, 2007b), will happen only a reduced percentage of times. The reduction in of dimensionality in clustering is achieved by using only the guaranteed top-*m* most frequent states as input points to the clustering procedure. These methods of combining local and central processing are paradigmatic examples of the path that distributed data mining algorithms should traverse.

2.4 Sensor Network Comprehension

This section of sensor network comprehension tries to extract knowledge in order to define dense regions of the sensor data space. Clustering examples in sensor networks can be used to search for *hot-spots* where sensors tend to produce data. In this settings, *grid-based* clustering represents a major asset as regions can be, strictly or loosely, defined by both the user and the adaptive process. The application of clustering to grid cells enhances the abstraction of cells as interval regions which are better interpreted by humans. Moreover, comparing intervals or grids is usually easier than comparing exact points, as an external scale is not required: intervals have intrinsic scaling. For example, when querying for the top *hot-spot* of a given sensor network, instead of achieving results such as "usually, sensor 1 is around 100.2 when sensor 2 is around 10.5", we would get "usually, sensor 1 is between 95 and 105 when sensor 2 is within 10.4 and 10.6". The comprehension of how sensors are interacting in the network is greatly improved by using *grid-based* clustering techniques for the data examples produced by sensors.

3 NETWORK COMPREHENSION BY CLUSTERING STREAMING SENSORS

Sensor network comprehension is also highly related with the interaction between sensors, in terms of similar behaviors or readings. Clustering streaming sensors is the task of clustering streaming data series produced by sensors on a wide sensor network. This process tries to extract knowledge about the similarity between data produced by different sensors. Most works on clustering analysis for sensor networks actually concentrate on clustering the sensors by their geographical position (Chan et al., 2005) and connectivity, mainly for power management (Younis & Fahmy, 2004) and network routing purposes (Ibriq & Mahgoub, 2004; Yoon and Shahabi, 2007). However, in this topic, we are interested in clustering techniques for data produced by the sensors, instead. Considering the dynamic behavior usually enclosed in streaming data, clustering streaming sensors should be addressed as an online and incremental procedure, in order to enable faster adaptation to new concepts and produce better models through time. However, centralized clustering strategies tend to be inapplicable as usual techniques have quadratic complexity on the number of sensors, and sensor networks grow unbounded.

The motivation for this is all around us. As networks and communications spread out, so does the distribution of novel and advanced measuring sensors. The networks created by current settings can easily include thousands of sensors, each one being capable of measuring, analyzing and transmitting data. From another point of view, given the evolution of hardware components, these sensors act now as fast data generators, producing information in a streaming environment. Clustering streaming time series has been already studied in various fields of real world applications. However, algorithms that were previously proposed to the task of clustering time series data streams tend to deal with data as a centralized multivariate stream (Rodrigues & Gama, 2007). They are designed as a single process of analysis, without taking into account the locality of data produced by sensors on a wide network, the transmission and processing resources of sensors, and the breach in the transmitted data quality. In fact, many motivating domains could benefit from (and some of them even require) a distributed approach, given their objective application or specialized setting.

3.1 Formal Setup

Sensor data streams usually consist of variables producing examples continuously over time. The basic idea behind this task is to find groups of sensors that behave similarly through time, which is usually measured in terms of time series similarities. Let be the complete set of n sensor streams and be the example containing the observations of all sensors at the specific time t. The goal of an incremental clustering system for streaming time series is to find (and make available at any time t) a partition P of those sensors, where sensors in the same cluster tend to be more alike than sensors in different clusters. In partitional clustering, searching for k clusters, the result at time t should be a matrix P of $n \times k$ values, where each is one if sensor belongs to cluster and zero otherwise. Specifically, we can inspect the partition of sensor streams in a particular time window from starting time s until current time t, using examples, which would give a temporal characteristic to the partition. In a hierarchical approach to the problem, the same possibilities exist, with the benefit of not having to previously define the target number of clusters, thus creating a structured output of the hierarchy of clusters.

3.2 Centralized Clustering of Streaming Sensors

Clustering streaming time series has been already targeted by researchers in order to cope with the tendentiously infinite amount of data produced at high speed. Beringer and Hüllermeier proposed an online version of *K-Means* for clustering parallel data streams, using a Discrete Fourier Transform approximation of the original data (Beringer & Hüllermeier, 2006). The basic idea is that the cluster centers computed at a given time are the initial cluster centers for the next iteration of *K-Means*, applying a procedure to dynamically update the optimal number of clusters at each iteration. Clustering On Demand *(COD)* is another framework for clustering streaming series which performs one data scan for online statistics collection and has compact multi-resolution approximations, designed to address the time and the space constraints in a data stream environment (Dai et al., 2006). It is divided in two phases: a first online maintenance phase providing an efficient algorithm to maintain summary hierarchies of the data streams and retrieve approximations of the sub-streams; and an offline clustering phase to define clustering structures of multiple streams with adaptive window sizes. Rodrigues et al. proposed the Online Divisive-Agglomerative Clustering *(ODAC)* system, a hierarchical procedure which dynamically expands and contracts clusters based on their diameters (Rodrigues et al., 2008b). It constructs a tree-like hierarchy of clusters of streams, using a top-down strategy based on the correlation between streams. The system also possesses an agglomerative phase to enhance a dynamic behavior capable of structural change detection. The splitting and agglomerative operators are based on the diameters of existing clusters and supported by a significance level given by the Hoeffding bound (Hoeffding, 1963).

However, if this data is produced by sensors on a wide network, the proposed algorithms tend to deal with them as a centralized multivariate stream. They process without taking into account the locality of data, the limited bandwidth and processing resources, and the breach in the quality of transmitted data. All of these issues are usually motivated by energy efficiency demands of sensor devices. Moreover, these algorithms tend to be designed as a single process of analysis without the necessary attention on the distributed setting (already addressed on some example clustering systems) which creates high levels of data storage, processing and communication.

3.3 Requirements for Distributed Procedures

Considering the main restrictions of sensor networks, the analysis of clusters of sensor streams should comply not only with the requirements for clustering multiple streaming series (Rodrigues and Gama, 2007) but also with the available resources and setting of the corresponding sensor network. If a distributed algorithm for clustering streaming sensors is to be integrated on each sensor, how can local nodes process data and the network interact in order to cluster similar behaviors produced by sensors far from each other, without a fully centralized monitoring process? If communication is required, how should this be done in order to avoid the known problems of data communication on sensor networks, prone to implosion and overlap? Sensor network comprehension is highly related with the relevance of this information, and its relationship with the geographical location of sensors.

Centralized models to perform streaming sensor clustering tend to be inapplicable as sensor networks grow unbounded, becoming overloaded if sensors are required to react to the definition of clusters, forcing the server to communicate with all sensors. Common sensor networks data aggregation techniques are based on the Euclidean distance (physical proximity) of sensors to perform summaries on a given neighborhood (Chan et al., 2005). However, the clustering structure definition of the series of data produced by the sensors may be completely orthogonal to the physical topology of the network (Rodrigues et al., 2009). Hence the need for completely distributed algorithms for clustering of streaming sensors. We can overview the features that act both as requirements for clustering streaming sensors and future paths to research in this area:

- The requirements for clustering streaming series must be considered, with more emphasis on the adaptability of the whole system;

- Processing must be distributed and synchronized on local neighborhoods or querying nodes;
- The main focus should be on finding similar sensors irrespectively to their physical location;
- Processes should minimize different resources (mainly energy) consumption in order to achieve high uptime;
- Operation should consider a compact representation of both the data and the generated models, enabling fast and efficient transmission and access from mobile and embedded devices.

The final goal is to infer a global clustering structure of all relevant sensors. Hence, approximate algorithms should be considered to prevent global data transmission.

3.4 Advantages of Distributed Procedures

Given the novelty of this research area, few previous works are actually related, even slightly, with this task. Therefore, we shall concentrate on the advantages that this task should bring to sensor network applications.

3.4.1 Advantages in Sensor Networks Processing

Distributed clustering of streaming sensors presents advantages for everyday processing in sensor networks. We can point out the implications in three areas: message forwarding, deployment quality and privacy preservation.

Sensor Networks Message Forwarding
One of the highest resources consuming tasks in sensor networks is communication. Moreover, information is usually forward through the network into a sink node. With sensors increasing in number, redundant information is also more

probable, so message forwarding will become a heavy resource leak. If a distributed clustering procedure is applied at each forwarding node, usual data aggregation techniques could be data-centric, in the sense that one node could decide not to transmit a message, or aggregate it with others, if it contains information which is quite similar to other nodes'.

Sensor Networks Deployment Quality

When sensor networks are deployed in objective areas, the design of this deployment is most of the times subject to expert-based analysis or template-based configuration. Unfortunately, the best deployment configuration is sometimes hard to find. Applying distributed clustering of sensors' data streams the system can identify sensors with similar reading profiles, while investigating if the sensors are in the same geographical cluster. If similar sensors, with respect to the produced data, are place in a dense, with respect to the geographical position, cluster of sensors, resources are being spoiled as less sensors would give the same information. These sensors could then be assigned to different positions in the network.

Privacy Preserving Clustering

The privacy of personal data is most of the times important to preserve, even when the objective is to analyze and compare with other people's data. Anonymization is the most common procedure to ensure this but experience as shown that it is not flawless. This way, centralizing all information in a common data server could represent a more vulnerable setup for security breaches. If we can achieve the same goal without centralizing the information, privacy should be easier to preserve. Furthermore, the system could achieve a global clustering structure without sharing sensible information between all nodes in the network (e.g. clustering using the fractal dimension (Barbará & Chen, 2000)).

3.4.2 Advantages in Specific Domains of Application

Although transverse to all sensor network applications, the advantages of distributed clustering of streaming sensors are better analyzed in specific real-world applications.

Electricity Demand Profiling

In electricity supply systems, the identification of *demand profiles* (ex: industrial or urban) by clustering streaming sensors' data decreases the computational cost of predicting each individual subnetwork load (Gama & Rodrigues, 2007b). This is a common scenario with thousands of different sensors distributed over a wide area. As sensors are naturally distributed in the electrical network, distributed procedures which would focus on local networks could prevent the dimensionality drawback.

Natural Phenomena Monitoring

A common problem in geoscience research is the monitoring of natural phenomena evolution. Several techniques are nowadays used to address these problems, and, given their increasing availability, sensor-based approaches are now *hot-topics* in the area. Sensor nodes can be densely deployed either very close or directly inside the phenomenon to be observed (Sun & Sauvola, 2006) (e.g. ocean streams or river flows, a twister or a hurricane, etc.). Sensors deployed in the objective area can monitor several measures of interest, such as water temperature, stream gauge and electricity resistance. Clustering the data produced by different sensors is helpful to identify areas with similar profiles, possibly indicating actual water or wind streams.

GPS Movement Tracks Monitoring

The Global Positioning System (GPS) is commonly used to monitor location, speed and direction of both people and objects. Identifying similar paths, for example, in delivery teams or traffic flow, is a relevant task to current enterprises and end-users (Moreira & Santos, 2005). Embedding these systems with context information is now a major research challenge to be able to improve results with real-time information (Zhang et al., 2007). However, the amount of data produced by each GPS receiver is so huge, and the allowed reply delay so narrow, that performing centralized clustering of GPS tracks is too expensive to perform. If each receiver is used to perform a distributed procedure for the clustering task, the same goal should be achieved faster and with better resource management.

Medical Patients Monitoring

In medical environments, clustering *medical sensor data* (such as ECG, EEG, etc.) is useful to determine association between signals (Sherrill et al., 2005), allowing better diagnosis. Detecting similar profiles in these measures among different patients is one way to explore uncommon conditions. Mobile and embedded devices could interconnect different patients and physicians, without revealing sensible information from patients while nevertheless achieving the goal of identifying similar profiles.

3.5 A General Setup for Distributed Clustering of Streaming Sensors

The main objective of a clustering system should be to be able to answer queries for the global clustering definition of the entire sensor network. If sensors are distributed on a wide area, with local sites being accessible from transient devices, queries could be issued at each local site, enabling fast answers to be sent to the querying device. However, current setups assume data is forwarded into a central server, where it is processed, being this the main answering device. This setup forces not only the data but also the queries to be transmitted across the network into a sink.

A general setup for distributed clustering of streaming sensors is proposed in the following,

taking into account the requirements and advantages previously enunciated. First we should be able to have some type of sketch of the stream being produced by each sensor, in order to reduce the computation of similarities between sensors. Then, we believe each sensor should communicate only locally with its neighbors, in order to reduce the amount of data being forwarded throughout the network. Finally, we must include mechanisms to prevent redundant communication, while monitoring for structural changes in the clustering definition of the entire network.

3.5.1 Sketching Streaming Sensors

Each sensor produces a stream of data, usually defined by one or more infinite time series. We want to define a clustering structure for the sensors, where sensors producing streams which are alike are clustered together. Hence, we shall define a similarity measure for such streams. However, we do not ever have access to the complete time series, and we would like to prevent the whole data to be transmitted in the network. We should consider approximate metrics, using simple sufficient statistics of the streams, or data synopsis, to achieve similar results. One way to summarize a data stream is by computing its sample mean and standard deviation, assuming some kind of data distribution. More complex strategies could include distribution distances based on the histograms of each sensor's data (e.g. relative entropy (Berthold & Hand, 1999)), where each sensor would have to transmit the frequency of each data interval to its neighbors, or using a approximations of the original data (Beringer & Hüllermeier, 2006). Overall, we should consider techniques that project each sensor's data stream into a reduced set of dimensions which suffice to extract similarity with other sensors. This estimates can be seen as the sensor's current overview of its own data, giving an indication of where in the data-space this sensor is included.

3.5.2 Local Approximations of the Global Clustering Structure

As each sensor is able to sketch its own data in a dimensionally-reduced definition, it is also able to interact with its neighbor nodes, in order to assess a local clustering of sensors. Overall, each sensor should include incremental clustering techniques which operate with distance metrics developed for the dimensionally-reduced sketches of the data streams. Our goal is to have at each local site a global clustering structure of the entire sensor network. To achieve this, at each time, each sensor should send to its neighbors its own estimate of the global clustering, instead of sending only its own sketch. Note that with this approach, each sensor keeps an approximate estimate of the global cluster centers. This estimate can be seen as the sensor's current overview of the entire network which, together with its own sketch, gives an indication of where in the entire network data-space this sensor is included. The key point is how to robustly update this definition using only neighborhood information.

3.5.3 Communication Management

Communication is one of the most resource-consuming procedures of sensor networks (Chan et al., 2005). If a central server is used to aggregate all data, each individual sketch must be forward through the network into a sink node. To enable each local site to have the global clustering structure of the entire network, the central server would have to reply with K values, largely increasing the global number of transmissions. If we transmit data only between neighbors, this would represent $2E$ communications of K values, where E is the number of links in the network, achieving an approximate clustering of the whole network at each node, with much less communication. On top of this, if the concept of the data being produced in the network is stable, then the clustering estimates will converge, and transmissions will become redundant. We should include mechanisms to allow each sensor to decide to which neighbors it is still valuable to send information. However, the world is not static. It is possible that, with time, the sketches of each sensor will change, adapting to new concepts of data. On a long run, the communication management strategy could prevent the system from adapting to new data. Overall, sensors should include change detection mechanisms that would trigger if the data changes, either univariatedly at each sensor, or in the global interaction of sensor data.

3.6 FUTURE DEVELOPMENTS

Although the physical topology of the network may be useful for data management purposes, the main focus should be on finding similar sensors irrespectively to their physical location. Also, minimizing different resources (mainly energy) consumption is a major requirement in order to achieve high uptimes for sensors. On top of this, a compact representation of both the data and the generated models must be considered, enabling fast and efficient transmission and access from mobile and embedded devices. Even though processing may be concentrated on local computations and short-range communication, the final goal is to infer a global clustering structure of all relevant sensors. Hence, approximate algorithms should be considered to prevent global data transmission. Given this, when querying a given sensor for the global clustering, we allow (and known beforehand that we will have) an approximate result within a maximum possible error with a certain probability. Each approximation step (local sketch, local clustering update, merging different cluster definitions, etc.) should be restricted by some stability bound on the error (Hoeffding, 1963). These bounds should serve as balancing deciders in the trade-off between transmission management and resulting errors.

Future research developments are requested to address the issues presented above, and surely researchers will focus on distributed data mining utilities for large sensor networks streaming data analysis, as sensors and their respective data become more and more ubiquitous and embedded in everyday life.

3.7 Sensor Network Comprehension

This section focused on clustering of the streams produced by sensors, which tries to extract knowledge about the similarity between data series produced by different sensors. This task relates with sensor network comprehension as clustering sensors finds groups of sensors that behave similarly through time. The distributed setup proposed in this section enables a transient user to query a local node for its position in the overall clustering structure of sensors, without asking the centralized server. For example, a query for a given sensor could be answered by "this sensor and sensors 2 and 3 are highly correlated", in the sense that when one's values are increasing the others' are also increasing, or "the answering sensor is included in a group of sensors that has the following profile or prototype of behavior". Hence, the comprehension of how sensors are related in the network is also greatly improved by using *distributed* sensor clustering techniques.

4 DIRECTIONS FOR FURTHER SENSOR NETWORK COMPREHENSION

The main idea behind these tasks is the following: some (or all) of the sensors enclosed in the network should perform some kind of processing over the data gathered by themselves or/and by their neighbors, in order to achieve an up-to-date definition of the entire network. However, different requirements need to be defined so that a clear

path in the development can be drawn. Distributed data mining appears to have most of the necessary features to address this problem. On one hand, the development of global frameworks that are capable of mining data on distributed sources is rising, taking into account the lack of resources usually encountered on sensor networks. Several parameters can then be controlled by the monitoring process in order to minimize energy consumption. On the other hand, given the processing abilities of each sensor, clustering results should be preferably localized on the sensors where this information becomes an asset. Information query and transmission should only be considered on limited situations. This trade-off between global and local knowledge is now the key point for knowledge discovery over sensor networks.

Sensor network comprehension is a wider concept than the two clustering task that were inspected in this chapter. Other tasks, both unsupervised and supervised, may yield additional elements for a global sensor network comprehension. Simple examples include: the extraction of rules for certain network events, which may reveal breaches of security in the current network topology; inspection of predictive errors across the network, which may reveal interactions between sensors not observed in unsupervised results; or the definition of a ranking of sensor activity, which may reveal unused or overloaded sensors in the network. Irrespectively of the technique, the main focus of any sensor network comprehension processes should be on using distributed processing of data and queries, and distributed data mining procedures, enabling fast answers and access from transient mobile devices.

Ubiquitous activities usually imply mobile data access and management, in the sense that even sensor networks with static topology could be queried by transient devices, such as PDAs, laptops or other embedded devices. In this setup, mining data streams in a mobile environment raises an additional challenge to intelligent systems, as model analysis and corresponding results need to be visualized in a small screen, requiring alternate multimedia-based human-computer interaction. Visualization techniques bounded to limited resources are required, especially for sensor network comprehension where data and model becomes more complex.

ACKNOWLEDGMENT

The work of Pedro P. Rodrigues is supported by the Portuguese Foundation for Science and Technology (FCT) under the PhD Grant SFRH/BD/29219/2006. Pedro P. Rodrigues and João Gama thank the Plurianual financial support attributed to LIAAD and the participation of Project ALES II under Contract POSC/EIA/55340/2004. Pedro P. Rodrigues and Luís Lopes are also partially supported by FCT through Project CALLAS under Contract PTDC/EIA/71462/2006.

REFERENCES

Aggarwal, C. C., Han, J., Wang, J., & Yu, P. S. (2003). A framework for clustering evolving data streams. In *VLDB 2003, Proceedings of 29th International Conference on Very Large Data Bases* (pp. 81–92). Morgan Kaufmann.

Agrawal, R., Gehrke, J., Gunopulos, D., & Raghavan, P. (1998). Automatic subspace clustering of high dimensional data for data mining applications. In *Proceedings of the ACM-SIGMOD International Conference on Management of Data* (pp. 94–105). Seattle, WA: ACM Press.

Akyildiz, I., Su, W., Sankarasubramaniam, Y., & Cayirci, E. (2002). A Survey on Sensor Networks. *IEEE Communications Magazine, 40*(8), 102–114. doi:10.1109/MCOM.2002.1024422

Barbará, D. (2002). Requirements for clustering data streams. *SIGKDD Explorations (Special Issue on Online, Interactive, and Anytime Data Mining), 3*(2), 23–27.

Barbará, D., & Chen, P. (2000). Using the fractal dimension to cluster datasets. In *Proceedings of the Sixth ACM-SIGKDD International Conference on Knowledge Discovery and Data Mining*, pages 260–264. ACM Press.

Beringer, J., & Hüllermeier, E. (2006). Online clustering of parallel data streams. *Data & Knowledge Engineering, 58*(2), 180–204. doi:10.1016/j.datak.2005.05.009

Bern, M., & Eppstein, D. (1996). *Approximation Algorithms for NP-hard Problems*, chapter 8. *Approximation* Algorithms for Geometric Problems (pp. 296-345). PWS Publishing Company.

Berthold, M., & Hand, D. (1999). *Intelligent Data Analysis - An Introduction*. Springer Verlag.

Bradley, P., Fayyad, U., & Reina, C. (1998). Scaling clustering algorithms to large databases. In *Proceedings of the Fourth International Conference on Knowledge Discovery and Data Mining* (pp. 9-15). AAAI Press.

Chakrabarti, D., Kumar, R., & Tomkins, A. (2006). Evolutionary clustering. In *KDD* (pp. 554–560).

Chan, H., Luk, M., & Perrig, A. (2005). Using clustering information for sensor network localization. In *First IEEE International Conference on Distributed Computing in Sensor Systems* (pp. 109–125).

Chen, Y., & Tu, L. (2007). Density-based clustering for real-time stream data. In *Proceedings of the 13th ACM SIGKDD International Conference on Knowledge Discovery and Data Mining* (pp. 133–142).

Cormode, G., Muthukrishnan, S., & Zhuang, W. (2007). Conquering the divide: Continuous clustering of distributed data streams. In *Proceedings of the 23rd International Conference on Data Engineering (ICDE 2007)* (pp. 1036–1045).

Culler, D. E., & Mulder, H. (2004). Smart Sensors to Network the World. *Scientific American.*

Dai, B.-R., Huang, J.-W., Yeh, M.-Y., & Chen, M.-S. (2006). Adaptive clustering for multiple evolving streams. *IEEE Transactions on Knowledge and Data Engineering, 18*(9), 1166–1180. doi:10.1109/TKDE.2006.137

Datta, S., Bhaduri, K., Giannella, C., Wolff, R., & Kargupta, H. (2006). Distributed data mining in peer-to-peer networks. *IEEE Internet Computing, 10*(4), 18–26. doi:10.1109/MIC.2006.74

Dempster, A., Laird, N., & Rubin, D. (1977). Maximum likelihood from incomplete data via the EM algorithm. *Journal of the Royal Statistical Society. Series A (General), 39*(1), 1–38.

Domingos, P., & Hulten, G. (2001). A general method for scaling up machine learning algorithms and its application to clustering. In *Proceedings of the Eighteenth International Conference on Machine Learning (ICML 2001)* (pp. 106-113).

Ester, M., Kriegel, H.-P., Sander, J., & Xu, X. (1996). A density-based algorithm for discovering clusters in large spatial databases with noise. In E. Simoudis, J.Han, & U. Fayyad (Eds.), *Second International Conference on Knowledge Discovery and Data Mining* (pp. 226–231). Portland, OR: AAAI Press.

Ferrer, F., Aguilar, J., & Riquelme, J. (2005). Incremental rule learning and border examples selection from numerical data streams. *Journal of Universal Computer Science, 11*(8), 1426–1439.

Gama, J., Fernandes, R., & Rocha, R. (2006). Decision trees for mining data streams. *Intelligent Data Analysis, 10*(1), 23–45.

Gama, J., Medas, P., Castillo, G., & Rodrigues, P. P. (2004). Learning with drift detection. In A. L. C. Bazzan,& S. Labidi (Eds.), *Proceedings of the 17th Brazilian Symposium on Artificial Intelligence (SBIA 2004)* (LNCS 3171, pp. 286-295). São Luiz, Maranhão, Brazil. Springer Verlag.

Gama, J., & Rodrigues, P. P. (2007a). Data stream processing. In J. Gama & M.M. Gaber (Eds.), *Learning from Data Streams - Processing Techniques in Sensor Networks* (pp. 25–39). Springer Verlag.

Gama, J., & Rodrigues, P. P. (2007b). Stream-based electricity load forecast. In J. N., Kok, J. Koronacki, R.L. de Mantaras, S. Matwin, D.Mladenic, & A. Skowron (Eds.), *Proceedings of the 11th European Conference on Principles and Practice of Knowledge Discovery in Databases (PKDD 2007)* (LNAI 4702, pp. 446-453). Warsaw, Poland. Springer Verlag.

Gay, D., Levis, P., von Behren, R., Welsh, M., Brewer, E., & Culler, D. (2003). The nesC Language: A Holistic Approach to Network Embedded Systems. In *ACM SIGPLAN Conference on Programming Language Design and Implementation (PLDI)* (pp. 1–11). ACM Press.

Gonzalez, T. F. (1985). Clustering to minimize the maximum inter-cluster distance. *Theoretical Computer Science, 38*, 293–306. doi:10.1016/0304-3975(85)90224-5

Guha, S., Meyerson, A., Mishra, N., Motwani, R., & O'Callaghan, L. (2003). Clustering data streams: Theory and practice. *IEEE Transactions on Knowledge and Data Engineering, 15*(3), 515–528. doi:10.1109/TKDE.2003.1198387

Guha, S., Rastogi, R., & Shim, K. (1998). CURE: An efficient clustering algorithm for large databases. In L. M. Haas & A. Tiwary, A., editors, *Proceedings of the 1998 ACM-SIGMOD International Conference on Management of Data* (pp. 73–84). ACM Press.

Halkidi, M., Batistakis, Y., & Varzirgiannis, M. (2001). On clustering validation techniques. *Journal of Intelligent Information Systems*, *17*(2-3), 107–145. doi:10.1023/A:1012801612483

Hoeffding, W. (1963). Probability inequalities for sums of bounded random variables. *Journal of the American Statistical Association*, *58*(301), 13–30. doi:10.2307/2282952

Hulten, G., Spencer, L., & Domingos, P. (2001). Mining time-changing data streams. In *Proceedings of the Seventh ACM SIGKDD International Conference on Knowledge Discovery and Data Mining* (pp. 97–106). ACM Press.

Ibriq, J., & Mahgoub, I. (2004). Cluster-based routing in wireless sensor networks: Issues and challenges. In *International Symposium on Performance Evaluation of Computer and Telecommunication Systems* (pp. 759-766).

Kargupta, H., Huang, W., Sivakumar, K., & Johnson, E. L. (2001). Distributed clustering using collective principal component analysis. *Knowledge and Information Systems*, *3*(4), 422–448. doi:10.1007/PL00011677

Klusch, M., Lodi, S., & Moro, G. (2003). Distributed clustering based on sampling local density estimates. In *Proceedings of the International Joint Conference on Artificial Intelligence* (pp. 485–490).

Moreira, A., & Santos, M. Y. (2005). Enhancing a user context by real-time clustering mobile trajectories. In *International Conference on Information Technology: Coding and Computing (ITCC'05)* (Vol. 2, p. 836). Los Alamitos, CA, USA: IEEE Computer Society.

O'Callaghan, L., Meyerson, A., Motwani, R., Mishra, N., & Guha, S. (2002). Streaming-data algorithms for high-quality clustering. In *Proceedings of the Eighteenth Annual IEEE International Conference on Data Engineering* (pp. 685–696). IEEE Computer Society.

Park, N. H., & Lee, W. S. (2004). Statistical grid-based clustering over data streams. *SIGMOD Record*, *33*(1), 32–37. doi:10.1145/974121.974127

Rodrigues, P. P., & Gama, J. (2007). Clustering techniques in sensor networks. In J. Gama & M. M. Gaber, (Eds.), *Learning from Data Streams - Processing Techniques in Sensor Networks* (pp. 125–142). Springer Verlag.

Rodrigues, P. P., & Gama, J. (2008). Dense pixel visualization for mobile sensor data mining. In *Proceedings of the 2nd International Workshop on Knowledge Discovery from Sensor Data* (pp. 50–57). ACM Press.

Rodrigues, P. P., Gama, J., & Lopes, L. (2008a). Clustering distributed sensor data streams. In W. Daelemans, B. Goethals & K. Morik (Eds.), *Proceedings of the European Conference on Machine Learning and Knowledge Discovery in Databases (ECMLPKDD 2008)* (LNAI 5212, pp. 282–297) Antwerpen, Belgium: Springer Verlag.

Rodrigues, P. P., Gama, J., & Lopes, L. (2009). Requirements for clustering streaming sensors. In A. R. Ganguly, J. Gama, O. A. Omitaomu, M. M. Gaber, & R. R. Vatsavai (Eds.), *Knowledge Discovery from Sensor Data* (pp. 33–51). CRC Press.

Rodrigues, P. P., Gama, J., & Pedroso, J. P. (2008b). Hierarchical clustering of time-series data streams. *IEEE Transactions on Knowledge and Data Engineering*, *20*(5), 615–627. doi:10.1109/TKDE.2007.190727

Sherrill, D. M., Moy, M. L., Reilly, J. J., & Bonato, P. (2005). Using hierarchical clustering methods to classify motor activities of copd patients from wearable sensor data. *Journal of Neuroengineering and Rehabilitation*, *2*(16).

Spiliopoulou, M., Ntoutsi, I., Theodoridis, Y., & Schult, R. (2006). Monic: modeling and monitoring cluster transitions. In *KDD* (pp. 706–711).

Sun, J.-Z., & Sauvola, J. (2006). Towards advanced modeling techniques for wireless sensor networks. In *Proceedings of the 1st International Symposium on Pervasive Computing and Applications* (pp. 133–138). IEEE Computer Society.

Tiny, O. S. (2000). The TinyOS Documentation Project. Retrieved from http://www.tinyos.org.

Wang, W., Yang, J., & Muntz, R. R. (1997). STING: A statistical information grid approach to spatial data mining. In M. Jarke, M. J. Carey, K. R. Dittrich, F. H. Lochovsky, P. Loucopoulos, & M. A. Jeusfeld (Eds.), *Proceedings of the Twenty-Third International Conference on Very Large Data Bases* (pp. 186–195). Athens, Greece. Morgan Kaufmann.

Yoon, S., & Shahabi, C. (2007). The clustered aggregation (CAG) technique leveraging spatial and temporal correlations in wireless sensor networks. *ACM Transactions on Sensor Networks, 3*(1). Article 3.

Younis, O., & Fahmy, S. (2004). HEED: A hybrid, energy-efficient, distributed clustering approach for ad hoc sensor networks. *IEEE Transactions on Mobile Computing, 3*(4), 366–379. doi:10.1109/TMC.2004.41

Zhang, K., Torkkola, K., Li, H., Schreiner, C., Zhang, H., Gardner, M., & Zhao, Z. (2007). A context aware automatic traffic notification system for cell phones. In *27th International Conference on Distributed Computing Systems Workshops (ICDCSW '07)* (pp. 48–50). IEEE Computer Society.

Zhang, T., Ramakrishnan, R., & Livny, M. (1996). BIRCH: An efficient data clustering method for very large databases. In *Proceedings of the 1996 ACM SIGMOD International Conference on Management of Data* (pp. 103–114). ACM Press.

Zhou, A., Cao, F., Yan, Y., Sha, C., & He, X. (2007). Distributed data stream clustering: A fast EM approach. In *Proceedings of the 23rd International Conference on Data Engineering* (pp. 736–745).

Chapter 7
Why General Outlier Detection Techniques Do Not Suffice for Wireless Sensor Networks

Yang Zhang
University of Twente, The Netherlands

Nirvana Meratnia
University of Twente, The Netherlands

Paul Havinga
University of Twente, The Netherlands

ABSTRACT

Raw data collected in wireless sensor networks are often unreliable and inaccurate due to noise, faulty sensors and harsh environmental effects. Sensor data that significantly deviate from normal pattern of sensed data are often called outliers. Outlier detection in wireless sensor networks aims at identifying such readings, which represent either measurement errors or interesting events. Due to numerous short-comings, commonly used outlier detection techniques for general data seem not to be directly applicable to outlier detection in wireless sensor networks. In this chapter, the authors report on the current state-of-the-art on outlier detection techniques for general data, provide a comprehensive technique-based taxonomy for these techniques, and highlight their characteristics in a comparative view. Furthermore, the authors address challenges of outlier detection in wireless sensor networks, provide a guideline on requirements that suitable outlier detection techniques for wireless sensor networks should meet, and will explain why general outlier detection techniques do not suffice.

INTRODUCTION

Advances in electronic processor technologies and wireless communications have enabled generation of small, low-cost sensor nodes with sensing, com-putation and short-range wireless communication capabilities. Each sensor node is usually equipped with a wireless transceiver, a small microcontroller, an energy power source and multi-type sensors such as temperature, humidity, light, heat, pressure, sound, vibration, motion, etc. A wireless sensor network (WSN) typically consists of a large num-

DOI: 10.4018/978-1-60566-328-9.ch007

ber of such low-power sensor nodes distributed over a wide geographical area with one or more possibly powerful sink nodes gathering information of others. These sensor nodes measure and collect data from the target area, perform some data processing, transmit and forward information to the sink node by a multi-hop routing. The sink node can also inform nodes to collect data by broadcasting a query to the entire network or a specific region in the network.

These small and low quality sensor nodes have severe limitations, such as limited energy and memory resources, communication bandwidth and computational processing capabilities. These constraints make sensor nodes more easily generate erroneous data. Especially when battery power is exhausted, probability of generating abnormally high or low sensor values will grow rapidly. On the other hand, operations of sensor nodes are frequently susceptible to environmental effects. The vision of large scale and high density wireless sensor network is to randomly deploy a large number of sensor nodes (up to hundreds or even thousands of nodes) in harsh and unattended environments. In such conditions, it is inevitable that some of sensor nodes will malfunction, which may result in erroneous readings. In addition to noise and sensor faults, abnormal readings may also be caused by actual events (e.g., once fire occurs, the readings of the temperature sensors around the region will intensively increase). These are potential reasons for generating abnormal readings in WSNs, often called *outliers*.

Coming across various definitions of an outlier, it seems that no universally accepted definition exists. The notion of outliers may even differ from one outlier detection technique to another (Zhang et al., 2007). Two classical definitions of an outlier include

(Hawkins, 1980) and (Barnett & Lewis, 1994). According to the former, "an outlier is an observation, which deviates so much from other observations as to arouse suspicions that it was

generated by a different mechanism'", where as the latter defines "an outlier is an observation (or subset of observations) which appears to be inconsistent with the remainder of that set of data". The term "outlier" can generally be defined as an observation that is significantly different from the other values in a data set.

Outlier Detection in WSNs

Based on the observation that sensor readings are temporally and geographically correlated, outliers in WSNs are sensor data significantly different from previous readings of own or neighboring nodes.

In WSNs, outliers are indication of either event or error:

- *Event*. In case of an event, the readings of the nodes in the event region are significantly different from those of nodes not in the event region. Examples of such events include fire, earthquake, chemical spill, etc. Decision making upon identification of events is of most importance.
- *Error*. This sort of outliers is also known as spurious data, erroneous readings or measurement faults. Compared to events, errors are more local. Due to the fact that such outliers influence the quality of data analysis and cause unnecessary communication overhead, they need to be identified and filtered or corrected if possible to prolong the network lifetime.

In the context of WSNs, outlier detection can assure high data quality and robustness of operation in harsh environments and in presence of noise and hardware failure. Filtering measurement errors can enhance the quality of sensor data analysis and reduce unnecessary communications overhead. Furthermore, identifying faulty sensor nodes that always generate outlier values may detect potential network attacks by adversaries. More importantly, outlier detection can lead to discovery of interesting events occurred in a specific region. Here, we exemplify the essence of outlier detection in several real-life applications of WSN.

- *Environmental monitoring*, in which sensors are deployed in harsh and unattended regions to monitor the environment. Outlier detection can identify when and where an event occurs and triggers an alarm upon detection.
- *Habitat monitoring*, in which endangered species can be equipped with small non-intrusive sensors to monitor their behavior. Outlier detection can indicate abnormal behaviors of the species and provide a closer observation about behavior of individual and groups of animals.
- *Health and medical monitoring*, in which small sensors are worn or attached to different positions of patients' body to monitor their well-being. Outlier detection showing unusual records can indicate whether the patient has potential diseases and allow doctors to take effective medical care.
- *Industrial monitoring*, in which machines are equipped with sensors to monitor their operation. Outlier detection can identify anomalous readings indicating possible malfunctioning or any other abnormality in the machines and allow for their corrections.
- *Target tracking*, in which moving targets are embedded with sensors to track them in real-time. Outlier detection can filter erroneous data to improve localization accuracy and to make tracking more efficiently and accurately.
- *Surveillance monitoring*, in which multiple sensitive and unobtrusive sensors are deployed in restricted areas. Outlier detection identifying position of the source of

anomaly can prevent unauthorized access and potential attacks by adversaries in order to enhance the security of these areas.

Contributions

The main objectives of this chapter are to:

- highlight the shortcomings of outlier detection techniques for general data and explain why they are not directly applicable for WSNs,
- provide a checklist and guideline on requirements that suitable outlier detection techniques for WSNs should meet.

We believe that both objectives are equally important as majority of recent proposed techniques for WSN have proved to:

- be optimization of outlier detection techniques for general data to lower down the complexity instead of proposing a completely new approach addressing challenges of WSNs,
- to focus only on one or two specific requirements of WSNs instead of the complete set of requirements, which has limited their applicability and generality.

To achieve our goals, in this chapter we first give an overview of the current state-of-the-art on outlier detection techniques for general data and provide a comprehensive technique-based taxonomy for them. We also discuss potential challenges and classification criteria for outlier detection in WSNs. Furthermore, we highlight the requirements for an outlier detection technique designed specially for WSNs and explain why outlier detection techniques for general data do not suffice for WSNs.

TECHNIQUE-BASED TAXONOMY FOR OUTLIER DETECTION TECHNIQUES FOR GENERAL DATA

There is a universally accepted assumption that number of anomalous data in a data set is considerably smaller than number of normal data. Thus, a straightforward approach to identify outliers is to construct a profile of normal pattern of data and then use this normal profile to detect outliers. Those observations whose characteristics significantly differ from the normal profile are declared as outliers (Tan et al., 2005). Based on their assumption on availability of pre-defined data, outlier detection techniques for general data can be classified into three basic categories, i.e., supervised, unsupervised and semi-supervised learning approaches (Tan et al., 2005). Both supervised and semi-supervised approaches require pre-classified normal or abnormal data to characterize all anomalies or non-anomalies in the training phase. The test data is compared against the learned predictive model for normal or abnormal classes. In many real-life applications, however, it is rarely possible to obtain the pre-classified data and also new types of normal or abnormal data may not be included in the pre-labeled data. Therefore, the focus of this chapter is on unsupervised outlier detection approaches that are more general as they do not need pre-labeled data.

Accuracy and execution time of outlier detection approaches depend on, among other things, type of data set. Therefore, data type is an important criterion for classification of outlier detection techniques. Common types of data sets are simple and complex data sets, of which the latter can be further categorized into high dimensional, mixed-type attributes, sequence, spatial, streaming and spatio-temporal data sets based on different semantics of data. These complex data sets pose significant challenges to the outlier detection problem.

Based on how probability distribution model is built, unsupervised approaches for simple data set are typically classified into the following categories (Markos & Singh, 2003):

- *Parametric* approaches, which assume that a single standard statistical distribution (e.g. normal distribution) can model the entire data set, and then directly calculate the parameters of this distribution based on means and covariance of the original data. Data that deviate significantly from the data model are declared as outliers.
- *Non-parametric* approaches, which make no assumption on statistical properties of data and instead identify outliers based on the full dimensional distance measure between data. Outliers are considered as data that are distant from their own neighbors in the data set. Non-parametric methods use user-defined parameters ranging from size of local neighborhood to threshold on distance measure.
- *Semi-parametric* approaches, which do not assume a standard data distribution for data, but instead map the data into a trained network model or a higher dimensional data space. On the basis of some unsupervised classification techniques such as neural network and support vector machine, semi-parametric methods further identify whether data deviate from the trained network model or are distant from other points in the higher dimensional data space.

Here we present a technique-based taxonomy, illustrated in Figure 1 to categorize unsupervised techniques for both simple and complex data types. The presented taxonomy can be used as to select a technique most suited to handle a specific data type.

To be able to highlight shortcomings of the outlier detection techniques for general data to

Figure 1. Taxonomy of outlier detection techniques

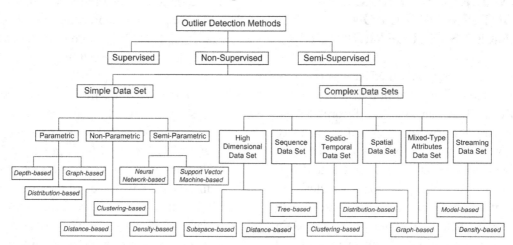

be directly applied to the WSN, in the following subsections we first give an overview and analysis of these techniques.

Outlier Detection Techniques for Simple Data Set

The simple data set has no complex semantics and is usually represented by low-dimensional real-valued ordering attributes (Hodge & Austin, 2003). Most existing outlier detection techniques are applicable to such simple data set. Outlier detection for simple data set has attracted techniques from statistics (i.e., distributed-, depth- and graph-based methods), data mining (i.e., clustering-, distance- and density-based methods), and artificial intelligent (i.e., neural networks-based and support vector machine-based methods).

Distribution-Based Methods

Distribution-based methods, as typical parametric methods, are the earliest approach to deal with the outlier detection problem. They assume that the whole data follow a statistical distribution (e.g., Normal, Poisson, Binomial) and make use of mathematics knowledge of applied statistics and probability theory to construct a data model.

These methods can fast and effectively identify outliers on the basis of an appropriate probabilistic data model.

Grubbs & Frank (1969) initially carry out the test on detecting outliers in a univariate data set. They assume that the whole data follows a standard statistical *t*-distribution and aim to identify one outlier at each iteration. Three most important fundamental textbooks concerning with outlier detection, i.e., (Hawkins, 1980), (Barnett & Lewis, 1994), and (Rousseeuw & Leroy, 1996) present classical definitions of distribution-based outliers. Barnett & Lewis (1994) and Rousseeuw & Leroy (1996) further address a comprehensive description and analysis of statistical outlier detection techniques. They discuss the problem of detecting outliers in univariate and multivariate data. For detecting univariate outliers, they assume that data points can be modeled by a statistical standard distribution. In this case, usually the Gaussian (normal) distribution is used. Three standard deviations are used as a threshold value to determine how significantly a point deviates from the data model. A simplified Z-score function more directly represents the degree of anomaly of each point. For detecting multivariate outliers, they usually assume a multivariate normal distribution and then use the Mahalanobis distance to detect multivariate outliers. Euclidean distance is another distance measure to be used but it cannot effectively capture the shape of the multivariate data distribution. Later on, Eskin (2000) proposes a mixture model approach to detect outliers in univariate data.

Based on their work, Hardin & Rocke (2004) propose a robust outlier detection approach that uses the minimum covariance determinant (MCD), which aims at alleviating the problem that the mean and standard deviation of the distribution may be extremely sensitive to outliers during the computation of Mahalanobis distance.

Depth-Based Methods

Depth-based methods exploit the concept of computational geometry (Preparata & Shamos, 1988) and organize data points into layers in multi-dimensional data space. Based on the definition of half-space depth (Tukey, 1997), also called as depth contours, each data point is assigned a depth and outliers are those points in the shallow layers with smaller depth value. These methods avoid the problem of fitting the data into a specific data distribution.

Rousseeuw & Leroy (1996) describe two basic depth-based outlier detection techniques for low dimensional data sets, i.e., minimum volume ellipsoid (MVE) and convex peeling. MVE uses the smallest permissible ellipsoid volume to define a boundary around the majority of data. Outliers are those points not in the densely populated normal boundary. Convex peeling uses peeling depth to map data points into convex hull layers in data space. Outliers are those points in the shallow convex hull layers with the lowest depth. Both MVE and convex peeling are robust outlier detection techniques that use the specific percentages of data points to define the boundary. Thus, these outlying points will not skew their boundary. The key difference between the two techniques is how many outliers are identified at a time. MVE

maintains all data points to define a normal boundary and then removes multiple outliers at once, while convex peeling builds convex hull layers and then peels away one outlier with the lowest depth at a time.

Based on their work, Ruts & Rousseeuw (1996) present an outlier detection approach using the concept of depth contour to compute the depth of points in a two-dimensional data set. The deeper the contour a data point fits in, the better chance of being an outlier. Johnson et al. (1998) further propose a faster outlier detection approach based on computing two-dimensional depth contours in convex hull layers. This approach only needs to compute the first k depth contours of a selected subset of points and it is robust against collinear points.

Graph-Based Methods

Graph-based methods use a powerful data image tool to map data into a graph to visualize single or multi-dimensional data spaces. Outliers are those points that are present in particular positions of the graph. These methods are suitable to identify outliers in real-valued and categorical data.

Laurikkala et al. (2000) propose an outlier detection approach for univariate data based on box plot, which is a simple single-dimensional graphical representation. Using box plot, points that lie outside the lower and upper threshold are identified as outliers. Also, these detected outliers can be ranked by the occurrence frequencies of outliers. Scatter plot (Panatier, 1996) is a graphical technique to detect outliers in two-dimensional data sets. It reveals a basic linear relationship between the axis X and Y for most of the data. An outlier is defined as a data point that deviates significantly from a linear model. In addition, spin plot (Valero-Mora et al., 2003) can be used for detecting outliers in 3-D data sets.

Clustering-Based Methods

Clustering-base methods are popular approaches to group similar data instances into clusters. Traditional clustering-based approaches, such as DBSCAN (Ester et al., 1996), BIRCH (Zhang, 1996), CURE (Guha, 1998) and TURN (Foss & Zaïane, 2002) are developed to optimize the process of clustering data with no specific interest in outlier detection.

New clustering-based outlier detection approaches can efficiently identify outliers as points that do not belong to clusters in a data set (Yu et al., 2002) or are clusters that are significantly smaller than other clusters (Jiang et al., 2001). These approaches do not require a priori knowledge about data distribution.

Yu et al. (2002) propose an outlier detection approach based on wavelet transformation, which has the multi-resolution property and can be extended to detect outliers in data sets with different densities. Jiang et al. (2001) propose to first partition the data into clusters, and then employ an outlier-finding process to identify outliers based on construction of a minimum spanning tree. He et al. (2003) introduce a new definition of a cluster-based local outlier, which takes size of a point's cluster and distance between the point and its closest cluster into account. Ren et al. (2004) propose a more efficient clustering-based local outlier detection approach, which combines detection of outliers with grouping data into clusters in a one-time process. Bohm et al. (2008) propose a robust clustering-based approach, which can be applied to a data set with non-Gaussian distribution to efficiently filter out the outliers.

Distance-Based Methods

Distance-based methods, as typical non-parametric methods, identify outliers based on the measure of full dimensional distance between a point and its nearest neighbor. Euclidean distance is commonly used as a similarity measure in distance-based methods. Outliers are points that are distant from the neighbors in the data set. These methods do not make any assumptions about data distribution and have better computational efficiency than depth-based methods, especially in large data sets.

Knorr & Ng (1998) propose three outlier detection algorithms, i.e., index-based, nested-loop and cell-based. The index-based algorithm is based on a priori constructed index structure and executes a range search with radius D for each point. If more than $M = (1-p)*N$ neighbors are found in a point's D-neighborhood, the search will stop and the point is declared as a non-outlier, otherwise it is an outlier. The nested-loop algorithm avoids the cost of preliminary construction of the index and instead partitions the entire set of points into blocks and then directly computes distance between each pair of points in the blocks. A point that has less than M neighbors within the distance D is declared as an outlier. The cell-based algorithm partitions the entire data set into cells and effectively prunes away a large number of non-outlier cells before finding out outliers.

Ramaswamy et al. (2000) provide a ranking of the top n outliers by measuring outlierness of data points. They define distance-based outliers to be the top n points with the maximum distance to their own k^{th} nearest neighbor. They then propose a partition-based algorithm to prune a significant number of partitions and efficiently identify the top n outliers in the rest of partitions from the data. Bay & Schwabacher (2003) propose an optimized nested-loops algorithm that has near-linear time complexity on mining the top n distance-based outliers. Barbara et al. (2006) propose a novel outlier detection technique based on statistical testing and distance calculation of Knorr & Ng (1998). The proposed technique can clean data with a large amount of outliers and also be effective in detecting further outliers even if some outliers are still in the presumed "cleaned" data.

Density-Based Methods

Density-based methods are proposed to take the local density into account when searching for outliers. These methods aim to effectively identify local outliers in data sets with diverse clusters. The computation of density still depends on full dimensional distance measures between a point and its nearest neighbors in the data set.

Breunig et al. (2000) introduce the notion of density-based local outliers. They assign a local outlier factor (LOF) value to each data point. LOF of a data point is calculated using the ratio of the local density of this point and the local density of its *MinPts* nearest neighbors. The single parameter *MinPts* of a point determines the number of its nearest neighbors in the local neighborhood. The LOF value indicates the degree of being an outlier depending on how isolated the point is with respect to the density of its local neighborhood. Points that have the largest LOF values are considered as outliers. Later on, many novel density-based approaches have been developed to further improve the effectiveness and efficiency of LOF.

Chiu & Fu (2003) present three enhancement schemes for LOF. The first two schemes are variants of the original LOF computation formulation. The third scheme uses a simple grid-based technique to prune away some non-outliers and then only computes the LOF values for the remaining points. Jin et al. (2001) propose an outlier detection approach, which can determine the top n local outliers having the maximal LOF values and use BIRCH (Zhang, 1996) to reduce the computation load of LOF for all points. Hu &

Sung (2003) propose an outlier detection approach for data sets with high density clustering and low density regularity. Papadimitriou et al. (2003) present a fast outlier detection approach called LOCI, which employs the concept of a multi-granularity deviation factor (MDEF) value to measure a point's relative deviation of its local neighborhood density from the average local neighborhood density in its neighborhood. A

point can be declared as an outlier by comparing its MDEF with a derived statistical value. Kim & Cho (2006) propose an outlier detection approach, which uses distance between a data point and its closest prototypes, i.e., small percentage of representative data from the original data, as the degree of outlierness.

Fan et al. (2006) and Kollios et al. (2003) combine distance-based and density-based approaches to identify outliers in a data set. Fan et al. (2006) introduce a novel outlier notion by considering both local and global features of the data set. The proposed approach uses TURN clustering technique (Foss & Zaïane, 2002) to identify outliers by consecutively changing the resolution of a set of data points. Kollios et al. (2003) propose a density-based biased sampling approach to detect DB-outlier based on kernel density estimator. These points can be used to represent the density of the whole data set and efficiently approximate the underlying probability distribution.

Neural Networks-Based Methods

Neural networks (NN) can autonomously model the underlying data distribution and distinguish the normal/abnormal classes. Neural networks do not require pre-labeled data to permit learning and can identify those data points that are not reproduced well at the output layer as outliers. These methods effectively identify outliers and automatically reduce the input features based on the key attributes.

Sykacek (1997) presents an outlier detection approach using the equivalent error bar to identify outliers in the trained network with multi-layer perception. Outliers are points that are residual outside the equivalent error bar depending on a pre-defined threshold. Harkins et al. (2002) present an outlier detection approach for large multivariate data sets based on the construction of a replicator neural network (RNN), which is a variant of the usual regression model. If some small number of input points are not reconstructed well and cause high reconstruction errors in the RNN, these points can be considered as outliers. Fu & Yu (2006) propose an outlier detection approach based on artificial neural network (ANN). Outlier detection can be performed in three ANNs by a modified Z-score.

Support Vector Machine-Based Methods

Support vector machine (SVM) based methods can distinguish between normal and abnormal classes by mapping data into the feature space. They do not require a pre labeled data set to determine a boundary region and can efficiently identify outliers as points that are distant from most of the other points or are in relatively sparse regions of the feature space.

Scholkopf et al. (2001) present an outlier detection approach, which uses a kernel function to efficiently map the original data into a vector space typically of high dimensions (feature space). Tax & Duin (1999) propose an outlier detection approach based on unsupervised SVM, called support vector domain description (SVDD).

The approach uses a Gaussian kernel function to map the whole data set to high dimensional feature space. It then can classify normal data by learning an optimal hypersphere, which is a sphere with minimum volume containing the majority of data points. Those points that lie outside this sphere are considered as outliers. Petrovskiy (2003) exploits the same idea of SVDD and presents an outlier detection approach by using fuzzy set theory.

Outlier Detection Techniques for Complex Data Sets

Outlier detection in complex data sets, such as high dimensional, mixed-type attributes, sequence, spatial, streaming and spatio-temporal data sets, are more difficult than in simple data set. These complex data sets have their own semantics of input data so that the above outlier detection techniques proposed for simple data set may not be well applicable. Thus, specific outlier detection techniques should be presented for these complex data sets.

Outlier Detection Techniques for High Dimensional Data Set

High dimensional data set contains a large number of data and each data point has a large number of attributes. In such high dimensional spaces, where the data is sparse, many outlier detection techniques designed for simple data sets will be susceptible to the problem of the curse of dimensionality. Specifically, convex hull or MVE becomes harder to discern, the notion of proximity is less meaningful and distance computation is computationally more expensive. Also, high dimensionality increases time complexity and makes it more difficult to accurately approximate distribution of underlying the data. Thus, several approaches have been proposed specially for detecting outliers in high dimensional data sets. They can generally be classified into subspace-based and distance-based methods. Subspace-based methods project the data into a low-dimensional subspace and declare a point as an outlier if this point lies in an abnormal lower-dimensional projection, where the density of the data is exceptionally lower than the average. These methods reduce the dimensions of data and efficiently identify outliers in high dimensional data sets.

Aggarwal & Yu (2001) propose a subspace-based outlier detection approach that observes density distribution of the projections, i.e., clusters in a low-dimensional subspace. An evolutionary search algorithm is used to determine low-dimensional projections since it can efficiently find hidden combinations of dimensions in which data is sparse and also speed up processing time compared to the naive brute-force search algorithm. Based on their work, Zhu et al. (2005) present an approach by incorporating directly user-defined example outliers. Points can be considered as outliers if they are in an extremely low-dimensional subspace.

Angiulli & Pizzuti (2002) design a distance-based method to efficiently find the top *n* outliers in large and high-dimensional data sets. This approach first determines an approximate subset including *k* candidate outliers and further determines the true outliers from these candidate outliers. Ghoting et al. (2006) present a fast distance-based outlier detection approach, which uses a divisive hierarchical clustering to effectively partition the data set into clusters using a distance similarity measure and further efficiently identifies outliers relying on a novel nested loop algorithm, which aims at finding a data point's *k* approximate nearest neighbors.

Outlier Detection Techniques for Mixed-Type Attributes Data Set

In some applications, data contains mixture of continuous (numeric) and categorical attributes. The latter usually has non-numeric and partial ordering values. This makes it difficult for most non-parametric approaches to use the notion of distance or density to measure the similarity between two data points in continuous data spaces. Moreover, distribution-based and neural network based approaches cannot work for mixed-type attributes data sets as they identify outliers only in numeric or ordinal data sets. The mixed-attribute data may influence the performance of detecting outliers if it is only simply disregarded. Thus, several approaches have been designed specially for detecting outliers in categorical or mixed-type attributes data sets using graph-based methods.

Otey et al. (2006) present an approach to identify outliers by taking into account the dependencies between continuous and categorical attributes. In a categorical attribute space, two data points are considered linked if they have at least one common attribute-value pair. Number of attribute-value pairs in common indicates the

strength of the associated link between these two points. A data point can be considered as an outlier if it has very few links or very weak links to other points. In a mixed attribute space, dependency between the values with mixed continuous and categorical attributes is captured by incrementally maintenance of covariance matrix. A data point can be considered as an outlier if number of its attribute-value pairs that are infrequent and its corresponding covariance are violated by the dependencies between the mixed attributes. Wei et al. (2003) and He et al. (2005) propose efficient approaches for detecting local outliers in categorical data. The former uses a hypergraph model to precisely capture distribution characteristics in a data subspace. The latter uses an entropy function to measure the degree of disorder of the rest of data set. A point is declared as an outlier if the entropy value after exchanging its label with each of the pre-defined outliers is decreased. Yu et al. (2006) propose an outlier detection approach for detecting centric local outliers in categorical/numerical data. A point can be considered as an outlier if its similarity relationship with its neighbors is lower than the similarity relationships among its neighbors' neighborhood.

Outlier Detection Techniques for Sequence Data Set

In the sequence data set, data is naturally represented as a sequence of individual entities. Also, two sequence data sets may not have the same size and their distributions are not a priori known. Therefore, it is difficult for traditional distance and density-based outlier detection techniques to define a standard notion of similarity to measure structural differences between two sequences. Thus, existing outlier detection techniques for sequence data sets exploits clustering-based and tree-based methods.

Budalakoti et al. (2006) introduce an approach, which efficiently clusters the sequence data into groups and finds anomalous subsequences that deviate from normal behaviors in a cluster. Sun et al. (2006) propose an approach on the basis of building a probabilistic suffix tree (PST), which exploits the theory of a variable-order Markov chain and uses a suffix tree as its index structure. Only nodes near the root of the tree need to be examined for identifying outliers. This approach uses the length of normalized probability as the sequence similarity measure to find the top *n* outliers in a sequence data set.

Outlier Detection Techniques for Spatial Data Set

Spatial data has non-spatial and spatial attributes. Spatial attributes contain location, shape, directions and other geometric or topological information. Spatial neighborhood is defined in terms of spatial relationship such as distance or adjacency. In traditional outlier detection techniques, distribution-based approaches work well for one-dimensional data sets and only consider the statistical distribution of non-spatial attribute values. They ignore the spatial relationships between data points. On the other hand, most non-parametric methods do not distinguish between spatial and non-spatial attributes, but use all dimensions to define the neighborhood based on concepts of distance, density and convex-hull depth. Existing spatial outlier detection approaches can be categorized into graphical and statistical approaches.

Shekhar et al. (2001) indicate that being a spatial outlier depends on the difference between an attribute value of the point and the average attribute value of its spatial neighbors. They propose an algorithm, which uses a single non-spatial attribute to compare the difference between spatial neighborhoods and identifies spatial outliers computationally efficiently by computing the global algebraic aggregate functions. The authors further consider the graph structure of the spatial data and exploit a graphical method for spatial outlier detection.

Lu et al. (2001) propose two iterative algorithms and a non-iterative algorithm to detect spatial outliers. Performance of these three algorithms depends on the choice of a neighborhood function and a comparison function. The neighborhood function refers to a summary statistic of attribute values of all the spatial neighbors of a data point. The comparison function is used to further compare the attribute value of this point with the summary statistic value of its neighbors. The non-iterative algorithm defines a different neighborhood function based on the median of the attribute values of the neighbors. They further detect spatial outliers with multiple attributes using Mahalanobis distance.

Kou et al. (2006) present two spatial weighted outlier detection algorithms, which consider the impact of spatial relationship on the neighborhood comparison. For a data point, each of neighbors in its spatial neighborhood is assigned a different weight in terms of their impact on the point. Chawla & Sun (2004) propose a spatial local outlier detection approach that computes the degree of outlierness of each point in a data set and considers the values of spatial autocorrelation and spatial heteroscedasticity, which are used to capture the effect of a data point on its neighborhood and the non-uniform variance of the data.

Outlier Detection Techniques for Streaming Data Set

Traditional outlier detection techniques work well for static data sets, where all data points are stationary. A data stream is a large volume of data that is arriving continuously and fast in an ordered sequence, and also data may be constantly added, removed, or updated. Thus, data stream can be viewed as an infinite sequence of data that continuously evolves over time. Outlier detection techniques for streaming data are categorized into model-based, graph-based, and density-based methods.

He et al. (2003) present a model-based approach to identify outliers in data streams by using frequent patterns, which represent common patterns of a majority of data points in data sets. The degree of outlierness for each point is measured by a frequent pattern outlier factor and *n* points that contain very few frequent patterns are considered as outliers. Yamanishi et al. (2006) detect outliers in non-stationary time series data based on a typical statistical autoregression (AR) model, which represents a statistical behavior of time series data. Muthukrishnan et al. (2004) define a new notion of an outlier in time series data streams based on a representation sparsity metric histogram. If the removal of a point from the time sequence results in a sequence that can be represented more briefly than the original one, then the point is an outlier. Pokrajac et al. (2007) propose an incremental density-based approach, which exploits the static iterated LOF algorithm to deal with each new point inserted into the data set and iteratively determines whether the point is an outlier. This technique can efficiently adapt to the update of the data profiles caused by insertion or deletion of data points.

Outlier Detection Techniques for Spatio-Temporal Data Set

Most existing spatial outlier detection techniques focus on detecting spatial outliers, which only consider the non-spatial attributes of data or the spatial relationships among neighbors. However, in many geographic phenomena evolving over time, the temporal aspects and spatial-temporal relationships existing among spatial data points also need to be considered.

Cheng & Li (2006) introduce a formal definition of a spatio-temporal outlier (ST-outlier), i.e., a spatial-temporal point whose non-spatial attribute values are significantly different from those of other spatially and temporally referenced points in its spatial or/and temporal neighborhoods. Considering the temporal aspects, the authors declare a point as a ST-outlier by checking if the point's attribute value at time T is significantly

different from the statistical attribute values of its neighbors at time $T-1$ and $T+1$. They further propose a four-step approach to detect ST-outliers, i.e., classification, aggregation, comparison and verification. Birant & Kut (2006) define a similar definition of ST-outlier and present a ST-outlier detection approach based on clustering concepts. This approach consists of three steps, i.e., clustering, checking spatial neighbors, and checking temporal neighbors. In the clustering step, DBSCAN clustering technique (Ester et al., 1996) has been improved in supporting temporal aspects and detecting outliers in clusters with different densities. As a result, potential outliers are those points that do not belong to any of clusters. The other two steps further verify these potential outliers.

GENERAL OUTLIER DETECTION TECHNIQUES DO NOT SUFFICE FOR WSNS

Up to now, we have presented a classification of outlier detection techniques for general data together with an overview and analysis of existing techniques. In this section, we address challenges of outlier detection in WSNs and present important classification criteria for these techniques. We then provide a checklist and guideline on requirements that suitable outlier detection techniques for wireless sensor networks should meet, and will explain why general outlier detection techniques do not suffice.

Challenges for Outlier Detection in WSNs

The context of sensor networks and the nature of sensor data make the design of an appropriate outlier detection technique challenging. Some of the most important challenges that outlier detection techniques designed for WSN should cope with are:

- *Resource constraints*. The low-cost and low quality sensor nodes have stringent resource constraints, such as limited energy, memory, computational capacity and communication bandwidth. Thus, a challenge for outlier detection in WSNs is how to minimize the energy consumption while using a reasonable amount of memory for storage and computational tasks.

- *High communication cost*. In WSNs, majority of the energy is consumed in radio communication rather than in computation. For a sensor node, the communication cost is often several orders of magnitude higher than the computation cost (Akyildiz et al., 2002). Thus, a challenge for outlier detection in WSNs is how to minimize the communication overhead in order to relieve the network traffic and prolong network lifetime.

- *Distributed streaming data*. Distributed sensor data coming from many different streams may dynamically change. Moreover, the underlying distribution of streaming data may not be known a priori, and also direct computation of probabilities is difficult (Gaber, 2007). Thus, a challenge for outlier detection in WSNs is how to process distributed streaming data online.

- *Dynamic network topology, frequent communication failures, mobility and heterogeneity of nodes*. The sensor network deployed in unattended environments over extended periods of time is susceptible to dynamic network topology and frequent communication failures. Moreover, sensor nodes may move among different locations at any point of time, and may have different sensing and processing capacities, even each of them may be equipped with different number and types of sensors. These dynamic characteristics of data, network and capabilities increase the complexity of designing an appropriate outlier detection technique for WSNs.

- *Large-scale deployment*. Deployed sensor networks can be large (up to hundreds or even thousands of sensor nodes). The key challenge for traditional outlier detection techniques is to maintain a high detection rate while keeping the false alarm rate low. This requires construction of an accurate normal profile that represents the normal behavior of sensor data (Tan et al., 2005). However, this is very difficult for large-scale sensor network applications.

- *Identifying outlier sources*. The sensor network is expected to provide the raw data sensed from the physical world and also detect events occurred in the network. Due to the fact that noise, harsh environmental effects, power exhaustion, events, and hardware failure are usual and common issues in WSNs, identification of what has caused outliers is not easy. Thus, a challenge for outlier detection in WSNs is identifying outlier sources and make a distinction between events and errors.

Thus, the main challenge faced by outlier detection techniques for WSNs is to satisfy the mining accuracy requirements while maintaining the resource consumption of WSNs to a minimum (Gaber, 2007). In another word, the question is how to process as much data as possible in a decentralized and online manner while keeping the communication overhead, memory and computational cost low (Ma et al., 2004).

CLASSIFICATION CRITERIA FOR OUTLIER DETECTION TECHNIQUES FOR WSNS

Several important classification criteria need to be considered for designing an optimal outlier detection technique for WSNs.

Input Sensor Data

Sensor data can be viewed as data streams, which refers to a large volume of real-valued data that is continuously collected by sensor nodes (Gaber et al., 2005). The type of input data determines which outlier detection techniques can be used to analyze the data (Chandola et al., 2007). Outlier detection techniques should consider the two following aspects of sensor data.

- *Attributes*. A data measurement can be identified as an outlier if it has anomalous values for its attributes (Tan et al., 2006). In univariate data, a single attribute presenting anomaly with respect to that attribute of other data points is an outlier. In multivariate data with multiple attributes, outlier detection depends on considering all attributes together, because sometimes none of the attributes individually may have an anomalous value (Sun, 2006).
- *Correlations*. There are two types of dependencies at each sensor node: (i) dependencies among the attributes of the sensor node data, and (ii) dependency between current and previous readings of the node and its neighboring nodes (Janakiram et al., 2006). Capturing attribute dependency helps improve the mining accuracy and computational efficiency. Capturing dependency between current and previous readings of a node and its neighboring nodes helps predict the trend of sensor readings and distinguish between errors and events.

Type of Outliers

Depending on the scope of data used for outlier detection, outliers may be local or global. Local models generated from data streams of individual nodes are totally different from global models (Subramaniam et al., 2006).

- *Local outliers*. Local outliers usually are identified at each individual sensor node. Two variations for local outlier identification exist in WSNs. One is that each node identifies the anomalous values only depending on its historical values. The other is that the node collects the readings of its neighboring nodes coupled with its own historical values to collaboratively identify the anomalous values. Compared with the first approach that lacks sufficient information, the latter approach takes advantage of the spatio-temporal correlations among sensor data and improves the accuracy and robustness of the outlier detection.
- *Global outliers*. Global outliers are identified in a more global perspective. Identification of global outliers can be done in different nodes depending on the network architecture (Chatzigiannakis et al., 2006). In a centralized architecture, all data is transmitted first to the sink node and the sink node identifies outliers. In aggregate/clustering-based architecture, the aggregator/clusterhead collects the data from nodes within its controlling range and then identifies outliers. Individual nodes should also be able to identify global outliers if they have a copy of global estimator model obtained from the sink node (Subramaniam et al., 2006).

Outliers Identity

Outliers may be indication of either events or errors. An event is defined as a particular phenomenon that changes the real-world state. An error refers to noise-related measurements or data coming from a faulty sensor.

- *Events*. This sort of outliers normally last for a relatively long time and change historical pattern of sensor data. They need to be distinguished from long segmental errors generated by faulty sensors. Sensor faults are likely to be stochastically unrelated, while event measurements are likely to be spatially correlated (Luo et al., 2006).

- *Errors*. Outliers caused by errors may occur frequently. The erroneous data is normally represented as an arbitrary change. Due to the fact that such errors influence the quality of data analysis, thus they need to be identified and corrected if possible since they may be still usable for data analysis after correction.

Degree of Being an Outlier

Outliers are measured by two units, i.e., scalar and outlier score (Chandola et al., 2007). The scalar unit is like a classification measure, which classifies each data measurement into normal or outlier class. The outlier score unit assigns an outlier score to each data measurement depending on the degree of which the measurement is considered as an outlier.

- *Scalar*. The output of scalar-based approaches is a set of outliers and a set of normal measurements. The scalar approach neither differentiates between different outliers nor provides a ranked list of outliers.
- *Outlier score*. The outlier score-based approaches provide a ranked list of outliers. An analyst may choose to either analyze the top *n* outliers having the largest outlier scores or use a cut-off threshold to select the outliers if their outlier scores are greater than a threshold.

Evaluation of Outlier Detection

Outlier detection techniques are required to maintain a high detection rate while keeping the false alarm rate low. False alarm rate refers to number of normal data that are incorrectly considered as

outliers. A receiver operating characteristic (ROC) curves (Lazarevic et al., 2003) is usually used to represent the trade-off between the detection rate and false alarm rate.

REQUIREMENTS FOR OUTLIER DETECTION TECHNIQUE FOR WSNS

Having seen challenges and classification criteria for outlier detection in WSNs, we identify the following requirements that an optimal outlier detection approach for WSNs should meet:

- It must process the data in a distributed manner to prevent unnecessary communication overhead and energy consumption and to prolong network lifetime.
- It must be an online technique for handling streaming and dynamically updated sensor data.
- It must have a high detection rate while keeping false alarm rate low.
- It should be unsupervised as the pre-classified normal or abnormal data is difficult to obtain in WSNs. Also, it should be nonparametric as there is no knowledge about distribution of the input sensor data.
- It should take multivariate data into account.
- It must be simple and computationally cheap.
- It must enable auto-configurability with respect to dynamic network topology or communication failure.
- It must scale well.
- It must consider dependencies among attributes of the sensor data as well as spatio-temporal correlations that exist among the observations of neighboring sensor nodes.
- It must effectively distinguish between erroneous measurements and events.

Table 1. Classification and comparison of general outlier detection techniques for WSNs (1)

Author/ Name of Technique	Technique-based on	Attribute			Requirements of outlier detection techniques for WSNs							
		Univariate	Multivariate		Use of correlation			Distributed	Online	Low computa-tional complexity	Distinguish Event/error	Scalability
			Moderate	High	Attribute	Spatial	Temporal					
Grubbs et al.	distribution	√										
Barnett et al.	distribution	√										
Eskin et al.	distribution	√										
Yamani. et al.	distribution		√									
MVE	depth		√									
Convex peeling	depth		√									
Box plot	graph	√										
Scatter plot	graph		√									
Yu et al.	clustering		√									
Jiang et al.	clustering		√									
He et al.	clustering		√							√		√
Ren et al.	clustering		√							√		√
Knorr et al.	distance		√			√						
Ramas. et al.	distance		√			√				√		√
Bay et al.	distance		√			√				√		√
LOF	density		√			√						
CF	density		√			√				√		√
Jin et al.	density		√			√				√		√
Hu et al.	density		√			√						
LOCI	density		√			√				√		√
Kim et al.	density		√			√						
RDF	density		√			√				√		√
Fan et al.	density		√			√				√		√
Kollios et al.	density		√							√		√
Harkins et al.	NN		√									
Fu et al.	SVM		√									

Shortcomings of General Outlier Detection Techniques for WSNs

General outlier detection approaches explained in the section of *Technique-based Taxonomy for Outlier Detection Techniques for General Data* have major drawbacks that make them not suitable to be directly applicable to outlier detection in WSNs. These shortcomings can be summarized as:

- They require all data to be accumulated in a centralized location and be analyzed offline, which causes too much energy consumption and communication overhead in WSNs.
- They have paid limited attention to reasonable availability of computational resources. They are usually computationally expensive and require much memory for data analysis and storage.
- They often ignore dependencies among the attributes of the data.
- They do not often distinguish between errors and events and regard outlier as errors, which results in loss of important hidden information about events.

Having the requirements of an appropriate outlier detection technique for WSN specified in the previous subsection, here we present a comparative overview on how the existing techniques satisfy these requirements. As it can be seen in Table 1 and Table 2, there is no general outlier detection technique, which satisfies all the requirements. This calls for an outlier detection technique specifically designed for WSN which is (i) not just an optimization of outlier detection techniques for general data to lower down the complexity, and (ii) does not only focus on one or two specific requirements of WSNs instead of the complete set of requirements.

CONCLUSION

In this chapter, we present a comprehensive technique-based taxonomy for contemporary outlier detection techniques for general data. We also highlight shortcomings of these techniques to be directly applicable to WSNs. Furthermore, we address challenges and important classification criteria for outlier detection in WSNs. Moreover, we provide a checklist and guideline on requirements that suitable outlier detection techniques for WSNs should meet. A comparative view on how the existing techniques satisfy these requirements is also presented, which clearly shows there is no general outlier detection technique that meet all these requirements.

REFERENCES

Aggarwal, C. C., & Yu, P. S. (2001). Outlier detection for high dimensional data, In *Proceedings of the ACM SIGMOD Conference on Management of Data* (pp. 37-47).

Akyildiz, I. F., Su, W., Sankarasubramaniam, Y., & Cayirci, E. (2002). Wireless sensor networks: a survey. *International Journal of Computer Networks*, 38(4), 393–422. doi:10.1016/S1389-1286(01)00302-4

Angiulli, F., & Pizzuti, C. (2002). Fast outlier detection in high dimensional spaces. *In Proceedings of Pacific-Asia Conference on Knowledge Discovery and Data Mining* (pp. 15-26).

Barbara, D., Domeniconi, C., & Rogers, J. P. (2006) Detecting outliers using transduction and statistical significance testing. In *Proceedings of the ACM SIGKDD conference on Knowledge Discovery and Data Mining* (pp. 55-64).

Barnett, V., & Lewis, T. (1994). *Outliers in statistical data*. New York: John Wiley Sons.

Table 2. Classification and comparison of general outlier detection techniques for WSNs (2)

Author/ Name of Technique	Technique-based on	Attribute			Requirements of outlier detection techniques for WSNs							
		Univariate	Multivariate		Use of correlation			Distributed	Online	Low computa-tional complexity	Distinguish Event/error	Scalability
			Moderate	High	Attribute	Spatial	Temporal					
Scholko et al.	SVM		✓									
Tax et al.	SVM		✓									
Petrovskiy	SVM		✓									
Aggarwal et al.	subspace			✓								
Zhu et al.	subspace			✓								
Shyu et al.	subspace			✓								
Angiulli et al.	distance			✓								
Ghoting et al.	distance			✓						✓		✓
Chaudhary et al.	distance		✓									
Otey et al.	graph		✓		✓	✓	✓	✓	✓	✓		✓
Wei et al.	graph		✓									
He et al.	graph		✓									
Yu et al.	graph		✓			✓						
Budalak et al.	clustering		✓									
Sun et al.	tree		✓									
Shekhar et al.	distribution	✓				✓						
Lu et al.	distribution	✓				✓						
Lu et al.	distribution		✓			✓						
Kou et al.	distribution		✓			✓						
Sun et al.	distribution		✓			✓						
He et al.	model		✓				✓		✓			
Yamani et al.	model		✓				✓		✓			
Muthukri. et al.	graph		✓				✓		✓			
Pokrajac et al.	density		✓				✓		✓			
Cheng et al.	clustering		✓			✓	✓					
Birant et al.	clustering		✓			✓	✓					

Bay, S., & Schwabacher, M. (2003). Mining distance-based outliers in near linear time with randomization and a simple pruning rule. In *Proceedings of the ACM SIGMOD Conference on Knowledge Discovery and Data* (pp. 29-38).

Birant, D., & Kut, A. (2006). Spatio-temporal outlier detection in large database. *Journal of Computing and Information Technology, 14*(4), 291–298.

Bohm, C., Faloutsos, C., & Plant, C. (2008). Outlier-robust clustering using independent components. In *Proceedings of the ACM SIG-MOD Conference on Management of Data* (pp. 185-198).

Breunig, M. M., Kriegel, H. P., Ng, R. T., & Sander, J. (2000). LOF: identifying density-based local outliers. In *Proceedings of the ACM SIGMOD Conference on Management of Data* (pp. 93-104).

Budalakoti, S., Cruz, S., Srivastava, A. N., Akella, R., & Turkov, E. (2006). Anomaly detection in large sets of high-dimensional symbol sequences. *The United States: California, National Aeronautics and Space Administration.*

Chandola, V., Banerjee, A., & Kumar, V. (2007). *Outlier detection: A survey* (Tech. Rep.). University of Minnesota.

Cheng, T., & Li, Z. (2006). A multiscale approach for spatio-temporal outlier detection. *Transactions in Geographic Information System, 10*(2), 253–263.

Chiu, A. L., & Fu, A. W. (2003). Enhancements on local outlier detection. In *Proceedings of International Database Engineering and Applications Symposium* (pp. 298-307).

Eskin, E. (2000). Anomaly detection over noisy data using learned probability distributions. In *Proceedings of International Conference on Machine Learning* (pp. 222-262).

Ester, M., Kriegel, H. P., Sander, J., & Xu, X. (1996). A density-based algorithm for discovering clusters in large spatial databases with noise. In *Proceedings of Knowledge Discovery and Data Mining* (pp. 226-231).

Fan, H., Zaiane, O. R., Foss, A., & Wu, J. (2006). A nonparametric outlier detection for effectively discovering top-n outliers from engineering data. In *Proceedings of Pacific-Asia Conference on Knowledge Discovery and Data Mining* (pp. 557-566).

Foss, A., & Zaïane, O. (2002). A parameterless method for efficiently discovering clusters of arbitrary shape in large datasets. In *Proceedings of International Conference on Data Mining* (pp. 179-186).

Fu, J., & Yu, X. (2006). Rotorcraft acoustic noise estimation and outlier detection. In *Proceedings of International Joint Conference on Neural Networks* (pp. 4401-4405).

Gaber, M. M. (2007). Data stream processing in sensor networks. In J. Gama & M. M. Gaber (Eds.), *Learning from data streams processing techniques in sensor network* (pp. 41-48). Berlin-Heidelberg: Springer.

Ghoting, A., Parthasarathy, S., & Otey, M. (2006). Fast mining of distance-based outliers in high dimensional datasets. *In Proceedings of the SIAM International Conference on Data Mining* (pp. 608-612).

Grubbs & Frank. (1969). Procedures for detecting outlying observations in samples. *Technometrics, 11*(1), 1–21. doi:10.2307/1266761

Guha, S., Rastogi, R., & Shim, K. (1998). CURE: An efficient clustering algorithm for large databases. In *Proceedings of the ACM SIGMOD Conference on Management of Data* (pp. 73-84).

Hardin, J., & Rocke, D. M. (2004). Outlier detection in the multiple cluster setting using the minimum covariance determinant estimator. *Journal of Computational Statistics and Data Analysis, 44,* 625–638. doi:10.1016/S0167-9473(02)00280-3

Harkins, S., He, H., Willams, G. J., & Baster, R. A. (2002). Outlier detection using replicator neural networks. In *Proceedings of International Conference on Data Warehousing and Knowledge Discovery* (pp. 170-180).

Hawkins, D. M. (1980). *Identification of outliers.* London: Chapman and Hall.

He, Z., Deng, S., & Xu, X. (2005). An optimization model for outlier detection in categorical data. In *Proceedings of International Conference on Intelligent Computing* (pp. 400-409).

He, Z., Xu, X., & Deng, S. (2003). Discovering cluster based local outliers. *Official Publication of Pattern Recognition Letters, 24*(9-10), 1651–1660.

He, Z., Xu, X., & Deng, S. (2003). Outlier detection over data streams. In *Proceedings of International Conference for Young Computer Scientists. Harbin, China.*

Hodge, V. J., & Austin, J. (2003). A survey of outlier detection methodologies. *International Journal of Artificial Intelligence Review, 22,* 85–126.

Hu, T., & Sung, S. Y. (2003). Detecting pattern-based outliers. *Official Publication of Pattern Recognition Letters, 24*(16), 3059–3068. doi:10.1016/S0167-8655(03)00165-X

Janakiram, D., Mallikarjuna, A., Reddy, V., & Kumar, P. (2006). Outlier detection in wireless sensor networks using Bayesian belief networks. In *Proceedings of Communication System Software and Middleware* (pp. 1-6).

Jeffery, S. R., Alonso, G., Franklin, M. J., Hong, W., & Widom, J. (2006). Declarative support for sensor data cleaning. In *Proceedings of International Conference on Pervasive Computing* (pp. 83-100).

Jiang, M. F., Tseng, S. S., & Su, C. M. (2001). Tw-phase clustering process for outliers detection. *Official Publication of Pattern Recognition Letters, 22*(6-7), 691–700. doi:10.1016/S0167-8655(00)00131-8

Jin, W., Tung, A. K. H., & Han, J. (2001). Mining top-n local outliers in large databases. In *Proceedings of the ACM SIGMOD Conference on Knowledge Discovery and Data* (pp. 293-298).

Johnson, T., Kwok, I., & Ng, R. T. (1998). Fast computation of 2-dimensional depth contours. In *Proceedings of the ACM SIGMOD Conference on Knowledge Discovery and Data* (pp. 224-228).

Kim, S., & Cho, S. (2006). Prototype based outlier detection. In *Proceedings of International Joint Conference on Neural Networks* (pp. 820-826).

Knorr, E., & Ng, R. (1998). Algorithms for mining distance-based outliers in large data sets. *International Journal of Very Large Data Bases* (pp. 392-403).

Kollios, G., Gunopulos, D., Koudas, N., & Berchtold, S. (2003). Efficient biased sampling for approximate clustering and outlier detection in large data sets. *International Journal of Knowledge and Data Engineering, 15*(5), 1170–1187. doi:10.1109/TKDE.2003.1232271

Kou, Y., Lu, C., & Chen, D. (2006). Spatial weighted outlier detection. In *Proceedings of SIAM International Conference on Data Mining* (pp. 613-617).

Laurikkala, J., Juhola, M., & Kentala, E. (2000). Informal identification of outliers in medical data. In *Proceedings of International Workshop on Intelligent Data Analysis in Medicine and Pharmacology*.

Lazarevic, A., Ozgur, A., Ertoz, L., Srivastava, J., & Kumar, V. (2003). A comparative study of anomaly detection schemes in network intrusion detection. In *Proceedings of SIAM Conference on Data Mining*.

Lu, C. T., Chen, D., & Kou, Y. (2003). Algorithms for spatial outlier detection. In *Proceedings of International Conference on Data Mining* (pp. 597-600).

Lu, C. T., Chen, D., & Kou, Y. (2003). Detecting spatial outliers with multiple attributes. *In Proceedings of International Conference on Tools with Artificial Intelligence* (pp. 122-128).

Luo, X., Dong, M., & Huang, Y. (2006). On distributed fault-tolerant detection in wireless sensor networks. *IEEE Transactions on Computers, 55*(1), 58–70. doi:10.1109/TC.2006.13

Ma, X., Yang, D., Tang, S., Luo, Q., Zhang, D., & Li, S. (2004). Online mining in sensor networks. In *Proceedings of International conference on network and parallel computing* (pp. 544-550).

Markos, M., & Singh, S. (2003). Novelty detection: A review-part 1: statistical approaches. *International Journal of Signal Processing, 83*, 2481–2497. doi:10.1016/j.sigpro.2003.07.018

Muthukrishnan, S., Shah, R., & Vitter, J. S. (2004). Mining deviants in time series data streams. In *Proceedings of International Conference on Scientific and Statistical Database Management*.

Otey, M. E., Ghoting, A., & Parthasarathy, S. (2006). Fast distributed outlier detection in mixed-attribute data sets. *International Journal of Data Mining and Knowledge Discovery, 12*(2-3), 203–228. doi:10.1007/s10618-005-0014-6

Palpanas, T., Papadopoulos, D., Kalogeraki, V., & Gunopulos, D. (2003). Distributed deviation detection in sensor networks. In *Proceedings of the ACM SIGMOD Conference on Management of Data* (pp. 77-82).

Panatier, Y. (1996) *Variowin: Software for spatial data analysis in 2D*. New York: Springer-Verlag Berlin Heidelberg.

Papadimitriou, S., Kitagawa, H., Gibbons, P. B., & Faloutsos, C. (2003). LOCI: fast outlier detection using the local correlation integral. In *Proceedings of International Conference on Data Engineering* (pp. 315-326).

Petroveskiy, M. I. (2003). Outlier detection algorithms in data mining system. *Journal of Programming and Computer Software, 29*(4), 228–237. doi:10.1023/A:1024974810270

Pokrajac, D., Lazarevic, A., & Latechi, L. J. (2007). Incremental local outlier detection for data streams. In *Proceedings of IEEE Symposium on Computational Intelligence and Data Mining* (pp. 504-515).

Preparata, F., & Shamos, M. (1988). *Computational geometry: An introduction*. New York: Springer-Verlag.

Ramaswamy, S., Rastogi, R., & Shim, K. (2000). Efficient algorithms for mining outliers from large data sets. In *Proceedings of the ACM SIGMOD Conference on Management of Data* (pp. 427-438).

Ren, D., Rahal, I., & Perrizo, W. (2004). A vertical outlier detection algorithm with clusters as by-product. In *Proceedings of International Conference on Tools with Artificial Intelligence* (pp. 22-29).

Rousseeuw, P. J., & Leroy, A. M. (1996). *Robust regression and outlier detection*. John Wiley and Sons.

Ruts, I., & Rousseeuw, P. (1996). Computing depth contours of bivariate point clouds. *Journal of Computational Statistics and data . Analysis, 23,* 153–168.

Schölkopf, B., Platt, J., Shawe-Taylor, J., Smola, A. J., & Williamson, R. C. (2001). Estimating the support of a high dimensional distribution. *Journal of Neural Computation, 13*(7), 1443–1471. doi:10.1162/089976601750264965

Shekhar, S., Lu, C. T., & Zhang, P. (2001). A unified approach to spatial outliers detection. *International Journal of GeoInformatica, 7*(2), 139–166. doi:10.1023/A:1023455925009

Subramaniam, S., Palpanas, T., Papadopoulos, D., Kalogeraki, V., & Gunopulos, D. (2006). Online outlier detection in sensor data using nonparametric models. *International Journal of Very Large Data Bases* (pp. 187-198).

Sun, P. (2006). *Outlier detection in high dimensional, spatial and sequential data sets.* Doctoral dissertation, University of Sydney, Sydney.

Sun, P., & Chawla, S. (2004). On local spatial outliers. In *Proceedings of International Conference on Data Mining* (pp. 209-216).

Sun, P., Chawla, S., & Arunasalam, B. (2006). Mining for outliers in sequential databases. In *Proceedings of SIAM International Conference on Data Mining* (pp. 94-105).

Sykacek, P. (1997) Equivalent error bars for neural network classifiers trained by bayesian inference. In *Proceedings of European Symposium on Artificial Neural Networks* (pp. 121-126).

Tan, P. N. Steinback. M., & Kumar, V. (2005). *Introduction to data mining.* Addison Wesley.

Tax, D. M. J., & Duin, R. P. W. (1999). Support vector domain description. *Official Publication of Pattern Recognition Letters, 20,* 1191–1199. doi:10.1016/S0167-8655(99)00087-2

Tukey, J. (1997). *Exploratory data analysis.* Addison-Wesley.

Valero-Mora, P. M., Young, F. W., & Friendly, M. (2003). Visualizing categorical data in ViSta. *Journal of Computational Statistics & Data Analysis, 43,* 495–508. doi:10.1016/S0167-9473(02)00289-X

Wei, L., Qian, W., Zhou, A., Jin, W., & Yu, J. X. (2003). HOT: hypergraph-based outlier test for categorical data. In *Proceedings of Pacific-Asia Conference on Knowledge Discovery and Data Mining* (pp. 399-410).

Yamanishi, K., & Takeuchi, J. (2006). A unifying framework for detecting outliers and change points from non-stationary time series data. *International Journal of Knowledge and Data Enginnering, 18*(4), 482–492. doi:10.1109/TKDE.2006.1599387

Yu, D., Sheikholeslami, G., & Zhang, A. (2002). Findout: finding outliers in very large datasets. *Journal of Knowledge and Information Systems, 4*(3), 387–412. doi:10.1007/s101150200013

Yu, J. X., Qian, W., Lu, H., & Zhou, A. (2006). Finding centric local outliers in categorical/numerical spaces. *Journal of Knowledge Information System, 9*(3), 309–338. doi:10.1007/s10115-005-0197-6

Zhang, T., Ramakrishnan, R., & Livny, M. (1996). BIRCH: an efficient data clustering method for very large databases. In *Proceedings of the ACM SIGMOD Conference on Management of Data* (pp. 103-114).

Zhang, Y., Meratnia, N., & Havinga, P. J. M. (2007). *A taxonomy framework for unsupervised outlier detection techniques for multi-type data sets.* The Netherlands: University of Twente, Technique Report.

Zhu, C., Kitagawa, H., & Faloutsos, C. (2005). Example-based robust outlier detection in high dimensional datasets. In *Proceedings of International Conference on Data Mining* (pp. 829-832).

Section 3
Clustering Sensor Network Data

Chapter 8
Intelligent Acquisition Techniques for Sensor Network Data

Elena Baralis
Politecnico di Torino, Italy

Tania Cerquitelli
Politecnico di Torino, Italy

Vincenzo D'Elia
Politecnico di Torino, Italy

ABSTRACT

After the metaphor "the sensor network is a database," wireless sensor networks have become an important research topic in the database research community. Sensing technologies have developed new smart wireless devices which integrate sensing, processing, storage and communication capabilities. Smart sensors can programmatically measure physical quantities, perform simple computations, store, receive and transmit data. Querying the network entails the (frequent) acquisition of the appropriate sensor measurements. Since sensors are battery-powered and communication is the main source of power consumption, an important issue in this context is energy saving during data collection. This chapter thoroughly describes different clustering algorithms to efficiently discover spatial and temporal correlation among sensors and sensor readings. Discovered correlations allow the selection of a subset of good quality representatives of the whole network. Rather than directly querying all network nodes, only the representative sensors are queried to reduce the communication, computation and power consumption costs. Experiments with different clustering algorithms show the adaptability and the effectiveness of the proposed approach.

INTRODUCTION

Smart sensors are small-scale mobile devices which integrate sensing, processing, storage and communication capabilities. They can programmatically measure physical quantities, perform simple computations, store, receive and transmit data. The lattice built by a set of cooperating smart sensors is called a sensor network. Because of the ambivalent role of

DOI: 10.4018/978-1-60566-328-9.ch008

each device, which acts simultaneously as a data producer and as a data forwarder, sensor networks provide a powerful infrastructure for large scale monitoring applications (e.g., habitat monitoring (Szewczyk et al., 2004), health care monitoring (Apiletti et al., 2006), condition maintenance in industrial plants and process compliance in food and drug manufacturing (Abadi et al., 2005)).

Querying the network entails the frequent acquisition from sensors of measurements describing the state of the monitored environment. To transmit the required information, sensors consume energy. Since sensors are battery-powered, network querying needs to be driven by three factors: (i) Power management, (ii) limited resources, and (iii) real-time constraints. While CPU overheads are very small (i.e., no significant processing takes place on the nodes), the main contributors to energy cost are communication and data acquisition from sensors (Deshpande et al., 2004). Thus, when querying a sensor network, the challenge is to reduce the data collection cost, in terms of both energy and bandwidth consumption. An important issue in this context is the reduction of energy consumption to maximize the longevity of the network.

SeReNe (Selecting Representatives in a sensor Network) (Baralis et al., 2007) is a framework which provides high quality models for sensor networks to efficiently acquire sensor data. Given sensor readings, the goal of SeReNe is to find and understand the relationships, both in the space and time dimensions, among sensors and sensor readings to select a subset of good quality representatives of the whole network. Rather than directly querying all network nodes, only the representative sensors are queried to reduce the communication, computation and power consumption costs. Many different approaches can be exploited to perform correlation analysis on sensor data. Correlation analysis in SeReNe is performed by means of clustering techniques. Furthermore, since a query optimizer aims to the identification of the execution plan with the least estimated cost, SeReNe may be profitability exploited by it. Given a set

of representatives sensors identified by SeReNe, the schedule that minimizes acquisition cost may be computed, for example, by means of a TSP solver (Li & Kernighan, 1971).

This chapter thoroughly describes different clustering techniques to efficiently discover spatial and temporal correlation among sensors and sensor readings. Clustering analysis has been validated by means of a large set of experiments performed on data collected from 54 sensors deployed in the Intel Berkeley Research lab between February 28th and April 5th, 2004. The experimental results show the effectiveness of sensor clustering in reducing energy consumption during data collection and extend a sensor's lifetime.

SENSOR NETWORK APPLICATIONS

Nowadays wireless sensor networks are being used for a fast-growing number of different application fields, with varying functional and operational requirements. Sensor network applications can be classified into two main classes: Habitat monitoring (Szewczyk et al., 2004) and surveillance applications (He et al., 2004). The habitat-monitoring applications (e.g., environment monitoring, highway traffic monitoring, habitat monitoring (http://www.greatduckisland. net/)) continuously monitor a given environment, while surveillance applications (e.g., health care monitoring, avalanche detection, condition maintenance in industrial plants and process compliance in food and drug manufacturing (Abadi et al., 2005)) alert the control system when a critical event occurs in an hostile environment or context. In the last case the alert needs to be detected with high confidence and as quickly as possible to allow the system to react to the situation. Some sensor network applications are described in the following.

Volcano area monitoring (Werner-Allen et al., 2006). A wireless sensor network has been deployed on Volcano Tungurahua, an active volcano in central Ecuador, to monitor volcano eruptions with low-frequency acoustic sensors (Werner-Allen et al., 2006). However, studying active volcanoes may address two different issues: (i) understanding long-term trends, or (ii) focusing on discrete events such as eruptions, earthquakes, or tremor activities. In both cases high data rates, high data fidelity, and large inter-node separations are required to perform an accurate monitoring. For the last constraint, sensors need to be able to transmit data to a long-distance. Since wireless sensor devices are characterized by low radio bandwidth, a high degree of redundancy is required to reduce the spatial distance between two sensors. Usually, the number of deployed sensors is greater than the number of strictly required sensors. Since the collected measures are highly correlated, an efficiently technique to gather correlated data may be exploited. Furthermore, the network needs to run for an extended period of time to study long-term trends. Hence, a power management technique needs to be exploited to minimize energy consumption and extend sensor lifetime.

Habitat monitoring (Szewczyk et al., 2004). A wireless sensor network has been deployed at the University of California's James Reserve in the San Jacinto Mountains of southern California to monitor habitat environment. This network continuously monitors the environment microclimate below and above ground, and animal presence in different locations. Since the monitored area is within 25 hectare area, a hierarchical network has been exploited to gather temperature, humidity, photo synthetically active radiation, and infrared thermopiles for detecting animal proximity. Furthermore, since the monitored area is densely deployed with sensing devices, collected measures are highly correlated, and to maximize the network's longevity an efficient gathering technique is required. However, since environment monitoring is performed for a long time and data are gathered from many sensors, a significant amount of bandwidth needs to be available

(if the bandwidth is not large, the probability of packet loss increases). Hence, sensors must last a (fairly) long time to reduce the cost of hardware resources and extend the sensor network lifetime. Furthermore, in long term environment sensing deployments, sensors are known to be failure prone (Szewczyk et al., 2004), i.e., nodes do not stop, but rather simply produce erroneous output. Hence, an important issue in this context is erroneous measure identification to avoid transmission and save energy during data collection.

Agricultural production monitoring (Burrel et al., 2004). A sensor network can be exploited in agricultural production (i) to identify the risk of frost damage to vines, (ii) to assess the risk of powdery-mildew outbreak (or to detect pests and irrigation needs), or (iii) to detect the presence of birds. A trial sensor network (it involved 18 motes) has been deployed in a local Oregon vineyard to collect different measures (e.g., temperature, lighting levels, humidity, presence of birds) for several weeks during the summer of 2002 (Burrel et al., 2004). By means of this deployment it has been possible to observe some correlation among sensor readings. There is a great variability across the vineyard during the day and less variation during the night, hence measurements are more correlated during the night and less during the day. Furthermore, there are different season issues (e.g., risk of frost damage to vines). For example, during the winter a wireless sensor network can be exploited in an agricultural production to gather frequent temperature readings and to alert the system only when a risk of frost damage is detected (i.e., temperature is lower than a given threshold). Hence, sensor readings are correlated both in time and space, and a more power-efficient technique would be necessary to efficiently collect the required information, thus extending the sensor network lifetime.

In the previous wireless sensor network scenarios each deployed sensor acquires measurements (e.g., temperature, light, humidity, fire), useful to monitor the physical phenomena, at discrete points. Each measurement is also characterized by a specific time and location of acquisition. Querying the sensor network entails the (frequent) acquisition of the appropriate sensor measurements. Since sensors are battery-powered, energy saving techniques are needed to extend the sensor network lifetime.

COMMUNICATION COST MODEL IN SENSOR NETWORKS

The main contributors to the energy cost are communication and data acquisition. While the data acquisition cost is obtained from the data sheets of sensors, the definition of the communication cost is more complex. It depends on the radio device[1] exploited on the sensor, on the data collection technique exploited to collect sensor data and on the network topology. The cost function becomes stochastic in presence of an unknown topology, or when the topology changes over time. As discussed in (Deshpande et al., 2004), we focus on networks with known topologies and unreliable communication. The unreliability issue is modeled by acknowledgment messages and retransmission. The sensor network can be represented by means of a network graph composed by a set of nodes (i.e., sensors) and a set of edges (Deshpande et al., 2004). Each edge between two nodes is characterized by a weight, which represents the average number of transmissions required to successfully complete the delivery. By considering p_{ij} and p_{ji} as the probabilities that a packet from i will reach j and vice versa, and by assuming that these probabilities are independent, the expected number of transmission and acknowledgment messages required to guarantee a successfully transmission between i and j is $\dfrac{1}{p_{ij} \cdot p_{ji}}$. This value (i.e., the edge weight) can be exploited to estimate the transmission cost required to query the network. The execution plan, also called schedule, is a list

of sensor nodes to be visited to collect sensor data. A plan usually begins and ends at the base station, which is the node that interfaces the query processor to the sensor network.

Given the sensor graph the schedule of sensor queries that minimizes acquisition cost can be easily computed on the network graph by means of a TSP solver (Li & Kernighan, 1971). The TSP solver algorithm selects the schedule which minimizes the communication cost and balances energy consumption. By tracing the minimum spanning tree through the network, energy consumption is not balanced among sensors. In any case, the schedule represents a single path through the network that visits all sensors and returns to the base station. The communication cost is computed by adding the weights of the edges traversed by the schedule.

ENERGY AWARE QUERY PROCESSING

Querying the network entails the (frequent) acquisition of the appropriate sensor measurements. Since sensors are battery-powered and communication is the main source of power consumption, an important issue in this context is energy saving during data collection.

The first approach in answering a sensor network query was based on (i) broadcasting the query to all sensors, (ii) appropriately scheduling measurement transmissions among sensors, and (iii) gathering all sensor answers in order to provide the best possible approximation of the considered phenomenon (Gehrke, 2004; Madden, 2003). Since a large number of sensors is queried, this approach is characterized by high communication cost and energy consumption.

One step further towards more energy-aware query processing was the reduction in the number of transmissions needed to answer a query. To achieve this goal, two different approaches have been proposed. The first one is based on the integration of statistical models of real-world processes into the query processing architecture of the sensor network. Statistical models can be exploited to perform approximate query answering (Chu et al., 2006; Deshpande et al., 2005). In the first phase, a statistical distribution of each considered phenomenon is independently inferred from the complete collection of sensor measurements. When the estimated accuracy is above a given threshold, the generated model is exploited to answer queries. Otherwise, the query is redirected to the network, according to the required accuracy. This approach is efficient. However, it does not work well when the topology of the network is dynamic and is sensitive to the presence of outliers. Other approaches (Chatterjea & Havinga, 2008; Chu et al., 2006; Tulone et al., 2006; Tulone et al., 2006) rely on the construction of simple predictive models which are kept synchronized on the base station and on sensors. When a query is submitted to the network, values are computed by means of the model, instead of collecting them through the network. Sensors transmit data only when the prediction does not satisfy an accuracy threshold. The efficiency of these systems relies on the effectiveness of the forecasting model, which is constrained to the limited resources of the sensors. Sensor memory limits the amount of data which can be used to compute the model and the computation usually requires the presence of a floating point unit (which is not available in all types of devices).

One step further towards query optimization was based on query similarity to support efficient multiple query executions (Trigoni et al., 2005; Xia et al., 2006). Query similarities have been exploited to merge them in a single query, which will be disseminated on the network. These approaches reduce the number of data requests submitted by the base station and the total cost is lower bounded by the cost of a single query. Efficient query execution strategies are still necessary, because this approach queries the entire network.

Correlation Aware Query Processing

Correlation analysis of sensor data may allow the definition of more effective strategies for data aggregation. It may be exploited to enhance bandwidth allocation and save energy consumption. To achieve this goal, different techniques have been proposed.

An effective technique focuses on electing a subset of nodes to represent the network (i.e., a snapshot of the network). To select a subset of relevant nodes, two different approaches have been proposed. The approach proposed in (Kotidis, 2005) picks the representative nodes by performing a continuous comparison among sensor measurements to detect similarities. This solution is implemented by means of a localized algorithm and a threshold value. For the election process, nodes need to exchange a set of messages with their neighbours to elect some of them as representatives of the surrounding environment. This approach can be enhanced by exploiting also temporal correlation among measures (Silberstein et al., 2006). Values are transmitted by a subset of sensors chosen as representatives of their neighbourhood if the value is changed since the last transmission (Silberstein et al., 2006). Both approaches are not able to detect correlation among faraway sensors, because only local similarities are considered.

A different strategy for node selection aims at approximating uniform random sampling (Bash et al. 2004). In this way, correlation among nodes is not analyzed. A second technique (Gupta, 2006) exploits the overlap of the sensing regions originated by each sensor. The choice of an optimal routing path is then mapped to the selection of the minimum number of sensors whose sensing regions cover the whole monitored area. This approach is effective for networks characterized by a high redundancy of sensors. The objective of this technique is a reduction in the redundancy of measures referring to the same location.

The proposed solutions are not able to deal with correlations among sensor measurements of faraway nodes. On the contrary, clustering techniques are capable of grouping together nodes which produce similar measures, regardless of their geographic position.

A parallel effort was devoted to the selection of the best plan to minimize the execution cost (Deshpande et al., 2004; Deshpande et al. 2005). Correlation among attributes is considered to identify the appropriate plan. If two attributes are correlated, the execution plan always considers the attribute whose acquisition cost is less. However, this approach queries the entire network.

In general, building a mathematical model to discover correlation on sensor data is an expensive computational task. As discussed in (Zhu et al., 2008) the coexistence of multiple sources causes the mathematical representation to be difficult, even in the case of a simple correlation model (Cristescu et al., 2004). Furthermore, finding the nearest neighbour (i.e., the most similar object) of an object may require computing the pair wise distance between all points. On the contrary, clustering algorithms can be much more efficient (Tan et al., 2005) in discovering the most correlated sensors. These algorithms provide a more scalable approach when a huge number of nodes is involved.

CLUSTERING SENSOR DATA

Clustering aims at grouping data into classes or clusters, so that objects within the same cluster are very similar, and objects in different clusters are very dissimilar. Many clustering algorithms have been proposed in literature. General features which characterize each approach are (i) the amount of domain knowledge to correctly set the input parameters, (ii) the ability to deal with clusters of different shapes, (iii) the ability to deal with noisy data, i.e. the sensitivity to outliers, missing or erroneous data. Different grouping strategies may

lead to a different organization of the network, and, consequently, to significant differences in power consumption and measure accuracy. Hence, we consider and compare four different algorithms to perform sensor data clustering. The selected algorithms can be classified into the following three categories: (i) Partitioning methods, (ii) density-based methods, and (iii) model-based methods. Partitioning and density-based methods require the definition of a metric to compute distances between objects in the dataset. In the SeReNe framework, distances between objects are measured by means of the Euclidean distance computed on normalized data. *Partitioning methods* subdivide a dataset of n objects into k disjoint partitions, where $k < n$. The general criterion to perform partitioning assigns objects to the same cluster when they are close, and to different clusters when they are far apart with respect to a particular metric. Partitioning methods are able to find only spherical-shaped clusters, unless the clusters are well separated, and are sensitive to the presence of outliers. K-Means (Juang & Rabiner, 1990) is a popular method which belongs to this category. *Density-based methods* are designed to deal with non-spherical shaped clusters and to be less sensitive to the presence of outliers. The objective of these methods is to identify portions of the data space characterized by a high density of objects. Density is defined as the number of objects which are in a particular area of the n-dimensional space. The general strategy is to explore the data space by growing existing clusters as long as the number of objects in their neighborhood is above a given threshold. DBSCAN (Ester et al., 1996) is the density-based method considered in the SeReNe framework. *Model-based methods* hypothesize a mathematical model for each cluster, and then analyze the data set to determine the best fit between the model and the data. These algorithms are able to correctly take into account outliers and noise by making use of standard statistical techniques. Expectation-maximization (EM) (McLachlan & Krishnan, 1997) is an algorithm which performs statistical modeling assuming a gaussian mixture distribution of the data. COBWEB (Fisher,1987) is a popular algorithm in the machine learning community which performs probability analysis of the data. An overview of the four algorithms is provided in the following.

K-Means (Juang & Rabiner, 1990) requires as input parameter *k*, the number of partitions in which the dataset should be divided. It represents each cluster with the mean value of the objects it aggregates, called centroid. The algorithm is based on an iterative procedure, preceded by a set-up phase, where *k* objects of the dataset are randomly chosen as the initial centroids. Each iteration performs two steps. In the first step, each object is assigned to the cluster whose centroid is the nearest to that object. In the second step centroids are relocated, by computing the mean of the objects within each cluster. Iterations continue until the *k* centroids do not change. K-means is effective for spherical-shaped clusters. Different cluster shapes are detected only if the clusters are well separated. Similar to other partitioning methods, K-Means is sensitive to outliers and requires the a-priori knowledge of the number of clusters.

DBSCAN (Ester et al., 1996) requires two input parameters, a real number *r*, and an integer number *minPts,* used to define a density threshold in the data space. A high density area in the data space is an n-dimensional sphere with radius *r* which contains at least *minPts* objects. DBSCAN is an iterative algorithm which iterates over the objects in the dataset, analyzing their neighborhood. If there are more than *minPts* objects whose distance from the considered object is less than *r*, then the object and its neighborhood originate a new cluster. DBSCAN is effective at finding clusters with arbitrary shape, and it is capable of identifying outliers as a low density area in the data space. The effectiveness of the algorithm is strongly affected by the setting of parameters *r* and *minPts.*

The *Expectation-Maximization* (McLachlan & Krishnan, 1997) algorithm is a general iterative procedure used in statistics to find maximum likelihood estimates of unknown parameters for which a likelihood function can be built. At the beginning, parameters are assigned with random values. Then, the algorithm iterates recomputing them until a convergence threshold is reached. Each iteration of the algorithm is composed by two phases. The first is the expectation (E) phase, where the parameters computed in the former iteration are used to update the expectation of the likelihood function. The second step is the maximization (M) phase, where the expected likelihood function determined in the E phase is maximized to determine new estimates of the unknown parameters. The found parameters are then used as the input for a new iteration of the algorithm. Consider a set of vectors, where each vector is interpreted as a sample picked by one of N gaussian distributions, i.e. the set of vectors is said to belong to a gaussian mixture. The EM algorithm may be exploited to group together vectors originated by the same distribution. It works by estimating (i) the mean and the standard deviation proper of each distribution (cluster), (ii) the sampling probability of each cluster, i.e. the probability that one of the N gaussian distributions is used as a source of data.

The *COBWEB* (Fisher, 1987) algorithm clusters a dataset in the form of a classification tree. A classification tree is a tree structure where the root node represents the whole dataset, the leaves represent single objects and internal nodes represent clusters. Each cluster is characterized by a probabilistic description. The tree is built incrementally with a single reading of the dataset. When new data is available, it is temporally added to each cluster to compute a metric called *category utility.* This measure evaluates the similarity of the data belonging to the same cluster and the dissimilarity among data belongs to other clusters. New data is expected to improve the overall *category utility.*

To achieve this goal four actions can take place: (i) New data is added to an existing class, (ii) a new cluster is created, (iii) an existing cluster is split, and (iv) two existing clusters are merged.

The previous algorithms require that the data are stored in a single repository. Hence, the training phase involves a high number of transmissions from the sensors to the sink, to collect enough data to partition the network. Furthermore, huge amounts of data require significant memory and good processing capabilities. To cope with data source distribution and to scale up data mining techniques, distributed data mining algorithms have been proposed in literature (Park & Kargupta, 2003). Generally, many algorithms for distributed data mining are based on algorithms which were originally developed for parallel data mining. However, the parallel versions of DBSCAN (Xu et al., 1999) and K-Means (Dhillon et al., 1999) still require a preprocessing step in which data is collected in a centralized storage unit.

Distributed clustering (Januzaj et al., 2003) is able to overcome this restriction by exploiting computation and memory resources which are disseminated across the network. Distributed clustering assumes that the objects to cluster are geographically distributed. In a first step, each local site independently computes a cluster model for its local data. Then, the central site tries to determine a global clustering model by exploiting local models. This step is generally difficult, since it is hard to cope with correlation among objects which are not located in the same site. Since the number of transmissions and the amount of data transmitted are reduced, this technique may decrease the energy consumption of the network during the model building phase.

THE SERENE FRAMEWORK

The SeReNe (Selecting Representatives in a sensor Network) framework allows the efficient acquisition of sensor data by means of intelligent techniques. The goal of SeReNe is to find and understand the relationships, both in the space and time dimensions, among sensors and sensor readings to select a subset of good quality representatives of the whole network. Rather than directly querying all network nodes, only representative sensors are queried to reduce the communication, computation and power consumption costs. Hence, the SeReNe framework is particularly suitable for monitoring applications.

Figure 1 shows the building blocks of the SeReNe framework. Given sensor readings, a sensor network model is created by means of two steps: (i) Correlation analysis and (ii) selection of representative sensors. Sensors can be physically (i.e., spatially) correlated even if they are not located nearby (e.g., a sensor, localized in (x_1, y_1) in area A, is correlated with a sensor in (x_2, y_2) in area B). Moreover, sensor readings may be correlated in time (e.g., the average temperature in a room at a given time during the day is correlated with the same measurement performed by the same sensor at the same time in different days). The correlation analysis block, based on clustering algorithms, discovers which sensors/measurements are correlated and when, and how strong is the correlation. To perform this step, during a short time window (which is application-dependent), data is gathered from all sensors. The more sensor data are analyzed, the more accurate the model will be. By means of clustering algorithms, correlated sensors/sensor readings are grouped in clusters, each of which can be represented by few sensor representatives, called R-Sensors, singled out in the second step. A selection algorithm selects the subset of sensors which (i) best represent the network state and (ii) correspond to the minimum communication cost. Finally, R-Sensors are exploited to efficiently query the sensor network. Each time a query is executed over the network, only R-Sensors are queried. The plan generation and optimization block of the SeReNe framework

Figure 1. SeReNe framework architecture

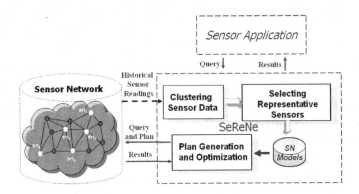

(see Figure 1) generates an energy saving transmission schedule among sensors which minimizes communication cost.

A query optimizer identifies the execution plan with the least estimated cost. Hence, SeReNe can be profitability exploited to this aim. Given a set of representatives sensors identified by SeReNe, the schedule among sensors that minimizes acquisition cost may be computed by means of a TSP solver (Li & Kernighan, 1971) running on the network graph represented by representative sensors. The schedule represents a single path through the network that visits all representative sensors and returns to the base station.

Since only R-Sensors are queried, the ability of learning the current state of the network is lost. A selective transmission strategy, running on smart sensors (Deshpande et al., 2005), allows acquiring only relevant changes in the network state (e.g., new nodes are added to the network) and dynamically adapting the sensor network model to the changes.

Definitions and Notation

In the following we provide the definition of some base concepts.

- **Definition 1.** *Measure.* A measure is an estimation of a physical variable expressed in a unit of measurement. An example of

measure is the temperature, which is usually expressed in degree Celsius.
- **Definition 2.** *Sensor node.* Let $M = \{M_1,\dots, M_k\}$ be a set of measures and f the reading frequency. A sensor is able to sense measures in M every $t = \frac{1}{f}$ time unit, also called epoch. Mica2Dot[2] is an example of sensor node.
- **Definition 3.** *Time band.* Let T be a sequence of time units and s a sensor node able to sense measures every $t \in T$. A time band is a continuous subset of T.
- **Definition 4.** *Time series.* Let T be a sequence of time units and s a sensor node able to sense a measure m. A time series is a sequence of observed values for m sampled by s every $t \in T$. The length of the time series is $|T|$.
- **Definition 5.** *Sensor reading.* Let s be a sensor node able to sense k measures. A sensor reading is an array of k values observed by s in time unit t.
- **Definition 6.** *Historical sensor readings.* Let S be a set of sensor nodes able to sense k measures, and T_{window} a time band. $\forall s \in S$ $\forall t \in T_{window}$ sensor readings are collected. Historical sensor readings are represented by means of a matrix H whose cardinality is $|T_{window}| \times |S| \times k$. Each value *(t, s, m)* in H is the observed value of measure

m, performed by sensor $s \in S$ at time $t \in T_{window}$.

Given historical sensor readings, collected in T_{window}, and an error bound τ, correlation both in space and time dimensions among sensors and sensor readings are analyzed. The objective is to single out a subset of sensors, denotes as R-Sensors, which best represent the network state in a validity window, denoted as T_{model}. $T_{model} \subseteq T_{window}$ is a time band in which representative sensors approximate the network within τ.

Correlation Analysis

Given historical sensor readings, we analyze two types of correlation: Physical correlation and time correlation. *Physical correlation* between two sensors depends on the similarity of the environment where the sensors are located. The following are both examples of physical correlations. (i) Two sensors located nearby sense similar values (e.g., sensors s_1 in room *1* and s_2 in room *2*, both at the second floor, sense the same value of temperature from 10 a.m. to 11 a.m. since they are in the shade and nobody is present). (ii) Far sensors, located in similar environments, sense correlated measurements (e.g., sensors s_1 in room *1* in building *A* and s_2 in room 2 in building *B* sense the same value of temperature from 12 a.m. until 1 p.m. since they are in the sun and many people are having lunch in these rooms). Sensor readings may be correlated *over time*. Two different cases need to be analyzed: (i) correlated phenomena and (ii) correlated measurements of the same environmental parameter. In the first case we may discover that two phenomena follow a similar pattern evolution. Hence, if we know the relationship between these parameters, it is not necessary to query the sensor network for both. In the second case we can find the variation pattern of the measurement (e.g., every hour), thus decreasing query frequency.

Since many phenomena can be characterized by a different correlation grade during the day,

the network state is analyzed at different times during the day to detect the best model for each time window (e.g., a given cluster set represents the network in the morning, a different grouping of sensors in the evening, and yet another one during the night). Furthermore, some environmental parameters could be correlated with different strength during different parts of the day (e.g., more correlated during the night and less during the day).

The probability theory (Chu et al., 2006; Deshpande et al., 2005; Deshpande et al., 2004) has been exploited to study correlation in sensor data. In the SeReNe framework, correlation analysis is based on clustering techniques. A cluster representation of the network is able to adapt appropriately to network topology changes or to the changed behavior of a node (see the Section Evolving the model).

The SeReNe framework focuses on clustering algorithms to study correlation analysis. Different clusterization sessions are performed to characterize sensor correlation appropriately. Each session considers a different set of measure combinations (e.g., temperature, humidity).

For each session two analyses are performed: (i) correlation over time and (ii) physical correlation. Clustering algorithms are exploited to perform both analyses. During the first phase, time series, collected by each sensor of the network, are clustered. For each sensor, the clustering algorithm returns a set of correlated sensor values. By plotting the clusterization results on a monodimensional diagram where the x-coordinate represents time evolution, a set of cyclic time bands is detected.

Since the network is composed by many sensors, clusterization results, obtained from all time series analyzed separately, need to be merged. Overlapped time bands are grouped together. For each group the largest time band is exploited to set the appropriate validity window (i.e., T_{model}). For each time band a sensor network model needs to be built. Hence, physical correlation is analyzed separately for each time band.

For the physical correlation analysis, a clusterization session is performed separately for each time unit in the time band. Each session analyzes the values observed by all sensors at a given time point and yields a set of sensor clusters. Each cluster set is evaluated by computing the overall cluster validity by means of the cohesion function, which is computed as the sum of the cohesion of individual clusters. The cluster set that maximizes the overall cohesion function will exploited for building the sensor network model in the corresponding time band. The selection of representative sensors is performed on this cluster set.

Given a time band, correlation analysis is performed separately for each time unit. Each clusterization session analyzes the values observed by all sensors at a given time unit and yields a set of sensor clusters. Thus, correlated measures collected at the same time unit by different sensors are clustered in the same group, independently of the spatial position of the sensing device. Each cluster set is evaluated by computing the overall cluster validity (Tan et al., 2006) by means of the cohesion function, which is computed as the sum of the cohesion of individual clusters. The cluster set that maximizes the overall cohesion function will exploited for building the sensor network model in the corresponding time band. The selection of representative sensors is performed on this cluster set.

Selection of Representative Sensors

This step consists in selecting from a set of correlated sensors a subset of representative sensors, denoted as R-Sensors. The subset may contain one or more sensors for each sensor group according to the required model accuracy (i.e., the error bound τ). The number n of representatives is set according to the required model accuracy. The number of representatives in each cluster is proportional to the number of cluster points. Reliable outliers are included in the R-Sensors.

Given a cluster of sensors, each of which senses k measures, we exploit the *Measure trend* selection strategy (Baralis et al., 2007) to single out the subset of sensors that better model correlated phenomena. This selection technique is based on the analysis of correlated phenomena. To represent both physical and temporal correlation among sensors and sensor readings, we consider (a) the physical clusterization in a given sampling time and (b) measurements collected during the considered time band. The best approximation of phenomenon i over the time band is represented by the average of the values collected by all sensors during the considered time band, denoted as $\overline{M_i}$. Let $\overline{O} = \left(\overline{M_1}, \ldots, \overline{M_k} \right)$, where k is the number of considered measures. For each sensor s_i we compute the average of each measure $\overline{M\left(s_i\right)} = \left(\overline{M_{1i}}, \ldots, \overline{M_{ki}} \right)$ over the time band. Representative sensors are the n nodes s_i nearest to $\overline{O}$ that correspond to the minimum communication cost. Distance is measured by

$$d\left(\overline{M\left(s_i\right)}, \overline{O}\right) = \sqrt{\left(\overline{M_{1i}} - \overline{M_1}\right)^2 + \ldots + \left(\overline{M_{ki}} - \overline{M_k}\right)^2}.$$

The communication cost to transmit data from s_i to s_j is computed by multiplying the energy required to send a data packet by the expected number of transmission and acknowledgement messages required to guarantee a successful transmission. In particular, to select R-Sensors, all sensor nodes are sorted according to the distance from $\overline{O}$. Hence, more accurate sensors appear at the head of the list. The top n sensors which minimize the communication cost are selected as R-Sensors. In particular, we scan the list in order and pick the sensors for which the communication cost is minimum, for every single sensor. If there are less than n sensors having minimum communication cost, we scan again the list and include sensors having the second larger communication cost, and so on, until n sensors have been found.

After selecting the representatives of a network state, the temporal interval in which these representatives approximate the network within the error bound τ has to be defined. The largest subset of contiguous sampling times in which the model provides a good approximation of a given phenomenon is identified. This time band is denoted as T_{model}. At first, we estimate the approximate value Mr_j of a measure j as the average on values collected by all representative sensors. The best approximation $\overline{M_j}$ is the average on the values gathered by querying all sensors. T_{model} is the largest subset of contiguous sampling times in which $\left| \overline{Mr_j} - \overline{M_j} \right| \leq \tau$.

Evolving the Model

Since only representative sensors are queried to monitor a given environment, two issues need to be considered: (a) Nodes which do not belong to the representative sensors do not send the collected measurement. Hence, the ability of learning the current state of the network is lost and the network model cannot be reliably adapted to changing patterns. To cope with this issue two different strategies can be exploited. Either all measures are periodically collected from all sensors, or smart sensors may be deployed. The first strategy is rather expensive, while the second is more efficient. For example, smart sensors (Deshpande et al., 2005) are able to exploit a selective transmission strategy to send their measures either when they are queried, or when the current measure exceeds a given (model dependent) threshold (locally stored on the sensor device). (b) The network topology may change over time because of three event types: (i) a new sensor is added to the network, (ii) a sensor is turned off, and (iii) new measures, gathered from a sensor, are very different from the previous ones. In this case, the model needs to be adapted to the new configuration. To cope with this issue we can either consider an incremental clustering algorithm to (incrementally) evolve the cluster model or

perform a new clusterization session, followed by a new representative selection.

Incremental clustering algorithms are able to update the clustering structure after insertion and/or deletion of new objects.

Among the considered algorithms, COBWEB is an incremental algorithm which processes data in a single scan. In general, single scan algorithms are able to change the clustering structure concurrently with the collection of new measures. However, these algorithms are only able to deal with the addition of new objects. In the case of a node failure, a portion of the network may become unreachable from the base station. Hence, we believe that, while single scan techniques may be adopted for less disruptive network changes, real fault tolerance is guaranteed only by a new clusterization session, followed by a new representative selection.

An incremental version of DBSCAN has been proposed in literature (Ester, 1998). This algorithm is able to deal with both insertion and deletion of new objects, and it can be proven that it yields the same results as the non-incremental DBSCAN algorithm. The integration of this approach in the SeReNe framework seems rather simple. Its use may substantially reduce the computational cost required to build a sensor network model in a dynamic situation.

EXPERIMENTAL RESULTS

We evaluated the performance of different clusterization techniques by means of a large set of experiments which analyze (i) the effectiveness in detecting sensor correlation over time and physical correlation, (ii) the support to selection algorithms in electing representative sensors, and (iii) the effectiveness in reducing dissipated energy. Correlation analysis was performed by means of different clustering algorithms such as DBSCAN (Ester et al., 1996), EM (McLachlan &

Krishnan, 1997), COBWEB (Fisher, 1987), and K-Means (Juang & Rabiner, 1990), all available in the machine-learning open-source environment WEKA (http://www.cs.waikato.ac.nz/ml/weka/).

Experimental Setting

Historical sensor readings are collected from 54 sensors deployed in the Intel Berkeley Research lab between February 28th and April 5th, 2004 (available at http://berkeley.intel-research.net/labdata/). The considered dataset contains 2.3 million sensor readings. Mica2Dot sensors collect temperature, humidity, light, and voltage values once every 31 seconds (epoch) by means of TinyDB in-network query processing system (Madden et al., 2003), built on the TinyOS platform (http://www.tinyos.net/). We also know the x and y coordinates of sensors expressed in meters. Experiments have been performed on an AMD Sempron(tm) 2400+ PC with 1666 MHz CPU and 512 Mb main memory, Linux operating system and WEKA version 3.5.2.

The analysis of historical sensor readings is preceded by a preprocessing phase, which aims at smoothing the effect of possibly unreliable measures performed by sensors. Preprocessing entails the following steps: 1) outlier detection and removal, and 2) standardization. Faulty sensors may provide unacceptable values for the considered measures. Anomalous values can be caused by transmission failure, faulty sensor readings, obstruction of sensors devices, low battery levels. We removed data outside the validity range for each measure (e.g., humidity<0 or humidity>100) and entire sensor readings when at least two measures are unacceptable. After this preprocessing step the dataset contains 1.7 million sensor readings. Several approaches have been proposed to deal with missing data, including non-parametric Expectation Maximization techniques (Davidson & Ravi, 2005) and association rules mining, to choose the most likely value to replace errone-

ous data (Jiang, 2007). These techniques could be easily integrated in the SeReNe framework to enhance its capability to handle missing data. The availability of a higher number of reliable values could result in an enhanced capability of detecting correlation over time and space.

Furthermore, for each sensed measure, values are normalized in the [0,1] interval. Different normalization processes are proposed in literature (Han & Kamber, 2006).

Correlation over Time

The analysis of historical sensor readings over time has been performed by considering separately each time series collected by each sensor. We performed the analysis for every combination of the collected measures (e.g., temperature, humidity and light). Hence, different clusterization sessions have been performed for each sensor.

For each session, the clustering algorithm returns a set of groups. Each group is a set of correlated sensor values. By plotting the clusterization results on a monodimensional diagram where the x-coordinate represents time evolution, we identified two/three cyclic time bands, which always correspond either to daytime, or to night time. The night band is shorter than the day band, and there is possibly a shorter time band between them. Overlapped time bands, identified by different clusterization sessions, are grouped together. For each group, the largest time band is considered to define the validity window (i.e., T_{model}). Hence, the largest time band for the night time and the largest time band for the daytime contribute to the corresponding ranges for T_{model}.

After a large set of experiments, the appropriate values of the clustering algorithm parameters have been devised. In particular, DBSCAN parameters are Epsilon in the range 0.08-0.1 and minPoints adapted to the sensor type. For COBWEB the Category utility is in the range 0.19-0.21, while for EM and K-Means the Cluster number is 2-3.

Figure 2. Spatial representation of physically correlated sensor clusters (4984 epoch)

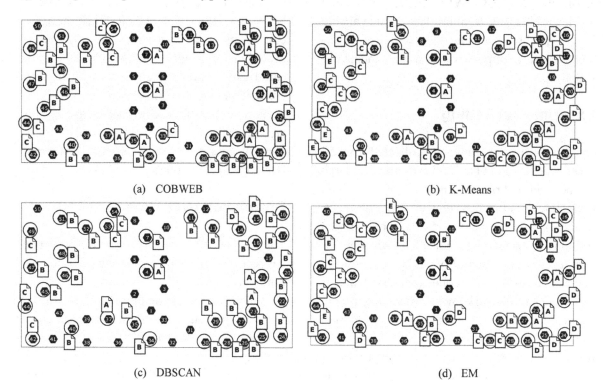

(a) COBWEB

(b) K-Means

(c) DBSCAN

(d) EM

Physical Correlation

To study physical correlation, for each time band, we have run separately experiments for each epoch. More specifically, each clusterization session analyzes all measures collected from all sensors at a given time point. Experiments performed on the day time band highlighted the following general trends. (i) During weekdays sensor readings are grouped in a single cluster. This effect may be due to air-conditioning in the lab. (ii) During holidays, 4 or 5 clusters have been identified, depending on the epoch. Experiments performed on the night time band highlighted 3 or 4 clusters according to the considered epoch.

Furthermore, the physical correlation analysis also addresses correlation among measures. We performed different clusterization sessions, each one considering a different set of measures. Some measures are strongly correlated (e.g., tempera-

ture and light). For example, when considering the temperature, light and humidity measures, the clusters are rather fragmented and the noise percentage is high, while this is not the case for temperature and light.

Figure 2 graphically shows the clusters of temperature and light measures, collected on Sunday, February 29nd 2004, around 5:35 p.m (4984 epoch), obtained with different clustering algorithms. For the DBSCAN algorithm, we set Epsilon to 0.08 and minPoints to 6%, for COBWEB we set acuity to 0.19, while for EM and K-Means we set the cluster number to 5. Four or five sub-areas can be identified inside the lab in which sensor readings are strongly correlated

To measure the agreement between the clustering results shown in Figure 2, we computed the Rand index. The Rand Index (Rand, 1971) computes the number of pair wise agreements between two partitions of a set. Hence, it may

Table 1. Rand Index between clustering results

	COBWEB	K-Means	EM	DBSCAN
COBWEB	-	0.90	0.89	0.90
K-Means	0.90	-	0.84	0.99
EM	0.89	0.84	-	0.83
DBSCAN	0.90	0.99	0.83	-

be exploited to provide a measure of similarity between the cluster sets obtained by two different clustering techniques (see Table 1).

Let O be a set of n objects, and X and Y two different partitions of set O to be compared. The Rand Index R is computed as

$$R = \frac{a+b}{\binom{n}{2}}$$

where a denotes the number of pairs of elements in O which are in the same cluster both in X and Y, and b denotes the number of pairs of elements in O which do not belong to the same cluster neither in X nor in Y. Therefore, the term $a + b$ is the number of pair wise agreements of X and Y, while $\binom{n}{2}$ is the number of different pairs of elements which can be extracted from O. The Rand Index ranges from 0 to 1, where 0 indicates that the two partitions do not agree for any data pair, and 1 indicates that the two partitions are equivalent.

Error! Reference source not found. shows the Rand Index value given by the pair wise comparison of the cluster sets shown in Figure 2. The following considerations hold. The partitions produced by K-Means and DBSCAN are almost equivalent. The EM clustering algorithm produces the most peculiar results, because it is characterized by the lowest Rand Index values when compared with the other techniques. On the contrary, COBWEB is characterized by a high level of agreement when compared with all the other techniques.

Sensor Network Model Validation

In the following we analyze the quality of the four clustering algorithms discussed in the Section Clustering sensor data. To build a sensor network model, we use as training data sensor readings collected during the day of February 29nd, 2004 in 12 hours of monitoring. Correlation, both in the time and space dimensions, among sensor and sensor readings has been studied by means of different clustering algorithms (i.e., DBSCAN, COBWEB, K-Means, EM). Analysis based on correlation over time identified a single time band which corresponds to the daytime. Representatives for the daytime sensor network model are selected from physically correlated sensor clusters related to the 4984 epoch. To select representative sensors by means of the measure trend strategy, we consider both the temperature and light measures together. The best epoch window T_{model} ranges from 4684 to 5495 epochs (i.e., 6 hours and 46 minutes, from 3:59 p.m. to 10:45 p.m.). When not otherwise specified, the number of representatives has been set to 50% of the network sensors. The generated sensor network model provides information for queries on both measures (i.e., temperature and light) either independently or jointly.

To estimate the error introduced by our model, we computed the mean square error (MSE) of our model in T_{model}, given by $\frac{1}{|T_{model}|} \sum_{t \in T_{model}} \left(\overline{M_{rt}} - \overline{M_t} \right)^2$.

Figure 3. Mean square error

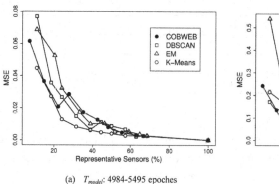 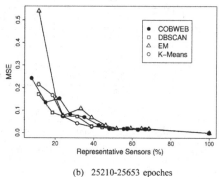

(a) T_{model}: 4984-5495 epoches   (b) 25210-25653 epoches

$\overline{M_{rt}}$ is the average measure computed by querying the representatives in a given epoch t and $\overline{M_t}$ is obtained by querying the whole network in the same epoch t^3. Figure 3 (a) shows the mean square error for each clustering algorithm (computed in T_{model}) when varying the percentage of representative sensors. For a low percentage (i.e., smaller than 40%) of representative sensors, COBWEB and K-Means provide more accurate models. However, when the percentage of representative sensors increases (i.e., greater than 45%) all algorithms provide models with similar accuracy.

To validate the effectiveness of our model, we applied it to querying the network during the same temporal interval of the following holidays. Representatives are queried in each epoch included in the time frame corresponding to T_{model}. Figure 3 (b) shows the mean square error for each clustering algorithm in the next holiday (March 7th 2004, 25210-25653 epoches), by varying the percentage of selected representatives. The mean square error is comparable to the value obtained on training data.

Figure 4 (a) shows the relative error distribution obtained by querying only the R-Sensors selected by means of each clustering algorithm with respect to querying the whole network. The relative error distribution has been computed by considering the

values collected during T_{model}. The four boxplots have close median values (i.e., approximately zero). The inter-quartile range for the COBWEB algorithm is the lowest, while for DBSCAN it is the highest. The error distribution variability of EM is very close to COBWEB. Hence, both algorithms provide good clusterization results. Furthermore, the extreme values of the relative error for any clustering algorithm is always smaller than sensor accuracy (i.e., 0.2^4).

Figure 4 (b) shows the relative error distribution obtained when the sensor network model is exploited to query the network in the following holidays. The relative error distribution, computed by considering values collected during the next holiday (March 7th 2004, 25210-25653 epoches), is comparable to the value obtained on training data. The inter-quartile range for the DBSCAN and EM algorithms is smaller than that of the COBWEB and K-Means algorithms. Since model-based algorithms (i.e., COBWEB and EM) search the best fit between the clusterization result and the data on which clustering is performed, their models are more accurate on training data than on test data. DBSCAN does not strongly adapt to the training data distribution. Hence, its model may become more accurate than other approaches on test data. Furthermore, the DBSCAN algorithm is rather robust to outliers with respect to, e.g., K-

Figure 4. Relative error distribution

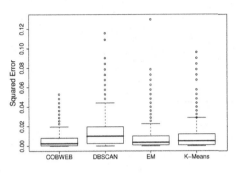 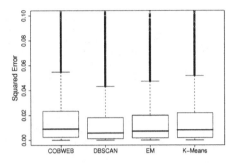

(a) T_{model}: 4984-5495 epoches   (b) 25210-25653 epoches

Means. We finally observe that the extreme values of relative error for all clustering algorithms are anyway always smaller than sensor accuracy.

Finally, Figure 5 shows the energy dissipated by querying only representative sensors compared with querying the whole network. It reports the energy dissipated by querying the network only for the temperature measure. Experiments have been performed by varying the percentage of selected representatives. For percentages of representative sensors larger than 50%, any clustering algorithm provides a model which is effective in reducing energy consumption during sensor network querying. Even if COBWEB is the clustering algorithm which provides the most accurate model on training data, for different ratio of representative sensors its model is less effective in reducing energy consumption. When considering jointly both model accuracy and energy consumption, the EM and DBSCAN algorithms seem to provide better quality sensor network models.

CONCLUSION

In this chapter we have discussed different clustering algorithms to analyze the correlation, both in the space and time dimensions, among sensors and sensor readings. Different algorithms (i.e., DBSCAN (Ester et al. 1996), EM (McLachlan & Krishnan, 1997), COBWEB (Fisher, 1987), and K-Means (Juang & Rabiner, 1990)) have been integrated in the SeReNe framework to generate good quality network models. These models may be exploited for querying sensor networks, thus reducing the data collection cost, in terms of both energy and bandwidth consumption. The effectiveness of the clustering algorithms in discovering sensor correlations has been validated on data collected by sensors located inside the Intel Berkeley Research lab. The experimental results show the effectiveness of the proposed approach in discovering both temporal and spatial correlation among sensors and sensor readings. The relative error affecting the result of query computation on the sensor representatives is often smaller than sensor accuracy and thus negligible. A good trade-off between model accuracy and energy consumption is provided by the EM and DBSCAN algorithms.

The proposed approach can be extended in a number of directions. This work was focused on efficient techniques for performing data collection. Representative sensors may also be exploited for network query optimization. Furthermore, to extend a network's lifetime, efficient turnover techniques for representative sensors should be exploited. Conventional scheduling policies such as Round Robin algorithm or LRU policies may be exploited to this purpose. Turnover techniques

Figure 5. Energy dissipation

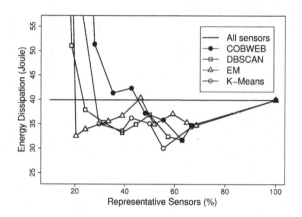

should be designed to balance the dissipated energy among different sensors in each clusters.

REFERENCES

Abadi, D. J., Madden, S., & Lindner, W. (2005). Reed: robust, efficient filtering and event detection in sensor networks. In *Proceedings of the 31st international conference on Very large data bases* (pp. 769-780). VLDB Endowment.

Apiletti, D., Baralis, E., & Bruno, G. G., & Cerquitelli, T. (2006). IGUANA: Individuation of Global Unsafe ANomalies and Alarm activation. In *3rd International IEEE Conference on Intelligent Systems,* (pp. 267-272). IEEE Press.

Baralis, E., Cerquitelli, T., & D'Elia, V. (2007). Modeling a sensor network by means of clustering. In *18th International Conference on Database and Expert Systems Applications*, (pp. 177-181). IEEE Press.

Bash, B. A., Byers, J. W., & Considine, J. (2004). Approximately uniform random sampling in sensor networks. In *DMSN* (pp. 32-39). ACM Press.

Burrell, J., Brooke, T., & Beckwith, R. (2004). Vineyard computing: Sensor networks in agricultural production. *IEEE Pervasive Computing / IEEE Computer Society [and] IEEE Communications Society, 1*(3), 38–45. doi:10.1109/MPRV.2004.1269130

Chatterjea, S., & Havinga, P. (2008). An adaptive and autonomous sensor sampling frequency control scheme for energy-efficient data acquisition in wireless sensor networks. *4th IEEE International Conference on Distributed Computing in Sensor Systems,* DCOSS 2008.

Chu, D., Deshpande, A., Hellerstein, J., & Hong, W. (2006). Approximate data collection in sensor networks using probabilistic models. In *Proc. of the 2006 Intl. Conf. on Data Engineering.*

Cristescu, R., Beferull-Lozano, B., & Vetterli, M. (2004). On network correlated data gathering. In *Proc. IEEE INFOCOM 2004.*

Davidson, I., & Ravi, S. S. (2005). Distributed pre-processing of data on networks of Berkeley motes using non-parametric EM. In *Proc. of SIAM SDM Workshop on Data Mining in Sensor Networks* (pp. 17-27).

Deshpande, A., Guestrin, C., & Madden, S. (2005). Using probabilistic models for data management in acquisitional environments. In *Proc. Biennial Conf. on Innovative Data Sys. Res* (pp. 317-328).

Deshpande, A., Guestrin, C., Madden, S., Hellerstein, J., & Hong, W. (2004). Model-driven data acquisition in sensor networks. In *VLDB* (pp. 588-599).

Deshpande, A., Guestrin, C., Madden, S., & Hong, W. (2005). Exploiting correlated attributes in acquisitional query processing. In *ICDE*, 2005.

Dhillon, I. S., & Modha, D. S. (1999). A data-clustering algorithm on distributed memory multiprocessors. *Large-scale parallel data mining* (pp. 245-260). Springer.

Ester, M., Kriegel, H. P., Sander, J., Wimmer, H. P., & Xu, X. (1998). Incremental Clustering for Mining in a Data Warehousing Environment. In *Proceedings of the 24rd International Conference on Very Large Data Bases* (pp. 323-333).

Ester, M., Kriegel, H. P., Sander, J., & Xu, X. (1996). A density-based algorithm for discovering clusters in large spatial databases with noise. In *Proceedings of the 2nd International Conference on Knowledge Discovery and Data Mining* (pp. 226-231).

Fisher, D. H. (1987). Knowledge acquisition via incremental conceptual clustering. *Machine Learning, 2*(2), 139–172.

Gehrke, J., & Madden, S. (2004). Query processing in sensor networks. *IEEE Pervasive Computing / IEEE Computer Society [and] IEEE Communications Society, 3*(1), 46–55. doi:10.1109/MPRV.2004.1269131

Gupta, H., Zhou, Z., Das, S.R., & Gu, Q. (2006). Connected sensor cover: Self-organization of sensor networks for efficient query execution. *IEEE/ACM Transactions on Networking, 14*(1), 55-67.

Han, J., & Kamber, M. (2006). *Data mining: Concepts and techniques*. The Morgan Kaufmann Series in Data Management Systems, Morgan Kaufmann Publishers.

He, T., Krishnamurthy, S., Stankovic, J., Abdelzaher, T., Luo, L., Stoleru, R., et al. (2004). Energy-efficient surveillance system using wireless sensor networks. *2nd International conference on Mobile systems, applications, and services* (pp. 270-283).

Januzaj, E., Kriegel, H. P., & Pfeifle, M. (2003). Towards effective and efficient distributed clustering. *Workshop on Clustering Large Data Sets (ICDM2003)*.

Jiang, N., (2007). A data imputation model in sensor databases (LNCS 4782, pp. 86-96).

Juang, B. H., & Rabiner, L. R. (1990). The segmental K-Means algorithm for estimating parameters of hidden Markov models. *IEEE Transactions on Acoustics, Speech, and Signal Processing, 9*(38), 1639–1641. doi:10.1109/29.60082

Kotidis, Y. (2005). Snapshot queries: Towards data-centric sensor networks. In *Proc. of the 2005 Intl. Conf. on Data Engineering* (pp. 131-142).

Li, S., & Kernighan, B. (1971). An effective heuristic algorithm for the TSP. *Operations Research, 21*, 498–516.

Madden, S., Franklin, M. J., Hellerstein, J. M., & Hong, W. (2003). The design of an acquisitional query processor for sensor networks. In *Proceedings of the 2003 ACM SIGMOD international conference on Management of data* (pp. 491-502). ACM Press.

McLachlan, G., & Krishnan, T. (1997). *The EM algorithm and extensions*. Wiley series in probability and statistics, John Wiley and Sons.

Park, B. H., & Kargupta, H. (2003). Distributed data mining. In N. Ye (Ed.), *The handbook of data mining* (pp. 341-348). Lawrence Erlbaum Associates.

Rand, W. M. (1971). Objective criteria for the evaluation of clustering methods. *Journal of the American Statistical Association, 66*, 846–850. doi:10.2307/2284239

Silberstein, A., Braynard, R., & Yang, J. (2006). Constraint chaining: On energy-efficient continuous monitoring in sensor networks. *SIGMOD '06: Proceedings of the 2006 ACM SIGMOD international conference on Management of data* (pp. 157-168). New York: ACM

Szewczyk, R., Mainwaring, A., Polastre, J., & Culler, D. (2004). An analysis of a large scale habitat monitoring application. In *Proceedings of the 2nd international conference on Embedded networked sensor systems* (pp. 214-226). ACM Press.

Tan, P. N., Steinbach, M., & Kumar, V. (2006). *Introduction to data mining*. Addison-Wesley.

Trigoni, N., Yao, Y., Demers, A., Gehrke, J., & Rajaraman, R. (2005). Multi-query optimization for sensor networks. In *Proceedings of the First IEEE International Conference on Distributed Computing in Sensor Systems (DCOSS 2005)*. Springer.

Tulone, D., & Madden, S. (2006). An energy-efficient querying framework in sensor networks for detecting node similarities. In *Proceedings of the 9th ACM international symposium on Modeling analysis and simulation of wireless and mobile systems* (pp. 191-300).

Tulone, D., & Madden, S. (2006). PAQ: Time series forecasting for approximate query answering in sensor networks. *Paper presented at the 3rd European Workshop on Wireless Sensor Networks (EWSN01906)*, Zurich, Switzerland.

Werner-Allen, G., Lorincz, K., Welsh, M., Marcillo, O., Johnson, J., Ruiz, M., & Lees, J. (2006). Deploying a wireless sensor network on an active volcano. *IEEE Internet Computing, 10*(2). doi:10.1109/MIC.2006.26

Xia, P., Chrysanthis, P., & Labrinidis, A. (2006). Similarity-aware query processing in sensor networks. *Parallel and Distributed Processing Symposium* (p. 8).

Xu, X., Jager, J., & Kriegel, H. P. (1999). A fast parallel clustering algorithm for large spatial databases. *Data Mining and Knowledge Discovery, 3*(3), 263–290. doi:10.1023/A:1009884809343

Zhu, Y., Vedantham, R., Park, S. J., & Sivakumar, R. (2008). A scalable correlation aware aggregation strategy for wireless sensor networks. *Information Fusion, 9*, 354–369. doi:10.1016/j.inffus.2006.09.002

ENDNOTES

[1] It may be obtained from the data sheet of the device.

[2] http://www.xbow.com/products/Product_pdf_files/Wireless_pdf/MICA2DOT_Datasheet.pdf

[3] For the sake of simplicity, in this work we only report diagrams for the temperature measure.

[4] Data available on user manuals accessible on http://www.xbow.com/Products/productsdetails.aspx?sid=84.

Chapter 9
Peer–to–Peer Data Clustering in Self–Organizing Sensor Networks

Stefano Lodi
University of Bologna, Italy

Gabriele Monti
University of Bologna, Italy

Gianluca Moro
University of Bologna, Italy

Claudio Sartori
University of Bologna, Italy

ABSTRACT

This work proposes and evaluates distributed algorithms for data clustering in self-organizing ad-hoc sensor networks with computational, connectivity, and power constraints. Self-organization is essential in environments with a large number of devices, because the resulting system cannot be configured and maintained by specific human adjustments on its single components. One of the benefits of in-network data clustering algorithms is the capability of the network to transmit only relevant, high level information, namely models, instead of large amounts of raw data, also reducing drastically energy consumption. For instance, a sensor network could directly identify or anticipate extreme environmental events such as tsunami, tornado or volcanic eruptions notifying only the alarm or its probability, rather than transmitting via satellite each single normal wave motion. The efficiency and efficacy of the methods is evaluated by simulation measuring network traffic, and comparing the generated models with ideal results returned by density-based clustering algorithms for centralized systems.

DOI: 10.4018/978-1-60566-328-9.ch009

INTRODUCTION

Distributed and automated recording of data generated by high-speed, high-volume information sources is becoming common practice in scientific research, environmental monitoring, as well as in medium sized and large organizations and enterprises. Whereas distributed core database technology has been an active research area for decades, distributed data analysis and mining have been dealt with only more recently (Kargupta & Chan, 2000; Zaki & Ho, 2000) motivated by issues of scalability, bandwidth, privacy, and cooperation among competing data owners.

A common scheme underlying all approaches is to first locally extract suitable aggregates, then send the aggregates to a central site where they are processed and combined into a global approximate model. The kinds of aggregates and combination algorithm depend on the data types and distributed environment under consideration, e.g., homogeneous or heterogeneous data, and numeric or categorical data.

Among the various distributed computing paradigms, self-administered, massive-scale networks, like sensor networks and peer-to-peer (P2P) computing networks, are currently the topic of large bodies of both theoretical and applied research. In P2P computing networks, all nodes (*peers*) cooperate with each other to perform a critical function in a decentralized manner, and all nodes are both users and providers of resources (Milojicic et al., 2002). In sensor networks, small devices equipped with a sensing unit and a transceiver and, possibly, a limited computing architecture, are deployed in an environment to be monitored continuously or at fixed intervals.

Applications of both computing paradigms are rapidly maturing. In data management applications, deployed peer-to-peer systems have proven to be able to manage very large databases made up by thousands of personal computers. Many proposals in the literature have significantly improved existing P2P systems from several view-points, such as searching performance and query expressivity, resulting in concrete solutions for the forthcoming new distributed database systems to be used in large grid computing networks and in clustering database management systems.

Sensor technology is overcoming many functional limitations of early devices. State-of-the-art sensors are equipped with memory and processors capable of executing moderately demanding algorithms, enabling the deployment of sensor networks capable of processing the data in-network, at least partially, without transmission to a sink or gateway node.

In light of the foregoing, it is natural to foresee an evolution and convergence of sensor and P2P networks towards supporting advanced data processing services, such as distributed data mining services, by which many nodes cooperatively perform a distributed data mining task. In particular, the data clustering task matches well the features of self-organizing networks, since clustering models mostly account for local information, and consequently carefully designed distributed clustering algorithms can be effective in handling topological changes and frequent data updates.

In this chapter, we describe an approach to cluster multidimensional numeric data that are distributed across the nodes of a sensor network by using the data partitioning strategies of multidimensional peer-to-peer systems with some revisions, namely without requiring any costly reorganization of data, which would be infeasible under the rigid energy constraints enforced in a sensor networks, and without reducing the performance of the nodes in message routing and query processing. We evaluate the data clustering accuracy of the proposed clustering approach by comparison to a well-known traditional density-based clustering algorithm. The comparisons have been done by conducting extensive experiments on the decentralized wireless sensor network and on the algorithms we have fully implemented.

Application Scenarios and Constraints

Technologies for the support of Earth science are rapidly evolving, as consensus on the significance of potential economic returns of research investments for environment preservation and monitoring grows. For applications in the latter domain, data-intensive technologies are particularly useful, as time is a critical constraint. For example, some important oceanic phenomena are irregular and their spatial and temporal extensions are limited. Near real-time, accurate pattern recognition and exploratory mining of the data describing such phenomena could be of great use for analysts and decision-makers. We list some potential applications and related research work in the sequel.

Harmful algal blooms are of concern for human and animal health and may also have negative economic impacts, such as an increase in water supply treatment and an adverse effect on aquatic ecosystems. For these reasons, response plans and procedures have been defined by public authorities (Queensland Department of Natural Resources and Mines, Environmental Protection Agency, Queensland Health, Department of Primary Industries, & Local Governments Association of Queensland, 2002). In (Schofield et al., 1999) it has been argued that harmful algal blooms forecasting systems could be an important support to traditional monitoring programs and accelerate response times; to this end, a combination of optical remote sensing, *in situ* moored technologies and pattern recognition has been proposed.

Another phenomenon motivating oceanographic and climatologic investigation is the upwelling of water masses near coasts. Coastal upwelling brings to sea surface cooler and lower water often containing a higher concentration of nutrients and hence favors larval growth; for this reason areas of coastal upwelling cover a large fraction of all fisheries production, and such areas are an active research topic. Anomalous coastal upwelling and algal blooms have been correlated to mass mortality of fishes and other sea organisms (Collard & Lugo-Fernández, 1999; Kudela et al., 2005). Automated recognition of coastal upwelling has been proposed by means of fuzzy clustering (Nascimento, Casimiro, Sousa, & Boutov, 2005).

Pattern recognition and data analysis have already been proposed and used for the support of oceanographic research, at a more general level and a larger scale. The proposed techniques are however difficult to implement at a smaller scale.

Ocean thermal analysis and detection of water masses, water fronts and eddies have received considerable attention, motivated by economic and security factors, and scientific interest in ocean circulation and biogeochemistry (Cummings, 1994; Oliver et al., 2004).

These studies are based on satellite imagery data, e.g., Advanced Very High Resolution Radiometer (AVHRR) data, which are processed by a variety of numerical and statistical pattern recognition techniques. Drawbacks of satellite images are the interference of clouds and the limitation to surface images.

The study of models and techniques for the forecast of medium-term and long-term weather changes is an important research activity in earth science, including climatology, oceanography, and geophysics. Such an effort is motivated, for example, by the scale of the economic consequences of global changes associated to irregular local climate phenomena, such as El Niño-Southern Oscillation (ENSO). To monitor the ENSO phenomenon, a large array of buoys has been deployed at equatorial latitudes as part of a multi-national effort to provide data for climate research and forecasting. The array is now complemented by the more extensive Argo project, aimed at deploying a global grid with 3° spacing at worst. Such arrays are relatively coarse-grained to monitor phenomena at small-space scale.

An alternative for small- and medium- scale phenomena is to deploy surface and depth sensors, connected as a multi-hop underwater acoustic sensor network. Such networks are a topic of active research and envisioned as a solution to extensive, reliable, cost-effective monitoring of aquatic environments (Akyildiz, Pompili, & Melodia, 2005; Cui, Kong, Gerla, & Zhou, 2006; Yang et al., 2002). Similarly to wireless radio sensor networks, the main constraint is energy consumption. A sensor consists of a battery powered acoustic modem, a processor, and memory, and is capable of switching between a low-power sleep mode and an active mode. Whenever an acoustic wakeup signal is received, the node switches automatically to active mode. With these assumptions, the design goal is to trade quality of service and energy consumption. Therefore, network topology is a design choice that must be made early.

In (Sozer, Stojanovic, & Proakis, 2000) two approaches to topology of underwater acoustic networks have been compared: The star topology, in which all nodes can communicate with a single master node, and the nearest-neighbor topology, in which every node can communicate only with its closest node, and communications to an arbitrary node are carried out via multi-hop routing. The results of the comparison show that the nearest-neighbor topology largely outperforms the star, by an increasing gap as the number of nodes increases. We assume therefore from now onwards that the underlying network is a multi-hop network with routing capabilities.

As far as the application level is concerned, we assume as reference design goal the deployment of a sensor network for oceanic monitoring, in one of the scenarios outlined above. For instance, we might want to gather signatures of a toxic alga and build a representation in a multidimensional space, including spatial coordinates, of its blooming, when the phenomenon is in its early stage, and send it to a land station. We also assume that state-of-the-art sensor processor and memory allow for computations complex enough to perform significant preprocessing work. Application design must include the decision whether some

form of distributed processing is feasible and useful, or rather the network acts mainly as a data collector, with all analytical processing delegated to a master node or a land station.

At least two factors may influence the decision. The first one is the type of mining that is applied to the data in the application. If the phenomenon at hand is relatively unknown, and the goal is to search for possible models in a broad spectrum, then the task most likely consists of time-consuming, iterative executions of different techniques. The entire relevant data, or a substantial portion of it, must be available, since it is not known in advance which technique will make regularities emerge. This approach is typical of scientific discovery and best supported by an architecture oriented to data collection. In a different scenario, it is known beforehand which model type describes an irregular phenomenon which must be detected early, by separating normal from anomalous model instances. In this case, models alone suffice and distributed processing might be appropriate.

In the latter case, another factor is the cost of sensor maintenance. If the time through which the application must remain functional is long, then distributed processing might help in lowering the total cost of the network. For example, if the achievement of the application's goal requires, or is made more likely by, the knowledge of models computed over a succession of time windows, spanning a large number of battery charge lifetimes. Accessibility of sensors is a sub-factor in evaluating costs. Maintenance of marine sensors will generally involve higher costs.

Oceanic monitoring and detection of water upwelling or algal blooms are examples, for which the fundamental design question, that is, whether distributed processing is feasible and useful, can be answered positively. Therefore, only partial models or aggregates are sent to other nodes, for further processing. Nodes will not send single observations to a master station via satellite communications, thereby sparing large amounts of energy.

Note that cases in which such distributed computation is made possible by simple algorithms are additive aggregates, like count, average and variance (Madden, Franklin, Hellerstein, & Hong, 2002). However, research in the field of distributed data mining have shown that the computation of a clustering model, and in general, of mining models, by combination of partial models is much harder.

Applications with a broader scope are also feasible and useful. For example, terrestrial applications of sensor networks in monitoring human activities can benefit from distributed data clustering as well. Sensors deployed in an urban environment are capable of detecting variations of the concentration of humans, vehicles, and facilities both spatially and over time. Whether collected data can be used off-line or as input of in-network mining, depends on factors already discussed above. Collected sensor data are useful for behavioral discovery and analysis, however, location- and time- related transient phenomena like traffic jams and slowdowns, saturation of parking slots, distribution of taxi cabs, concentration of pollutants emitted by car exhausts are best detected by a distributed algorithm, especially if the monitoring task must be put into execution repeatedly or continuously. Data clustering in environment observation systems is another example of potential terrestrial application of clustering in sensor networks. Data records of movements of landslides can be grouped to highlight masses in three dimensions which have changed their speed or physical features. Similarly, flood forecasting can benefit from the monitoring groups of multidimensional data of atmospheric variables.

Overview of the Methods

The proposed approach exploits non-parametric density-based clustering to cluster the data in sensor network with small energy consumption. In density-based clustering, membership of an object in a cluster is influenced by a small neighborhood of the object in the data space. In fact, such a clustering typically estimates the density at every object and clusters together the objects which are located on the slopes of a local maximum; non-parametric density estimation weights objects decreasingly with distance, thus allowing for ample opportunities to economize on object visits, by ignoring far objects which contribute negligibly. If allocation of objects to nodes preserves locality, only a limited number of nearby nodes has to be queried, thereby reducing transmission costs further.

In summary, the proposed approach has the following characteristics:

- Data objects are allocated to nodes in a way that preserves locality, that is, near objects in the data space are stored in near nodes.
- The execution proceeds by sequential steps, with nodes executing in parallel the same step.
- In parallel, the nodes gather density information from near nodes, using a small amount of inter-node communication.
- Subsequently, the nodes determine in parallel the cluster memberships of owned objects, based on density information obtained in the previous step, by connecting every object to the best near object with higher density.
- The objects in a tree of the resulting forest constitute a cluster.

The amount of communication needed depends mainly on the amount of detail on density that is gathered in the first parallel step. In this chapter we propose and evaluate experimentally two approaches, KDE and MV, which differ greatly in the amount of density information which is transmitted in the network.

BACKGROUND

Data Clustering

Data clustering is a fundamental problem in exploratory analysis and data mining.

The goal of data clustering is to divide a collection of objects, usually represented as a multidimensional vector of observations, into clusters, ensuring that objects within a cluster are more similar to each other than they are to an object belonging to a different cluster (Jain, Murty, & Flynn, 1999). Cluster can be disjoint, or overlapping, and their union may cover the collection of patterns or not. In the latter case, objects not belonging to any cluster are exactly the objects recognized as outliers of the collection. In the sequel, we will focus on disjoint clusters and no outlier recognition will be performed.

Algorithms for clustering data are often classified into two broad classes: Hierarchical and partitional (Jain et al., 1999). Hierarchical algorithms create a sequence of nested partitions.

Agglomerative hierarchical algorithms start with the partition of singleton clusters and iteratively compute the next partition in the sequence by joining two most similar clusters. Divisive hierarchical algorithms start with the singleton partition and iteratively compute the next partition in the sequence by splitting a cluster.

Partitional algorithms create a single clustering of the collection of objects. In most cases, the number of cluster is a parameter of the algorithm, and must be chosen by other means. The most popular partitional algorithms for numerical data try to minimize a squared error function, that is, the summation, over all objects, of the squared distance between an object and the mean of its cluster. The popular k-means algorithm is an efficient heuristic to minimize a squared error function for clustering data (Hartigan & Wong, 1979).

An important separate subclass of partitional algorithms is constituted by density-based methods. The idea underlying all density-based approaches is that similarity is expected to be high in densely populated regions of the given data set. Consequently, searching for clusters may be reduced to searching for dense regions of the data space separated by regions of relatively lower density. Popular methods in the class have been investigated in the context of non-parametric density estimation (Silverman, 1986) and data mining (Ankerst, Breunig, Kriegel, & Sander, 1999; Ester, Kriegel, Sander, & Xu, 1996; Hinneburg & Keim, 1998; Xu, Ester, Kriegel, & Sander, 1998). See figures 1 and 2.

Distributed Clustering

The data clustering problem has been investigated also in the distributed setting. Research on the topic has been motivated mainly by two scenarios. In case different data owners wish to cooperate, e.g., in a joint fraud detection project, competition among data owners makes them reluctant to share their data, as the resulting gain would not compensate for the loss of competitive advantage; moreover, privacy regulations could forbid the transfer of data outside the space controlled by a data owner. In general, the total size of the local datasets collected by networked sites may be such that collecting the datasets on a single site is too slow or expensive, due to the limited bandwidth available.

Distributed data mining problems are classified into *homogeneous* and *heterogeneous*. In the homogeneous case, the global dataset is horizontally partitioned, that is sites store data on the same attributes; it is vertically partitioned in the heterogeneous case, that is, the features for which sites store values may vary from site to site (Kargupta, Park, Hershberger, & Johnson, 2000). Both in the homogeneous and heterogeneous case, the majority of algorithms is based on the collection and aggregation of partial, local models at a facilitator node. The algorithms presented so far for the clustering problem likewise build small local cluster descriptions which are aggregated

Figure 1. Two clusters drawn from two normal distributions with distribution means (-2, 0) and (2,0) and unit variance. Each local dataset consists of the data in a single quadrant. A cross marks the mean of the data in a region.

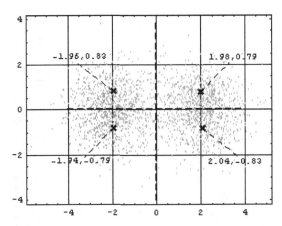

centrally with limited communication overhead (Johnson & Kargupta, 1999; Kargupta, Huang, Sivakumar, & Johnson, 2001; Klusch, Lodi, & Moro, 2003; Merugu & Ghosh, 2003; Tasoulis & Vrahatis, 2004).

We show by a simple example why the problem of distributed clustering under communication constraints is not a trivial one. To avoid data transfers, we could use the following technique: first find all clusters locally; represent each by its vector mean; cluster the vector means using a distributed protocol, for example, transfer the means to a chosen site and apply a centralized partitional algorithm. Assume the data are distributed in two clusters, drawn from two normal distributions with distribution means (-2, 0) and (2,0) and unit variance, the data in each of the four quadrants are stored at four different sites, and we know the number of clusters in advance. The distance between the means is most likely smaller for the pairs of means calculated on the parts of the same cluster, and the partitional algorithm would easily find the correct clusters. Suppose now the clusters are drawn from two normal distributions with distribution means (-2, 0) and (2,0) and covariance 0.95. The space is partitioned into four regions separated by the straight lines

$y = 0$ and $y = x$, and the data in each of the four regions are stored at four different sites. In this case, the distance between the means is roughly the same for all local datasets, being 2.17 for the means within the left cluster and 2.21 for the means within the right cluster, whereas the distances between means of the facing upper and lower parts are 2.03 and 2.01, respectively. Note that, if the clusters were properly aligned along their major axis, then distances between the means of the parts of different clusters facing each other would be even smaller, whereas the distances of the means within one cluster would not change.

Difficulties with distributed clustering arise not only in partitional homogeneous clustering. Johnson and Kargupta (1999) show that single linkage hierarchical clustering requires careful analysis of the distance bounds satisfied by local hierarchies to attain efficiency and accuracy.

Clustering Based on Non-Parametric Density Estimation

Among density-based methods, clustering based on non parametric kernel density estimation has been investigated for its generality and theoretical soundness (Hinneburg & Keim, 1998; Klusch

Figure 2. Two clusters from drawn from two normal distributions means (-2, 0) and (2,0), unit variance, and covariance 0.95. The dashed lines separate the regions allocated to the sites. A cross marks the mean of the data in a region.

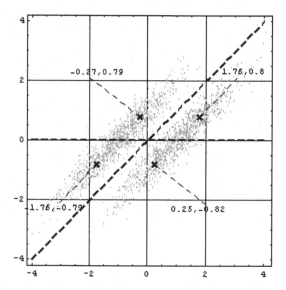

et al., 2003; Silverman, 1986). In the following, we describe the general features and techniques of clustering algorithms based on kernel density estimation.

Assume a set $S = \{O_i \mid i = 1, ..., N\} \subset R^d$ of objects. Kernel estimates capture the idea that the value of an estimated probability density function must increase with the fraction of data objects falling in the neighborhood of the point of estimate. The contribution of each data object O_i on the point of estimate $O \in R^d$ is quantified by a suitable kernel function $K(x)$, i.e., a real-valued, non-negative function on R^d having unit integral over R^d. In most cases, kernel functions are symmetric and non-increasing with x. When the argument of the kernel is the vector difference between O and O_i, the latter property ensures that the contribution of any data object O_i to the estimate at some $O \in R^d$ is larger than the contributions of farther data objects in the same direction. A *kernel estimate* $\varphi(O):R^d \rightarrow R_+ \cup \{0\}$ is defined as the normalized sum over all data objects O_i of the differences between O and O_i, scaled by a factor h, called

smoothing or *window width*, and weighted by the kernel function K:

$$\phi(O) = \frac{1}{Nh^d} \sum_{i=1}^{N} K(\frac{O - O_i}{h}) \qquad (1)$$

Prominent examples of kernel functions are the standard multivariate normal density $(2\pi)^{-d/2}e^{-(1/2)x.x}$, and the uniform kernel $(c_d)^{-1}I_{[0,\infty)}(1 - x.x)$, where c_d is the volume of the d-dimensional unit sphere and $I_{[0,\infty)}$ is the indicator function of $[0,\infty)$.

A kernel estimate is therefore a sum of exactly one "bump" placed at each data object, dilated by h. The parameter h controls the smoothness of the estimate. Small values of h result in fewer bumps to merge and a larger number of local maxima. Thus, the estimate is more sensitive to slight local variations in the density. As h increases, the separation between regions having different local density tends to blur and the number of local maxima decreases, until the estimate is unimodal.

A widely accepted objective criterion to choose h is to minimize the mean integrated square error

Figure 3. Example dataset

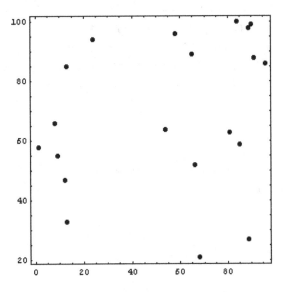

(MISE), that is, the expected value of the integrated squared pointwise difference between the estimate and the true density of the data. An approximate minimizer is given by

$$h_{opt} = A(K)N^{-1/(d+4)} \tag{2}$$

where $A(K)$ depends also on the dimensionality d of the data and the unknown true density. In particular, if the data are distributed according to the unit multivariate normal density

$$A(K) = \left(\frac{4}{2d+1}\right)^{1/(d+4)} \tag{3}$$

For an unknown distribution, the window width h is set to the optimal value for normal data multiplied by the root of the average marginal variance of the data:

$$h = h_{opt}\sqrt{d^{-1}\sum_{j=1}^{d} s_{jj}} \tag{4}$$

where s_{jj} is the data variance on the j-th dimension (Silverman, 1986). In the following we refer to the simple example dataset depicted in Figure 3; the coordinate values of its objects are shown in Figure 4. The dataset contains two visible clusters, at the left and right hand side, spanning the vertical coordinate. Both clusters can be subdivided into smaller clusters, two in the left cluster and three in the right cluster.

In Figure 5 and Figure 6, three-dimensional plots of multivariate normal kernel estimates of the example data are shown, with h set to 0.5 h_{opt} and 2 h_{opt} respectively, and h_{opt} given by Equation 2, Equation 3, and Equation 4. At 0.5 h_{opt} the finer structure of the dataset emerges; whereas at 2 h_{opt} even the coarse structure is obscured. In Figure 7, optimal smoothing is used and the overall clustered structure is recognizable.

In practice, often the summation in a kernel estimate need not be extended to all objects.

For all points O and $\gamma > 0$ and some $\alpha \in [0,1]$ depending on O, γ, S, and K,

$$\frac{1}{Nh^d} \sum_{\substack{1 \le i \le N \\ i:\|O-O_i\|\le\gamma h}} K\left(\frac{O-O_i}{h}\right) = \alpha\,\phi(O). \tag{5}$$

Figure 4. Coordinate values for the example dataset

O_{1-4}: **(13,85) (1,58) (9,55) (24,94)**

O_{5-8}: **(12,47) (68,21) (13,33) (8,66)**

O_{9-12}: **(89,27) (85,59) (54,64) (84,100)**

O_{13-16}: **(66,52) (90,99) (96,86) (58,96)**

O_{17-20}: (81,63) (89,98) (91,88) (65,89)

Figure 5. Plot of the multivariate normal kernel estimate of the example data. The smoothing parameter h is set to half its optimal value

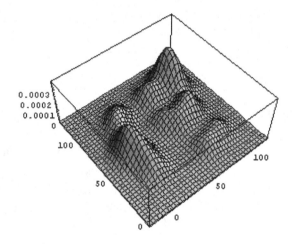

Figure 6. Plot of the multivariate normal kernel estimate of the example data. The smoothing parameter h is set to twice its optimal value

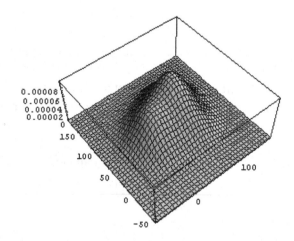

Figure 7. Plot of the multivariate normal kernel estimate of the example data. The smoothing parameter h is set to its optimal value.

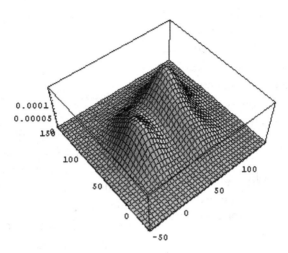

The support of many kernels is the unit sphere, so that $\alpha = 1$ for $\gamma > 1$. For the normal kernel, if $\gamma > 3$ and $\{O_i \in S: \|O - O_i\| \leq \gamma h \}$ is nonempty, then $\alpha \approx 1$.

Kernel estimates have therefore a strong locality property. If the data set is centralized, the property allows for a substantial efficiency gain provided a spatial access method to retrieve $\{O_i \in S: \|O - O_i\| \leq \gamma h\}$ efficiently is available. Similarly, if the data set is distributed, then the availability of a distributed multidimensional index is crucial to avoid massive data transfers.

Once the kernel estimate of a data set has been computed, the strategy to cluster its objects is to detect disjoint regions of the data space populated by objects with large estimated density and group all data objects of each region into one cluster. The different approaches to construct the regions determine different clustering schemes.

Two main approaches have been proposed. In the first approach, the density is zeroed where its value is smaller than a threshold. The support of the resulting density is split into maximal connected subsets. An example of this approach is given by the density-based clusters extracted by the DBSCAN algorithm (Ester et al., 1996):

Each cluster collects all data objects included in a maximal connected region where the value of a uniform kernel estimate exceeds a threshold. Another approach searches for maximal regions covered by a single bump in the estimate. A first example is the approach of Koontz, Narendra and Fukunaga (1976), as generalized in (Silverman, 1986): Each data object O_i is connected by a directed edge to the data object O_j that maximizes the average steepness of the density estimate between O_i and O_j, over all O_j within a distance threshold possibly depending on O_i, and satisfying $\varphi(O_i) < \varphi(O_j)$. Usually, the distance threshold is a small multiple of the smoothing parameter h. Clusters are then defined by the connected components of the resulting graph. We call this approach a *density tree clustering*. A density tree clustering of the example dataset is shown in Figure 8, with the edges between an object and the object which maximizes the average steepness, and a contour plot of the kernel estimate; the distance threshold for searching for maximizers is set to twice the optimal smoothing. More recently, Hinneburg and Keim (1998) proposed the DENCLUE algorithm which is based on a different approach. In their proposal, two types of cluster are defined. Center-defined clusters are based on the idea that every sufficiently large local maximum corresponds to a cluster including all data objects which can be connected to the maximum by a continuous, maximally steep path in the graph of the estimate. An arbitrary-shape cluster is the union of center-defined clusters having their maxima connected by a continuous path whose density exceeds a threshold. In either case, we call the approach a *density gradient clustering*.

Another fundamental device in non-parametric density estimation is the density histogram (Scott, 1992). After choosing d smoothing parameters h_1, $h_2,..., h_d \in R_+$, the d-dimensional space is partitioned into a regular mesh of *bins*

$$B_k = [k^{(1)}h_1, (k^{(1)}+1)h_1) \times \cdots \times [k^{(d)}h_d, (k^{(d)}+1)h_d),$$

where $k^{(i)}$ is the i-th component of $k \in Z^d$. Let v_k be the count of objects in B_k, that is, $v_k = \# S \cap B_k$. The *density histogram estimate* is the function

taking the value $v_k /(N h_1 h_2 ...h_d)$ at all points of bin B_k:

$$H(x) = \frac{1}{Nh_1 h_2 \cdots h_d} \sum_{k \in Z^d} \nu_k I_k(x),$$

where I_k is the indicator function of B_k.

The histogram is rarely used in multivariate analysis for $d > 1$, due to the difficulty of presenting and perceiving the structure of its graph. It can be nevertheless used as a basis for clustering, in the same way the kernel estimate is used, as the long as the method to construct high density regions does not require the continuity of the estimate. For example, the density-tree clustering approach, once a representative object is chosen for every nonempty bin, creates a tree of the bins and successively a partition of the bins, inducing a clustering of the objects.

Figure 8. Plot of the density tree clustering of the example data, with contours of the estimate, h set to its optimal value. A tree edge joins an object to the object which maximizes the average steepness of the density estimate. The maximum distance to search for the maximizer is set to twice the optimal smoothing.

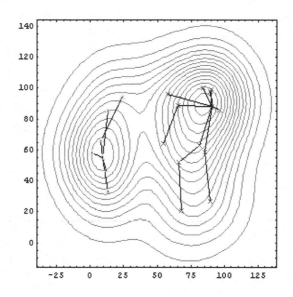

Sensor Networks

We exclude from the scope of this work simple sensor networks in which sensed data are periodically gathered at a single point, or sink, for external transmission and processing. We focus instead on a new emergent sensor network which has been recently investigated in the literature, namely, a sensor network that, in addition to sensing data, operates as a distributed database with the capability to answer queries and provide mining models to the user. In such networks, the data are stored and indexed at the sensors, and the keys are distributed across the network.

Using this strategy, sensors will no longer transmit raw data. Users will inject queries or tasks in general whenever needed. Fundamental tasks in this kind of network are routing and data management. Routing is needed whenever a requested data item must be retrieved. Solutions for routing in wireless ad-hoc networks exist in the literature (Intanagonwiwat, Govindan, Estrin, Heidemann, & Silva, 2003; Karp & Kung, 2000; Li, Kim, Govindan, & Hong, 2003) but they do not satisfy sensor network constraints. Geographic routing, for instance, has dead-ends problems and requires high energy consumption in order to find out the physical position of the devices. Regarding data management, some solutions are based on distributed hash tables, for instance in GHT (Ratnasamy et al., 2003) each piece of data is converted, using a hash function, into a physical position. Then the data is stored at the sensor which is closest to that position and geographic routing is used to navigate to the position.

Usually GPSR (Karp & Kung, 2000) is used for low level routing. In (Monti & Moro, 2008; Monti, Moro, & Lodi, 2007; Monti, Moro, & Sartori, 2006; Moro & Monti, 2006) an infrastructure for ad-hoc and sensor networks is proposed, and its peculiarity is to integrate multi-dimensional data management and routing in a cross-layer fashion without using GPS and message broadcast operations.

DENSITY-BASED CLUSTERING IN SENSOR NETWORKS

Efficiency of Sensor Networks for Clustering Data

The goal of a sensor network data mining system is to compute models of the global dataset, as efficiently and accurately as possible. As to efficiency, the system should satisfy some specific requirements, which we list in the sequel. We group them into three categories: overall efficiency, efficiency under network evolution, and load balancing. Some of these desiderata hold for other types of massive-scale systems, e.g., for P2P multidimensional query answering systems (Ganesan, Yang, & Garcia-Molina, 2004).

- **Overall efficiency**
  - **Locality** When clustering data by similarity, the objects that are similar to each other must be visited and grouped together. A data partitioning scheme that preserves locality reduces the number of nodes to be visited, thus improving efficiency.
  - **Transmission complexity** The transceiver of a sensor is by far the component with the largest power consumption, in transmission mode. The battery lifetime of a sensor node therefore depends largely on transmission time, and the total number of messages transmitted over the network to cluster the entire dataset should be as small as possible.
- **Efficiency under network evolution** Sensor nodes are essentially unattended. Therefore, the network is expected to be unstable due to node failures and the intrinsic volatility of wireless connections. A network that reacts fast to topology changes is more likely to preserve the accuracy of its answers.

- ◦ **Indexes** The size of local and network indexes on the data should be small.
- ◦ **Links** The number of connecting links between neighboring nodes should be small, to minimize the number of update actions needed for the connection and disconnection of nodes.
- **Load balance**
  - ◦ **Data** The number of stored objects should be the same at every node. Notably, such objects include information needed to store partial models of clustering.
  - ◦ **Forwarded messages** Even if the transmission time is short overall, if some nodes transmit much longer than others, they will fail sooner and the network's view of its environment will lose homogeneity, thereby deteriorating its accuracy, due to missing data. Ideally, energy consumption should be equal at all nodes to minimize the variance of node survival, hence the frequency of forwarded messages should be equal at all nodes.

The relative weight of the requirements above in the techniques to solve mining problems may of course vary widely. As clustering is based on data similarity, if all $O(n^2)$ similarity values between objects are computed, then the locality of data partitioning is of minor importance. However, most efficient approaches to the clustering problem are local in that they avoid the computation of all $O(n^2)$ similarity values; exploiting the fact that the membership of an object o in a cluster is not affected significantly by objects far from it, only objects in the spatial neighborhood of o are visited. If a local clustering approach is used in a distributed system with many nodes, then depending on the size of the neighborhood and the extent data locality is preserved, the system may access between zero and a large number of nodes per visited object. Data locality is therefore crucial to the efficiency of any massive distributed clustering system subject to tight constraints on communication costs. In the following sections, we examine how partitioning schemes used by traditional centralized spatial access methods have been adapted to the P2P and sensor network domains and how such schemes attempt to preserve data locality.

Data Partitioning Schemes

Many multidimensional access methods (Gaede & Günther, 1998) supporting search operations partition recursively the search space into hyper-rectangular disjoint regions, contiguous at *split hyper-planes*, thereby creating a tree of regions. Other than at random, the selection of split hyper-planes follows one of two strategies, which we may term *mass-based partitioning* and *volume-based partitioning*.

In mass-based partitioning, every split hyper-plane is selected in such a way that the number of objects in any sub-region exceeds the number of objects in any other sub-region by at most one. A prominent multidimensional access method using mass-based partitioning is the *adaptive k-d-tree*.

In volume-based partitioning, every split hyper-plane is such that all generated new sub-regions enclose equal volumes. An example of multidimensional access method using volume-based partitions is the *bintree*.

In an adaptive *k*-d-tree (Bentley & Friedman, 1979) in *k* dimensions, the initial region, that bounds all the data, is recursively partitioned into two hyper-rectangular sub-regions contiguous at a $(k-1)$-dimensional hyper-plane, having constant value on a coordinate called *discriminator*. The *discriminator* is selected as the coordinate for which the spread of values, for example measured by variance, is maximum. The constant value for the discriminator is chosen as its median over the data in the region. This strategy is adaptive in that data distribution determines the splits.

An alternative strategy for choosing a discriminator is to alternate coordinates cyclically. Numbering coordinates as $0, 1, ..., k-1$, the discriminator for the initial region is 0; the discriminator for its sub-regions is 1; the discriminator of a region at depth i in the tree is i, until depth k is reached, for which the discriminator is reset to 0, and so on. That is, in general, the discriminators for both sub-regions of a region having discriminator i equal $(i+1) \bmod k$. This strategy is used in the original, non-adaptive, k-d-tree (Bentley, 1975).

In a bintree (Tamminen, 1984), the split hyperplane divides a region into two sub-regions of equal volume and the choice of split coordinates follows the alternating strategy of the non-adaptive k-d-tree.

Prominent P2P systems supporting multidimensional queries use one of the data partitioning strategies above as well. An example of a mass-based partitioning system is the MURK (Multi-dimensional Rectangulation with K-d-trees) network (Ganesan et al., 2004). Examples of systems using volume-based partitioning are the CAN (Content-Addressable Network) overlay network (Ratnasamy, Francis, Handley, Karp, & Schenker, 2001) and G-Grid (Moro & Ouksel, 2003).

Independent of partitioning strategy, the properties of MURK and CAN networks can be summarized in terms of intervals as follows. Assume the data objects are elements of R^k and the coordinates are numbered $0, 1, ..., k-1$. A *zone* is the product of k half-open intervals $[x_0, y_0) \times ... \times [x_{k-1}, y_{k-1})$. We say a zone is *contiguous* to another zone if the projections of the two zones on one coordinate, the *partitioning coordinate*, are a partition of some half-open interval, and their projections on the remaining k-1 coordinates are equal. If j is the discriminator, and $[x_j, y_j)$, $[y_j, z_j)$ are the projections of the two contiguous zones on j, the value y_j is the *separator* of the contiguous zones. Then, the network represents a collection of zones satisfying the following properties:

- Exactly one zone, called the root zone, is contiguous to no zone
- For every non-root zone
  - there is exactly one contiguous zone
  - the union with its contiguous zone
    - is an element of the collection
    - has $(j+k-1) \bmod k$ as partitioning coordinate
- A zone is owned by a peer if and only if it is not a proper superset of another

A MURK network (Ganesan et al., 2004) of peers distributes the data to the peers as they enter the network according to an allocation strategy which is a strict analog to k-d-tree partitioning. In the following, we will say a zone is an *immediate neighbor* of another zone if the projections of the two zones on one coordinate are a partition of some half-open interval, and for each of the remaining k-1 coordinates, the length of the intersection of the projections of the two zones on the coordinate is not zero.

Initially, one peer holds the entire dataset in a single hyper-rectangular zone. In general, the peers in the network own one hyper-rectangular zone each, corresponding to a leaf in a k-d-tree. A peer joining the network randomly selects a zone; using routing, it a sends split request to the owner peer. Upon receiving the request, the owner partitions its zone into two sub-zones, containing the same number of objects each, transfers the ownership of one of the sub-zones and all enclosed objects to the new peer, and retains the other with all enclosed objects. The partition is computed, as in k-d-tree partitioning, by creating two sub-zones contiguous at a $(k-1)$-dimensional hyper-plane, having a coordinate of constancy. The coordinate is selected cyclically. With every zone a fixed partitioning coordinate is stored. At partition time, the coordinate which follows it in a fixed ordering is stored with the sub-zone. The routing tables of all the peers owning the zones which are neighbors to the affected zone are updated.

Every peer maintains a routing table, which consists of a list of pointers to its immediate neighbors, and a list of the boundaries of their zones. Using this information, queries are forwarded from the peer where they originate to the destination peer by greedy routing. If a multi-dimensional query q originates at peer p, routing cost is defined as the minimum L_1 distance from any point in p's zone to q. Initially, the query message is routed from p to an immediate neighbor which decreases routing cost most; the process is repeated until the peer owning the zone containing q is reached.

A CAN overlay network (Ratnasamy et al., 2001) is a type of distributed hash table by which (*key,value*) pairs are mapped to a multi-dimensional toroidal space by a deterministic hash function. The allocation policy of the toroidal hash space is similar to that of a MURK network with some differences. Zones are partitioned into equal sub-zones; therefore, CAN uses volume-based partitioning. When the ownership of a sub-zone is transferred to an entering peer, (*key,value*) pairs hashed to the sub-zone are transferred to the peer. Finally, there is provision for the departure of

peers. A peer leaving the network hands over its zone and the associated (*key,value*) pairs to one of its neighbors. Both at peer arrival and departure, the routing tables of all the peers owning the zones which are neighbors to the affected zone are updated.

G-Grid (Moro & Ouksel, 2003), in comparison to CAN, introduces new concepts and techniques, such as region nesting, region split based on region bucket size, preservation of data locality using an appropriate linearization function, and a learning capability, not essential for its functioning, according to which each node builds locally and gradually the knowledge of the P2P networks in order to improve the efficiency of queries and updates.

In this paper, we will be concerned with sensor networks for clustering data using mass-based and volume-based partitions, which follow MURK and CAN approaches to network construction and zone definition and allocation. However, when using volume-based partitioning, we will allocate zones to peers directly in the data space. Notably, distributed hashing is not a suitable technique for

Figure 9. Mass-based partitioning of an example dataset. The partitioning coordinate is selected by an alternating strategy. The separator value is such that the numbers of objects in a zone exceeds the number of objects in a contiguous zone at most by one.

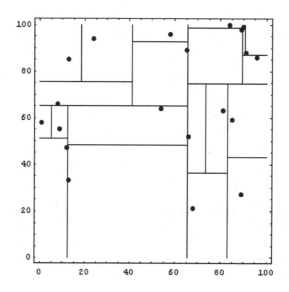

distributed data management underlying a system for clustering data. Hashing scatters neighboring objects, so that, in general, objects controlled by a network node are far in the space of origin, and neighborhood queries are very costly. For the example dataset of Figure 3, mass-based partitions and volume-based partitions are depicted in Figure 9 and Figure 10. In the latter figure, the dataset was first partitioned in the x coordinate, i.e., with a vertical line.

Distributed Partitioning Schemes in Sensor Networks

In data centric sensor networks, which this work refers to, the network acts like a distributed database managing and indexing sensed data in order to efficiently perform in-network tasks, such as routings and searches. It is worth to notice that a distributed index in such kind of sensor networks is essential to perform unicast routings, namely avoiding to broadcast/flood each message to the entire network. In other words, without an index the only way to find a certain record in a file is to sequentially scan the file, and analogously in a

sensor network this corresponds to broadcast the search to all sensors, causing network congestions and expensive energy consumptions.

Basically routing is necessary whenever a data sensed must be transmitted elsewhere in the network, including an external machine, proactively or reactively according to periodic tasks or queries submitted to the network system.

For instance, in oceanographic observations sensors are currently used for gathering data by following currents, such as salinity, oxygen, temperature, fluorescence, optical attenuation, etc. Data are collected both on the water surface, by means of float sensors, and deeply until 2000 dbar[1] employing glider sensors, and then transmitted via satellite. If sensors communicated each other forming networks as mentioned above they could aggregate and fuse data in order to elaborate models and transmit only synthetic information, rather than huge amount of raw data, reducing of orders of magnitude the energy spent for satellite communications.

There are several approaches to map the scheme partitions described in the preceding subsection in sensor networks. In (Li et al., 2003) and (Xiao

Figure 10. Volume-based partitioning of the example dataset. The partitioning coordinate is selected by an alternating strategy. The separator value is such that contiguous zones are equal in size.

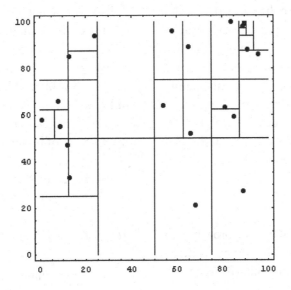

& Ouksel, 2005) data space partitions in sensor networks follow the physical positions of nodes, which means that each sensor manages one or more data space partitions that better corresponds to geographical position of the sensor. Of course, this implies that in both solutions sensors must know both their geographical positions and the physical coordinates delimiting the network geographic area. This means that all devices or a good percentage be equipped with a Global Positioning System (GPS) technology and the network area should not change because the distribution of data partitions over sensors depends on physical coordinates. Moreover, GPS could be costly and may not work if weather conditions are not good or if the network is too dense because of limitations in the precision. W-Grid (Monti & Moro, 2009; Monti & Moro, 2008; Moro & Monti, 2006) overcomes the preceding limitations by means a novel decentralized technique for the assignment of virtual coordinates to nodes that guarantees independence between data distribution and physical node positions, moreover, by construction, the network produces at least two disjoint paths between each couple of nodes. This innovation drastically reduces network traffic, while guaranteeing robustness and efficiency.

Our data clustering methods work in any network structure and exploit the data space partitions already present in the network in order to extract the model, namely the set of clusters. When the network receives a mining task each sensor computes the density of its data space partitions and the clustering by communicating with sensor neighbors according to the principle of density-based algorithms described above and following the methods illustrated in the following subsection.

Clustering the Data

In this section, we present two approaches to density-based clustering in a sensor network. In both approaches, the computation of clusters is based on density-tree clustering. Both approaches can be used with a distributed partition independent of the splitting strategy.

The first approach, KDE, relies entirely on kernel density estimation. The method constructs a forest of trees on the data objects, with every object linked to an object of higher density, and builds clusters from the connected components of the forest. Initially, a master node determines the optimal smoothing parameter and distributes it to all nodes; such computation involves only additive aggregates and can therefore be performed efficiently. In parallel, every node approximates the kernel estimate of its objects by collecting the objects within a small, predefined number of hops, and computing the estimate using objects from the collection only. Because of the locality properties of mass-based and volume-based partitioning, the retrieved collection approximates the object content of balls centered at the node's objects. Next, every node in parallel obtains from the nodes within a small number of hops their local density maxima. Finally, every node associates every owned object to a maximum collected at the previous step, if its density is higher, choosing the one reachable with the steepest line; if none has higher density than the object, then the object becomes a tree root. This guarantees that a polyline from every object is always monotone in density at its vertices, and ends at a high density object; each resulting connected component consists of objects on the slopes of a single maximum (cf. Figure 8).

The algorithm of the KDE method is described in Figure 11.

The parameter H_e allows to exploit the locality property of kernel estimates. Limiting the set of objects retrieved to compute the estimate by limiting the number of hops to H_e is a heuristic for the retrieval of the objects in

$$\{O_j : \| O_i - O_j \| \leq \gamma h\},$$

(cf. the left member of Equation 5).

The parameter H_c specifies the distance threshold in density tree clustering. To obtain meaningful clusters, its value must be of the same order of magnitude of h.

With the running example data, assume the mass-based partitioning of Figure 9 and $H_e = H_c = 1$, that is, queries are routed only to immediate neighbors. At Step 2, the node containing object $O_3 = (9,55)$ routes a query to its immediate neighbors and receives their data objects as result, that is, objects O_2, O_5, O_8, and O_{11}. The kernel estimate when summation is extended to such objects equals 92.21×10^{-6}. As a comparison, the exact estimate, computed over all objects, is 111.3×10^{-6}. At Step 3, the node now has to decide which data object maximizes the ratio between the estimate value at the object and its distance from O_3. It routes a query to its immediate neighbors, which select and return their respective object maximizing the local density. Upon receiving the local maxima, the node maximizes the ratio and selects the object to link to O_3. In our case, the returned estimates are 86.3×10^{-6}, 84.62×10^{-6}, 86.84×10^{-6}, 52.18×10^{-6}, for O_2, O_5, O_8, and O_{11}, respectively. None of these objects has a higher density value, thus O_3 is linked to no object. Consider now the node containing O_2. Its immediate neighbors are the three nodes containing O_3, O_5, O_8. After executing Step 2 and Step 3, at Step 4 only O_3 and O_8 must be evaluated, since O_5 has a smaller density than O_2 itself.

The ratios are 10.79×10^{-6} and 8.17×10^{-6} in favor of O_3. O_2 is therefore linked to O_3 (as expected; cf. the centralized density tree clustering depicted in Figure 8).

The second approach, MV (Mass divided by Volume), directly exploits the data space partitions generated by the data management subdivision among sensors (or nodes) as described in the previous section. The data are not purposely reorganized or queried to compute a density estimate to perform a clustering. In this case, the density value at data objects in a zone is set as the ratio between the number of objects in the zone and the volume of the zone.

The method constructs a forest of trees on the data objects, like the KDE method, and builds clusters from the connected components of the forest. There are two major differences, however. First, every object is linked to a regional *representative* object of higher density. The representative object of a region is an object having maximum kernel density in the region; to this end the kernel estimate is computed using the region's objects only. Second, the representative object is still chosen according to the ratio between density and distance, but density is the mean density of the region containing the representative object.

Initially, every node in parallel computes its region's representative object and geometric mean. Such computation involves only local objects. In

Figure 11. The KDE method

1. At the master:
   a. Collect from every region assigned to a node its object count, object sum, and square sum to globally choose a window width h according to **Error! Reference source not found.**, **Error! Reference source not found.**, and **Error! Reference source not found.**.
   b. Send h to every node.
2. At every node: For every local data object O_i, compute the kernel estimate value $\varphi(O_i)$, from the local data set and the remote data objects that are reachable by routing a query message for at most H_e hops, where H_e is a parameter.
3. At every node: Query the location and value of all local maxima of the estimate located within other regions, having a value greater than $\varphi(O_i)$, routing a query message for at most H_c hops.
4. At every node: associate each local data point to the maximum, received at Step 3, which maximizes the ratio between the value of the maximum and its distance from the point.

parallel, every node collects the mean density of the regions within a small, predefined number of hops, and computes the mean density of its region as a ratio between count and volume, and adds a corrective contribution based on the collected mean densities. Such computation involves transmitting only a few floating point numbers. Next, every node in parallel obtains the representatives and mean region densities from the nodes within a small number of hops. Finally, every node associates every owned object to one of the collected representatives, if its density is higher, choosing the one reachable with the steepest line; if none has higher density than the object, then the object becomes a tree root. Similarly to the KDE method, a polyline from every object is monotone at its vertices, however, in this case *mean* density is considered.

The algorithm of the MV method is described in Figure 12.

With the running example data again, assume the volume-based partitioning of Figure 10 and $H_e = H_c = 1$. At Step 2, the node containing $O_3 = (9,55)$ sends its geometric mean $(9.375, 56.25)$ and its mean density 0.0128 to its immediate neighbors, that is the peers containing objects O_2, O_5, O_8, and one peer owning an empty region. From such

peers it receives their geometric means, $(3.125, 56.125)$, $(6.25,37.5)$, $(6.25,68.75)$, $(18.75,62.5)$, and mean densities, 0.0128, 0.0032, 0.0064, 0.0. The local density is updated by adding the ratios between each received density and the distance between the corresponding geometric mean and the local geometric mean. At Step 3, the peer performs a computation similar to Step 3 and Step 4 of the KDE approach, using region mean densities instead of density maximum values, and computing the ratios with distances to the object that maximizes the region's kernel estimate. Note that such estimate is computed entirely locally, and therefore has zero communication costs.

In the approach above, the volume of information sent over the network at Step 2.a for computing densities is very small, consisting of $n (d + 1) f (H_e)$ floating point numbers, where $f (H_e)$ is the average number of nodes within H_e hops of a node and n is the number of nodes in the network. The costs for computing the density tree are analogous.

Note that for a kernel estimate to be accurate, the objects within at least h units from the point of estimate must be retrieved. If the set $\{O_i \in S: \|O - O_i\| \leq \gamma h\}$ is covered by more than one region, data objects must transferred and the total volume

Figure 12. The MV method

---
1. At every node:
    a. Set the local mean region density m as the object count in the node's region divided by its volume.
    b. Set the local representative x^* of the node's region to the object that maximizes a kernel estimate computed on the node's region data only.
    c. Set the local geometric mean g of the region to the point that divides in half the region's intervals on all coordinates.
2. At every node:
    a. Send to every node within H_e hops the local geometric mean g and the local mean region density m and receive its local geometric mean γ and its local mean region density μ. Collect the received pairs of means into M.
    b. For all pairs (γ, μ) in M set $m = m + \mu / \|x^* - \gamma\|$.
3. At every node:
    a. Get the local representative ξ^* and local mean region density μ of all regions within H_c hops. Collect the received pairs into R.
    b. Associate each local data object O to the local representative ξ^{**} maximizing, over all regions within H_c hops, the ratio between the region's local mean density and its distance from O: $\xi^{**} = \arg\max_{\xi^*}\{\frac{\mu}{\|O - \xi^*\|} : (\xi^*, \mu) \in R\}$.
---

of messages for computing densities in the KDE method is expected to be significant. Therefore, MV is expected to outperform in efficiency the KDE method.

It is therefore natural to raise a question about the accuracy of MV compared to KDE.

MV estimates density in a way similar to a particular type of histogram, the percentile histogram (Scott, 1992). It is well-known that histograms are not very accurate in estimating the modes of a density. In the neighborhood of a mode the estimate is almost flat, however the neighborhood may cover more than a region. Due to the variance of the histogram over the regions, more maxima can be detected by the density tree. Therefore, compared to KDE, MV is not expected to perform as well as far as efficacy is concerned. To mitigate this drawback, at Step 2.b the estimated density in a region is computed with weighted contributions from near regions.

To be able to answer the question experimentally, a centralized algorithm for clustering the data should be taken as a reference. For its generality and accuracy, the natural candidate would be the DENCLUE (Hinneburg & Keim, 1998) algorithm. However, DENCLUE contains some approximations that are fundamental for its efficiency in real cases, but inessential when the goal is to find a term of reference to objectively measure the efficacy of other approaches. We therefore chose to take the density gradient clustering method with a normal kernel, coupled with numerical optimization techniques to guarantee convergence in the search for a local maximum, as a term of reference.

Experiments

The main goal of the experiments described in this section is to compare the accuracy of the clusters produced by the two approaches presented in this paper, namely their efficacy, as a function of the network costs, that is their efficiency as clustering algorithms. To determine the accuracy of cluster-

ing, we have compared the clusters generated by each solution as a function of the number of hops, with the clustering computed by a centralized system according to the density-based algorithm mentioned in the preceding section. Limiting the number of hops means that the computed estimate is an approximation of the true estimate, which should be computed by routing queries to the entire network.

As a measure of clustering accuracy, we have used the Rand index (Rand, 1971). Given a dataset $S = \{O_1,..., O_N\}$ of N data objects and two data clusterings of S to be compared, X and Y, the Rand index is determined by computing the following variables:

- a = The number of objects in S that are in the same partition in X and in the same partition in Y,

- b = The number of objects in S that are not in the same partition in X and not in the same partition in Y,

- c = The number of objects in S that are in the same partition in X and not in the same partition in Y,

- d = The number of objects in S that are not in the same partition in X but are in the same partition in Y.

The sum $a + b$ can be thought of as the number of agreements between X and Y, and $c + d$ as the number of disagreements between X and Y. The Rand index $R \in [0, 1]$ is the fraction of agreements on the total number of pairs of objects:

$$R = \frac{a + b}{a + b + c + d}$$

In our case, one of the two data clusterings is always the one computed by a centralized density-based algorithm, namely the one corresponding to a distributed density-based algorithm with no hop limitation.

We have conducted extensive experiments by implementing in Java the two distributed density-based clustering algorithms described in Section 4, together with the sensor network infrastructure for the management of multi-dimensional data, for the execution of physical message routings and queries among sensors. Experiments have been executed by a desktop workstation equipped with two Intel dual-core Xeon processors at 2.66GHz and 2GB internal memory.

Two generated datasets and a real dataset of two-dimensional real vectors have been used in our experiments. The first dataset, S_0 shown in Figure 13, contains 2500 data vectors generated from 5 similar normal densities. The second dataset S_1, shown in Figure 14, contains the same number of vectors, but the 5 normal densities are dissimilar. Each of the two large groups, which has a deviation of 70, is 4.5 times bigger than each of the three small groups, which are very close in mean, with a deviation of 10. The third dataset, S_2, shown in Figure 15, is a subset of the El Niño dataset published in the UCI KDD Archive (Hettich & Bay, 1999). The dataset contains observations collected by the Tropical Atmosphere Ocean (TAO) array of buoys, deployed by the Tropical Ocean Global Atmosphere (TOGA) program to improve the detection and prediction of climate variations, such as El Niño-Southern Oscillation (ENSO). The attributes of the data include date, latitude, longitude, zonal winds, meridional winds, relative humidity, air temperature, and sea surface temperature. We selected sea surface temperature and humidity of 2500 observation collected in the eastern Pacific Ocean from December 1997 to February 1998; in this period, the magnitude of the 1997-1998 El Niño, the strongest of the century, was maximum. The subset contains two visible clusters.

The experiments have been performed on the three datasets S_0, S_1, and S_2 for both method KDE and MV, and each experiment has been executed over the two data structures distributed in an ad-hoc wireless network of 500 sensors:

- one data structure is generated by partitions of regions balancing the mass, namely balancing the number of data objects in each region (in short mass-based partitions)
- the other data structure is generated by partitions of regions balancing the volume, namely the Euclidean space occupied by each region (in short volume-based partitions).

Figure 13. First Data Set S_0

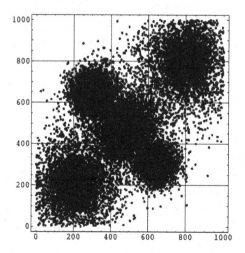

In particular, each experiment compares the clustering achieved by the two distributed data structure as the number of hops, to compute the density, varies from 1 to 5, keeping the number of hops to compute the clustering unchanged to 4 and 5, for the first two and the third data set respectively. For each experiment we have analyzed

(i) how the Rand index improves as the number of hops increases, i.e., how efficacy improves, and (ii) the efficiency measured by counting the number of messages among sensors generated by both the computation of the density estimate and the clustering.

Figure 14. Second Data Set S1

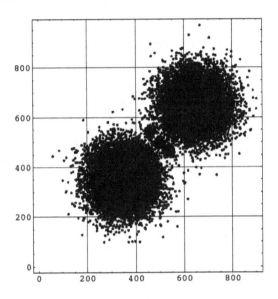

Figure 15. Third Dataset S_2 Sea surface temperature and humidity collected by the TAO array of buoys in the eastern Pacific Ocean from December 1997 to February 1998

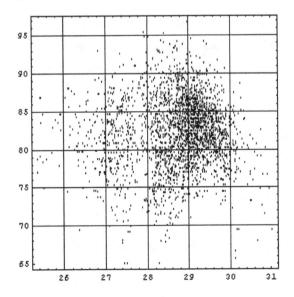

Results

The efficacy achieved in the first two data sets are described in Figure 16 and Figure 17, while the corresponding efficiency evaluations are depicted in Figure 18 and Figure 19. Both aspects of the third dataset are illustrated in Figure 20, Figure 21, Figure 22, and Figure 23.

As far as the first two data sets are concerned, Figure 16 depicts the efficacy of clustering computing the density according to the technique MV applied to the two data sets and, in each data set, to the two kinds of distributed structures. For both data sets, the volume-based partition structure is better than the mass-based one, moreover the clustering quality of the latter decreases as the

Figure 16.

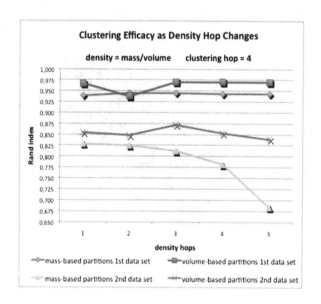

Figure 17.

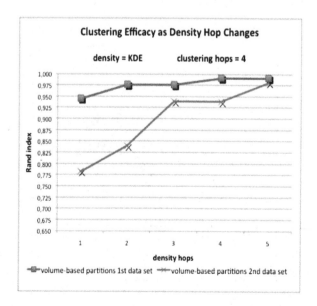

number of density hops increases. This behavior, which is more evident in the second data set (Figure 16), is due to the fact that mass-based partitions produce highly irregular spatial regions, causing distortions in the computation of the MV density as the number of hops increases, in fact the larger the number of regions involved, the larger the number of errors.

On the contrary, volume-based partitions generate much more regular regions; in fact, ideally, if volume-based partitions were so small, such that each one contained at most one data object, then the MV density computation would approximate

the KDE technique, which is the solution with the best efficacy as shown in Figure 17. Another expected result is that it is slightly more difficult to correctly cluster the second data set S_1 than S_0, both for KDE and MV techniques in both distributed data structures.

An important result is that the clustering quality of the MV technique is very similar to KDE (Figure 16), particularly for the first data set S_0 where the difference from volume-based partitions is less than 0.01, moreover for one density hop the MV quality is also slightly superior with a very high value of 0.97. It is also interesting to notice

Figure 18.

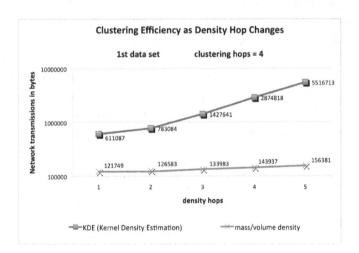

Figure 19.

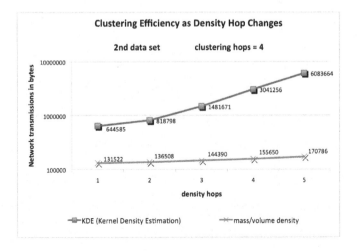

that the quality of MV on the second data set S_1 is better than KDE for the first two hops achieving more than 0.85; moreover, a low number of hops means lower network costs, but above all the network cost generated by KDE, as depicted in Figure 18 and Figure 19 (log scale), the first and second data set respectively, is on average 20 times greater than MV.

These evaluations are also confirmed in the third dataset, even if the best clustering quality, with 5 density hops, is a bit lower than the quality of the first two data sets (see Figure 20 and Figure 21). Again in the volume-based structure, the clustering quality of the MV density is very similar to the one produced by KDE, for instance with 5 hops they are greater than 0.81; instead in the mass-based structure the two density approaches show some differences in their best values, which are 0.72 and 0.89 with MV and KDE respectively.

The network costs in the third dataset are analogous to the first two data sets, even if they are slightly higher because the clustering, whose cost is constant with respect to the density, has been performed with one more hop, as depicted in Figure 22 and Figure 23 for volume-based and mass-based structures respectively.

To perform coherent comparisons, we have not reported in the graphics of Figure 20 and Figure 21 the clustering quality with higher density hops, but it is worth mentioning that the structure with volume-based partitions achieves very close results to the first two data sets, in fact with 6 hops the Rand index is 0.91 and 0.97 using the MV and KDE density respectively.

Finally, it is worth mentioning that network diameter is 31, and each sensor, according to the number of hops employed in the simulations, contacts at most between 50 and 75 neighbors, namely at most only between 10% and 15% of the network.

In conclusion, the MV technique guarantees a higher index and better quality clustering than KDE for a low number of hops, also with irregular distribution of data, and it cuts the network costs of more than one order of magnitude.

Figure 20.

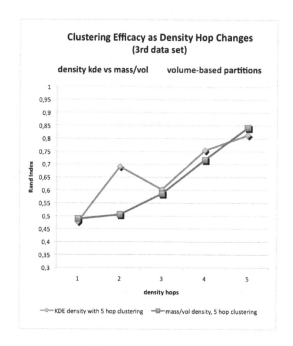

Figure 21.

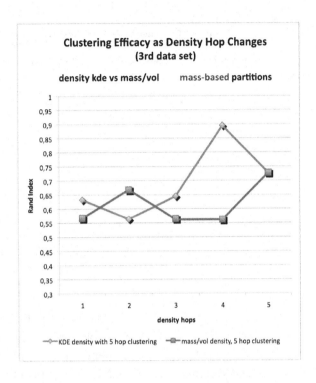

Figure 22.

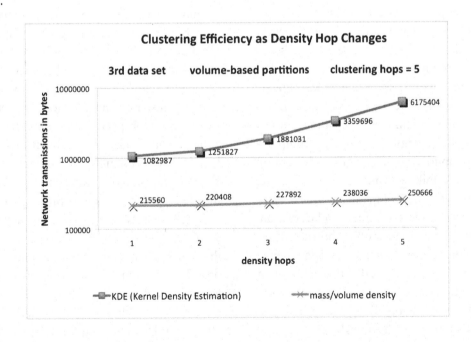

Figure 23.

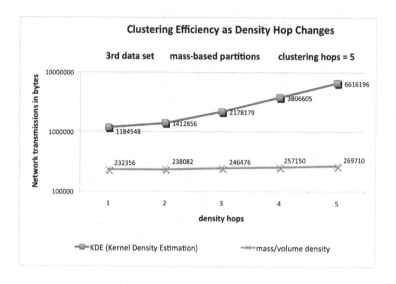

Related Work

In the last few years a major research effort in data mining has been devoted to distributed systems. However, the proposed solutions initially did not take into consideration additional constraints that sensor networks imply with respect to distributed systems of computers. Recently, sensor technology has evolved considerably and, as a consequence, distributed data mining in sensor networks has attracted a growing interest (Pedersen & Jul, 2005). In these works, sensors interact as peers to solve a distributed learning problem, e.g., regression, classification, facility location, subject to constraints of energy consumption, computational throughput, available memory, connection volatility, node failure, which are characteristic of sensor and P2P environments.

As of writing, there are three studies on data clustering in massive-scale, self-administered networks. In (Klampanos & Jose, 2004; Klampanos, Jose, & van Rijsbergen, 2006) the problem of P2P information retrieval is addressed by locally clustering documents residing at each peer and subsequently clustering the peers by a single-pass algorithm which assigns each new peer to the closest cluster, or initiates a new peer cluster with

it, depending on a distance threshold. Although the approach results in a global clustering of the documents in the network, these works do not compare directly to ours, since their main goal is to show how simple forms of clustering can be exploited to reorganize the network for improving query answering effectiveness.

In (Li, Lee, Lee, & Sivasubramaniam, 2006) the PENS algorithm is proposed to cluster data residing in P2P networks with a CAN overlay, according to a density-based criterion. First, the DBSCAN algorithm is run locally by each peer. Then, a cluster expansion check is performed to detect, for each peer, which neighboring CAN zones contain clusters which can be merged to local clusters contained in the peer's zone. The check is performed bottom-up in the virtual tree induced by CAN's zone-splitting mechanism. Finally, clusters are merged by appropriately selected arbiters in the tree. The authors show that the communication requirements of their approach is linear in the number of peers. Similarly to the methods we have considered in our analysis, this work assumes a density-based model of clustering. However, clusters emerge by cutting the data space following contours defined by a density threshold, as in the DBSCAN algorithm, whereas the algorithms considered in the present paper utilize a density-tree criterion, similar to the one proposed in (Silverman, 1986) to define center-based clusters.

In (Nowak, 2003) the author proposes a distributed version of the Expectation-Maximization algorithm to construct parametric Gaussian mixture models in a sensor network computing environment. The effectiveness of the approach is demonstrated by convergence analysis and simulation experiments. A possible drawback of the proposed approach is the need to select the model order, i.e., the number of mixture components, in advance. However, approaches to automatic order selection are suggested by the author.

CONCLUSION AND FUTURE WORK

Although a consistent effort in data mining research deals with distributed computation, there is comparatively little work that has been published on mining in sensor networks, despite this kind of network is spreading in many commercial, scientific and social application domains. Equipping small wireless devices with data mining algorithms means to bring in everyday life pervasive intelligent systems capable of making predictions anticipating behaviors and complex phenomena. This will lead to an amazing variety of network applications, such ambient intelligence, logistics, precision agriculture, environmental monitoring, robotics, industrial processing, vehicular traffic control, body sensor networks, wellness systems etc.

These potentials lead also to several research challenges, among which the development of new in-network data mining algorithms and techniques, beyond the data clustering we have presented in this paper, able to efficiently produce accurate results according to typical limits of this technology, such as connectivity, computation, memory and energy consumption.

As far as our future work is concerned, we are interested in studying the behavior of MV in new multi-dimensional distributed data structures in order to reach almost 100% of accuracy also for highly complex data distributions.

REFERENCES

Akyildiz, I. F., Pompili, D., & Melodia, T. (2005). Underwater acoustic sensor networks: Research challenges . *Ad Hoc Networks*, *3*, 257–279. doi:10.1016/j.adhoc.2005.01.001

Ankerst, M., Breunig, M., Kriegel, H.-P., & Sander, J. (1999). OPTICS: Ordering points to identify the clustering structure. In *ACM SIGMOD Record, Proceedings of the 1999 ACM SIGMOD International Conference on Management of Data*, *28*(2), 49-60.

Babcock, B., Babu, S., Datar, M., Motwani, R., & Widom, J. (2002). Models and issues in data stream systems. In *Proceedings of the Twenty-First ACM SIGMOD-SIGACT-SIGART Symposium on Principles of Database Systems* (pp. 1-16). New York: ACM.

Bentley, J. L. (1975). Multidimensional binary search trees used for associative searching. *Communications of the ACM*, *18*, 509517. doi:10.1145/361002.361007

Bentley, J. L., & Friedman, J. H. (1979). Data structures for range searching. *ACM Computing Surveys*, *11*, 397409. doi:10.1145/356789.356797

Collard, S. B., & Lugo-Fernández, A. (1999). *Coastal upwelling and mass mortalities of fishes and invertebrates in the northeastern Gulf of Mexico during spring and summer 1998* (OCS Study MMS 99-0049). New Orleans, LA: U.S. Department of the Interior, Minerals Management Service, Gulf of Mexico OCS Region. Retrieved August 4, 2008, from http://www.gomr.mms.gov/PI/PDFImages/ESPIS/3/3207.pdf

Cui, J.-H., Kong, J., Gerla, M., & Zhou, S. (2006). The challenges of building mobile underwater wireless networks for aquatic applications. *IEEE Network, 20*(3), 12–18. doi:10.1109/MNET.2006.1637927

Cummings, J. A. (1994). Global and regional ocean thermal analysis systems at Fleet Numerical Meteorology and Oceanography Center. In *OCEANS '94, 'Oceans Engineering for Today's Technology and Tomorrow's Preservation.' Proceedings, Vol. 3*, (pp.III/75-III/81). Brest, France: IEEE Press.

Ester, M., Kriegel, H.-P., Sander, J., & Xu, X. (1996). A density-based algorithm for discovering clusters in large spatial databases with noise. In E. Simoudis, J. Han, & U. M. Fayyad (Eds.), *Proceedings of the Second International Conference on Knowledge Discovery and Data Mining* (KDD-96) (pp. 226-231). AAAI Press.

Gaede, V., & Günther, O. (1998). Data structures for range searching. *ACM Computing Surveys, 30*, 170–231. doi:10.1145/280277.280279

Ganesan, P., Yang, B., & Garcia-Molina, H. (2004). One torus to rule them all: multi-dimensional queries in P2P systems. In *Proceedings of the 7th International Workshop on the Web and Databases* (pp. 19-24). New York: ACM.

Hartigan, J. A., & Wong, M. (1979). Algorithm AS136: A k-means clustering algorithm. *Applied Statistics, 28*, 100–108. doi:10.2307/2346830

Hettich, S., & Bay, S. D. (1999). The UCI KDD archive. Irvine, CA: University of California, Department of Information and Computer Science. Retrieved July 22, 2008, from http://kdd.ics.uci.edu/databases/el_nino/tao-all2.dat.gz

Hinneburg, A., & Keim, D. A. (1998). An efficient approach to clustering in large multimedia databases with noise. In R. Agrawal, P. Stolorz, & G. Piatetsky-Shapiro (Eds.), *Proceedings, The Fourth International Conference on Knowledge Discovery and Data Mining* (pp. 58-65). Menlo Park, CA: AAAI Press.

Intanagonwiwat, C., Govindan, R., Estrin, D., Heidemann, J., & Silva, F. (2003). Directed diffusion for wireless sensor networking. *IEEE/ACM Transactions on Networking, 11*, 2-16.

Jain, A. K., & Dubes, R. C. (1988). *Algorithms for clustering data.* Englewood Cliffs, NJ: Prentice-Hall.

Jain, A. K., Murty, M. N., & Flynn, P. J. (1999). Data clustering: A review. *ACM Computing Surveys, 31*, 264–323. doi:10.1145/331499.331504

Johnson, E., & Kargupta, H. (1999). Collective, hierarchical clustering from distributed heterogeneous data. In M. Zaki & C. Ho (Eds.), *Large-Scale Parallel KDD Systems*, Volume 1759 of Lecture Notes in Computer Science (pp. 221-244). Berlin/Heidelberg, Germany: Springer.

Kargupta, H., & Chan, P. (Eds.). (2000). *Distributed and parallel data mining.* Menlo Park, CA / Cambridge, MA: AAAI Press / MIT Press.

Kargupta, H., Huang, W., Sivakumar, K., & Johnson, E. L. (2001). Distributed clustering using collective principal component analysis. *Knowledge and Information Systems, 3*, 422–448. doi:10.1007/PL00011677

Kargupta, H., Park, B.-H., Hershberger, D., & Johnson, E. (2000). Collective data mining: a new perspective toward distributed data mining. In H. Kargupta & P. Chan (Eds.), *Advances in distributed and parallel knowledge discovery* (pp. 131-174). Menlo Park, CA / Cambridge, MA: AAAI Press / MIT Press.

Karp, B., & Kung, H. (2000). GPSR: Greedy perimeter stateless routing for wireless networks. In *MobiCom '00: 6th Annual International Conference on Mobile Computing and Networking,* (pp. 243-254). New York: ACM.

Klampanos, I. A., & Jose, J. M. (2004). An architecture for information retrieval over semi-collaborating peer-to-peer networks. In *Proceedings of the 2004 ACM Symposium on Applied Computing* (pp. 1078-1083). New York: ACM.

Klampanos, I. A., Jose, J. M., & van Rijsbergen, C. J. K. (2006). Single-pass clustering for peer-to-peer information retrieval: The effect of document ordering. In *INFOSCALE '06. Proceedings of the First International Conference on Scalable Information Systems.* New York: ACM.

Klusch, M., Lodi, S., & Moro, G. (2003). Distributed clustering based on sampling local density estimates. In *Proceedings of the 19th International Joint Conference on Artificial Intelligence* (pp. 485-490). Acapulco, Mexico: AAAI Press.

Koontz, W. L. G., Narendra, P. M., & Fukunaga, K. (1976). A graph-theoretic approach to non-parametric cluster analysis. *IEEE Transactions on Computers, C-25,* 936–944. doi:10.1109/TC.1976.1674719

Kudela, R., Pitcher, G., Probyn, T., Figueiras, F., Moita, T., & Trainer, V. (2005). Harmful algal blooms in coastal upwelling systems. *Oceanography (Washington, D.C.), 18*(2), 184–197.

Li, M., Lee, G., Lee, W.-C., & Sivasubramaniam, A. (2006). PENS: An algorithm for density-based clustering in peer-to-peer systems. In *INFOSCALE '06. Proceedings of the First International Conference on Scalable Information Systems.* New York: ACM.

Li, X., Kim, Y., Govindan, R., & Hong, W. (2003). Multidimensional range queries in sensor networks. In *SenSys '03: Proceedings of the 1st International Conference on Embedded Networked Sensor Systems* (pp. 63-75). New York: ACM.

Lodi, S., Moro, G., & Sartori, C. (in press). Distributed data clustering in multi-dimensional peer-to-peer networks. In H. T. Shen & A. Bouguettaya (Eds.), Conferences in Research and Practice in Information Technology (CRPIT): Vol. 103. Proceedings of the Twenty-First Australasian Database Conference (ADC2010). Brisbane, Australia: Australian Computer Society.

Madden, S., Franklin, M. J., Hellerstein, J. M., & Hong, W. (2002). TAG: A tiny aggregation service for ad-hoc sensor networks. *ACM SIGOPS Operating Systems Review, 36, Issue SI (Winter 2002), OSDI '02: Proceedings of the 5th symposium on Operating Systems Design and Implementation, SPECIAL ISSUE: Physical Interface,* 131-146.

Merugu, S., & Ghosh, J. (2003). Privacy-preserving distributed clustering using generative models. In X. Wu, A. Tuzhilin, & J. Shavlik (Eds.), *Proceedings of the 3rd IEEE International Conference on Data Mining.* Los Alamitos, CA: IEEE Computer Society.

Milojicic, D. S., Kalogeraki, V., Lukose, R., Nagaraja, K., Pruyne, J., Richard, B., et al. (2002). *Peer-to-peer computing* (Technical Report HPL-2002-57R1). HP Lab.

Monti, G., & Moro, G. (2008). Multidimensional range query and load balancing in wireless ad hoc and sensor networks. In K. Wehrle, W. Kellerer, S. K. Singhal, & R. Steinmetz (Eds.), *Eighth International Conference on Peer-to-Peer Computing* (pp. 205-214). Los Alamitos, CA: IEEE Computer Society.

Monti, G., & Moro, G. (2009). Self-organization and local learning methods for improving the applicability and efficiency of data-centric sensor networks. *Sixth International ICST Conference on Heterogeneous Networking for Quality, Reliability, Security and Robustness, QShine/AAA-IDEA 2009*, (LNICST 22, pp. 627-643). Berlin/Heidelberg, Germany: Springer.

Monti, G., Moro, G., & Lodi, S. (2007). W*-Grid: A robust decentralized cross-layer infrastructure for routing and multi-dimensional data management in wireless ad-hoc sensor networks. In M. Hauswirth, A. Montresor, N. Shahmehri, K. Wehrle, & A. Wierzbicki, *Seventh IEEE International Conference on Peer-to-Peer Computing* (pp. 159-166). Los Alamitos, CA: IEEE Computer Society.

Monti, G., Moro, G., & Sartori, C. (2006). W^R-Grid: A scalable cross-layer infrastructure for routing, multi-dimensional data management and replication in wireless sensor networks. In G. Min, B. Di Martino, L. T. Yang, M. Guo, & G. Ruenger (Eds.), *Frontiers of High Performance Computing and Networking – ISPA 2006 Workshops* (LNCS 4331, pp. 377-386). Berlin/Heidelberg, Germany: Springer.

Moro, G., & Monti, G. (2006). W-Grid: A cross-layer infrastructure for multi-dimensional indexing, querying and routing in ad-hoc and sensor networks. In A. Montresor, A. Wierzbicki, & N. Shahmehri (Eds.), *IEEE Int. Conference on P2P Computing* (pp. 210-220). Los Alamitos, CA: IEEE Computer Society.

Moro, G., & Ouksel, A. M. (2003). G-Grid: A class of scalable and self-organizing data structures for multi-dimensional querying and content routing in P2P networks. *Agents and Peer-to-Peer Computing* (LNCS 2872, pp. 123-137). Berlin/Heidelberg, Germany: Springer.

Nascimento, S., Casimiro, H., Sousa, F. M., & Boutov, D. (2005). Applicability of fuzzy clustering for the identification of upwelling areas on sea surface temperature images. In B. Mirkin & G. Magoulas (Eds.), *Proceedings of the 2005 UK Workshop on Computational Intelligence* (pp. 143–148). London, United Kingdom. Retrieved August 6, 2008, from http://www.dcs.bbk.ac.uk/ukci05/ukci05proceedings.pdf

Nowak, R. D. (2003). Distributed EM algorithms for density estimation and clustering in sensor networks. *IEEE Transactions on Signal Processing, 51*, 2245–2253. doi:10.1109/TSP.2003.814623

Oliver, M. J., Glenn, S., Kohut, J. T., Irwin, A. J., Schofield, O. M., & Moline, M. A. (2004). Bioinformatic approaches for objective detection of water masses on continental shelves. *Journal of Geophysical Research*, 109.

Pedersen, R., & Jul, E. (Eds.). (2005). *First International Workshop on Data Mining in Sensor Networks*. Retrieved April 18, 2007, from http://www.siam.org/meetings/sdm05/sdm-Sensor-Networks.zip

Queensland Department of Natural Resources and Mines. Environmental Protection Agency, Queensland Health, Department of Primary Industries, & Local Governments Association of Queensland (2002). *Queensland Harmful Algal Bloom Response Plan*. Version 1. Retrieved from http://www.nrw.qld.gov.au/water/blue_green/pdf/multi_agency_hab_plan_v1.pdf

Rand, W. M. (1971). Objective criteria for the evaluation of clustering methods. *Journal of the American Statistical Association, 66*, 846–850. doi:10.2307/2284239

Ratnasamy, S., Francis, P., Handley, M., Karp, R., & Schenker, S. (2001). A scalable content-addressable network. In *Proceedings of the 2001 Conference on Applications, Technologies, Architectures, and Protocols for Computer Communications* (pp. 161–172). New York, NY: ACM.

Ratnasamy, S., Karp, B., Shenker, S., Estrin, D., Govindan, R., & Yin, L. (2003). Data-centric storage in sensornets with GHT, a geographic hash table. *Mobile Networks and Applications, 8,* 427–442. doi:10.1023/A:1024591915518

Schofield, O., Grzymski, J., Paul Bissett, W., Kirkpatrick, G. J., Millie, D. F., & Moline, M. (1999). Optical monitoring and forecasting systems for harmful algal blooms: Possibility or pipe dream? *Journal of Phycology, 35,* 1477–1496. doi:10.1046/j.1529-8817.1999.3561477.x

Scott, D. W. (1992). *Multivariate density estimation. Theory, practice, and visualization.* New York: Wiley.

Silverman, B. W. (1986). *Density estimation for statistics and data analysis.* London: Chapman and Hall.

Sozer, E. M., Stojanovic, M., & Proakis, J. G. (2000). Underwater acoustic networks. *IEEE Journal of Oceanic Engineering, 25,* 72–83. doi:10.1109/48.820738

Steinbach, M., Karypis, G., & Kumar, V. (2000). A comparison of document clustering techniques. In *Proceedings of the ACM SIGKDD Workshop on Text Mining.* Retrieved from http://www.cs.cmu.edu/~dunja/KDDpapers/

Tamminen, M. (1984). Comment on quad- and oc-trees. *Communications of the ACM, 27,* 248–249. doi:10.1145/357994.358026

Tasoulis, D. K., & Vrahatis, M. N. (2004). Unsupervised distributed clustering. In M. H. Hamza (Ed.), *IASTED International Conference on Parallel and Distributed Computing and Networks* (pp. 347-351). Innsbruck, Austria: ACTA Press.

Xiao, L., & Ouksel, A. M. (2005). Tolerance of localization imprecision in efficiently managing mobile sensor databases. In U. Cetintemel & A. Labrinidis (Eds.), *Proceedings of the 4th ACM International Workshop on Data Engineering for Wireless and Mobile* (pp. 25- 32). New York: ACM.

Xu, X., Ester, M., Kriegel, H.-P., & Sander, J. (1998). A distribution-based clustering algorithm for mining in large spatial databases. In *Proceedings 14th International Conference on Data Engineering* (pp. 324-331). Los Alamitos, CA: IEEE Computer Society.

Yang, X., Ong, K. G., Dreschel, W. R., Zeng, K., Mungle, C. S., & Grimes, C. A. (2002). Design of a wireless sensor network for long-term, in-situ monitoring of an aqueous environment. *Sensors, 2,* 455–472. doi:10.3390/s21100455

Zaki, M. J., & Ho, C.-T. (Eds.). (2000). *Large-Scale Parallel Data Mining* (LNCS 1759). Berlin/Heidelberg, Germany: Springer.

ENDNOTE

[1] The pressure from the weight of 1 meter of water

Section 4

Query Languages and Query Optimization Techniques for Warehousing and Mining Sensor Network Data

Chapter 10
Intelligent Querying Techniques for Sensor Data Fusion

Shi-Kuo Chang
University of Pittsburgh, USA

Gennaro Costagliola
Università di Salerno, Italy

Erland Jungert
Swedish Defense Research Agency, Sweden

Karin Camara
Swedish Defense Research Agency, Sweden

ABSTRACT

Sensor data fusion imposes a number of novel requirements on query languages and query processing techniques. A spatial/temporal query language called ΣQL has been proposed to support the retrieval of multimedia information from multiple sources and databases. This chapter investigates intelligent querying techniques including fusion techniques, multimedia data transformations, interactive progressive query building and ΣQL query processing techniques using sensor data fusion. The authors illustrate and discuss tasks and query patterns for information fusion, provide a number of examples of iterative queries and show the effectiveness of ΣQL in a command-action scenario.

INTRODUCTION

Sensor data fusion is an area of increasing importance that requires novel query languages and query processing techniques for the handling of spatial/temporal information. Sensors behave quite differently from traditional database sources. Most sensors are designed to generate information in a temporal sequence. Sensors such as video camera

and laser radar also generate large quantities of spatial information. Therefore, the query language and the query processing techniques must be able to handle sources that can produce large quantities of streaming data, which due to the imperfections of the sensors also result in uncertain information generated within short periods of time.

Another aspect to consider is that user's queries may be modified to include data from more than one sensor and therefore require fusion of multiple

DOI: 10.4018/978-1-60566-328-9.ch010

sensor information. In our empirical study we collected information from different types of sensors, including, among others, laser radar, infrared video (similar to video but generated at 60 frames/sec) and CCD digital camera. In a preliminary analysis of the above described sensor data, it is found that data from a single sensor yields poor results in object recognition. Object recognition can be significantly improved if the query is modified to obtain information from other types of sensors, while allowing the target to be partially hidden. In other words, one (or more) sensors may serve as a guide to the other sensors by providing status information such as position, time and accuracy, which can be incorporated in multiple views and formulated as constraints in the refined query.

Existing query processing techniques are not designed to handle sensors that produce large quantities of streaming data within short periods of time. With existing query languages such as SQL, it is also difficult to systematically refine the query to deal with information fusion from multiple sensors and distributed databases. To support the retrieval and fusion of multimedia information from multiple sources and distributed databases, a spatial/temporal query language called ΣQL has been proposed (Chang et al., 2004). ΣQL is based upon the σ-operator sequence and in practice expressible in a syntax similar to SQL. ΣQL allows a user to specify powerful spatial/temporal queries for both multimedia data sources and multimedia databases, eliminating the need to write separate queries for each. A ΣQL query can be processed in the most effective manner by first selecting the suitable transformations of multimedia data to derive the multimedia static schema, and then processing the query with respect to the selected multimedia static schema.

The main contribution of this chapter is to provide a systematic approach of intelligent querying consisting of fusion techniques, multimedia data transformations and interactive progressive query building processing techniques for sensor data fusion. The chapter is organized as follows. Section 2 presents background and related research. The basic concept of the dual representation of the σ-query is explained in Section 3. The usage of the various types of operators is discussed in Section 4. The techniques of sensor data fusion are explained in Section 5. Section 6 recalls the architecture of the system. In section 7 tasks and query patterns for information fusion are illustrated and discussed. In Sections 8 we provide a number of examples of iterative queries, while section 9 illustrates the use of ΣQL with some examples taken from a scenario and where also interactive progressive query building is included. Section 10 concludes the chapter.

BACKGROUND AND RELATED RESEARCH

Sensor data fusion poses some special problems. First of all, there is no general solution to the problem of sensor data fusion for an unlimited number of different types of sensors. The problem is usually restricted to a limited number of object types observed from a specific perspective by a limited number of sensors (White, 1998). One such example is to select sensors that are looking only at ground objects, primarily vehicles, from a top view perspective where the sensors are carried by a flying platform such as a helicopter. By studying this restricted problem in detail, we may be able to understand better how to deal with complex queries for sensor data fusion. For a more general view on sensor data fusion, see (Hall & Llinas, 2001).

As explained in the preceding section, sensor data fusion requires a query language that supports multiple sensor sources and the systematic modification of queries. In early research on query modification, queries are modified to deal with integrity constraints (Stonebraker, 1975). In query augmentation, queries are augmented by adding constraints to speed up query processing (Grafe, 1993). In query refinement (Vélez et al., 1997)

Figure 1. An experimental prototype of the ΣQL system

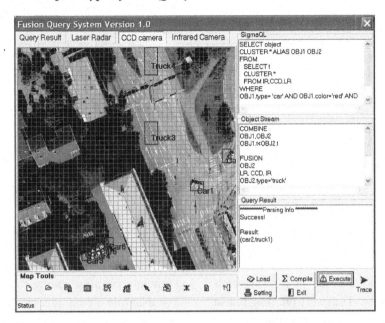

multiple term queries are refined by dynamically combining pre-computed suggestions for single term queries. In (Chakrabarti et al., 2000), query refinement techniques were applied to content-based retrieval from multimedia databases. It is worth noting that there also are some query techniques employed in information retrieval in sensor networks (Gehrke & Madden, 2004). However, these techniques mainly focus on the problems about data collection from sensor nodes and preservation of energy more than information fusion. In our approach the refined queries are manually created to deal with the lack of information from a certain source or sources, and therefore not only the constraints can be changed, but also the source(s). This approach has not been considered previously in database query processing because usually the sources are assumed to provide the complete information needed by the queries.

In our previous research, a spatial/temporal query system called the ΣQL System was developed to support the retrieval and fusion of multimedia information from real-time sources and databases (Chang at al., 2004; Chang et al.,

2002a). Figure 1 shows the experimental prototype of the ΣQL System (Li et Chang, 2003), in which a textual query interface is provided for users based on a SQL-like query language called ΣQL, which allows the users to specify powerful spatial/temporal queries for multiple data sources.

To investigate the usability of the ΣQL system, experimental results are analyzed and summarized in Figure 2. Figure 2(a) shows system CPU/memory usage during processing of a complex spatial/temporal query. Depending on the workload, query processing can be divided into two phases, marked as Phase A and Phase B:

- **Phase A:** The characteristics of this phase is low workload on CPU/memory for the system (CPU Usage<30%; Memory Usage < 10%). The system is almost idle because it is accepting user's input, checking syntax of query or doing other light-load tasks.
- **Phase B:** CPU/memory becomes much busier because the system is executing query and visualizing result.

Figure 2. Experimental result of sigma query system

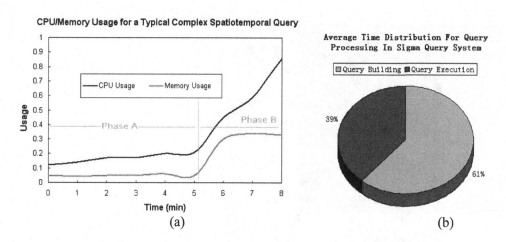

Figure 2 (b) shows the statistical result from the user's point of view. The *query building* phase (corresponding to Phase A) begins from inputting a query to submitting a query, followed by the *query execution* phase (corresponding to Phase B). Results from the view of the system (Figure 2 (a)) and the view of the user (Figure 2 (b)) both consistently lead to the following conclusion: for the ΣQL system the users spend more time building a query than executing a query. In other words *query building* is the most significant phase in the processing of complex spatial/temporal queries. On the other hand the system is almost idle while the user is composing a complex query.

This explains why optimization for *query execution* does not significantly improve the usability of the system. *Query execution* is actually just a small part of the entire query processing task. The query interface plays a crucial role in this task because it is a tool to help a user build complex queries, which is the most time consuming part of the query processing task.

To reduce complexity, in our approach complicated queries are broken into several separated iterations, which is illustrated in (Chang et al., 2006). It is worth noting that in many cases, users are not completely clear on what they are really trying to find before starting or they may change

their interest after seeing the result of a query. For example a commander may want to find all tanks in a battle field. Once tanks are shown on the map, objects of threat near these tanks probably will become the target of the next query. In other words queries could be generated step by step and coupled together, making the queries more and more complex. This is another reason to make the query building interactive and progressive.

THE DUAL REPRESENTATION OF THE ΣQL QUERY LANGUAGE

Let us now recall the main concepts from (Chang et al., 2004). ΣQL is a spatial/temporal query language for information retrieval from multiple heterogeneous sources and databases. Unlike SQL, which does not explicitly deal with spatial/temporal queries, ΣQL is designed with that purpose in mind. Unlike SQL, which deals only with databases, ΣQL is designed to deal with both static sources (databases) and dynamic sources of streaming sensor data, and furthermore these sources may be distributed. Its strength is its simplicity: the query language is based upon a single operator - the σ-operator. Yet the concept is natural and can easily be mapped into an SQL-like query

Figure 3. An example of extracting three time slices from a source

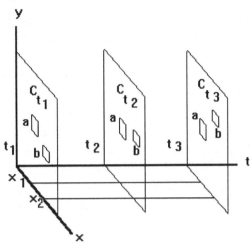

language. The σ-query is useful in theoretical investigation, while the SQL-like query language is easy to implement and is a step towards a user-friendly visual query language. An example is illustrated in Figure 3. Each *time slice* C_{ti} consists of objects with the same time value. In turn, the time slices form the *source*, also called *universe*. To extract three pre-determined time slices at times t_1, t_2, t_3 from the source R, the σ-query in mathematical notation is: $\sigma_t (t_1, t_2, t_3)$ R.

The meaning of the σ-operator in the above query is "select", i.e. we want to select the time dimension and three slices along this dimension. The subscript t in σ_t indicates the selection of the time dimension. In the SQL-like language the ΣQL query is expressed as:

```
SELECT t
CLUSTER t₁, t₂, t₃
FROM R
```

A new keyword "CLUSTER" is introduced so that the parameters such as t_1, t_2, t_3 for the σ-operator can be listed. The word "CLUSTER" indicates that objects belonging to the same subset of the universe (a cluster) must share some common characteristics (such as having the same

time value, being similar to one another, etc.) Clustering is a technique used in pattern classification to form subsets of similar objects. A cluster may have a sub-structure specified in another (recursive) query. Clustering is a natural concept when dealing with spatial/temporal objects that are specifiable only through similarity to other objects. The result of a ΣQL query is a string that describes the relationships among the clusters. This string is called a *cluster-string*, which will also be discussed further in Section 4.

The dual representation of ΣQL means that a query can be formulated as an SQL-like query (Chang et al., 1998) or as a sequence of *generic* operators (the σ-operators introduced above) and *specialized* operators (the φ-operators to be discussed in the following section). Translation from one representation to the other is quite straightforward.

The operators may handle both qualitative and quantitative information. Primarily, the operators allow operations on a sensor-data-independent level, i.e. sensor data should be transformed into information structures at high abstraction levels that are sensor independent. To accomplish this, the queries are expressed in terms of operator sequences where the transformations of the sensor data are carried out stepwise by the operators. The operators reduce the dimensions of the multi-dimensional search space to which each successive operator is applied. Intuitively, the reduced search space is also another *cluster*. Thus, as successive operators are applied, the clusters become more and more refined.

In contrast to this refinement of the search space, the *fusion* and related operators take multiple clusters as input and fuse the information to determine a belief value that may support a certain hypothesis such as whether different observations in the clusters correspond to the same object. Furthermore the fusion and related operators can handle uncertain information through the use of belief values.

OPERATOR CLASSES

The operators in ΣQL can be categorized with respect to their functionality. The two main classes are the *transformational operators* (the σ-operators) and the *fusion operators* (the φ-operators). In this section the two main operator classes are discussed according to their input, output and functionality.

σ-operators

A σ-operator is defined as an operator to be applied to any multi-dimensional source of objects in a specified set of intervals along a dimension. The operator projects the source along that dimension to extract clusters (Chang et al., 1998). Each cluster contains a set of objects or components whose projected positions fall in one of the given intervals along that dimension. As an example, let us write a σ-expression for extracting the video frame sequences in the time intervals $[t_1\text{-}t_2]$ and $[t_3\text{-}t_4]$ from a video source VideoR. The expression is $\sigma_{time}([t_1\text{-}t_2], [t_3\text{-}t_4])$ VideoR where VideoR is projected along the time dimension to extract clusters (frames in this case) whose projected

positions along the time dimension are in the specified intervals.

In case of uncertainty, the components of the clusters may be associated with various probabilities or belief values. Input and output data may be of either qualitative or quantitative type, although generally the later type is of main interest. Thus input data will be accessed from either a raw-data source such as a sensor, or from a structured data source such as a database, or from some internal source such as *qualitative strings* that are strings consisting of object descriptions projected along certain dimension(s) (Chakrabarti et al., 2000) . The output data correspond to clusters in relational representations that in practice may be available as qualitative strings of various types (Chang et al., 1998). The general syntax can thus be expressed in the following way:

```
σ_dimension (<intervals>){<source> | <cluster>
| < cluster_strings> | …} ⇒
{<cluster> | < cluster_strings> | …}
```

A variety of σ-operators can be defined (Jungert, 1999a). Many of these operators are common in most spatial applications. Examples are the determination of various attributes and spatial relations, such as 'northwest-of', 'to the left of', etc. For simple inputs, these operators can be described as:

```
σ_attribute (<attribute-values>)<cluster_
strings> ⇒
<relational_strings_of_(<attribute > <
object> <attribute value>)>
σ_relation ((<relational_value>)< cluster_
strings> ⇒
{(<relation>(<relation_value> <object>-i
<object>-j))}
```

As an example to find a pair of objects such that the blue object precedes (<) the red object along the spatial dimension V, the σ-operator instance is:

```
σ_color(blue, red)(V:A < B) = (V:(color A
blue) < (color B red))
```

where (V:A<B) is a cluster string.

In case of uncertainty the input and output to the σ-operators may include an attribute corresponding to a specific belief value. The σ_{type}- and the σ_{motion}- operators may include this attribute. The σ_{type}-operator is concerned with matching between objects found in a sensor image and objects stored in a library database and where both objects are described in the same terms that may be either qualitative or quantitative. Traditionally matching was regarded as a part of information fusion. Generally, however, the σ_{type}-operator and its result, i.e. a set of objects and their corresponding normalized belief values, can be expressed as follows when the input to the operator is a single cluster:

```
σ_type (type_value) < cluster_strings> ⇒
< cluster_strings_of_ (<type_value>
<object>-i <nbv>-i)>
```

where *<nbv>* corresponds to the normalized belief value that in practice becomes an attribute to the actual object. An instance of this is:

```
σ_type (car)(U:A < B < C) = (U:(car A 0.7) <
(car B 0.1) < C)
```

φ-operators

The φ-operators are more complex because they are concerned with sensor data fusion. Consequently these operators require more complex expressions as well as input data in different time periods from multiple sensors.

The φ_{fusion}-operator performs sensor data fusion from heterogeneous data sources to generate fused objects. Fusion of data from a single sensor in different time periods is also allowed. The output of the φ_{fusion}-operator is some kind of high level, qualitative representation of the fused object, and may include object type, attribute values and status values. The output may also include a normalized belief value for each fused object.

The output from the fusion operator may serve as the answer to a query. This result may consist of a list of objects each having a belief value. The object with the highest belief value is the most likely answer to the query and thus should come first in the list. The general description of the fusion operator is therefore:

```
φ_fusion(cluster | cluster-string) =>
{(type, obj, nbv)|(type, obj, nbv)-list}
```

SENSOR DATA FUSION

In sensor data fusion (Chang et al., 1996), queried objects from the different sensors need be associated to each other in order to determine whether they are the same object registered at different times and at different locations. Tracking is another problem that is closely related to the sensor data fusion problem. In tracking the problem is to verify whether different object observations represent the same or different objects. Another problem of concern is the uncertainties of the sensor data. Consequently, there is a demand for a well-structured and efficient general framework to deal smoothly with a large number of different query types with heterogeneous input data. Some of these aspects are further discussed by (Horney et al., 2004).

For certain sensors the objects can only be determined with respect to their type but rarely with respect to their identity. Therefore classification of the objects is necessary. This is a process that can be carried out in a matching algorithm that should display a result that includes not only the type of the object but a normalized belief value, *nbv*, associated to the observed object type. A number of attributes and state variables can be extracted from the sensor data where the actual types of attributes depend on the actual sensor types. Among the most important state variables are orientation, type and position, direction of motion, speed and acceleration. Most attributes and state variables, such as position and orientation, may be determined either in quantitative or in qualitative terms. In ΣQL reasoning is generally performed on qualitative information, basically represented in terms of Symbolic Projection. For this reason it is an advantage to use qualitative input to the fusion process as well.

A query demonstrating the usage of ΣQL with the sensor sources video and laser-radar can now be provided. Laser radars use laser beams to scan and process the signal echoed from targets, to create a virtual picture of the area. Given the input information from the laser-radar, Figure 4(a), and the video camera, Figure 4(b) the query is as follows: *is there a moving vehicle present in the given area and in the given time interval?* In this query the laser-radar data can be used as an index to the video. In this way most of the computational efforts can be avoided since the vehicles can be identified in almost real time in a laser-radar image. However, in the laser-radar used here, it cannot be determined whether an identified vehicle is moving or not. Consequently once a vehicle has been identified in a laser-radar image, we need to determine whether it is moving by analyzing a small number of video frames taken in a short time interval. This is possible to accomplish because the location of the vehicle at a certain time is known from the laser-radar information, which is illustrated in the Figure 4(a) and 4(b). The three images illustrate a situation where a car is first about to enter a parking lot (the two video frames) and at a later time the car has entered the parking lot (the laser-radar image).

The query is logically split into two parts, one looking for the vehicles in the laser-radar image and another looking for the vehicles in the video frames during the same time interval as the laser-radar image. The result of these two sub-queries are fused by applying the fusion operator $\varphi_{type,position,direction}$ which includes the fusion

procedure with the voting scheme. The fusion is applied with respect to type, position and direction including also their belief values.

The query demonstrating the problem can thus be expressed as:

$$\varphi_{pe,position,direction}$$
$$(\sigma_{motion}(moving)\ \sigma_{type}(vehicle)$$
$$\sigma_{xy}(*)$$
$$\sigma\ (T)_{T\ mod\ 10\ =\ 0\ and\ T>t1\ and\ T\ <t2}$$
$$\sigma_{media\_sources}\ (video)\,media\_sources$$
$$\sigma_{ype}\ (vehicle)\ \sigma_{xyz}(*)$$
$$\sigma\ (T)_{\ T>t1\ and\ T<t2}$$
$$\sigma_{media\_sources}(laser\_radar)\ media\_sources)$$

In processing this query it is assumed that it is reasonable to just pick out every tenth of the video frames for analysis. However it is quite possible that the query will work well even for larger gaps between the frames. The main problem is to keep track of the co-ordinates, which require knowledge about the speed and altitude of the platform. A GPS sensor system is required for monitoring this. This is a problem of the video while the laser-radar has a built-in GPS system and is therefore simpler to deal with from this perspective.

The φ-operator performs the fusion process during the execution of the query. Each vehicle identified in any of the images during the query is subject to the determination of these state variables and attributes and their belief values. Fusion is then completed with respect to all the possible combinations of the identified vehicles. The proposition that takes place in the fusion process for this query can thus be formulated as: *can observation A be associated to observation B?*

A large number of data fusion techniques exist. A requirement of particular importance in relation to query language is that the fusion method must be quick, primarily to avoid waiting times during the query process. The method used here will not be discussed further; the reader is instead referred to the method described in (Folkesson et al., 2006).

Figure 4. (a) A laser radar image of a parking lot with a moving car (encircled). (b) Two video frames showing a moving white vehicle (encircled) while entering a parking lot

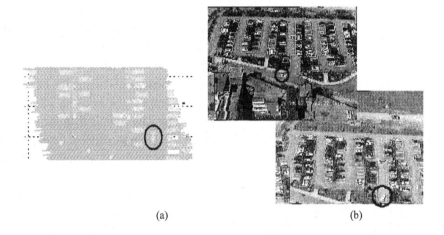

(a)   (b)

A SYSTEM ARCHITECTURE FOR SENSOR DATA FUSION

Figure 5 illustrates a generic system architecture for sensor data fusion. A *user* interacts with a user interface to produce a σ-*query*. This can be done directly or through a domain specific virtual environment customized to the user's level of expertise and described in the *user awareness subsystem*. So far, two independent visual user interfaces has been designed and implemented for this purpose, see (Camara & Jungert, 2007; Chang et alt., 2006). Once a σ-*query* has been formulated, it can be compiled and its correctness and feasibility checked. For a σ-*query* to be executable all the required operators must have been implemented in the system. The knowledge of what type of queries the system can execute is given in a knowledge base formed by the *Meta Database* and the *Applied operators*.

The Meta Database contains a set of tables describing which operators are implemented in the system and how they can be used. The Applied operators are the set of algorithms that implement the operators. Once a query has been successfully compiled the *Sigma Query Engine* executes it against the *Distributed Multimedia Database* or directly against the *sensors input*.

During the execution it applies *input filtering, indexing and data transformation* required by the application of the operators in the query. The sensors are controlled directly by the user through the user interface. The execution of a query produces (*Fused) Knowledge* that can then be used to modify the virtual environment in which the user operates, providing useful feedback through appropriate *result visualizations*. If the initial results are uninformative then the Reasoner guided by the user creates a more elaborate query by means of some rule and returns the query to the query processor. The query processor executes it and returns a more informative answer. Rules may be initially specified by the user and subsequently learned by the Reasoner (Chang et al., 2006).

One of the important characteristics of this architecture is that the same query language can be used to access any data source. This is possible due to the fact that the Meta Database and the Applied Operators hide the information to be processed, providing the ΣQL processor with a general data model, on which programmers base their queries.

With respect to the initial definition, ΣQL has evolved into a query language for multiple sensor data sources with capabilities not only for sensor data fusion but also for *sensor data independence*.

Figure 5. A system architecture including means for sensor data fusion

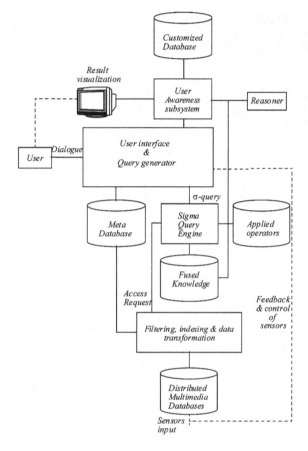

TASKS AND QUERY PATTERNS FOR INFORMATION FUSION

In what follows we describe the typical tasks for information fusion, which can be grouped into three types. The reasoning process as carried out here depends on whether the belief values that are output from any user query has got a value that is uninformative. Three different types of task queries have been identified. As explained in Section 5, these task queries are repeated structures called query patterns. By using query path schemata, many repeated structures can be revealed during the process of query building. For example, the query "finding a red car followed by a truck in 10 minutes" could be built by a similar way with the query "find a yellow bus followed by a bicycle in 30 minutes".

The first type of tasks, or query patterns, is concerned with how to improve the result of the original query by considering other aspects of the sensor data sources. The second type is concerned with how to associate different object instances with each other, and the third group is concerned with queries that need to inspect the dependency tree, that is the proposed approach to query refinement (Chang et al., 2002b), from the last user query. All these three task types are related in that they will result in new elaborate queries that are part of the iterative information fusion process. Here we also demonstrate the three task types with some relevant examples of elaborate ΣQL queries.

Type I: Tasks/Query Patterns

These tasks require the generation of new and elaborate ΣQL queries that basically depends on certain conditions among the spatial objects normally found in sensor data. This type of Query Patterns requires retrieving similar objects to get a more accurate and complete result, for example:

Sensor data independence here means that an end-user should be able to apply queries without being aware of which sensors that are involved in the process of answering the queries. This is especially important as the use of a particular sensor depends on weather and light conditions and it is difficult for a user to determine, which is the best sensor at the given circumstances. Furthermore, the various sensors may be of different types and generate heterogeneous sensor data images. For this reason, a large number of algorithms for sensor data analysis (Horney et al., 2004) must be available to the system as well.

1) Are there any other objects in the proximity of the retrieved object that are of similar or of equal type as the retrieved object? *Proximity* refers to the AOI and *similarity* to the ontology; the Reasoner creates new elaborate queries from the templates.

2) Have there been any other objects in the proximity of the retrieved object that are of similar or of equal type as the retrieved object? The Reasoner creates new elaborate queries from the templates.

3) Is the present background (context) information of proper type relative to the type of the retrieved object? Context refers to the geographical background in the AOI and the Reasoner creates new elaborate queries from the templates. As a consequence the query processor responds to the question whether it is possible that the primary object can be found in an area with the actual geographic background.

4) Has the object type been previously observed in this background context? This query is similar to 3 but requires involvement of earlier output instances from ΣQL, i.e. the IOI must be involved as well.

5) Is the retrieved object partly hidden by some other object that is part of the context? The context refers to the local background that must be in high resolution. This allows the Reasoner to create a new elaborate query.

6) Are there any other object types in the proximity of the retrieved object that may have any *impact* of some kind on the found object? This query is similar to task type 6 but here the object may have a location that is too close to some other object which will have some more or less serious consequences on the primary object.

Type II: Tasks/Query Patterns

This task type requires generally the invocation of a particular function that basically is concerned with the resolution of the association problem that may occur in single cases, that is between a pair of registered objects, or as a part of a tracking task, including a time sequence of the registered objects.

7) Can the retrieved object be associated to an earlier single observation? This query type requires the solution of the association problem.

8) Can the retrieved object be part of an existing track. This task type is a recursive variation of query type 8 that includes the application of the association problem.

Type III: Tasks/Query Patterns

This task type requires only investigation of the result of the various sub-queries of user defined queries that corresponds, in most cases, to an inspection of the content of the dependency tree.

9) Is it possible that the quality of the background information may have influenced the query result? For instance, the question could be whether there are any missing data in the AOI. This could be determined from a particular ΣQL query. For a lot of missing data the risks of not finding a particular object increases.

10) Did the result of some of the sensors (data sources) contradict each other? Does not require any new ΣQL query; only an inspection of the dependency tree.

11) Did any sensor (data sources) contribute to the result in any extreme way? This refers to the single belief values from the various sensor related sub-queries; consequently just a check of the dependency tree is required.

12) Which sensors (data sources) were used to answer the query? Does not require any new ΣQL query; only an inspection of the dependency tree.

13) Are there any attributes or status values of the retrieved object that *in particular* could have diverted the outcome of the primary query? This means that the attribute did not in any case contribute to the query result. On the contrary it could have contributed to another outcome of the query.

EXAMPLES OF ITERATIVE QUERIES

In this section a set of iterative queries generated by the Reasoner will be shown. These queries are examples of the three task types introduced in Section 7. Most of these queries require two iterations although it is possible to refine query formulation into three or even four iterations, depending upon how the query formulation is done by the user. In the query examples *relation* refers to topological relations between a pair of spatial objects of which AOI is also considered being a spatial object.

Example 1: The following ΣQL query corresponds to Query 1 of Task Type 1. Initially, the user tries to find "trucks" in a certain area of interest. This corresponds to the light grey colored ΣQL query. If the results are uninformative, the user may guide the Reasoner to apply a more elaborate query as shown below. After the elaborate query is processed, the user is satisfied with the results. He can then tell the Reasoner to remember the rule under a certain user-assigned task description, such as "objects similar to trucks". For an explanation of ΣQL query language and syntax, see (Chang et al., 2004).

Query:*Are there any other objects in the proximity of the retrieved object that are of similar or of equal type as the retrieved object?*

```
Select object_k.type, object_k.position, ob-
ject_j.type, object_j.position
cluster * alias object_k
from PerceptionSource
where Inside(AOI, object_k)
```

```
    and object_j.t = object_k.t
    and distance(object_k, object_j) < δ
        and similar(object_k, object_j)
    and object_j in Select objedct_i.type,
object_i.position
        cluster * alias object_i
        from PerceptionSource
        where Inside(AOI, object_i)
            and object_i.t = t_given
            and object_i.type = 'truck'
```

In the above *similar* function, similarity is determined by means of the ontology. That is, an object similar to a given object can be defined either as the parent or a sibling of the given object considering the structure of the object part of the ontology. The *distance* function is simply defined as the Euclidean distance between the objects.

Example 2: The following ΣQL query corresponds to Query 2 of Task Type 1. As in the previous example, initially the user tries to find "trucks" in a certain area of interest. This corresponds to the light grey colored ΣQL query. The user may then decide to guide the Reasoner to apply a more elaborate query as shown below to find similar objects in previous time periods. After the elaborate query is processed, the user is satisfied with the results. He can then tell the Reasoner to remember the rule under a certain user-assigned task description, such as "objects similar to trucks in previous time periods".

Query:*Have there been any other objects in the proximity of the retrieved object that are of similar or of equal type as the retrieved object?*

```
Select object_k.type, object_k.position,
object_k.t, object_j.type, object_j.position
cluster * alias object_k
from PerceptionSource
where Inside(AOI, object_k)
        and t_start < object_i.t < object_k.t_given
        and distance(object_k.position, ob-
ject_j.position) < δ
        and similar(object_k.type, object_j.
```

```
type)
      and object, in Select objedct_i.type,
object_i.position
          cluster * alias object_i
          from PerceptionSource
          where Inside(AOI, object_i)
              and object_i.t = t_given
              and object_i.type = 'truck'
```

Example 3: task type query 3

Query: *Is the present background (context) information of proper type relative to the type of the retrieved object?*

```
Select object_j.type, object_j.position, ob-
ject_p.type
cluster * alias object_p
from ContextSource
where Inside(object_p, object_j)
      and properbackground(object_p.type,
object_j.type)
      and object_j in Select objedct_i.type,
object_i.position
          cluster * alias object_i
          from PerceptionSource
          where Inside(AOI, object_i)
              and object_i.t = t_given
              and object_i.type = 'bus'
```

Example 4: task type query 4

Query: *Has this object type been previously observed in this background context?*

```
Select object_p.type, object_i.type, objec-
t_i.position, object_i.t
cluster * alias object_p
from ContextSource
where object_j.type = object_p.type
  and object_i.type = object_k.type
  and t_start < object_i.t < object_k.t
  and object_i in
    Select object_m.type, object_m.t, ob-
ject_m.position
```

```
cluster * alias object_m
from PerceptionSource
where Inside(AOI, object_m)
  and t_start < object_m.t < t_given
  and object_m.type = 'bus'
  and object_j in
    Select object_i.type
    cluster * alias object_i
    from ContextSource
    where Overlap(AOL, object_i)
      and object_k in
        Select objedct_n.type, object_n.
position
          cluster * alias object_n
          from PerceptionSource
          where and Overlap(AOI, ob-
ject_n)

              and object_n.t = t_given
              and object_n.type = 'bus'
```

Example 5: task type query 5

Query: *Is the retrieved object partly hidden by some other object that is part of the context?*

```
Select object_p.type, object_p.position, ob-
ject_j.type, object_j.position
cluster * alias object_p
from PerseptionSource
where Partlyoverlap(AOI, object_p)
  and ((object_p.type = 'terrain-feature')
      or (object_p.type = 'building')
      or (object_p.type = 'natural-ob-
ject))
  and ((Partlyoverlap(object_p, object_j))
      or (Beside(object_p, object_j)))
  and object_p.t = object_j.t
  and object_j in Select type
          cluster * alias object_i
          from PerceptionSource
          where Inside(AOI, object_i)
              and object_i.t = t_given
              and object_i.type = 'bus'
```

Example 6: task type query 6

Query:*Are there any other object types in the proximity of the retrieved object that may have any impact of some kind on the found object?*

```
Select object_k.type, object_k.position, ob-
ject_j.type, object_j.position
cluster * alias object_k
from PerceptionSource
where Inside(AOI, object_k)
      and distance(object_k.position, ob-
ject_j.position) < δ
      and non-coexistant(object_k, object_j)
      and object_k.t = object_j.t
      and object_k.type = 'tank'
      and object_j in Select type
            cluster * alias object_i
            from PerceptionSource
            where Inside(AOI, object_i)
                  and object_i.type = 'bus'
                  and object_i.t = t_given
```

An illustration to this query may be a bus with refugees close to a tank. The possible impact could be that it is prohibited to fire at a tank that is close to a bus.

Example 7: task type query 7

Query:*Can the retrieved object be associated to an earlier single observation?*

```
Select object_k.type, object_k.t, object_k.
position, object_j.type,
      object_j.t, object_j.position
cluster * alias object_k
from PerceptionSource
where Inside(AOI, object_k)
      and distance(object_k.position, ob-
ject_j.position) < δ
      and t_start < object_k.t < object_j.t
      and associate(object_k, object_j)
      and object_j in Select object_i.type,
object_i.position
            cluster * alias object_i
            from PerceptionSource
```

```
      where Inside(AOI, object_i)
            and object_i.t = t_given
            and object_i.type = 'truck'
```

In this query it is assumed that the function *associate* associates two objects to each other, i.e. the two object observations are the same although observed at different positions and at different times.

Example 8: task type query 8, "Can the retrieved object be part of an existing track", is a recursive variation of query type 7 and will not be further elaborated.

Example 9: task type query 9

Query:*Is it possible that the quality of the background information may have influenced the query result?*

```
Select object_k.sensor, object_k.background,
object_j.type, object_j.position
cluster * alias object_k
from DependencyTree
where
      and missingdata(AOI, object_k.back-
ground) > 50
      /* more than 50% of the background
data may be missing inside AOI*/
      and object_j in Select object_i.type,
object_i.position
            cluster * alias object_i
            from PerceptionSource
            where Inside(AOI, object_i)
                  and object_i.t = t_given
                  and object_i.type = 'bus'
```

The measure of data quality here is based on the level of missing data. However, other definitions of quality can be thought of.

Example 10: task type query 10

Query:*Did the final result contradict the result from any of the sensors (data sources)?*

```
Select object_k.type, object_k.sensor, ob-
ject_j.type, object_j.position
```

```
cluster * alias object_k
from DependencyTree
where object_k.position = object_j.position
     and object_k.type ≠ object_j.type
     and object_k.t = object_j.t
     and object_j in Select object_i.type,
object_i.position
          cluster * alias object_i
          from PerceptionSource
          where Inside(AOI, object_i)
               and object_i.t = t_given
               and object_i.type = 'truck'
```

Example 11: task type query 11

Query: *Did any sensor (data source) contribute to the result in any extreme way?*

```
Select object_k.type, object_k.sensor
cluster * alias object_k
from DependencyTree
where (or (qualitative-difference(object_k.
belief-value, object_j.belief-value) =
     'large'
     and object_k.position = object_j.posi-
tion
     and object_k.t = object_j.t)
     (object_k.t = object_j.t
     and object_k.position = object_j.posi-
tion
     and object_k.type ≠ object_j.type))
     and object_j in Select object_i.type,
object_i.position
          cluster * alias object_i
          from PerceptionSource
          where Inside(AOI, object_i)
               and object_i.t = t_given
               and object_i.type = 'truck'
```

Example 12: task type query 12

Query: *Which sensors (data sources) where used to answer the query?*

```
Select object_k.sensor
cluster * alias object_k
```

```
from DependencyTree
where object_k.position = object_j.position
     and object_k.type = object_j.type
     and object_k.t = object_j.t
     and object_j in Select type
          cluster * alias object_i
          from PerceptionSource
          where Inside(AOI, object_i)
               and object_i.t = t_given
               and object_i.type = 'truck'
```

A SCENARIO

In dealing with queries it should be enough for a user to have a general understanding of the problems associated with the sensor data, i.e., the user should not have to bother with sensor data uncertainty or with the fact that all sensors cannot measure all possible attributes. Thus sensor data independence has to be achieved. To achieve this, the user should not work with concepts related to the sensors, but instead with what is relevant to the user. The basic questions that should be answered in a spatial/temporal environment is *where?, when?* and *what?*. We call the concepts that are the answers to these questions *area of interest* (AOI), *time-interval of interest* (IOI) and *object types*. Hence, a user must indicate the AOI in a map and supply the IOI by supplying the start and end points in time. IOI can be more advanced by only setting the starting time and thus allowing continuous repetition, i.e. answering the same query several times but with different IOIs that are consecutive over time. Object types can in its simplest form just be chosen from a list that mirrors the actual object ontology. The elementary queries are also described in (Silvervarg, 2004; Silvervarg, 2005a; Silvervarg, 2005b). This basically indicates how queries can be applied and below we will give a more illustrative example of how this can be done.

To illustrate the use of ΣQL we have chosen some examples from a scenario by Camara and

Jungert (2007). This illustration will also demonstrate a possible user interface with a visual query capability and techniques for extendable query development. The purpose of this scenario driven concept is also to demonstrate how the system can work.

The scenario was implemented and executed with the simulation platform MOSART (Horney et al., 2006). Other tools used are explained in detail in Silvervarg and Jungert (2006), and Camara and Jungert (2007). Here only an overview of the scenario will be presented.

At midnight of November 12, 2005, the UN troops have received rumours that there will be a demonstration larger than usual the next day, but the rumours are unconfirmed. With the help of ΣQL the area is monitored and incoming data are used to search for vehicles in the area. A simple query is applied:

```
AOI: A rather large area covering the
vicinity of the peak, but none of the
nearby towns.
Time: From midnight and continuously for-
ward.
Object type: Vehicles
```

The result of this query can be seen in Figure 6, which, in red, shows the path of a vehicle approaching the demonstration area.

At half past two in the morning the pass of three vehicles are registered by the southern stationary ground sensor net in an east-westerly direction. As the observation post has not detected any vehicles on the big road going south the conclusion is that the vehicles have stopped between the sensor network and the big road, see Figure 7. The vehicles have been classified by the net as one large vehicle such as truck or bus, and two small vehicles such as passenger cars.

At dawn, 6.12, the observation post reports that a roadblock has been setup during the night north-east of the observation post (see Figure 8). It consists of one truck and three passenger cars. The conclusion is that the vehicles south-east of the observation post that arrived during the night is some kind of a roadblock. Together they are probably meant to keep the UN troops locked

Figure 6. The result returned from ΣQL at 2:37

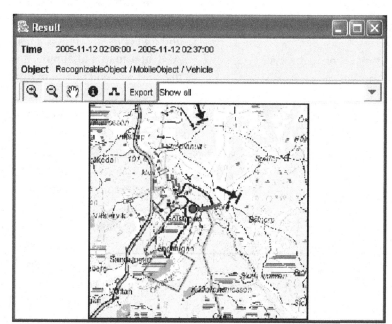

in to not be able to patrol the area. Since the UN troops has flying reconnaissance available that will be used instead of forcing a confrontation CARABAS (Hellsten et al., 1996) is requested to fly across the area to find further information about the situation. CARABAS is a synthetic aperture radar, SAR, (Carrara et al., 1995). SAR data consist of high-resolution reflected returns of radar-frequency energy from terrain that has been illuminated by a directed beam of pulses generated by the sensor. By supplying its own source of illumination, the SAR sensor can acquire data day or night without regard to cloud cover.

When the detections from CARABAS are delivered they are found to be too many. Thus it is not possible to take a closer look at them all (>100 observations). With the help from ΣQL vehicles on roads can be found. The query to ΣQL was:

```
AOI: The area covered by CARABAS, but ex-
cluding the nearby towns.
Time: The time for the CARABASE detec-
tions.
Object type: Vehicles.
Condition: Vehicles on road.
```

The result to this query including the extension concerning the condition "vehicles on road" can be seen in Figure 9. The result contains a number of false hits that somehow need to be eliminated

Figure 7. The estimated location of the vehicles that arrived at 2:30 as demonstrated by the operational picture in the scenario

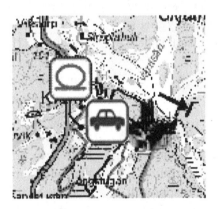

Figure 8. The known situation at 6.12

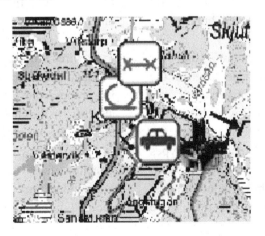

Figure 9. The vehicles on roads found in the CARABAS data, presented by ΣQL

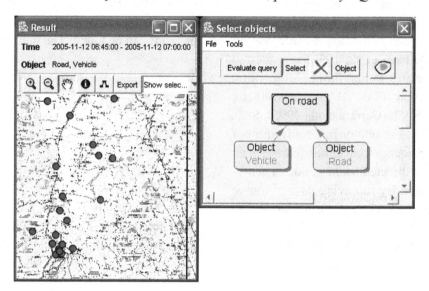

but that is subject to the remaining part of the scenario where some other services are used to deal with this problem. Consequently this is outside the scope of the work demonstrated here in this work.

DISCUSSION

In this chapter we described a query system with multiple sensor data sources, which requires a method for sensor data fusion. In our approach the queries are manually created, and then modified, to deal with the lack of information from a certain source or sources, and therefore not only the constraints can be changed, but also the source(s).

An experimental ΣQL query processing system has been implemented by researchers at the University of Pittsburgh, the University of Salerno and the Swedish Defence Research Agency, to demonstrate the feasibility of applying the proposed techniques to data from various types of sensors, including laser-radar, infrared video (similar to video but generated at 60 frames/sec) and CCD digital camera. The users have

successfully tested a number of queries, ranging from simple queries to complex ones for fusion, and systematic usability study is currently being conducted. Having established the feasibility of the techniques, we now discuss a number of issues for further research. The implementation of the query system has been made so that each component can be run on a separate computer or all of them on the same one. All the databases have been implemented using MySQL and all the code is implemented in Java, thus making it possible to run the query system on both Linux and Windows platforms. So far, no efforts have been made to package all the pieces together, thus there is, at this time, no easy way to download and try out the implementation.

The sensors in the above experiment are limited to the three pre-specified types of image sensors. To handle a large number of different sensors, we propose the following extension (Chang et al., 2002b): the characteristics, applicable ranges and processing algorithms of these sensors are stored in a knowledge base, which enables the system to deal with new sensors in a sensor data independent way. The incorporation of domain-specific information into the knowledge base makes this approach extendible to other multimedia applications.

The fusion method is discussed in (Folkesson et al., 2006). Examples of other fusion methods that can be used are Basian networks (Jensen, 1996) and Dempster-Schafer (Yager et al., 1994). The proposed information structure is an information flow structure that works in parallel with the queries and allows acquisition and determination of the information necessary to carry out the sensor data fusion process. It is not only necessary to determine the objects, their state variables and attributes requested by the query but also the belief values associated to them. This will put a heavy burden on the user to judge the result of the queries with respect to the belief values returned by the query system based on the uncertainty of the sensor information, because there will always be uncertainties in data registered by any sensor. How to replace the manual query refinement process by a semi-automatic or fully automatic query refinement process is of great importance from a user's point of view and will be further investigated.

Regarding the issue of generality of the ΣQL language, it is at least as powerful as SQL because an SQL query can be regarded as an ΣQL query with the clause "CLUSTER *". Since ΣQL can express both spatial and temporal constraints individually using the SELECT/CLUSTER construct and nested sub-queries, and sensor data sources are by nature spatial/temporal, there is a good fit. Its limitation is that constraints simultaneously involving space and time cannot be easily expressed, unless embedded in the WHERE clause. Although such constraints may be infrequent in practical applications, further investigation is needed in order to deal with such complex constraints.

Finally, the qualitative methods used by the σ-operators are developed to support indexing and efficient inference making by transforming the information acquired from the heterogeneous data sources into a unified spatial/temporal structure. Such a unified structure is desirable because

generic reasoning techniques can be applied independently of the original sensor data structures. Thus generic σ-operators based on qualitative methods can be designed and implemented to support qualitative structure such as Symbolic Projection, which is discussed further in (Chang et al., 1996) where a number of alternative qualitative approaches can be found as well.

REFERENCES

Camara, K., & Jungert, E. (2007). A visual query language for dynamic processes applied to a scenario driven environment. *Journal of Visual language and Computation. Special Issue on Human-GIS Interaction, 18*(3), 315–338.

Carrara, W. H., Majewski, R. M., & Goodman, R. S. (1995). *Spotlight synthetic aperture radar: Signal processing algorithms*. Artech House.

Chakrabarti, K., Porkaew, K., & Mehrotra, S. (2000). Efficient query refinement in multimedia databases. In *Proceedings of the 16th International Conference on Data Engineering. San Diego, California, February 28 – March 3, 2000*.

Chang, S.-K., Costagliola, G., & Jungert, E. (2002b). Multi-sensor information fusion by query refinement. *Recent Advances in Visual Information Systems* (LNCS 2314, pp. 1-11).

Chang, S.-K., Costagliola, G., Jungert, E., & Orciuoli, F. (2004). Querying distributed multimedia databases and data sources for sensor data fusion. *IEEE Transactions on Multimedia, 6*(5), 687–672. doi:10.1109/TMM.2004.834862

Chang, S.-K., Dai, W., Hughes, S., Lakkavaram, P., & Li, X. (2002a). Evolutionary query processing, fusion and visualization. In *Proceedings of the 8th International Conference on Distributed Multimedia Systems. San Francisco Bay, California, September 26-28, 2002* (pp. 677-686).

Chang, S.-K., & Jungert, E. (1996). *Symbolic projection for image information retrieval and spatial reasoning*. London: Academic Press.

Chang, S.-K., & Jungert, E. (1998). A spatial/temporal query language for multiple data sources in a heterogeneous information system environment. *International Journal of Cooperative Information Systems, 7*(2-3), 167–186. doi:10.1142/S021884309800009X

Chang, S.-K., Jungert, E., & Li, X. (2006). A progressive query language and interactive reasoner for information fusion. *Journal of Information Fusion, 8*(1), 70–83. doi:10.1016/j.inffus.2005.09.004

Folkesson, M., Grönwall, C., & Jungert, E. (2006). A fusion approach to coarse-to-fine target recognition. In B. V. Dasarathy (Ed.), *Proceedings of SPIE -6242- Multisensor, Multisource Information Fusion: Architectures, Algorithms, and Applications*.

Gehrke, J., & Madden, S. (2004). Query processing in sensor networks. *Pervasive Computing*, Jan-March, 46-55.

Grafe, G. (1993). Query evaluation techniques for large databases. *ACM Computing Surveys, 25*(2), 73–170. doi:10.1145/152610.152611

Hall, D. L., & Llinas, J. (Eds.). (2001). *Handbook of multisensor data fusion*. New York: CRC Press.

Hellsten, H., Ulander, L. M. H., Gustavsson, A., & Larsson, B. (1996). Development of VHF CARABAS-II SAR. In . *Proceedings of the Radar Sensor Technology SPIE., 2747*, 48–60.

Horney, T., Ahlberg, J., Jungert, E., Folkesson, M., Silvervarg, K., Lantz, F., et al. (2004). An information system for target recognition. In *Proceedings of the SPIE Conference on Defense and Security, Orlando, Florida, April* (pp. 12-16).

Horney, T., Holmberg, M., Silvervarg, K., & Brännström, M. (2006). MOSART Research Testbed. *IEEE International Conference on Multisensor Fusion and Integration for Intelligent Systems. 3-6 Sept. 2006* (pp. 225-229).

Jensen, F. V. (1996). *An introduction to Bayesian networks*. New York: Springer Verlag.

Jungert, E. (1999a). A qualitative approach to reasoning about objects in motion based on symbolic projection. In *Proceedings of the Conference on Multimedia Databases and Image Communication (MDIC'99). Salerno, Italy, October 4-5* (pp. 89-100).

Jungert, E., Söderman, U., Ahlberg, S., Hörling, P., Lantz, F., & Neider, G. (1999). Generation of high resolution terrain elevation models for synthetic environments using laser-radar data. In *Proceedings of SPIE Modeling, Simulation and Visualization for Real And Virtual Environments. Orlando, Florida, April 7-8, 1999* (pp. 12-20).

Li, X., & Chang, S.-K. (2003). An interactive visual query interface on spatial/temporal data. In *Proceedings of the 10th international conference on Distributed Multimedia Systems. San Francisco, September 8-10, 2003* (pp. 257-262).

Silvervarg, K., & Jungert, E. (2004). Visual specification of spatial/temporal queries in a sensor data independent information system. In *Proceedings of the 10th International Conf. on Distributed Multimedia Systems. San Francisco, California, Sept. 8-10* (pp. 263-268).

Silvervarg, K., & Jungert, E. (2005a). Uncertain topological relations for mobile point objects in terrain. In *Proceedings of 11th International Conference on Distributed Mutimedia Systems, Banff, Canada, September 5-7* (pp. 40-45).

Silvervarg, K., & Jungert, E. (2005b). A visual query language for uncertain spatial and temporal data. In *Proceedings of the Conference on Visual Information systems (VISUAL). Amsterdam, The Netherlands* (pp. 163-176).

Silvervarg, K., & Jungert, E. (2006). A scenario driven decision support system. In *Proceedings of the eleventh International Conference on Distributed Multimedia Systems. Grand Canyon, USA, August 30 – September 1, 2006* (pp. 187-192).

Stonebraker, M. (1975). Implementation of integrity constraints and views by query modification. In *Proceedings of the 1975 ACM SIGMOD international conference on Management of data San Jose. California, 1975* (pp. 65-78).

Vélez, B., Weiss, R., Sheldon, M. A., & Gifford, D. K. (1997). Fast and effective query refinement. In *Proceedings of the 20th ACM Conference on Research and Development in Information Retrieval (SIGIR97). Philadelphia, Pennsylvania.*

White, F. E. (1998). Managing data fusion systems in joint and coalition warfare. In *Proceedings of EuroFusion98 – International Conference on Data Fusion, October 1998, Great Malvern, United Kingdom* (pp. 49-52).

Yager, Fedrizzi & Kacprzyk (eds.) (1994). *Advances in Dempster-Shafer theory of evidence.* New York: Wiley & Sons.

Chapter 11
Query Optimisation for Data Mining in Peer-to-Peer Sensor Networks

Mark Roantree
Dublin City University, Ireland

Alan F. Smeaton
Dublin City University, Ireland

Noel E. O'Connor
Dublin City University, Ireland

Vincent Andrieu
Dublin City University, Ireland

Nicolas Legeay
Dublin City University, Ireland

Fabrice Camous
Dublin City University, Ireland

ABSTRACT

One of the more recent sources of large volumes of generated data is sensor devices, where dedicated sensing equipment is used to monitor events and happenings in a wide range of domains, including monitoring human biometrics and behaviour. This chapter proposes an approach and an implementation of semi-automated enrichment of raw sensor data, where the sensor data can come from a wide variety of sources. The authors extract semantics from the sensor data using their XSENSE processing architecture in a multi-stage analysis. The net result is that sensor data values are transformed into XML data so that well-established XML querying via XPATH and similar techniques can be followed. The authors then propose to distribute the XML data on a peer-to-peer configuration and show, through simulations, what the computational costs of executing queries on this P2P network, will be. This approach is validated approach through the use of an array of sensor data readings taken from a range of biometric sensor devices, fitted to movie-watchers as they watched Hollywood movies. These readings were synchronised

DOI: 10.4018/978-1-60566-328-9.ch011

with video and audio analysis of the actual movies themselves, where we automatically detect movie highlights, which the authors try to correlate with observed human reactions. The XSENSE architecture is used to semantically enrich both the biometric sensor readings and the outputs of video analysis, into one large sensor database. This chapter thus presents and validates a scalable means of semi-automating the semantic enrichment of sensor data, thereby providing a means of large-scale sensor data management which is a necessary step in supporting data mining from sensor networks.

INTRODUCTION

We are currently witnessing a groundswell of interest in pervasive computing and ubiquitous sensing which strives to develop and deploy sensing technology all around us. We are also seeing the emergence of applications from environmental monitoring to ambient assisted living which leverage the data gathered and present us with useful applications. However, most of the developments in this area have been concerned with either developing the sensing technologies, or the infrastructure (middleware) to gather this data and the issues which have been addressed include power consumption on the devices, security of data transmission, networking challenges in gathering and storing the data, and fault tolerance in the event of network and/or device failure. If we assume these issues can be solved, or can at least be addressed successfully, we are then left to develop applications which are robust, scalable and flexible, and at such time the issues of efficient high-level querying of the gathered data becomes a major issue.

The problem we address in this chapter is how to manage, in an efficient and scalable way, and most importantly in a way that is *flexible* from an application developer or end user's point of view, large volumes of sensed and gathered data. In this, we have a broad definition of sensor data and we include raw data values taken directly from sensor devices such as a heart rate monitor worn by a human, as well as *derived* data values such as the frame or time offsets of action sequences which appear in a movie. In the case of the former there would be little doubt that heart rate moni-

tor readings are sensor values, whereas the latter still corresponds to data values, taken from a data stream, albeit with some intermediate processing (audio-visual analysis in this case). We now describe the motivation for our work.

Motivation and Contribution

To design a scalable system to manage sensor data, it is first necessary to enrich the data by adding structure and semantics in order to facilitate manipulation by query languages. Secondly, in order to improve efficiency, the architecture should be suitably generic to make it applicable to other domains. Specifically, it should not be necessary to redesign the system or write new program code when new sensor devices are added. Finally, when the number of sensor devices increases to very large numbers, the system should be capable of scaling accordingly.

The contribution of the research work reported here is the development of an architecture that is both generic, and has the capability to scale to very large numbers. In this respect, our XSENSE architecture facilitates the addition of new sensor devices by requiring that the knowledge worker or user provides only a short script with structural information regarding the sensor output. Scalability is provided in the form of a Peer-to-Peer (P2P) architecture that classifies sensors into clusters, but otherwise contains no upper limit on the numbers of sensors in the network.

The chapter is structured as follows: in §2, a description of sensor networks is provided and in particular the sensor network we use in our experiments, together with the issues involved in this specific domain; in §3, we describe our solution to problems of scale and processing by way of an architecture that transforms raw data and provides semantically rich files; in §4, we provide scalability by removing the centralised component and replacing it with a Peer-to-Peer Information System; in §5, we demonstrate good query response times for distributed queries; a discussion on related research is provided in §6, and finally in §7 we offer some conclusions.

SENSOR NETWORK BACKGROUND

In previous work (Rothwell *et al.*, 2006), we reported a study conducted to investigate the potential correlations between human subject responses to emotional stimuli in movies and observed biometric responses. This was motivated by the desire to extend our approach to film analysis by capturing real physiological reactions of movie viewers. Existing approaches to movie analysis use audio-visual (AV) feature extraction coupled with machine learning algorithms to index movies in terms of key semantic events: dialogue, exciting sequences, emotional montages, etc. However, such approaches work on the audio-visual signal only and do not take into account the visceral human response to viewed content. As such, they are intrinsically limited in terms of the level of semantic information they can extract. However, integrating and combining viewer response with AV signal analysis has the potential to significantly extend such approaches toward really useful semantic-based indexing.

For the study, we created a controlled cinema-like environment and instrumented both this and movie watchers, in a variety of ways. We also performed our AV analysis on all films watched by our viewers and synchronised these analysis results with the captured biometric responses. The instrumented environment, termed the "CDV*Plex*", was designed to replicate a true cinematic experience as closely as possible. It corresponded to an air-conditioned windowless room with comfortable seating for up to 4 people, in which a Dolby 5.1 surround sound system, DVD player and large-screen digital projector were installed.

We gathered a total of 6 biometric sensor data feeds from each of our participants watching each of our movies, via 3 different sensor devices, as follows:

- **Polar S610i™ heart-rate monitor.** This consists of a fabric band which fits around a person's chest and detects and logs their *heartrate*, sampled every few seconds.
- **BodyMedia SenseWear®.** This sensor array is worn around the upper arm and measures and logs the following: *galvanic skin response*, a measure of skin conductivity which is affected by perspiration; *skin temperature*, which is linearly reflective of the body's core temperature activities; *heat flux* which is the rate of heat being dissipated by the body; *subject motion* using an in-built 3-axis accelerometer.
- **Smart Chairs.** Each of the chairs used had a specially designed foam-based pressure sensor (Dunne *et al.*, 2005) integrated into its backrest to record changes in viewer posture.

The participants were 43 staff and student volunteers from across the university. In total, 37 full length feature films of 10 different genres (e.g. Action/Adventure, Animation, Documentary, Horror, etc.) were shown, resulting in over 500 hours of recorded biometric data from the set of sensors.

As outlined in (Rothwell *et al.*, 2006), this gathered data, when combined with automatically detected movie events (see section 3.1), is potentially a hugely valuable resource for modelling and integrating human responses with automatic content structuring and indexing. Unfortunately, the value of this resource is significantly reduced in the absence of a seamless and efficient means

Figure 1. Class model for the sensor network

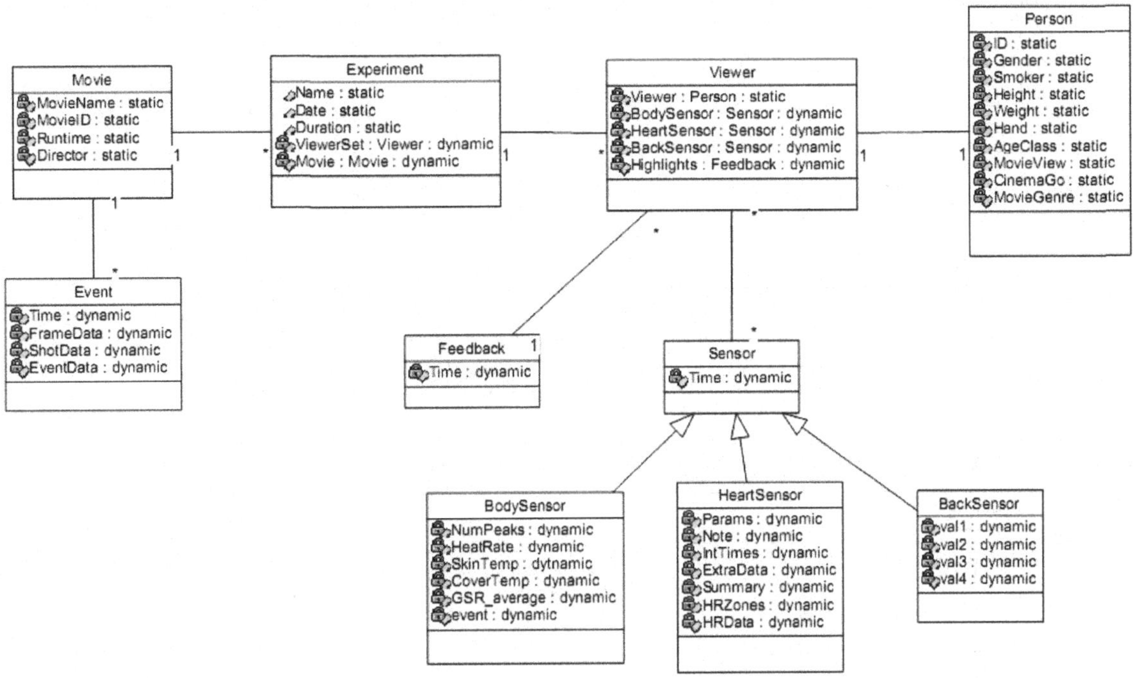

to perform semantic queries against this repository. In this chapter, we report our work on using XML for the semantic enrichment of this gathered sensor data.

Describing the Network

In order to employ any data management utility for a large volume of information such as is the case here, a user needs a compact yet detailed description of the overall system components, in our case the sensors, their data, and how they inter-operate. In Figure 1, the Sensor Network is represented as a UML class diagram. The principal class is the Experiment class, where one instance is created for each experiment. Each experiment requires a single movie (Movie class) and multiple viewers (Viewer class). The Movie has a set of static attributes associated with that movie, together with dynamic data (captured by the Event class) that is generated as the movie is processed by our movie

event detection software (Lehane & O'Connor, 2006). Each Viewer is associated with a Person class that captures static information about the viewer, and four other classes containing dynamic information: 3 types of Sensor class and a single Feedback class.

One property of these experiments that cannot be captured in Figure 1 is the time dependency across all of the classes containing dynamic information. All experiments are time-related and the classes Sensor, Feedback and Event, are bound together using the start time of the movie.

Calibration and Normalisation Issues

Sensor networks generally have two forms of data: static and dynamic. Static data is not generated by sensors or from the video analysis process. It refers to information regarding the movie, the experiment or an individual person (a viewer). Static data generated during experiments includes Personal

Figure 2. Sensor data graph

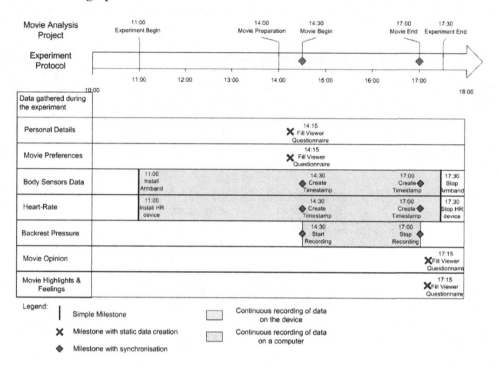

Details, Movie Preferences and Movie Opinions. Dynamic data includes movie semantics regarding scenes, shots, events and Movie Highlights & Feelings. Sensor data involves the generation of biometric sensor data, heart rate and backrest pressure on the sensors in the chairs.

Two important points are clear from Figure 2: there is a strict timing relationship across sensor sources, and some experimental data will contain anomalies, e.g. before watching the movies, participants logged up to 3 hours of biometric data in order to establish their baseline biometric values but the duration of this baseline varied considerably from person to person and from movie to movie. Thus, sensor output is influenced by the users' actions in many cases.

There is a synchronisation activity carried out at the beginning and end of each movie showing for each of the sensors' measurements. A synchronisation process is also required to link information concerning viewers' reactions and movies' events. These are the events that have

been identified using the semantic analysis of audio and video contents of the shots (described later in §3). Movie data includes all data related to a movie and is independent of Viewer data. Viewer data includes all data related to a viewer during an experiment for a movie and is generated during the experiments. They are three sources of sensor data: body sensor data generated by the armband, heart-rate measured by the HR device and the backrest pressure as measured on the chairs. Thus, one of the issues for the sensor network is how to facilitate the calibration of sensor data by associating timing events across many sensor output streams.

For the purposes of working through our P2P XSENSE architecture we use the CDV*Plex* data consisting of human biometric readings, described here, though there are many other applications and domains for which data enrichment, efficient querying and data mining are needed. These include sensor values gathered from environmental monitoring of water or of air quality, data readings

Figure 3. XSENSE processing architecture

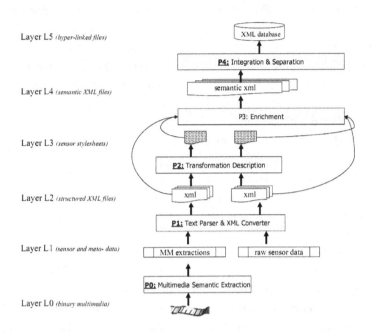

from people-movement in an urban setting (CCTV and other people-counting), or data readings from network usage. In the next section we introduce our XSENSE Enrichment Architecture.

XSENSE ENRICHMENT ARCHITECTURE

The XSENSE Architecture illustrated in Figure 3 comprises six layers, with each pair of layers joined by a processor performing a key activity in the enrichment process. Layer 0 contains raw multimedia files that use processor P0 (described in the next section) to extract meaningful content. The files generated are textual and contain timing and event data that is quite similar to raw sensor data. At Layer 1, raw text files (both sensor and multimedia metadata files) are converted by Process P1 into basic XML files and, using Process P2, stylesheets are automatically generated to enrich the basic XML files. By Layer 4, the output from the previous two layers are combined (by Process P3) to form semantically rich sensor data files. At

this point, sensor files are autonomous with no relationship across files (indicating for example, that they were used in the same experiment or that their times are synchronised). The final process (P4) adds relationship semantics to link sensor files at the global layer.

Extracting Multimedia Semantics

In order to extract semantics from the video content, a process corresponding to P0 in Figure 3, we employ a movie indexing framework capable of automatically detecting three different kinds of semantic events — dialogue between two or more characters, exciting sequences (e.g. car chases, fight scenes, etc) and emotionally laden musical sequences (Lehane & O'Connor, 2004; Lehane et al., 2005; Lehane & O'Connor, 2006). The approach relies on a series of AV feature extraction processes designed to mimic well-known film creation principles. For example, when filming a dialogue sequence, the director needs to ensure that the audience can clearly interpret the words being spoken and uses a relaxed filming style with little camera movement, large amounts of shot repetition and clearly audible speech (Bordwell & Thompson, 1997). Conversely, when shooting an exciting part of a film, the director uses fast-paced editing combined with high amounts of movement (Dancyger, 2002). Emotionally laden events, on the other hand, are shot with a strong musical soundtrack, usually combined with slower paced editing and filming style (Bordwell & Thompson, 1997; Lehane *et al.*, 2006).

We extract a set of AV features to detect the presence of these characteristics. We characterise editing pace by detecting the rate of occurrence of *shot boundaries* using a standard colour histogram technique. To measure onscreen motion, we use MPEG-7 *motion intensity* for local motion (e.g. character movement) (Manjunath *et al.*, 2002) as well as a measure of global *camera movement* (Lehane *et al.*, 2005). A support vector machine based classifier is used to classify the audio track into one of: *speech, music, silence* and *other audio*. A set of Finite State Machines (FSMs) are then employed to detect parts of a film where particular combinations of features are prominent. For example, in order to detect dialogue events, we detect temporal segments which contain various combinations of speech shots, still cameras and repeating shots. Similar rules are employed for the other event types. An evaluation over ten films of very different genres and origins, in (Lehane *et al.*, 2005), found that 95% of Dialogue events, 94% of Exciting events and 90% of Musical events as judged by multiple viewers were detected by the system, which indicates the usefulness of the system for detecting semantically important sequences in movies.

Processors for Generating Sensor Semantics

Each of the layers in the XSENSE Architecture is joined by processors that operate to build the final sensor database. In this section, we provide a more detailed description of the functionality of

each processor. The innovation in the XSENSE architecture is its generic nature: it was designed to accommodate a heterogeneous collection of sensors. By providing basic scripts and the XSENSE Term Database (a small terminology database), XSENSE can integrate most or all sensor data formats.

Example 1. Raw Sensor Data

```
File: EVK Action.evtkey
176 183 177 179 181
423 431 425 427 429
```

P1: Text to XML Conversion. At layer 1, raw sensor files contain no semantic information and at this point, it is only possible to add structural information to the sensor file. Example 1 illustrates a subset of a multimedia event file. The output from this process is a basic XML file with structural semantics but not the real content semantics as required by a query processor. Firstly, a naming convention (located in the XSENSE Term Database) is applied to sensor data files to enable the parser to recognize the type of sensor file. For example, a file that contains action events is renamed to EVT_Action.events. Example 2 illustrates the same sensor data after structural information has automatically been added.

Example 2. XML Structure (output from P1)

```
<?xml version=\1.0" encoding=\UTF-8"?>
<document>
    <event>
        <startShot>176</startShot>
        <endShot>183</endShot>
        <keyShot-group>
            <keyShot><value>177</val-
ue></keyShot>
            <keyShot><value>179</val-
ue></keyShot>
            <keyShot><value>181</val-
```

```
ue></keyShot>
        </keyShot-group>
    </event>
```

The XSENSE System uses ServingXML schema (ServingXML Project, 2007) to provide a generic process for incorporating new types of sensor devices. A ServingXML service describes the sequence of tasks that create a set of rules to coordinate the activities of the parser. The activity for the knowledge worker who is charged with managing and manipulating such sensor data is to provide a short script describing the file structure. The output for the next layer is a set of XML data files of the type shown in Example 2.

P2: Transformation Description. The aim of this process is to create an XSLT stylesheet describing the transformations necessary to create a semantically enriched XML sensor file. Each sensor file has its own time format or an implicit time interval (e.g. a reading every second). In this situation, this XSLT stylesheet describes how to transform timing information into the system time format and how to normalize times. In XSENSE, there are currently four transformation categories.

- **Semantic Transformation.** Rules for file naming and deriving information from the content.
- **Structural Transformation.** Rules for changing entity or attribute names and for changing information groupings.
- **Normalisation.** Rules for content format normalization (i.e. date format), generating IDs and synchronising time information.
- **File Transformation.** Rules to merge several XML files, divide XML files, and generating new files.

The output from this process is the stylesheet for this sensor type. Once created, the stylesheet will be reused for all sensors of this type, eliminating P2 from the next iteration of the process.

P3: Semantic Enrichment. The aim of the formatting process is to transform a basic XML file into a semantically enriched file by applying the XSLT stylesheet. Processors P1 and P2 generated the structured XML file and XSLT stylesheet respectively. The ServingXML service is updated by process P3 in order to facilitate the XSLT transformations and build the enriched data files. When a sensor that has been processed previously is detected by the system, the ServingXML service updates are not required.

Example 3. Enriched XML File

```
<?xml version=\1.0" encoding=\UTF-8"?>
<events>
    <event type=\Action">
        <startShot id=\176"/>
        <endShot id=\183"/>
        <keyShot id=\177"/>
        <keyShot id=\179"/>
        <keyShot id=\181"/>
    </event>
```

Example 3 illustrates how the sensor data used in the previous two examples can now be queried using the standard XPath language. The Saxon XSLT Processor (Saxon Project, 2007) manages all transformations for this process.

P4: Integration and Separation. At this point, XML files contain information relating to a single sensor file. The aim of this process is to add information from other sensors' files, to merge several XML files or to generate new files. Those transformations are once again described in XSLT stylesheets. This is necessary as most sensor networks contain sensors that while physically autonomous, operate in relation to other sensors to deliver the overall monitoring process.

In our sample domain, events files of different types are merged and sorted according to the time they occur in the movie. Sensor files are edited to ensure that links referring to the same experiment, heart monitor, film event, etc., are preserved. This

Table 1. Creation of the Experiment database

	Input	Output	Time
P1 P3	171 sensor files (316MB) 171 raw XML files	171 raw XML files 188 enriched files (651MB)	23 min
P4 (incl. manual editing)	188 enriched files (651MB)	221 linked files (651MB)	1h 27min (+ 27 min)
Total	171 sensor files (316MB)	221 linked files (651MB)	1h 50min

process ensures that users need not be aware of implicit relationships across the sensor network: instead they are explicitly created inside sensor documents. This provides an added advantage when sensor networks are very large and require a distributed architecture to manage scalability. In Appendix A, we provide a subset of the files created for the sensor data used in the examples in this section. This file is automatically created after user input (script data) to the ServingXML process.

Building Sensor and Movie Databases

The sensor network comprises 33 experiments, covering 29 movies, and a total of 171 sensor outputs, creating 316MB of raw data. The XSense architecture requires that sensor files follow a specific naming convention and that binary files be converted to text files.

When all of the sensor data for a particular experiment is generated, this causes the creation of an experiment XML file (by processor P4). This experiment XML file contains timing information for both the movie and experiment. However, some information such as movie id, title and language are entered manually before they are stored in the XML (eXist) database. Table 1 presents the times for the creation of the database (approx 1 hour and 50 minutes).

Movie Database. The analysis of the 29 films generated 268 semantic files, however, processor 4 merges all XML documents generated for the same film. Unlike sensor files from the movie-

viewing experiments, movie analysis files are small (less than 1MB) and can be merged. The generated movie file must be edited to add the movie id, the movie title and language before files are inserted into the eXist XML database. Table 2 presents the timing for each processor. As both processes run in parallel, the movie database is created before the sensor database.

BUILDING A DISTRIBUTED SENSOR NETWORK

As many sensor networks comprise very high volumes of sensors and data, a centralised approach to data management would gradually reduce in performance. In prior work (Bellahsène & Roantree, 2004), we developed a distributed information system architecture that operated over a Peer-to-Peer (P2P) network. The XPeer architecture was effectively a logical IS architecture that provides the scalable benefits of P2P systems with the information management functionality of traditional database systems. In this section, we describe how XSENSE was extended with P2P concepts to facilitate scalable data management for data generated for a larger sensor network.

In order to optimise the management of peers on a single machine, each peer is a thread and not a JVM process. Previously, each peer occupied its own JVM causing memory problems and an upper limit on the number of peers. With the current version of XPeer, we can launch all the CDVPlex peers simultaneously.

Table 2. Creation of the Movie database

	Input	Output	Time
P1 P3	268 text files (2.3MB) 268 raw XML files	268 raw XML files 326 enriched files	8 min
P4 (incl. manual editing)	326 enriched files (44.2MB)	29 movie files (24.7MB)	36 min (+ 30 min)
Total	268 text files (2.3MB)	29 movie files (24.7MB)	44 min

The current version of the XPeer System contains 268 peers:

- 221 Data Peers where each Data Peer represents a single sensor output (as a single XML document) and can have multiple XML documents as input.
- 39 Cluster Peers
  - 1 Cluster per movie (33 Clusters)
  - 1 Cluster for Body sensor data
  - 1 Cluster for Smart chair data
  - 1 Cluster for Heart rate sensor data
  - 1 Cluster for Chest belt sensor data
  - 1 Cluster for male viewer data
  - 1 Cluster for female viewer data
- 6 Repository Peers (arbitrary number).
- 2 Query Peer (for 2 Data Mining clients).

For an open sensor network, the Query Peers and Cluster Peers must access the repository to retrieve information about themselves as peers are generic. This requires Repository Peers which must be launched first. With a low number of peers, it takes only a few seconds for the other peers to be set up but with a large number of peers running on the same machine (since a peer is a thread), it means they share the processor resources. There is strong contention for execution time and thus, it takes some time for the system to startup.

Definition 1. $T_L = RP + C \times n + D \times m$. Where T_L: System Launch Time, RP: Repository Peer time, C: Cluster Peer, time (depending on the number n of Cluster Peers), and D: Data Peer time (depending on the number m of individual Data Peers).

In Definition 1, we provide a formal definition of the time required for system startup.

- The Repository Peers require approximately 10 seconds. On a positive note, this is independent of the number of peers.
- Cluster Peers require approximately 7.5 seconds each to set up.
- Data Peers and Query Peers take about 0.1 second each to set up.

Example 4. In the *Computer 1* configuration, described in Figure 5, the launch time would be:

$$T_L = 10s + 18 \times 7.5s + 88 \times 0.1s = 153.8s$$

There is a limitation on the number of peers running on the same machine because sometimes, if too many threads are launched (over 100), they need so much time for set up that we have JXTA connections timeouts. This is a machine processing power limitation. Systems with a large number of peers on a single machine should use a multi-core machine (dual core, quad core), but in a real-world P2P scenario, the distribution of peers should ensure that this is not necessary.

Our main reason for using a P2P network in our work is to address scalability which is an important characteristic of sensor networks. By their very definition, sensor networks have to be large

Figure 4. Mining sensor data in the CDVPlex archive

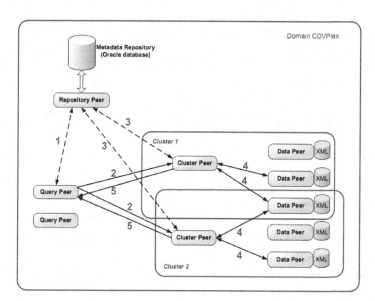

in scale in order to be useful. In our experiments we chose to use as heterogeneous a collection of sensor types (chair, heart rate, movement, and viewer data) as we could in order to replicate the real-world scenario where we will use networks of heterogeneous sensors and to configure these so that our experiments can easily scale up. A second, beneficial, reason for using P2P is that P2P also assumes that peers may join and leave at any time. This is ideal for a sensor network where sensors die and new ones are deployed without bringing down the network. Scale and come/go are proven features of P2P however this comes at a price, that price being that a P2P network is low-level, with no data management or DM functionality, which creates the opportunity for our contribution.

However, in addition to just re-using the proven P2P framework for managing data, in this case sensor data, we also include semantic enrichment as part of our contribution, where numbers/values become XML. This provides XML streams (formerly sensor values) that can be queried using standard languages such as XPath. In order to mine the distributed sensor system, each mining

query must be expressed in XPath together with some keywords. If no keywords are provided, the system will auto-extract keywords from the XPath query although the effect of this will vary based on the XPath query. For example, in Q1 in the following section, the keywords extracted from this query will do little to reduce the search space. Figure 4 provides an illustration of the query process where a QueryPeer receives a query and several keywords with which to target the appropriate cluster.

1. The Query Peer accesses the repository for the IDs of the clusters matching the query keyword.
2. The Query Peer queries those clusters.
3. The Cluster Peer requests (from the repository) the IDs of the independent Data Peers matching the query keyword.
4. Data Peers each process the XML query and can each process multiple XML documents.
5. The Cluster Peer aggregates the individual Data Peers' results and sends the entire result back to the Query Peer.

Figure 5. Network configuration

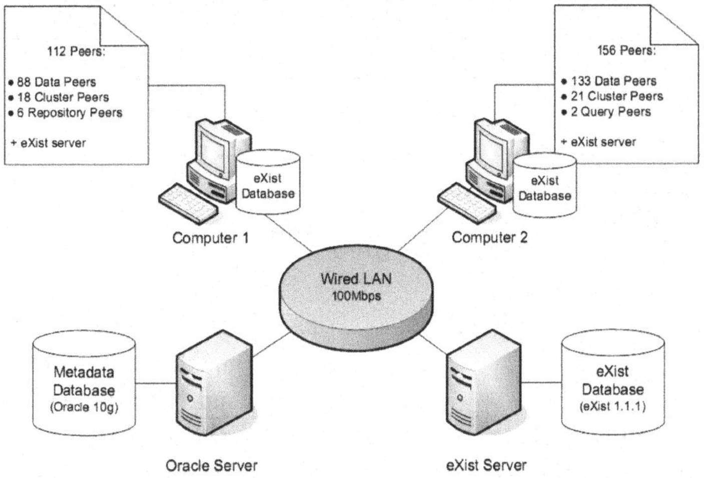

QUERY SETUP AND RESULTS

One of the advantages of the architecture and implementation we have developed for managing sensor data is the support it gives to data mining operations. In this section, we describe how we mine the distributed sensor network for personal health data. To provide a baseline execution time for queries, we begin by recording query times over a basic peer-to-peer network, without using any of our metadata features for clustering groups of sensors. We then run a more detailed series of queries using the metadata service and clusters generated by knowledge workers. While we expect some performance penalty as query peers must connect to, and retrieve data from, the metadata service, the reduction in search space should provide significant improvement in execution time for most queries. In all cases the experiment setup is identical: experiments ran on two 3.2GHz Pentium IV machines, each with 1GB of RAM using the Windows XP Pro operating system, and the P2P Query Processor was implemented using Java Virtual Machine (JVM) version 1.5.

Basic Peer-to-Peer Setup

In this experiment, we have an unstructured peer-to-peer environment. There are no clusters, meaning there is no need for a repository connection. The entire network is queried and all of the Data Peers will process the query (a total of 221 Data Peers and 1 or 2 Query Peers).

We use the same mining queries for all experiments although in some cases, these are revised for subsequent data mining interactions.

- **Q1**. The user is mining for device types used across different experiments. In this case, it will search all experiments.
- **Q2**. In this query, the values generated by the device are sought.
- **Q3**. In this query, the user is searching for experiment start times in order to normalise across all experiments in the query set. Different start times will cause peaks at unexpected times.
- **Q4**. This query searches for patterns resulting from different interval times used by sensors to generated readings.
- **Q5**. This query revised the previous query to seek interval patterns for specific peaks

Table 3. Response times in a basic P2P configuration

#	Query	Times
Q1	//@deviceId/string()	7.36s
Q2	//measurement/@name/string()	9.08s
Q3	//startDateTime	10.43s
Q4	//interval	12.65s
Q5	//interval[MinHR='66']	7.78s

(heart rate readings greater than 66) in output data.

Query Times. As with all experiments, a query is executed 10 times with average response times displayed in Table 3.

Response time variations are due to the size of some XML documents and the size of the query itself. This now provides a baseline by which we can evaluate the performance of our own data mining queries.

Using the Super-Peer Architecture

In the XSENSE-P2P architecture, individual data peers, which can have multiple XML documents, are grouped into clusters, with keywords, and a Repository peer handles the database containing the cluster configuration. The Cluster peer must communicate with a Repository peer to target the relevant peers and this will have an impact on execution times. The network configuration used for this experiment is displayed in Figure 5.

Sensor Queries. In this section we report on queries executed on clusters of sensors: all body sensors, all heart rate monitors, all smart chair sensors and all chest belt sensors. In each case, a single cluster is queried. Table 4 shows a number of XML queries processed on different sensor clusters and their execution times. In each case, keywords accompany the XPath query to influence the identification of appropriate clusters.

The response time largely depends on the number of Data Peers queried: we obtain the fastest results for the Chest Belt cluster (10 Data Peers) and the slowest for Body Sensor (75 Data Peers) and Heart Rate (70 Data Peers) clusters. As expected, the response time also increases with the amount of sensor data generated at each Data Peer. However, it is clear that clustering has greatly improved the response times for queries due to the classification process.

Depending on cluster size, the response time for these queries is 3 to 15 times faster than without using the Super-Peer architecture. Appendix B shows a detailed query timings chart for a sample //measurement/@name/string() query on the body sensors cluster: we can see that communication between Cluster Peers and the Repository Peer is very fast (the time from *SendFromClusterToRepository* to *ReceivedByClusterFromRepository* is 0.12 - 0.067 = 0.053 second) and performed once per query process. Overall, communication with a repository takes a relatively small amount of time while cluster targeting allows a great improvement in times.

Random Queries. Table 5 reports on queries executed on smaller and more random clusters. The second column, C(P), shows the number of Cluster Peers receiving the query and the number of Data Peers involved in the (cluster) query. With these very small clusters (4 to 5 Data Peers), the response is very fast as the search space is highly selective.

This clearly shows the benefits from the super-peer architecture: without this architecture, all the documents would have been queried and it would have taken more than 5 seconds. In this experi-

Table 4. Sensor Cluster Response Times for a Single Cluster

Cluster Type (Data Peers Num.)	#	Query	Times
Body Sensor (75)	Q1	//@deviceId/string()	2.653s
	Q2	//measurement/@name/string()	2.699s
	Q3	//startDateTime	3.291s
	Q4	//interval	3.262s
Heart Rate (70)	Q5	//@deviceId/string()	2.340s
	Q6	//measurement/@name/string()	2.463s
	Q7	//startDateTime	3.775s
	Q8	//interval[MinHR='66']	2.666s
	Q9	//interval[MinHR='66']/MaxHR	2.323s
	Q10	//interval[MaxHR/number()$>$200]	2.292s
Smart Chairs (33)	Q11	//@deviceId/string()	1.583s
	Q12	//measurement/@name/string()	1.770s
Chest Belt (10)	Q13	//@deviceId/string()	0.517s
	Q14	//measurement/@name/string()	0.454s
	Q15	//Params	0.564s
	Q16	//Params/Length	0.485s
	Q17	//VO2max	0.501s
	Q18	//measurement/value[@time/string()='75000']	2.851s

Table 5. Random and Multiple Cluster Response Times

Keyword	Clusters Type	C(P)	#	Query	Times
Shrek	Shrek movie	1(5)	Q19	//@deviceId/string()	0.266s
			Q20	//measurement/@name/string()	0.269s
			Q21	//title	0.297s
Harry.*	Harry Potter movies	3(14)	Q22	//@deviceId/string()	0.438s
			Q23	//measurement/@name/string()	0.484s
			Q24	//title	0.469s

ment we only have 221 Data Peers but it is clear that without our logical super-peer architecture, there is a limit to the number of sensors that can be added to the system without employing a cut-off point for searching, and thus, risking missing relevant data.

Fine Tuning Response Times

Despite the fact that we employ a super-peer architecture for our mining experiments, we limited our setup to a single data server. By analysing detailed query timing charts and tables, we can determine where the process incurs delays by finding large horizontal gaps (as time is displayed on the horizontal axis). The chart displayed in Appendix B, shows detailed query timings for a

//measurement/@name/string() query on the body sensors cluster. It contains a large horizontal gap between the *processingXMLQuery* checkpoint and the *XMLQueryProcessed* checkpoint. This demonstrates to us that the reason we had a delay in our response times was due to a data server bottleneck. When we roll this out to a real-world scenario, there will be a large number of data servers. We then examined the effect of this by using 3 data servers instead of a single server. We employ the same Super-Peer architecture and the same machines as in Fig. 5 but now, we use the additional XML servers as shown in Fig. 6.

Results. We can see in the chart displayed in Appendix C that the large gap between the *processingXMLQuery* checkpoint has disappeared. The eXist database querying is now almost instan-

Figure 6. Network configuration

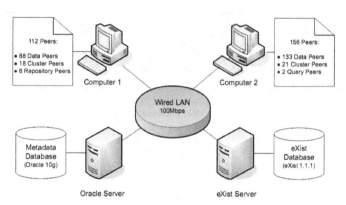

taneous, because the XML querying processing power has been improved by a factor of 3. All the limitations we have in our experiments are processing power limitations because we simulate a large P2P network on only 3 workstations.

For this cluster (body sensors cluster, the biggest one, containing 75 Data Peers), a series of 30 queries was processed for both 1-server configuration and 3-server configuration, showing that the average response time is 480 milliseconds faster with 3 servers than with 1. We can see on the chart displayed in Appendix C that all the curves are extremely close to each other. This means that the limitations we encounter are processing power limitations due to running 268 peers on 2 machines. A real world distributed system will avoid these limitations as peers will be spread over many machines.

RELATED RESEARCH

In this section, we examine related research under two different criteria: their ability to provide generic frameworks for sensor management and querying, and their ability to generate meaningful semantics for complex sensor data (e.g. multimedia). We first begin with a brief review of other work which takes a P2P approach to managing data.

Starting with a non-sensor network data application, a peer-to-peer network approach was used in a distributed caching system for OLAP queries (Kalnis *et al.*, 2002). This was called the PeerOLAP architecture and it contains a group of peers each of which pose OLAP queries to data warehouses. The cache associated with each peer can be shared, as can computational capabilities, and as a result, queries addressed to a specific peer can be either processed locally or propagated to neighbours. This is a typical peer-to-peer approach.

Edutella (Nejdl *et al.*, 2002) is a semantic web project which implements an RDF-based metadata infrastructure for P2P-networks. This is done in order to enable interoperability between heterogeneous JXTA applications. The network underlying Edutella is a hypercube of super-peers, which is a variation of P2P, to which peers are directly connected. Each super-peer acts as a mediator to route the query to some of its neighbour super-peers according to a strategy that exploits the hypercube topology to guarantee a worst-case logarithmic time for reaching the relevant super-peer. SomeWhere (Adjiman *et al.*, 2005) is a semantic web oriented P2P mediation, based on propositional logic for defining ontologies, mappings, and queries.

While sensor networks are now quite popular, there is not a great deal of published work in the area of semantic enrichment for sensor networks, nor in the area of evaluating performance over enriched sensor sources. Instead, most of the published work covers the use of XML for reasons of interoperability (Rose *et al.*, 2006), or evaluates query processing using simple methods operating at the network level (Kotidis, 2006).

Rose *et al.* (2006) provide a template for incorporating non-XML sources into an XML environment. They tackle the same issue as that faced by knowledge workers in sensor networks: converting data to a usable format. Their approach is similar to XSENSE in that they use their Data Format Description Language to generate the XML representation of data. This is similar to our usage of ServingXML to meet the same goals although their approach requires a lot more user input as descriptions can often be quite lengthy. The key contribution of their work is that no conversion of sensor data is necessary as they create a view definition to interpret the raw data. On the negative side, they provide only a template system that has not been applied to any domain (instead they provide some use-case descriptions), and no query response times are possible. In this respect, their query optimiser could face problems when converting between the view definition and the physical data.

In (Kotidis, 2006; Whitehouse *et al.*, 2006), the authors process and query raw streams of sensor data and avoid conversion to XML. This has its benefits as the construction times for XML repositories (both centralised and distributed) are often reported to be quite large, and we have reported similar issues here. In (Whitehouse *et al.*, 2006), the approach is to enrich raw data into semantic streams and process these streams as they are generated. Their usage of constraints on the data streams provides a useful query mechanism with possibilities for optimisation. However, this work is still theoretical and contains no evidence of experiments or query times. In (Kotidis, 2006), they employ the concept of proximity queries where network nodes monitor and record interesting events in their locality. While their results

are positive in terms of cost, queries are still at a relatively low level (no common format for query expression), and it is difficult to see how this type of proximity network can be applied in general terms due to the complexity of the technologies involved.

Kawashima *et al.* (2006) provide semantic clusters within their sensor network. This is a similar approach to our work, where we cluster related groups of sensors. They also adopt a semi-automated approach and are capable of generating metadata to describe sensors and thus, support query processing. However, their object-oriented approach is likely to lead to problems with interoperability and this could be exacerbated through the lack of common query language. While this can be addressed with a canonical layer (probably using XML) for interoperability, it is likely to have performance related issues.

None of the related work which we have discovered has been able to allow querying of large amounts of sensor data for the purpose of explicitly supporting the type of complex queries characteristic of data mining applications. Our work, as reported here, achieves this.

It is interesting at this point to consider what the longer term roadmap for our work might be and where it is likely to develop. Within the research community there is a strong belief in the prospects for semantic web (SW) technology across a wide range of applications. The original motivation for developing SW technologies was to address the dissatisfaction with (text-based) search on a WWW scale, where ambiguities due to word polysemy, for example, still cause such frustration in search. To illustrate, when searching for documents containing the work *"Jaguar"* we might retrieve pages about a car, an animal or an operating system! The motivation behind the semantic web is that a more fine-grained and semantically rich and accurate representation of a web page, a document, a picture, a video, any type of digital object, and likewise for a user's query, will result in more accurate information management. This principle can be extended to other data besides multimedia documents, and can be extended to include sensor data and sensor networks.

Within the broad Semantic Web research community there is work on-going on extending SW principles to sensor data. For example, the Open Geospatial Consortium (OGC) have developed and adopted several Sensor Web Enablement Standards including *SensorML* for describing sensor systems and processes to help in their discovery and location, *TransducerML* for describing real-time streaming of data to/from sensor networks and the *Sensor Planning Service*, a web service acting as an intermediate layer between a client and an environment for managing sensor networks. These developments, and others, provide a target area for our own work to migrate towards and to build upon.

CONCLUSION

As sensor networks become more pervasive, the volumes of data generated will pose challenges to managing the stream of data values in an efficient, scalable, and yet useful manner. This data will include data readings from sensing devices, such as the biometric sensing devices we have used in this chapter. It will also include data values derived from analysis of raw sensed data, such as the highlights and event detection results from an audio-visual data stream which we have illustrated in this chapter through our analysis of movie video. While many differing solutions exist for efficiently managing the raw data values, our focus here is on allowing semantic enrichment of the raw data to take place in order to facilitate data mining activities. To address this, we have presented and used the XSENSE architecture, extended with Peer-2-Peer concepts and implementation, to realise scalable management of sensor data at a semantic level. We have demonstrated this in operation using a dataset of 33 experiments covering 29 movies and a total of 171 sensor outputs, realising a data volume of 316 MB of raw data. Our implementation has been tested using a collection of different *semantic* query types, and response time performance, on a standard desktop machine, has been good.

Our contribution is in addressing the problem of having very long query response times when sensor data is gathered, enriched using our XSENSE architecture into XML streams, and then deployed on a P2P architecture. We have demonstrated that cluster formation, together with keywords, stored in the system repository, significantly optimise query response time. The approach we took is that the XPath expression is stripped of keywords, and these are used to match to 1 or more clusters and by querying only those clusters, we reduce query processing time. We also demonstrated that this works best in a truly distributed setup (P2P) as when we break the system over multiple machines, the multiple CPUs improve times further. All this yields a much faster turnaround time in querying sensor data. Our current work is thus on constructing a layer of DM primitives using XPath and Java which will in turn support analytics and data mining.

There are many directions in which we intend to pursue further work. As clusters of sensors inside specified domains have similar properties to tree-based systems, we intend to optimise distributed XSENSE queries by extending techniques previously developed for XML trees (O'Connor *et al.*, 2005). We also intend to examine issues of real-time ingestion of sensor data, and real-time querying. Finally, we will also address questions of how to handle instances of sensors dropping out of the sensing process, and then returning later, usually done in order to save power consumption on the device.

REFERENCES

Adjiman, P., Chatalic, P., & Goasdou, F. Rousset, M., & Simon, L. (2005). Somewhere in the Semantic Web. *3rd Workshop on Principles and Practice of Semantic Web Reasoning* (LNCS 3703, pp. 1-16).

Bellahsène, Z., & Roantree, M. (2004). Querying Distributed Data in a Super-Peer based Architecture. *15th International Conference on Database and Expert System Applications* (LNCS 3180, pp. 296-305).

Bordwell, B., & Thompson, K. (1997). *Film Art: An Introduction*. McGraw-Hill.

Dancyger, K. (2002). *The Technique of Film and Video Editing. History, Theory and Practice.* Focal Press.

Dunne, L. E., Brady, S., Smyth, B., & Diamond, D. (2005). Initial development and testing of a novel foam-based pressure sensor for wearable sensing. *Journal of Neuroengineering and Rehabilitation, 2*(1), 4. doi:10.1186/1743-0003-2-4

Kalnis, P., Ng, W., Ooi, B., Papadias, D., & Tan, K. (2002). An Adaptive Peer-to-Peer Network for Distributed Caching of OLAP Results. In SIGMOD Record, *ACM SIGMOD International Conference on the Management of Data, 31*(2), 25-36. ACM Press.

Kawashima, H., Hirota, Y., Satake, S., & Imai, M. (2006). MeT: A Real World Oriented Metadata Management System for Semantic Sensor Networks. *3rd International Workshop on Data Management for Sensor Networks (DMSN)* (pp. 13-18).

Kotidis, Y. (2006). Processing Proximity Queries in Sensor Networks. *3rd International Workshop on Data Management for Sensor Networks (DMSN)* (pp. 1-6).

Lehane, B., & O'Connor, N. E. (2004). Action Sequence Detection in Motion Pictures. *The International Workshop on Multidisciplinary Image, Video, and Audio Retrieval and Mining.*

Lehane, B., & O'Connor, N. E. (2006). Movie Indexing via Event Detection. *7th International Workshop on Image Analysis for Multimedia Interactive Services.*

Lehane, B., O'Connor, N. E., & Murphy, N. (2005). Dialogue Scene Detection in Movies. *International Conference on Image and Video Retrieval (CIVR)* (pp. 286-296).

Lehane, B., O'Connor, N. E., Smeaton, A. F., & Lee, H. (2006). A System for Event-Based Film Browsing. *TIDSE 2006, 3rd International Conference on Technologies for Interactive Digital Storytelling and Entertainment* (LNCS 4326, pp. 334-345).

Manjunath, B. S., Salember, P., & Sikora, T. (2002). *Introduction to MPEG-7, Multimedia content description language.* John Wiley and Sons Ltd.

Nejdl, W., et al. (2002). EDUTELLA: a P2P networking infrastructure based on RDF. *11th International World Wide Web Conference* (pp. 604-615). ACM Press.

O'Connor, M., Bellahsene, Z., & Roantree, M. (2005). An Extended Preorder Index for Optimising XPath Expressions. *3rd XML Database Symposium* (LNCS 3671, pp 114-128).

Rose, K., Malaika, S., & Schloss, R. (2006). Virtual XML: A toolbox and use cases for the XML World View. *IBM Systems Journal, 45*(2), 411–424.

Rothwell, S., Lehane, B., Chan, C. H., Smeaton, A. F., O'Connor, N. E., Jones, G. J. F., & Diamond, D. (2006). The CDVPlex Biometric Cinema: Sensing Physiological Responses to Emotional Stimuli in Film. *Pervasive 2006 - the 4th International Conference on Pervasive Computing.*

Saxon Project. (2007). Retrieved from http://saxon.sourceforge.net

Serving, X. M. L. Project (2007). Retrieved from http://servingxml.sourceforge.net

Whitehouse, K., Zhao, F., & Liu, J. (2006). Semantic Streams: a Framework for Composable Semantic Interpretation of Sensor Data. *3rd European Workshop on Wireless Sensor Networks (EWSN)* (LNCS 3868, pp. 5-20).

APPENDIX

A. Semantic File for Film Happiness.xml

```
<movie movieId=\tt0147612" langCode=\eng">
    <title>Happiness</title>
    <movieStructure>
        <scene id=\0">
            <cluster id=\0">
                <shot id=\0">
                    <keyFrame id=\0"/>
                </shot>
                ...
            </cluster>
            ...
        </scene>
        ...
    </movieStructure>
    <events>
        <event id=\0" type=\Dialogue">
            <startShot id=\0"/>
            <endShot id=\3"/>
            <keyShot id=\2"/>
            <keyShot id=\1"/>
            <keyShot id=\99999999"/>
        </event>
        ...
        <event id=\18" type=\Action">
            <startShot id=\176"/>
            <endShot id=\183"/>
            <keyShot id=\177"/>
            <keyShot id=\179"/>
            <keyShot id=\181"/>
        </event>
        ...
    </events>
    <shots>
        <shot id=\0">
            <startFrame id=\0"/>
            <keyFrame id=\0"/>
            <endFrame id=\30"/>
            <startTime>0</startTime>
            <endTime>1001</endTime>
            <motionIntensity>0</motionIntensity>
```

```
        <percentCameraMovement>0.2</percentCameraMovement>
        <percentSilence>0</percentSilence>
        <percentSilenceMusic>1</percentSilenceMusic>
        <percentSpeech>0</percentSpeech>
        <percentMusic>0</percentMusic>
        <percentOtherAudio>0</percentOtherAudio>
    </shot>
</shots> </movie>
```

B. Timings Chart using a Single-Server Configuration

The graph (Figure 7) displays the time (X-axis) required to process various numbers of peers (Y-axis). Significant gaps such as that between processingXMLQuery and XMLQueryProcessed represent delays caused by CPU contention.

C. Timings Chart using the 3-Server Configuration

The graph (Figure 8) displays the time (X-axis) required to process various numbers of peers (Y-axis). The elimination of major gaps between processing steps is achieved by distributing sensor peers.

Figure 7.

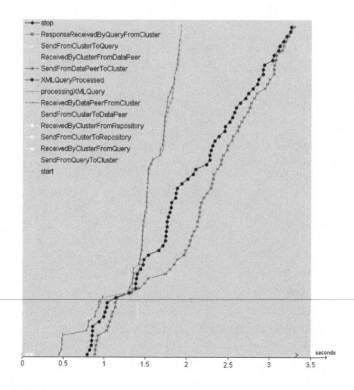

Figure 8.

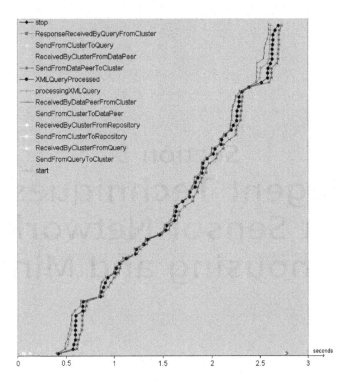

Section 5
Intelligent Techniques for Efficient Sensor Network Data Warehousing and Mining

Chapter 12

Geographic Routing of Sensor Data around Voids and Obstacles

Sotiris Nikoletseas
University of Patras, Greece

Olivier Powell
University of Geneva, Switzerland

Jose Rolim
University of Geneva, Switzerland

ABSTRACT

Geographic routing is becoming the protocol of choice for many sensor network applications. Some very efficient geographic routing algorithms exist, however they require a preliminary planarization of the communication graph. Planarization induces overhead which makes this approach not optimal when lightweight protocols are required. On the other hand, georouting algorithms which do not rely on planarization have fairly low success rates and either fail to route messages around all but the simplest obstacles or have a high topology control overhead (e.g. contour detection algorithms). This chapter describes the GRIC algorithm which was designed to overcome some of those limitations. The GRIC algorithm was proposed in (Powell & Nikoletseas, 2007a). It is the first lightweight and efficient on demand (i.e. all-to-all) geographic routing algorithm which does not require planarization, has almost 100% delivery rates (when no obstacles are added), and behaves well in the presence of large communication blocking obstacles.

INTRODUCTION

We consider the problem of routing in ad hoc wireless networks of location aware stations, i.e. the stations know their location in a coordinate system such as the Euclidean plane. This problem is commonly called *geographic routing*. An emerging technology which has recently attracted considerable research efforts towards improving geographic routing algorithms is that of wireless sensor network. This text falls within this context and is concerned with the specific requirements imposed on geographic routing

DOI: 10.4018/978-1-60566-328-9.ch012

by such networks. The constraints imposed on the engineering of algorithms dedicated to sensor networks are stringent and routing algorithms are no exceptions to this rule: routing algorithms for sensor networks should be realistic, lightweight, on demand and efficient. Routing algorithms for sensor networks should be *realistic* and usable in real world scenarios, including urban scenarios with large communication blocking obstacles such as walls or buildings. Routing algorithms should also accommodate themselves with areas of low node density, also called routing holes. Of course, the success and performance of routing algorithms should not be dependent on the assumptions made on the model used as an approximation to real world wireless networks. They should be *lightweight*, which is notably understood in terms of topology maintenance overhead and, more generally, in terms of protocol message exchange overhead. Routing algorithms for sensor networks should also be *on demand*, which means they should be all-to-all (as opposed to all-to-one and one-to-all algorithms) without relying on solution which are obviously not lightweight as they require the storing of routing tables or other expensive and difficult to maintain and update information. *Efficiency* is measured in terms of success rate (the probability that a message will reaches its destination) and hop-stretch (i.e. the path length should be short), which somehow encompasses at a high level many other metrics, such as energy consumption, latency and traffic.

Problem definition 1. *The problem we are considering is that of routing messages in a network of wireless and localized nodes with regions of low node density and with large emission blocking obstacles (such as walls, buildings, lakes, etc...). The geographic routing algorithms we allow ourselves to consider should be lightweight, fast, low cost, on demand and realistic.*

Geographic Routing for Sensor Nets and Applications

Recent advances in micro-electromechanical systems (MEMS) and wireless networking technologies have enabled the development of very small sensing devices called sensor nodes (Rabaey, Ammer, da Silva, Patel, & Roundy, 2000; Warneke, B., Last, Liebowitz, & Pister, 2001; Akyildiz, Su, Sankarasubramaniam, & Cayirci, 2002). Unlike traditional sensors that operate in a passive mode, sensor nodes are smart devices with sensing, data-processing and wireless transmission capabilities. Data can thus be collected, processed and shared with neighbors. Sensor nodes are meant to be deployed in large *wireless sensor networks* instrumenting the physical world. Originally developed for tactical surveillance in military applications, they are expected to find growing importance in civil applications such as building, industrial and home automation, infrastructure monitoring, warehousing and data mining, agriculture and security. In particular, sensor networking can be very useful in automated warehousing, in the sense that tiny wireless sensory devices (possibly together with RFIDs) can be attached to the items kept at the warehouse. The sensors can then exchange data towards collaboratively answer queries injected in the system, like how many items of this type are there, what is the actual condition of item with a given id etc. Needless to say, such a data exchange is usually done in an environment in which wireless communication and data routing is obstructed, e.g. by walls, narrow corridors, voids due to physical failures of some devices etc. The theme of this chapter is exactly how to efficiently route sensory data in the presence of obstacles. Their range of applications makes sensor networks heterogeneous in terms of size, density and hardware capacity. Their great attractiveness follows the availability of low cost, low power and miniaturized sensor nodes pervading, instrumenting and reality augmenting the physical world *without requiring human intervention*: sensor networks are self-organizing, self repairing and highly autonomous. Those properties are implemented by designing protocols addressing the specific requirements of sensor networks. Common to all sensor net architecture and applications is the need to propagate data inside the network. Almost every scenario implies multi-hop data transmission because of low power used for radio transmission mostly to limit depletion of the scarce energy resources of battery powered sensor nodes. As a consequence, routing comes as a fundamental primitive of every fully fledged sensor network protocol, whatever the application, the topology and the hardware composing the sensor net. Zhao and Guibas (Zhao & Guibas, 2004a) claim correctly that routing protocols for sensor nets need to respond to specific constraints.

The most appropriate protocols [for sensor networks] are those that discover routes on demand using local, lightweight, scalable techniques, while avoiding the overhead of storing routing tables or other information that is expensive to update such as link costs or topology changes.

Geographic routing is a very attractive solution for sensor networks. The most basic geographic routing algorithms are the greedy geographic routing algorithms from (Finn, 1987; Takagi & Kleinrock, 1984), where messages are propagated with minimization of the remaining distance to the message's destination used as a heuristic for choosing the next hop relay node or maximization of the projected progress towards destination along the line from source to destination respectively. Fundamental (and some more advanced) geographic routing techniques and limitations are covered in most text books on sensor nets, e.g. (Karl & Willig, 2005; Zhao & Guibas, 2004a). The greedy routing algorithm discovers routes on demand and is all-to-all (as opposed to all-to-one), it uses only location information of neighbor nodes, it scales perfectly well, requires no routing tables

and adapts dynamically to topology changes. The only requirement, which is a strong one, is that sensor nodes have to be *localized*, i.e. they should have access to coordinates defining their position. The major problem of the greedy approach is that messages are very likely to get trapped inside of local minimums (nodes which have no neighbors closer to the destination than themselves). For this reason early geographic routing has received much attention and has been the subject of abundant research, motivated by the gain in momentum of ad hoc wireless networking. Ingenious geographic routing techniques have been developed and will be discussed in more detail in section 2.

Location Awareness

Geographic routing is based on the assumption that nodes of the communication network are localized, i.e. they are aware of their position in a coordinate system. A natural question in the context of sensor networks is thus to ask how reasonable it is to assume sensor nodes to be localized. It is usually accepted that the use of global positioning system (GPS) or similar technology on every node is not reasonable for sensor networks. Among the reasons why GPS may not be appropriate is the increased cost, weight, and energy depletion, not to mention the fact that GPS does not work indoors. Therefore, one may ask to what extent the strong hypothesis of localized nodes restricts the application of geographic routing techniques to a specialized niche of sensor networks. First of all, it should be acknowledged that the localization of nodes comes at a cost. Although probably high, this cost can nevertheless be kept reasonable through the use of one of the localization techniques for sensor networks often using GPS-like technology to localize a few beacon nodes followed by a distributed protocol to localize the bulk of sensor nodes as described in general books on sensor networks such as (Karl & Willig, 2005; Beutel, 2005; Zhao & Guibas, 2004b), recent papers such as (Leone, Moraru, Powell, & Rolim, 2006) or in the extensive survey in (Hightower & Borriello, 2001). Those techniques come at the price of an overhead in terms of protocol complexity which would diminish the attractiveness of geographic routing if it was not for the fact that in many applications localization is inevitable since, as explained by Karl and Willig (2005):

...in many circumstances, it is useful and even necessary for a node in a wireless sensor network to be aware of its location in the physical world. For example, tracking or event-detection functions are not particularly useful if the [sensor net] cannot provide any information where an event has happened.

As a consequence, geographic routing is not confined to a specialized niche but on the contrary it is becoming a protocol of choice for sensor networks as advocated by Seada, Helmy and Govindan (2001):

... [geographic routing] is becoming the protocol of choice for many emerging applications in sensor networks, such as data-centric storage (Ratnasamy et al, 2002) and distributed indexing (Rao, Ratnasamy, Papadimitriou, Shenker, & Stoica, 2003).

It thus seems reasonable that geographic routing will remain a sustainable approach for many sensor net applications and that the ongoing research will keep improving the state of the art to a point where localization will be made available with high precision at low protocol overhead for sensor networks. Interestingly, it may be noticed that geographic routing can even be used when nodes are not localized by using *virtual coordinates* as was proposed in (Rao, Ratnasamy, Papadimitriou, Shenker, & Stoica, 2003).

BACKGROUND

As explained in the previous section, the early greedy geographic routing algorithms proposed in (Finn, 1987; Takagi & Kleinrock, 1984) are very appealing solutions for routing in wireless sensor networks. The major limitation of those greedy geographic routing algorithms is the so called *routing hole problem* (Karl & Willig, 2005; Zhao & Guibas, 2004a; Ahmed, Kanhere, & Jha, 2005) where messages get trapped by local minimum nodes which have no neighbors closer to the destination of the message than themselves. The incidence of routing holes increases as network density diminishes and the success rate of the greedy algorithms drops very quickly with network density. On the contrary, under conditions likely to be fulfilled in high density networks it is guaranteed that greedy geographic routing succeeds. More precisely, (Xing, Lu, Pless, & Huang, 2004) proves that in a unit disc graph, greedy geographic routing has 100% success rate if the density condition of being 0.5-covered is fulfilled. I.e., greedy routing guarantees delivery if the sensing covered region with sensing radius 0.5 times the communication radius is convex. In real networks, this density condition is likely not to be fulfilled since low density regions are statistically likely to appear even in relatively dense deployments. Furthermore, in the case where the network has "artificial holes" because of obstacles such as a pond, a building, a vehicle or a lake, the sensing covered region has no chance of being convex, whatever the density. Improvements on the initial greedy approach to geographic routing have mainly focused on increasing the success rate (the probability that a message is successfully routed) while keeping the "lightweight" advantages of the greedy type of routing algorithms, although recent research efforts have also been proposed to deliver efficient geographic routing algorithms with respect to other metrics (Stojmenovic & Datta, 2004; Kuruvila, Nayak, & Stojmenovic, 2006). Most geographic routing algorithms can be classified into one of the following two groups: those that *guarantee delivery* (i.e. the success rate is 100% if the communication graph is connected) (Bose, Morin, Stojmenovic, & Urrutia, 1999; Karp & Kung, 2000) and *probabilistic approaches* that achieve certain performance trade-offs, such as (Chatzigiannakis, Dimitriou, Nikoletseas, & Spirakis, 2004; Chatzigiannakis, Dimitriou, Nikoletseas, & Spirakis, 2004; Chatzigiannakis, Nikoletseas, & Spirakis, 2002; Chatzigiannakis, Nikoletseas, & Spirakis, 2005; Kuruvila, Nayak, & Stojmenovic, 2006). Alternatively, it is also possible to run topology discovery algorithms to discover routing holes and obstacles and to use this information to route messages around them. An example is the *boundhole* algorithm from (Fang, Gao & Guibas, 2006) which uses the *tent* rule to discover local minimum nodes and then to "bound" the contour of routing holes. Furthermore, with this approach the information gained during the contour discovery phase may be used for other applications than routing. For example, this information can be used for path migration, information storage mechanisms or identification of regions of interest (Fang, Gao & Guibas, 2006). On the downside, this approach has a high overhead when compared to greedy geographic routing (Finn, 1987; Takagi & Kleinrock, 1984) or the GRIC routing algorithm presented in section 3.2. It also seems fragile towards topology instability such as node mobility or temporary link failure. One other interesting approach, which seems to combine robustness and reasonable message overhead, is the loop-free hybrid single-path and flooding solution proposed in (Stojmenovic & Lin, 2001). This solution is quite effective. It handles the local minimum problem by using a controlled flooding mechanism. There is one possible limitation, which is the fact that this algorithm was not designed to take into consideration the case of obstacles. Indeed, while the message overhead due to the controlled flooding necessary to get out of the local minimums induced by areas

of low network density is reasonable, the case of obstacles is different since the whole region contained "inside" the obstacle will need to be flooded with messages. This area can get quite large when the obstacle size increases.

Probabilistic Approaches

Guaranteed delivery algorithms have received plenty of attention in the literature because, when applicable, they are very powerful algorithms. A taxonomy of geographic routing algorithms is provided in (Stojmenovic, 2002), focusing on guaranteed delivery algorithms but with a review of many other solutions. Some other solutions include the PFR (Chatzigiannakis, Dimitriou, Nikoletseas, & Spirakis, 2004; Chatzigiannakis, Dimitriou, Nikoletseas, & Spirakis, 2006) VTRP (Antoniou, Chatzigiannakis, Mylonas, Nikoletseas, & Boukerche, 2004), the LTP protocols (Chatzigiannakis, Nikoletseas, & Spirakis, 2002; Chatzigiannakis, Nikoletseas, & Spirakis, 2005) or the CKN protocol from (Chatzigiannakis, Kinalis, & Nikoletseas, 2006). Those protocols employ randomization. They are very lightweight and are also general enough so that they may be used in a more general context than that of localized networks. On the down side, they do not guarantee message delivery. PFR is a probabilistically limited flooding algorithm (it is thus multi-path), VTRP proposes to use variable transmission range capable hardware to bypass routing holes in low density areas and LTP is a combination of a modified greedy algorithm with limited backtracking as a rescue mode. Those algorithms are quite interesting and offer an attractive alternative, even in the presence of low density regions. A feature of interest to this text is the behavior of those protocols in the presence of large communication blocking obstacles. The effect of obstacles on those three protocols was studied in (Chatzigiannakis, Mylonas, Nikoletseas, 2006). One negative result found in (Chatzigiannakis, Mylonas, Nikoletseas, 2006) is that all three protocols fail in the case of a large wall obstacle. In terms of obstacle avoidance, the GRIC algorithm presented in section 4 thus prevails over PFR, VTRP and LTP (as well as over all other "lightweight" geographic routing algorithms we know of). In the case of no obstacle, we shall show that GRIC is more efficient than LTP (and therefore than PFR, as follows from results in Chatzigiannakis, Mylonas, Nikoletseas, 2006). VTRP is more efficient than GRIC in the case of networks of very low densities with no obstacles by taking advantage of additional hardware properties, more precisely, it "jumps over" routing holes by augmenting the transmission range when faced with a blockage situation (whereas GRIC does not assume transmission range variation to be possible). It may be noticed that although adaptive transmission range is a reasonable assumption (although maybe not for all possible hardware configuration), this is of little help if the obstacle is a wireless communication blocking obstacle. More recently, (Chatzigiannakis, Kinalis, & Nikoletseas, 2006) proposes a routing protocol that implicitly copes with simple obstacles and node failures by "learning", i.e. gaining limited local knowledge of the actual network conditions, and by using this information to optimize data propagation. By planning routes a few hops ahead and by varying the transmission range it can therefore bypass routing holes (i.e. areas of low density but through which radio transmission is possible if the transmission range is increased) and some small and simple obstacles. However, increasing the transmission range will be of little help in the case of physical obstacles (through which radio propagation is impossible) or large obstacles which cannot be "jumped over" by increasing the transmission range.

Routing with Guaranteed Delivery

Arguably, most geographic routing algorithms are inspired by the GFG geographic routing algorithm proposed in (Bose, Morin, Stojmenovic, & Urrutia, 1999). GFG uses a greedy routing algorithm until the message gets trapped in a local minimum. When the message gets trapped in a local minimum, the algorithm switches to a *rescue mode*, escapes the local minimum, and then switches back to greedy mode. This algorithm has the very strong property of guaranteeing message delivery as long as the graph is connected. The rescue mode used in GFG to escape local minimums is called the FACE algorithm. The FACE algorithm from (Bose, Morin, Stojmenovic, & Urrutia, 1999) is a modification of the COMPASS-II algorithm from (Kranakis, Singh, & Urrutia, 1999). FACE can be seen as an optimization of COMPASS-II, in the sense that the length of the path along which messages are routed by FACE is smaller than for compass-II. One requirement for the FACE (and the COMPASS-II) algorithm to succeed is that the underlying communication graph should be planar. One of the important contributions of (Bose, Morin, Stojmenovic, & Urrutia, 1999) is to propose a simple and distributed algorithm to extract a planar subgraph of the initial communication graph, since the former is typically not planar. This has the enormous advantage of making the protocol usable for real world networks, a point that was not addressed by the otherwise very ingenious COMPASS-II algorithm of (Kranakis, Singh, & Urrutia, 1999). Other major contributions to guaranteed delivery geographic routing algorithms include the GPSR algorithm of (Karp & Kung, 2000). GPSR is very similar to (Bose, Morin, Stojmenovic, & Urrutia, 1999). One major difference is that (Karp & Kung, 2000) proposes a distributed algorithm to extract the relative neighborhood graph (a planar subgraph of the communication graph) instead of the Gabriel graph used in (Bose, Morin, Stojmenovic, & Urrutia, 1999). (Karp & Kung, 2000) also includes a detailed experimental validation of GPSR, including MAC layer considerations in the simulations. Other major contributions to geographic routing are the GOAFR and GOAFR+ algorithms from (Kuhn, Wattenhofer, & Zollinger, 2003; Kuhn, Wattenhofer, Zhang, & Zollinger, 2003). Those algorithms are inspired by GFG/GPSR. One major contribution of (Kuhn, Wattenhofer, & Zollinger, 2003; Kuhn, Wattenhofer, Zhang, & Zollinger, 2003) is to come up with a formal analysis of the GOAFR/GOAFR+ algorithms in terms of performance and path lengths. FACE routing (as well as algorithms using it such as GFG/GPSR/ GOAFR) is very attractive because it guarantees delivery, however it involves a number of issues. The most important issue is the fact that it requires a preliminary planarization phase, a procedure which may not be achievable easily in many scenarios. Other issues include dependency on the unit disc graph model that can only be erased by introducing high topology maintenance overhead, sensitivity to localization errors, increased path length, requirement of symmetric links and limitation to 2-dimensional networks. Issues, and how they motivate the need for an alternative approach to geographic routing, will be discussed in more detail in section 3.1.

Planarization

One limitation of GFG and GPSR is that the distributed planarization algorithms proposed rely heavily on the assumption that the initial communication graph is a disc graph. That is, the assumption is made that two nodes can communicate if and only if they are at distance at most r, for some fixed constant communication radius r. This assumption is a simplistic approximation of real world communication graphs which is very convenient for simulation and analysis purposes, but it can be very different from real communication networks. Depending on the usage of this simple model, this can be a source of erroneous, incomplete or imprecise findings. In the case of the planarization algorithms proposed in (Bose, Morin, Stojmenovic, & Urrutia, 1999; Karp & Kung, 2000), the disc graph hypothesis is oversimplifying and the planarization protocols fail in most real world network configurations.

To improve the matter, (Barriere, Fraigniaud, Narayanan, & Opatrny, 2003) proposes a dis-

tributed planarization algorithm that works in a more general model than the unit disc graph. More precisely, it works in quasi-unit disc graphs where two nodes can never communicate if their distance is greater than R, can always communicate if their distance is smaller than r, and there are no conditions for nodes at a distance between r and R from one another. The algorithm proposed by (Barriere, Fraigniaud, Narayanan, & Opatrny, 2003) introduces virtual edges to reduce the quasi-disc graph to a case similar to the original disc graph. For their solution to work, it has to be that the ratio $\dfrac{R}{r}$ is smaller than $\sqrt{2}$.

In (Kim, Govindan, Karp, & Shenker, 2005b) a series of pitfalls associated with geographic routing which extends and deepens the microanalysis of (Seada, Helmy, & Govindan, 2001) are identified. Those pitfalls follow from the fact that the communication graph of real world wireless sensor networks is not so well approximated by quasi-disc graphs, which suggests that the solution of (Barriere, Fraigniaud, Narayanan, & Opatrny, 2003) needs further improvements.

Following the findings of (Kim, Govindan, Karp, & Shenker, 2005b), the same authors propose in (Kim, Govindan, Karp, & Shenker, 2005a) the cross-link detection protocol (CLDP). CLDP extracts a planar subgraph of the initial communication graph. To our knowledge, it is the first planarization algorithm to work on arbitrary communication graphs. In fact, CLDP does not necessarily produce a completely planar (embedding of a) communication graph, but it is proven that GFG/GPSR guarantees message delivery on the subgraph constructed using CLDP. Unfortunately, CLDP introduces very large topology maintenance overhead and many messages have to be propagated through the network to construct the (quasi) planar subgraph.

In (Leong, Liskov, & Morris, 2006) *Greedy Distributed Spanning Tree Routing* (GDTSR) is proposed. GDTSR is not a planar graph based routing algorithm, but it is also a guaranteed delivery routing algorithm. It is shown in (Leong, Liskov, & Morris, 2006) that GDTSR has a much lower communication overhead than CLDP. GDTSR has the originality of considering a new way of routing messages while in rescue mode, i.e. as a replacement to FACE routing, based on the use of convex-hull spanning trees.

In (Kim, Govindan, Karp, & Shenker, 2006), the lazy cross-link removal (LCR) algorithm is proposed to extract a planar subgraph more efficiently than CLDP. When compared to CLDP or GDTSR, LCR significantly reduces the topology maintenance overhead.

In a different repertoire, (Moraru, Leone, Nikoletseas, & Rolim, 2007) proposes to reduce the routing path length of GFG by using a marking mechanism of nodes where previous routing required the use of the rescue mode. Those nodes will be avoided for further routing. After convergence, the technique is shown to improve performance if the network density is high enough. Another interesting approach is proposed in (Frey & Gorgen, 2006) in the context of sensing covered networks. Nodes are organized into geographic clusters and routing is performed along the virtual edges of a virtual overlay graph.

LIGHTWEIGHT GEOGRAPHIC ROUTING AROUND OBSTACLES

Issues, Controversies, Problems

GFG, GPSR and GOAFR (+) type of algorithms are based on extracting a planar subgraph of the total communication graph. The initial distributed algorithms proposed in (Bose, Morin, Stojmenovic, & Urrutia, 1999; Karp & Kung, 2000) rely heavily on the assumption that the communication graph is a unit disc graph, which is violated in real world networks. The fact that virtual edges and quasi disc graphs can replace disc graphs (Barriere, Fraigniaud, Narayanan, &

Opatrny, 2003), c.f. section 2.2.1, is an important improvement. However, as shown by the micro-analysis of (Seada, Helmy, & Govindan, 2001), deviation from those assumptions may lead to unrecoverable errors for GFG based algorithms. Unfortunately, it seems that real world scenarios are likely to often times deviate from the quasi disc graph model. (Kim, Govindan, Karp, & Shenker, 2005b) identifies such a scenario (which is a typical indoor sensor network deployment with a common hardware platform). One of the reason why quasi disc graphs are not likely to be the solution to all problems is that the hypothesis that sensor nodes can communicate to neighbor nodes whenever their distance is beyond a given threshold seems not to be true in many real networks, c.f. (Zhou, He, Krishnamurthy, & Stankovic, 2004) for a study of radio irregularity in sensor networks providing evidence on this. As a consequence, the series of improvements to planarization such as those proposed in CLDP, GDTSR and LCR are likely to be the only currently available solution to offer guaranteed delivery geographic routing for real world applications. Those algorithms, although innovative and reasonable, have a few drawbacks. One limitation is the necessity of having bidirectional links, which may not be the case, particularly at the MAC layer and may also limit some optimization techniques (e.g. a link leading to a low energy node may be "virtually" removed while the symmetric link leading to a high energy node should be kept for energy optimization purposes). A more serious limitation is the topology maintenance overhead. Although LCR implies consistently lower topology maintenance overhead than CLDP and GDTSR, it is still fair to say that the overhead would be prohibitive for many dynamic scenarios, e.g. if the nodes are mobile. In short, LCR makes GFG/GPSR robust, but it somehow diminishes their property of being lightweight. On the other side, lightweight guaranteed delivery protocols such as the one based one controlled flooding from (Stojmenovic & Lin, 2001) are not capable of (efficiently) bypassing large communication blocking obstacles. The same holds for probabilistic based algorithms such as PFR, VTRP or LTP, as was shown in (Chatzigiannakis, Mylonas, Nikoletseas, 2006). Summarizing, there is a need for a lightweight geographic routing algorithm capable of bypassing large communication blocking obstacles. The GRIC algorithm proposed in (Powell & Nikoletseas, 2007b; Powell & Nikoletseas, 2007a; Nikoletseas & Powell, 2008). answers those requirements and is presented in the following section.

Solutions and Recommendations

From our previous discussion, it becomes clear that wireless networking, in particular sensor networks, would greatly benefit from the existence of a geographic routing solution which is both lightweight and capable of bypassing large (communication blocking) obstacles. To our knowledge, such solutions are scarce (c.f. section 2), and the most efficient is the *Geographic Routing* around *obstaCles* (GRIC) algorithm (Powell & Nikoletseas, 2007b; Powell & Nikoletseas, 2007a; Nikoletseas & Powell, 2008). The main idea of this algorithm is to use movement directly towards the destination to optimize performance appropriately combined with an inertia effect that forces messages to keep moving along the "current" direction so that it will closely follow obstacle shapes in order to efficiently bypass them. In order to route messages around more complex obstacles a "right-hand rule" component is added to it. The right-hand rule is borrowed from maze solving algorithms (algorithms one can use to find the way out of a maze). The right-hand rule is a well known "wall follower" technique to get out of a maze (Hemmerling, 1989) which is used by some of the most successful geographic routing algorithms known so far (COMPASS-II, GFG, GPSR, etc... c.f. section 2). However, whereas GFG and other similar algorithms use a strict implementation of the right hand rule, the GRIC algorithm is only inspired by it. The rationale

behind this is that the right-hand rule, originally developed to find a path out of a maze, only works for planar graphs. The GRIC algorithm wants to avoid the overhead of planarization and should run on the complete (non planarized) communication graph. As a consequence, GRIC needs a "soft" version of the right-hand rule. The rest of this paper is dedicated to the detailed description of GRIC, as well as the demonstration of its efficiency.

Overview of GRIC

The algorithm was implemented and its performance evaluated against those of other representative algorithms (greedy, LTP, and FACE) in (Powell & Nikoletseas, 2007a). The evaluation focused on three performance measures: success rate, distance travelled and hop count. The impact on performance of several types of obstacles (both convex and concave) and representative regimes of network density was studied.

Comparison

When compared to other geographic routing algorithms, GRIC features (1) no topology maintenance overhead, (2) low constraints on other layers (physical and MAC) of the protocol stack since it runs on arbitrary directed communication graphs and (3) the capacity of routing messages to their destination with high success rate, even in the presence of large communication blocking obstacles. Also, it may be added (4) that GRIC is fairly easy to implement, which is important for hard to debug networks like sensor networks. To our knowledge, the combination of those features makes GRIC unique. Some previous algorithms like GFG or the BOUNDHOLE algorithm have high success rates, however none of them works on directed communication graphs and they all suffer either from a high topology maintenance overhead (for example, in practice GFG has to be combined with a complicated and high overhead planarization protocol like LCR). Other algorithms are lightweight (i.e. have low topology maintenance overhead), like the probabilistic algorithms discussed in section 2.1 or the controlled flooding algorithm of (Stojmenovic & Lin, 2001) but none of them is capable of efficiently routing messages around large obstacles.

Strengths of the Approach

GRIC has the advantage of being many-to-many and on demand, i.e. no routing tables have to be stored, no flooding is required to establish a gradient and no interests (or other topology information) have to be propagated for the routing to be successful. This is possible because the nodes are localized, which is arguably part of the strengths inherent to all geographic routing algorithms. It is in large part responsible for the attractiveness of this kind of protocols. GRIC has a few attractive properties. *In terms of algorithm design*, it combines the lightweight and simplicity of the greedy type of algorithms. *In terms of overhead*, it completely avoids the topology maintenance phase (usually planarization) required by the GFG (Bose, Morin, Stojmenovic, & Urrutia, 1999) and GPSR (Karp & Kung, 2000) algorithms and their successors, c.f. section 3.2. *In terms of performance*, it was found that in most cases GRIC outperforms previous lightweight algorithms. Furthermore, its performance with respect to the path length is very close to the length of an optimal path in the case of no global knowledge, e.g. when the presence of obstacles is not known to the algorithm in advance but only detected when reaching the obstacles. It may also be worth pointing out that this algorithm seems to be resistant to temporary link failure, as suggested by the fact that the randomized version of GRIC actually performs better than the deterministic one, c.f. section 4.3. Because it is the only lightweight algorithm to efficiently bypass large and complex obstacles without requiring a preliminary high overhead topology discovery or maintenance phase, it makes GRIC an innovative and important geographic routing algorithm, complementing other approaches such as GFG/GPSR/LCR combinations.

Sensor Network Model

This subsection describes the sensor net model we assume, following (Powell & Nikoletseas, 2007a). The assumptions we make are fairly weak and general. We consider sensor networks deployed on a two dimensional surface and we allow messages to piggy-back $O(1)$ bits of information which is being used for the decision making of the routing algorithm (in fact, the only required information is the position of the last node visited by the message, the position of the final destination of the message and the value of a mark-up flag, c.f. subsection 4). The most important part of the model is the communication model. We assume that each node is aware of its own position, has access to the list of its *outbound* neighbors and to their positions, and that it can reliably send messages to its neighbors, i.e. GRIC runs on a directed communication graph where nodes are localized and aware of their neighbors as well as their positions. This high-level combinatorial model may or may not be realistic and reasonable for sensor networks, depending on what it is being used for. In the case of the GRIC routing algorithm, it seems a quite reasonable and realistic model. Indeed, the GRIC algorithm is a *network layer* algorithm. The network layer relies on an underlying *data link* layer which in turn relies on a MAC and physical layer. Real world sensor networks implementation of this algorithm would therefore have to implement all levels of the protocol stack, and it is reasonable to assume that the link layer provides the level of abstraction we will be using. We discuss in more detail a realistic and simple full layer implementation of our algorithm (including physical, MAC and data link layers) in the next subsection.

Lower Layers of the Protocol Stack

For completeness we discuss the assumptions we make for every layer of the protocol stack upon which the network layer relies: the physical, the MAC and the link layers. Because the algorithm under consideration is situated at the network layer level, it is not strongly dependent on the physical layer, i.e. the type of nodes. Highly limited piconodes forming very large smart dust nets (Akyildiz, Su, Sankarasubramaniam, & Cayirci, 2002; Warneke, B., Last, Liebowitz, & Pister, 2001; Estrin, Govindan, Heidemann, & Kumar, 1999) composed of nodes smaller than a cubic centimeter, weighing less than 100 grammes and with costs well under a dollar such as envisioned in (Rabaey, Ammer, da Silva, Patel, & Roundy, 2000; Warneke, B., Last, Liebowitz, & Pister, 2001) but with probably strong limitations in terms of resource and features are perfectly acceptable. On the other hand, less futuristic sensor nodes currently available on the market from original equipment manufacturers are fine too. Those nodes are typically the size of a matchbox (including battery) and with a cost of more than a hundred dollars. We assume that each sensor node is a fully-autonomous computing and communicating device, equipped with a set of sensors (e.g. for temperature, motion, or metal detector). It has a small CPU, some limited memory and is strongly dependant on its limited available energy (battery). Each sensor can communicate wirelessly with other sensors of the network which are within communication range. We do not assume that sensor nodes can vary their transmission range.

We assume that a MAC layer protocol such as the S-MAC protocol from (Ye, Heidemann, & Estrin, 2002) or an IEEE 802.11-like MAC protocol is running on each sensor node. The MAC layer protocol is in charge of neighborhood discovery and it provides the link layer protocol with a list of neighbors with which the sensor node is capable of symmetric communication.

At the link layer, we assume a protocol that provides each node with a list of reliable *outbound* wireless communication links (i.e. at the link layer we do not need to assume the communication link to be symmetric, which is a feature of our algorithm). This means that the link layer protocol can remove some of the links provided at the MAC layer (e.g. those links which cannot be made reliable at reasonable costs, for example by including some quality metric on the links).

Georouting algorithms all make a further assumption: sensor nodes are localized for example using one of the currently available localization techniques (Beutel, 2005; Zhao & Guibas, 2004c; Karl & Willig, 2005; Hightower & Borriello, 2001). We assume that the link layer protocol adds this information to the list of links: the position of the sensor to which messages are sent if a given wireless link is chosen.

The GRIC routing protocol situates itself at the network layer. It runs on each individual node and the input to the algorithm is the position of the node on which it runs, as well as the list of links (including the position of the node to which those links lead to) provided by the link layer protocol.

GRIC is assumed to be capable of reliably sending messages to any of its outbound neighbor. The reliability issue (collision avoidance, acknowledgments, retransmissions, etc...) is not taken into account by GRIC but by the lower layer protocols (link layer and MAC layer). Reliable transmission can be achieved through many possible techniques, like time division multiple access schemes (TDMA) at the MAC layer, acknowledgement with retransmission or error correcting codes at the data-link layer or, alternatively multipath redundancy could be added on top of GRIC. Because many different options exist, our assumption of reliable communication links is realistic but out of the scope of the current report.

While the high level and abstract model assumed in this work is a good high level approximation of a real communication network, one thing which may not be very realistic is to assume stability of the communication graph. I.e., wireless links may fluctuate. Temporary congestion, collisions or other non constant environmental pattern may imply that the communication graph evolves over time. In other words, links go up or down. A link may be available at a given point in time, and become unavailable shortly after. A protocol which is not robust to such changes may be hard to use in practice or impose very heavy constraints on underlying (physical, MAC and data-link) protocols. As a consequence, it is necessary that an efficient geographic routing algorithm does not rely on the simplification assumptions of the model. One very interesting strength of the GRIC algorithm is precisely that it does not depend strongly on the stability of the underlying communication graph. Even more surprising, it actually performs better in the presence of limited link instability, as was experimentally observed in (Powell & Nikoletseas, 2007a).

THE GRIC ALGORITHM

Like the guaranteed delivery geographic routing algorithm GFG/GPSR/GOAFR (Bose, Morin, Stojmenovic, & Urrutia, 1999; Karp & Kung, 2000; Kuhn, Wattenhofer, & Zollinger, 2003; Kuhn, Wattenhofer, Zhang, & Zollinger, 2003), GRIC uses two different routing modes: a normal mode called *inertia mode* and a rescue mode called *contour mode*. The inertia mode is used when the message makes progress towards the destination, and the contour mode when it is going away from the destination, typically to go around an obstacle. The two major ingredients which make this approach successful are the use of the *inertia principle* and the use of the so called *right-hand rule*. The inertia principle is borrowed from physics and is used to control the trajectory of messages inside the network. Informally, messages are "attracted" to their destination (like a

celestial body is attracted in a planet system) but have some inertia (i.e. they also have an incentive to follow the "straight line"). The right-hand rule is a "wall follower" technique to get out of a maze (Hemmerling, 1989) and is used to route messages around complex obstacles. This informal idea is used to ensure that messages follow the contour of obstacles. The successful implementation of both of these abstract principles uses a virtual *compass device* (described in subsection 4.2) which treats a message's destination as the north pole. The *compass device*, the *inertia principle* and the *right-hand rule* enable the computation of an ideal direction in which the message should be sent (c.f. figure 1). Since there may not be a node in this exact direction, the message is actually sent to the node maximizing progress towards the ideal direction.

Routing with Inertia

In this subsection, we give a high level description of the *inertia routing mode*. Intuitively, the inertia mode performs *greedy routing* with an adjunct *inertia conservation* parameter β ranging in [0,1]. When a node routes a message in *inertia routing mode*, it starts by computing an *ideal direction vector* v_{ideal}. Once this ideal direction vector is computed, the message is sent to the node maximizing progression in the ideal direction. We next describe how to compute v_{ideal} and how to choose the node with best progression towards the ideal direction. To compute the ideal direction, the routing node needs to know its own position p, the position p' of the node from which the message was received and the position p'' of the message's destination, c.f. figure 1(a).

The position p' is assumed to be piggy-backed on the message, therefore, access to p, p' and p'' is consistent with the sensor network model described in subsection 3.3. First, the node p computes $v_{prev} = \overrightarrow{p'p}$, which is a vector pointing in the last direction travelled by the message, as

Figure 1. Computation of the ideal direction and the compass device

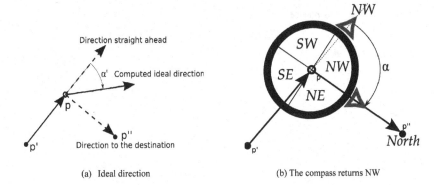

(a) Ideal direction   (b) The compass returns NW

well as $v_{dest} = \overrightarrow{pp''}$, which is a vector pointing in the direction of the message's destination. Through elementary trigonometry an angle α is computed, such that α is the unique angle in [$-\pi$, π] with $R_\alpha \cdot v_{prev}^T = v_{dest}^T$, where v^T denotes vector transposition and where R_α is the following rotation matrix:

$$R_\alpha = \begin{pmatrix} \cos(\alpha) & -\sin(a) \\ \sin(a) & \cos(\alpha) \end{pmatrix}$$

The node can now compute the *ideal direction* in which to send the message. For the *inertia mode*, the ideal direction is defined as $v_{ideal} = R_{\alpha'} \cdot v_{prev}$, where $R_{\alpha'}$ is a rotation matrix with α' defined by

$$\alpha' = -\beta\pi \ \text{if} \pm < -\beta\pi$$
$$\alpha' = \beta\pi \quad \text{if} \ \alpha > \beta\pi$$
$$\alpha' = \alpha \qquad \text{otherwise}$$

The ideal direction v_{ideal} is thus obtained by applying a rotation to the previous direction v_{prev} towards the destination's direction v_{dest}, however, the maximum rotation angle allowed is bounded, in absolute value, by $\beta\pi$. This implies that the ideal direction is somewhere between the previous direction and the direction of the message's destination.

Remark 1. *To enhance intuitive understanding of the* inertia routing mode, *it may be useful to notice that setting β=1 implies that $v_{ideal} = v_{dest}$*. In other words the ideal direction is always towards the message's destination and inertia routing *is equal to the simple* greedy geometric routing. *At the other extreme, setting β=0 implies that $v_{ideal} = v_{prev}$*, i.e. the message always tries to go straight ahead: there is maximal inertia. In (Powell & Nikoletseas, 2007a), it was found through simulations that setting $\beta = \dfrac{1}{6}$ was a good choice for practical purposes.

After the ideal direction has been computed the message is sent to the neighbor maximizing progress towards the ideal direction. Maximal progress is defined to be reached by the node m (amongst outbound neighbors of the node currently holding the message) such that $progress(m) := \langle v_{ideal} \mid pos(m) - p \rangle$ is maximized, where $\langle \cdot \mid \cdot \rangle$ is the standard scalar product and $pos(m)$ is the position of m.

Inertia routing is already quite an improvement over simple greedy routing, as can be seen on figure 2 (a) where a message is successfully routed from a point a=(0,10) to a point b=(20,10) in the presence of an obstacle (a stripe).

We do not give detailed explanations of the simulation context in which the plots of figure 2 where obtained and refer the reader to (Powell

Figure 2. Inertia routing with two different obstacles

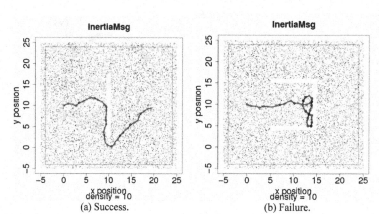

(a) Success.   (b) Failure.

& Nikoletseas, 2007a). Our intention in showing figure 2 is to show that although inertia routing successfully avoids a simple obstacle (the stripe of figure 2(a)), it fails for the more complicated U shape obstacle of figure 2(b). This routing failure is the motivation for adding a *contour mode* to which the algorithm can switch when it needs to route messages around complicated obstacles.

Routing Around Obstacles

We will now describe the *contour mode* of the algorithm, as well as the mechanism which permits to switch between *contour mode* and *inertia mode*. The *contour mode*, as well as the *switching mechanism* both makes use of a *virtual compass* indicating the direction travelled by the message during its last hop relatively to its final destination.

The Compass Device

When a node generates a message, we take the convention of saying that the message has travelled its last hop towards the north. Otherwise, we use the points p, p' and p'' to compute the angle α between the last direction travelled by the message v_{prev} and the direction to the message's destination v_{dest} as was done in subsec-

tion 4.1, c.f. figure 1. Seeing v_{dest} as the north, the virtual compass device returns a value in north-west, north-east, south-west or south-east depending on the quadrant in which α lies: SW if $\alpha \in [-\pi, -\pi/2]$, NW if $\alpha \in [-\pi/2, 0]$, NE if $\alpha \in [0, \pi/2]$ and SE otherwise. See figure 1(b) for an intuitive explanation of the compass device, while a rigorous description of the computation steps implementing the virtual compass is given in procedure 1 (compass).

Procedure 1: Compass

$v_{last} \leftarrow \overrightarrow{p'p}$

$v_{north} \leftarrow \overrightarrow{pp''}$

$\alpha \leftarrow$ angle from to(v_{prev}, v_{dest})

if $\alpha \in [-\pi, -\pi/2]$ then

    return SW

else if $\alpha \in [-\pi/2, 0]$ then

    return NW

else if $\alpha \in [0, \pi/2]$ then

    return NE

else {Comment: in this case $\alpha \in [\pi/2, \pi]$}

    return SE

The Right and Left Hand Rules to Route around Obstacles

The inertia routing mode manages to route messages around some simple obstacles, but not around more complicated obstacles. In order to improve it, it is necessary to detect when a message is trying to go around an obstacle and make the algorithm switch to the *contour mode* when appropriate. The main idea in the contour mode is to choose either the *right-hand rule* or the *left-hand rule*, and to go around the obstacle according to this rule. The right-hand rule means that the message will go around the obstacle by keeping it on its right. When a node receives a message, it makes a call to the *compass* procedure to get the direction of the message and uses a to flag which is piggy-backed on the message to decide if the right-hand rule (or the left-hand rule) should be applied. If the flag is down, the algorithm looks at the compass. If the compass points to the north, nothing special is done (it means the message is making progress towards its destination, and we stay in the inertia mode). But if the compass points south, it means that the message is actually going away from its destination and the algorithm interprets this as the fact that the message is being routed around an obstacle. To acknowledge this fact, GRIC raises the contour flag and tags it with E or W if the compass points to the southeast or the south-west respectively. Once the flag is up, the message is considered to be trying to go around an obstacle using the right-hand rule or the left-hand rule if the flag is tagged W or E respectively. See procedure 2 (raise flag) for a detailed explanation.

Procedure 2: Raise Flag (Raise and Mark the Flag)

```
{By assumption the flag is down}
if Compass == SW then
     Raise flag
     Tag flag with the W value
```

```
else if Compass == SE then
     Raise flag
     Tag flag with the E value
else
     Let the flag down
```

Once the message is considered to be trying to go around an obstacle using the right-hand rule (i.e. if the flag is up and tagged W), the algorithm will not change its mind until the message reaches a node where the compass returns the value NW (or NE if the left-hand rule was used). See procedure 3 (lower flag) for a rigorous description.

Procedure 3: Lower Flag

```
{By assumption the flag is up}
if (flag is tagged with W and compass ==
NW) then
     put the flag down
else if (flag is tagged with E and com-
pass == NE) then
     put the flag down
else
     Let the flag up and don't change the
tag
```

To summarize the discussion so far, when a node receives a message two different cases may occur: either the flag is down or it is up. If the flag is down (the message is not currently trying to go around an obstacle), GRIC runs procedure 2 (raise flag) to see if the message has just started going around an obstacle and flag and tag it appropriately. On the other hand, if the flag is up, it means the message was going around an obstacle, procedure 3 (lower flag) is used to see if the message should be considered as having finished going around the obstacle and the flag can be brought down. The position of the flag, together with the position of the compass is used to determine if the message is to be routed according to the *inertia* or the *contour* mode.

271

Mode Selection

The flagging mechanism described in the previous subsection enables a node to know if a message is being routed around an obstacle by looking at the position of the flag (which is piggy-backed on the message). When the flag is down, the *inertia mode* is used. However, when the flag is up, the *contour mode* (which we shall describe below) *may* have to be used. We next describe informally the mechanism used to decide whether the contour mode or normal mode should be used *when the flag is up*. Therefore, suppose the flag is up, and in order to simplify the discussion, suppose it is tagged with the E value (if the tag is W, a symmetric case is applied). Note that by case assumption, it is implied that the compass either returned NW, SW or SE, since otherwise procedure 3 (lower flag) would have put the flag down. The message is therefore currently trying to go around an obstacle using the left-hand rule described in the previous subsection, trying to keep the obstacle on its left. Two different cases have to be considered. **In the first case**, the compass points SE. Recalling the definition of the *ideal direction* v_{ideal} and of the previous direction v_{prev} from subsection 4.1, it is easy to see that since by case assumption the compass points to south-east, v_{ideal} is obtained by applying to v_{prev} a rotation to the left (i.e. counter-clockwise), which is equivalent to saying angle α defined in subsection 4.1 takes a value in *[0,π]* and therefore that angle α' takes a value in *[0,π/6])*. In other words, the *inertia routing mode* tries to make the message turn left, which is consistent with the idea of the left-hand rule: routing the message along the perimeter of the obstacle while keeping it to the left of the message's path. There is therefore no need to switch to the rescue *contour mode*. **The second case** which may occur is when the compass points either to the SW or to the NW. A reasoning similar to the previous one shows that in this case the inertia routing mode will make the message turn right

and angle α' will be in $\alpha \in [-\pi/6, 0]$. According to the left-hand rule idea, the obstacle should be kept on the left of the message. However turning right, according to the inertia routing mode, will infringe the idea of keeping the obstacle to the left. Instead, the message would turn right and get away from the perimeter of the obstacle. In this case, it is therefore required to call the rescue mode: the *contour mode*. Rigorous description of the above discussion is given in procedure 4 (mode selector).

The Contour Mode

By case assumption, the flag is up. One can also suppose without loss of generality that the tag on the flag is E, since the case of a W tag is similar. We have seen that procedure 4(mode selector) calls the contour mode when the inertia mode is going to make the message *turn right* (or *left* if the tag is W), which is equivalent to saying that $\alpha \in [-\pi, 0]$ and $\alpha' \in [-\pi/6, 0]$. However, the left-hand rule would advise turning to the left to stay as close as possible to the obstacle and to keep it on the left of the message path. Therefore, in order to stay consistent with the left-hand rule idea, we define $\alpha_2 = -sign(a)(2\pi - |\alpha|)$, thus α_2 is an angle such that $sign(a_2) = -a$ and such that $R_{\alpha_2} \cdot v_{prev}^T = R_\alpha \cdot v_{prev}^T = v_{prev}^T$. In order to give some inertia to the message, define α'_2 from α_2 in a similar way as α' was defined from α, and the ideal direction v_{ideal}, is defined by $v_{ideal}^T = R_{\alpha'_2} \cdot v_{prev}^T$, where $\alpha'_2 = \beta a_2$, where β is the inertia conservation parameter of subsection 4.1. Putting all things together, GRIC is formally described by (the non-randomized version of) Algorithm 1.

Procedure 4: Mode Selector

```
if the flag is up
  if (the flag is tagged with E and com-
  pass ∈ {NW,SW}) then return contour mode
```

```
else if (the flag is tagged with W and
compass ∈ {NE,SE}) then return contour
mode
else
    return inertia mode
```

Algorithm 1. GRIC, running on the node n_which is at position p_n

1. **if** the flag is down **then**
2. call **Raise flag** (procedure 2)
3. else
4. call **Lower flag** (procedure 3)
5. *mode*:= **Mode selector** (procedure 4)
6. **if** *mode* == *inertia mode* **then**
7. $\gamma=\alpha'$ (Where α' is the angle defined in subsection 4.1)
8. else
9. $\gamma = \alpha'_2$ (Where α'_2 is the angle defined in subsection 4.2)
10. $v_{ideal}^T = R_\gamma \cdot v_{prev}^T$
11. Let V be the set of neighbors of n
12. **If** running the non-random version of GRIC **then**
13. Send the message to the node $m \in V$ maximizing $\left\langle v_{ideal} \mid \overrightarrow{p_n p_m} \right\rangle$, where p_m is the position of m
14. **else** if running the random version of GRIC **then**
15. Let V' be an empty set.
16. **for all** $v \in V'$ **do**
17. add v to V' with probability 0.95
18. Send the message to the node $m \in V'$ maximizing $\left\langle v_{ideal} \mid \overrightarrow{p_n p_m} \right\rangle$, where p_m is the position of m

Randomization

In (Powell & Nikoletseas, 2007a), it is shown experimentally that GRIC succeeds in routing messages around local minimum and around obstacles with high probability (where the probability is taken over randomly generated networks). However, in unfavorable cases it may fail. It was observed experimentally that adding a small random disturbance to the GRIC algorithm improves its behavior. Intuitively, if the message gets blocked in a local minimum and starts looping, a small random perturbation may be sufficient to make the message escape the routing loop. Instead of considering V to be the set of neighbors of n, we consider V to be the *subset* of neighbors of n where each neighbor is added to V with probability 1-ε, where ε is a small constant. Each time a node has to take a routing decision, the neighborhood is randomly selected, thus the communication graph is dynamic and changes over time. For practical purposes experiments have shown the choice of $\varepsilon = 0.05$ to be good. Formally, this idea is implemented by introducing the if-else cases on lines 12 and 14 of algorithm 1 to differentiate between random and non-random routing modes. One important thing to notice is that the random version of GRIC, when used in a simulation environment with a static communication graph, can also be interpreted as simulating temporary link failure in the network. In other words, the static communication graph becomes dynamic because links go up and down randomly over time. One important finding from (Powell & Nikoletseas, 2007a), as previously stated, is that not only does GRIC behave well in the presence of limited link instability, it actually performs better. Even more important is the fact that this robustness to random perturbation to the communication graph means that GRIC will be easy to use in real world scenarios where the availability of wireless links over time may fluctuate for various reasons, either for physical reasons or for MAC layer protocol reasons, such as message collision, node switching to sleep mode, loss of synchronization between neighbors, etc...

PERFORMANCE EVALUATION

In this section, we summarize the main experimental results concerning the GRIC algorithm. For a full report of simulation results, we refer the reader to the original paper (Powell & Nikoletseas, 2007a). For evaluation purposes, the following numerical experiments were designed, which we describe from a high level point of view. To start with, sensor nodes are randomly uniformly deployed in a square region of 30×30 meters. A unit disc graph model is used to construct a communication graph (that is, two nodes are linked in the graph if and only if their distance is less than 1). A start point a is selected on the left of the network, at the position $x=0$ and $y=10$, and a destination point b at position $x=20$ and $y=10$ is selected on the right of the network (see for example figures 2 and 3). A message is then routed from point a to point b according to the GRIC algorithm. The routing is considered successful if the message reaches point b, while it is considered to be a failure if it gets "out of bounds"; i.e. if it gets within 1 meter of the border of the network, or if a maximum number of propagation hops has been reached without the message reaching destination; i.e. messages have a time to leave (TTL) counter, and they are dropped if the timer expires. In order to get meaningful results, the

simulation is repeated many times (1000 times), and the success probability is defined as being the ratio of positive outcomes on the total number of trials. This simulation is repeated for different network densities. For example, on figure 4(a), we can see that the success rate of the randomized version of the GRIC algorithm is around 0.7 for a density of 2 nodes per square meter, and very close to 1 for a density of 4 nodes per square meter. Finally, for purposes of performance comparison a similar experiment is repeated with a few other representative routing algorithms.

GRIC Without Obstacles

Figure 4 summarizes results when no obstacles are present. Looking at figure 4(a), we see success rates for different routing algorithms. One of them is the FACE routing algorithm from (Kranakis, Singh, & Urrutia, 1999), which is a guaranteed delivery protocol; i.e. the algorithm never fails, as long as the communication graph is connected (which is not always the case when the node density is low). The main conclusion of Figure 4(a) is that, without obstacles, the success rate of the randomized GRIC algorithm is almost as high as that of the FACE algorithm, i.e. GRIC has *almost guaranteed delivery*, which is a very nice result. One may also observe that the perfor-

Figure 3. GRIC routing with two different obstacles

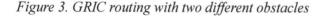

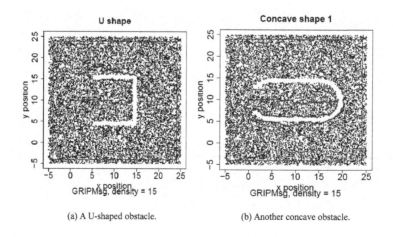

(a) A U-shaped obstacle.

(b) Another concave obstacle.

mance of the simple inertia routing algorithm is quite good too, whereas LTP and greedy routing are not competitive.

An important and usual metric is the hop count, i.e. the "speed" at which a message reaches its destination, since it captures many other network statistics (latency, energy or generated traffic). On figure 4(b), the average hop distance required to reach destination is plotted, for each density and for each routing protocol. As can be seen, GRIC reaches destination fast, except for the lowest densities (densities below 2), where the network is likely to be disconnected anyway (and it is thus impossible to route to destination), as can be seen from the fact that even the guaranteed delivery FACE algorithm fails often for such low densities.

GRIC with a U-Shaped Obstacle

Next, we present similar simulation results, except that this time a U-shaped obstacle is added in between the start and destination points, similarly as can be seen on figure 5. This time, only GRIC and FACE are capable of bypassing such an obstacle. We can see that the performance of the randomized version of GRIC becomes good if the network density is high enough (roughly, densities above 4 nodes per square meter). By

comparing the randomized (p=0.95) and non randomized (p=1) versions of GRIC, we see that the link instability does indeed augment the success rate, but that this comes at the cost of increasing the hop-count (the randomized version of GRIC has a random walk-like behavior), particularly for lower densities.

For an intuitive interpretation of network densities, we refer the interested reader to (Powell & Nikoletseas, 2007a), where it is argued that densities of 3 are low, densities of 4.5 are medium and densities of 6 or higher are high densities. Based upon that informal interpretation of network densities, one could summarize the simulation findings about GRIC in the presence of a large and difficult obstacle like the one of figure 5(a) by saying that GRIC behaves well only if the density is at least between low and medium, and that its performance diminishes for lower network densities.

The behavior of GRIC on some other obstacles, as well as more extensive simulation details and analysis are available in the original paper (Powell & Nikoletseas, 2007a).

CONCLUSION

We have presented in detail the GRIC algorithm originally described in (Powell & Nikoletseas,

Figure 4. No obstacles

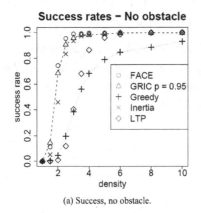

(a) Success, no obstacle.

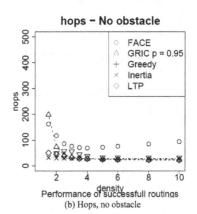

(b) Hops, no obstacle

Figure 5. U-shaped obstacle

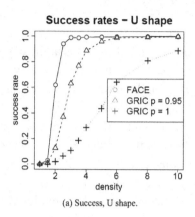

(a) Success, U shape.

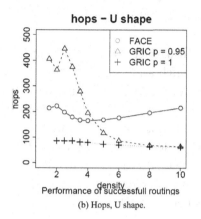

Performance of successfull routings

(b) Hops, U shape.

2007a). The GRIC algorithm has almost all of the advantages of the simple and lightweight greedy algorithm, while avoiding its main weakness: the inability to avoid even the simplest routing holes. In fact, GRIC is capable of bypassing large and difficult concave and communication blocking obstacles. As a consequence, it seems that *GRIC should always be preferred to the greedy routing algorithm.*

GRIC also offers a viable alternative to the FACE family of routing algorithms, such as GFG and its successors, but with a slightly different application niche: while GFG needs to amortize the cost of the inevitable planarization phase, it is perfectly suitable for stable networks, and also works for all network densities, including very low ones. On the other hand, *when networks are more dynamic GRIC may be a preferable alternative,* unless there are large obstacles combined with a very low network density.

We have also provided an extensive literature survey of significant competing geographic routing algorithms, with the purpose of justifying the need for lightweight and obstacle avoiding geographic routing algorithms.

REFERENCES

Ahmed, N., Kanhere, S. S., & Jha, S. (2005). The holes problem in wireless sensor networks: a survey. *SIGMOBILE Mob. Comput. Commun. Rev.*

Akyildiz, I. F., Su, W., Sankarasubramaniam, Y., & Cayirci, E. (2002). Wireless sensor networks: a survey. *Computer Networks.*

Antoniou, T., Chatzigiannakis, I., Mylonas, G., Nikoletseas, S., & Boukerche, A. (2004). A new energy efficient and fault-tolerant protocol for data propagation in smart dust networks using varying transmission range. In *Annual Simulation Symposium (ANSS).* ACM/IEEE.

Barriere, L., Fraigniaud, P., Narayanan, L., & Opatrny, J. (2003). Robust position-based routing in wireless ad hoc networks with irregular transmission ranges. *Wireless Communications and Mobile Computing.*

Beutel, J. (2005). Location management in wireless sensor networks. In *Handbook of Sensor Networks: Compact Wireless and Wired Systems.* CRC Press.

Bose, P., Morin, P., Stojmenovic, I., & Urrutia, J. (1999). Routing with guaranteed delivery in ad hoc wireless networks. In *Discrete Algorithms and Methods for Mobile Computing and Communications*.

Chatzigiannakis, I., Dimitriou, T., Nikoletseas, S., & Spirakis, P. (2004). A probabilistic forwarding protocol for efficient data propagation in sensor networks. In *European Wireless (EW) Conference on Mobility and Wireless Systems beyond 3G*.

Chatzigiannakis, I., Dimitriou, T., Nikoletseas, S., & Spirakis, P. (2006). A probabilistic forwarding protocol for efficient data propagation in sensor networks. *Journal of Ad hoc Networks*.

Chatzigiannakis, I., Kinalis, A., & Nikoletseas, S. (2006). Efficient and robust data dissemination using limited extra network knowledge. In *IEEE Conference on Distributed Computing in Sensor Systems (DCOSS)*.

Chatzigiannakis, I., Mylonas, G., & Nikoletseas, S. (2006). Modeling and evaluation of the effect of obstacles on the performance of wireless sensor networks. In *Annual Simulation Symposium (ANSS)*. ACM/IEEE.

Chatzigiannakis, I., Nikoletseas, S., & Spirakis, P. (2002). Smart dust protocols for local detection and propagation. In *Principles of Mobile Computing (POMC)*. ACM.

Chatzigiannakis, I., Nikoletseas, S., & Spirakis, P. (2005). Smart dust protocols for local detection and propagation. *Journal of Mobile Networks (MONET)*.

Estrin, D., Govindan, R., Heidemann, J., & Kumar, S. (1999). Next century challenges: Scalable coordination in sensor networks. In *Mobile Computing and Networking*. ACM.

Fang, Q., Gao, J., & Guibas, L. (2006). Locating and bypassing holes in sensor networks. *Mobile Networks and Applications*.

Finn, G. G. (1987). Routing and addressing problems in large metropolitan-scale internetworks (Tech. Rep.). Information Sciences Institute.

Frey, H., & Gorgen, D. (2006). Geographical cluster-based routing in sensing-covered networks. *IEEE Transactions on Parallel and Distributed Systems*, *17*(9), 899–911. doi:10.1109/TPDS.2006.124

Greenstein, B., Estrin, D., Govindan, R., Ratnasamy, S., & Shenker, S. (2003). DIFS: A distributed index for features in sensor networks. *Ad Hoc Networks*.

Hemmerling, A. (1989). *Labyrinth Problems: Labyrinth-Searching Abilities of Automata*. B.G. Teubner, Leipzig.

Hightower, J., & Borriello, G. (2001). Location systems for ubiquitous computing. *Computer*.

Karl, H., & Willig, A. (2005). *Protocols and Architectures for Wireless Sensor Networks*, chapter 10.1.2 Aspects of topology control algorithms. Wiley.

Karp, B., & Kung, H. T. (2000). GPSR: greedy perimeter stateless routing for wireless networks. In *Mobile Computing and Networking*.

Kim, Y.-J., Govindan, R., Karp, B., & Shenker, S. (2005a). Geographic routing made practical. In *Networked Systems Design & Implementation*.

Kim, Y.-J., Govindan, R., Karp, B., & Shenker, S. (2005b). On the pitfalls of geographic face routing. In *Foundations of mobile computing*.

Kim, Y.-J., Govindan, R., Karp, B., & Shenker, S. (2006). Lazy cross-link removal for geographic routing. In *Embedded Networked Sensor Systems*.

Kranakis, E., Singh, H., & Urrutia, J. (1999). Compass routing on geometric networks. In *Canadian Conference on Computational Geometry*.

Kuhn, F., Wattenhofer, R., Zhang, Y., & Zollinger, A. (2003). Geometric ad-hoc routing: of theory and practice. In *Principles of Distributed Computing*.

Kuhn, F., Wattenhofer, R., & Zollinger, A. (2003). Worst-case optimal and average-case efficient geometric ad-hoc routing. In *Mobile ad hoc Networking & Computing*.

Kuruvila, J., Nayak, A., & Stojmenovic, I. (2006). Progress and location based localized power aware routing for ad hoc and sensor wireless networks. *International Journal of Distributed Sensor Networks*.

Leone, P., Moraru, L., Powell, O., & Rolim, J. (2006). A localization algorithm for wireless ad-hoc sensor networks with traffic overhead minimization by emission inhibition. In *Algorithmic Aspects of Wireless Sensor Networks (ALGOSENSORS)*.

Leong, B., Liskov, B., & Morris, R. (2006). Geographic routing without planarization. In *NSDI'06: Proceedings of the 3rd conference on 3rd Symposium on Networked Systems Design & Implementation* (pp. 25–25). Berkeley, CA, USA, 2006. USENIX Association.

Moraru, L., Leone, P., Nikoletseas, S., & Rolim, J. D. P. (2007). Near optimal geographic routing with obstacle avoidance in wireless sensor networks by fast-converging trust-based algorithms. In *Q2SWinet '07: Proceedings of the 3rd ACM workshop on QoS and security for wireless and mobile networks* (pp. 31–38). New York: ACM.

Nikoletseas, S., & Powell, O. (2008). Obstacle avoidance algorithms in wireless sensor networks. *Encyclopedia of Algorithms*. Springer.

Powell, O., & Nikoletseas, S. (2007a). Simple and efficient geographic routing around obstacles for wireless sensor networks. In *WEA 6th Workshop on Experimental Algorithms*, Rome, Italy. Springer Verlag.

Powell, O., & Nikoletseas, S. (2007b). *Geographic routing around obstacles in wireless sensor networks* (Technical report). Computing Research Repository (CoRR).

Rabaey, J. M., Ammer, M. J., da Silva, J. L., Patel, D., & Roundy, S. (2000). Picoradio supports ad hoc ultra-low power wireless networking. *Computer*.

Rao, A., Ratnasamy, S., Papadimitriou, C., Shenker, S., & Stoica, I. (2003). Geographic routing without location information. In *Mobile Computing and Networking*.

Ratnasamy, S., Karp, B., Yin, L., Yu, F., Estrin, D., Govindan, R., & Shenker, S. (2002). GHT: a geographic hash table for data-centric storage. In *Wireless Sensor Networks and Applications*. ACM.

Seada, K., Helmy, A., & Govindan, R. (2001). On the effect of localization errors on geographic face routing in sensor networks. In *Information Processing in Sensor Networks*.

Stojmenovic, I. (2002). Position-based routing in ad hoc networks. *Communications Magazine*.

Stojmenovic, I., & Datta, S. (2004). Power and cost aware localized routing with guaranteed delivery in unit graph based ad hoc networks. *Wireless Communications and Mobile Computing*.

Stojmenovic, I., & Lin, X. (2001). Loop-free hybrid single-path/flooding routing algorithms with guaranteed delivery for wireless networks. *IEEE Transactions on Parallel and Distributed Systems*, *12*(10). doi:10.1109/71.963415

Takagi, H., & Kleinrock, L. (1984). Optimal transmission ranges for randomly distributed packet radio terminals. *Communications, IEEE Transactions on [legacy, pre - 1988]*, *32*(3), 246–257.

Warneke, B., Last, M., Liebowitz, B., & Pister, K. S. J. (2001). Smart dust: communicating with a cubic-millimeter computer. *Computer*.

Xing, G., Lu, C., Pless, R., & Huang, Q. (2004). On greedy geographic routing algorithms in sensing-covered networks. In *MobiHoc '04: Proceedings of the 5th ACM international symposium on Mobile ad hoc networking and computing* (pp. 31–42). New York: ACM Press.

Ye, W., Heidemann, J., & Estrin, D. (2002). An energy-efficient MAC protocol for wireless sensor networks. In *INFOCOM.*

Zhao, F., & Guibas, L. (2004a). Geographic Energy-aware Routing. In *Wireless Sensor Networks, an Information Processing Approach.* Elsevier.

Zhao, F., & Guibas, L. (2004b). *Wireless Sensor Networks, an Information Processing Approach.* Elsevier.

Zhao, F., & Guibas, L. (2004c). Localization and Localization Services. In *Wireless Sensor Networks, and Information Processing Approach.* Elsevier.

Zhou, G., He, T., Krishnamurthy, S., & Stankovic, J. A. (2004). Impact of radio irregularity on wireless sensor networks. In *Mobile systems, applications and services.*

Chapter 13
Sensor Field Resource Management for Sensor Network Data Mining

David J. Yates
Bentley University, USA

Jennifer Xu
Bentley University, USA

ABSTRACT

This research is motivated by data mining for wireless sensor network applications. The authors consider applications where data is acquired in real-time, and thus data mining is performed on live streams of data rather than on stored databases. One challenge in supporting such applications is that sensor node power is a precious resource that needs to be managed as such. To conserve energy in the sensor field, the authors propose and evaluate several approaches to acquiring, and then caching data in a sensor field data server. The authors show that for true real-time applications, for which response time dictates data quality, policies that emulate cache hits by computing and returning approximate values for sensor data yield a simultaneous quality improvement and cost saving. This "win-win" is because when data acquisition response time is sufficiently important, the decrease in resource consumption and increase in data quality achieved by using approximate values outweighs the negative impact on data accuracy due to the approximation. In contrast, when data accuracy drives quality, a linear trade-off between resource consumption and data accuracy emerges. The authors then identify caching and lookup policies for which the sensor field query rate is bounded when servicing an arbitrary workload of user queries. This upper bound is achieved by having multiple user queries share the cost of a sensor field query. Finally, the authors discuss the challenges facing sensor network data mining applications in terms of data collection, warehousing, and mining techniques.

DOI: 10.4018/978-1-60566-328-9.ch013

INTRODUCTION

Applications for mining sensor network data vary in scale from monitoring and controlling microscopic manufacturing equipment, to implementing an earthquake early warning system for a country like Japan. Sensor network data mining requires bridging the gap between low-level data that is acquired in sensor fields and high-level knowledge that is useful to real-world applications. Our research describes a new approach to sensor field resource management, which is necessary but not sufficient to bridge this gap.

There are many performance metrics of interest in sensor networks for data mining. We focus on two that are common to the vast majority of applications that mine sensor network data:

1. The *accuracy* of the data acquired by the mining application from the sensor networks; and
2. the total *system end-to-end delay* incurred in the sequence of operations needed for an application to obtain sensor data.

Although almost all sensor network applications have performance requirements that include accuracy and system delay, their relative importance may differ between applications. We therefore define the *quality* of the data provided to data mining applications to be a combination of accuracy and delay. Measuring the quality of sensor network data is important for data mining applications since the data quality determines the level of confidence in the knowledge extracted from the mining process. As in most systems, improved quality usually comes at some *cost*. For current wireless sensor networks, the most important component of cost typically is the energy consumed in providing the requested data. In turn this is dominated by the energy required to transport messages through the sensor field. This cost versus quality trade-off has recently been an active area of research (Boulis et al., 2003; Hu et al., 2006; Sharaf et al., 2004; Son et al., 2005; Tilak et al., 2002; Yu et al., 2004).

To perform our research, we construct a model for a system that acquires and mines sensor network data. We then develop novel policies for caching sensor network data values in sensor field gateway servers, and then retrieving these values via cache lookups. We also propose a new objective function for data quality that combines accuracy and delay. So that we can compare data quality across different data mining applications and systems, this objective function normalizes data quality for a given system to values in the range [0,1]. Finally, we use our system model to assess the impact of several factors on data quality and query cost performance:

- Our caching and lookup policies;
- the relative importance of data accuracy and system end-to-end delay; and
- the manner in which the sensed data values in the environment change.

This assessment evaluates seven different caching and lookup policies by implementing them in a simulator based on CSIM 19 (Schwetman, 1990, 2001).

The remainder of this chapter is organized as follows. In the next section we provide the background for sensor network data mining, outlining the three data management phases common to such applications and the resource requirements for each phase. This chapter's main focus is on the first phase (sensor network data acquisition). In the subsequent section we describe our approach to acquiring and caching sensor network data. This approach addresses the problem of balancing data quality and the cost of acquiring sensor network data. We evaluate the performance of our approach and discuss our results, including their implications for sensor field resource management. We also identify the challenges of sensor network data mining, including showing how processing sensor data streams is fundamentally different from processing static data in traditional data mining applications. Finally, we point out potential applications and techniques for mining sensor network data after data is acquired and cached using our approach. The last two sections suggest possible future research directions and conclude the chapter.

BACKGROUND

Sensor network data mining has attracted much attention from the research community in recent years. Data mining refers to non-trivial extraction of implicit, previously unknown, and potentially useful knowledge from data (Fayyad et al., 1996). Data mining techniques are used for easy, convenient, and practical exploration of very large collections of data for organizations and users, and have been applied in marketing, finance, manufacturing, biology, and many other domains (e.g., predicting consumer behavior, detecting credit card fraud, or clustering genes that have similar biological functions). Note that these traditional data mining applications usually take static data as input and then generate patterns and knowledge as output. However, data collected from sensor networks are live data streams that are dynamic. Such a difference poses new challenges and requirements for managing data in sensor network data mining applications.

Almost all sensing systems that acquire and mine data have three main data management phases and three corresponding subsystems:

1. *Data acquisition.* This subsystem may include one or more sensor fields consisting of sensor nodes that communicate with one or more base stations. Sensor network data is aggregated at sensor network data servers within this subsystem. While sensor network data is being gathered from sensor fields, the remaining stages in the data mining process

need to produce accurate and timely output based on dynamically changing data.

2. *On-line data processing.* This subsystem includes a sensor network data mining system, and may also include a sensor network data processing system. On-line data processing components obtain their input data from sensor network data servers (or gateways) that accept requests for sensor data and generate responses to these requests.

3. *Data warehousing and mining.* This subsystem is the sink for all important sensor network data (and perhaps data from other sources too). A data warehouse, managed by a data warehouse management system, is the main component within this subsystem.

Figure 1 shows the architecture of such a system deployment with one sensor field, one data server, a data mining system, a data processing system, and one data warehouse. Components that handle data in this figure are shown in rectangles labeled with their abbreviation. For example, the sensor network data server (**SNDS**) appears at the bottom of this figure, and is shown as part of the sensor field. Note that if the sensor network data server is augmented with storage, it can store and cache sensor field data values. On the right hand side of the figure, the data warehouse management system (**DWMS**) manages data that is archived after it has been processed by the other components.

Consider the flow of data within the sensor network data mining architecture. The sensor field is the data source. The flow of data between components is indicated by the "data flow" arrows that mostly point from left to right in this figure. Before data is sent along any of these arrows, it can be filtered, transformed, or selected according to specifications sent via the "command or control" arrows. There are three possible data flows from the **SNDS**. One or more of these data flows can be active at the same time. The upper arrow sends data directly to the sensor network

Figure 1. Architecture for sensor network (sensornet) data mining

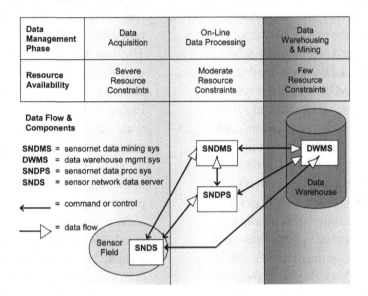

data mining system (**SNDMS**). The middle arrow delivers data to a sensor network data processing system (**SNDPS**). Finally, the lower arrow delivers sensor field data directly from the sensor field to the data warehouse via the **DWMS**. Which of the flows is delivering data and how the data is being processed by each component depends on the application.

There are two additional dimensions to this architecture which are described by the labels across the top of Figure 1. Both of these dimensions mirror the flow of data from left to right in the figure. The top row describes the phases of data management, which from left to right are:

1. Data acquisition;
2. On-line data processing that is performed on sensed or streamed data flows; and
3. Data warehousing and (traditional) data mining.

For a sensor network data mining system to work effectively, it is also important to consider the resource constraints within each of the data management phases. The resources available in

the devices used to perform each major function are summarized in the second row of Figure 1. For example, sensors that perform data acquisition in the sensor field usually only have kilobytes of RAM, and CPU clock speeds that are measured in MHz. In contrast, servers used to warehouse and mine data can have gigabytes of RAM, and one or more CPUs running at 1 GHz or faster. As one might expect, devices that perform on-line data processing typically have memory and CPU specifications that fall between these two extremes.

Our chapter focuses on the first data management phase, data acquisition, and its resource constraints. In the following subsection, we briefly introduce the process of data acquisition and describe how one might cache sensor data to conserve resources in wireless sensor networks.

Data Acquisition and Caching in Wireless Sensor Networks

The caching approaches we propose are designed to be general since they make no assumptions about whether the sensor network architecture uses a structured or unstructured data model. In other words, our approaches are independent of the database model for the sensor network. The database could implement a structured schema that extends a standard like the Structured Query Language (SQL). The TinyDB and Cougar systems both advocate this approach (Demers et al., 2003; Madden et al., 2005). However, the schema could also be modified while the system is running, e.g., as in IrisNet (Gibbons et al., 2003). The database model might also expose a low-level interface to the sensing application. Directed Diffusion (Intanagonwiwat et al., 2003) does this by allowing applications to process attributes or attribute-value pairs directly. Kairos (Gummadi et al., 2005), Tenet (Gnawali et al., 2006), Regiment (Newton et al., 2007), and DSN (Chu et al., 2007) do this by introducing task-oriented or declarative programming languages for applications to acquire and process sensor network data.

Consider the impact of adding a cache to the sensor network data server in Figure 1. Figure 2 shows such a system in which a cache is added to the internal architecture of the **SNDS**, on the "border" between the sensor field(s) and the backbone network that delivers sensor data for on-line data processing. There are two possible data paths that can be traversed in response to a query from the backbone network:

- For a *cache miss*, a query is sent to the sensor field by the **SNDS**, incurring a cost. The data path for a cache miss is shown by the light grey arrows in Figure 2. To update the cache, each sensor data value v_i is copied into a *cache entry*. A cache entry associates with location l_i, the most recent value observed at this location, and its timestamp into the tuple $\langle\ l_i,\ v_i,\ t_i\ \rangle$. We say that the *system delay*, S_d, is the time between an application query arriving at the point labeled $Query_m$ in Figure 2 and the corresponding reply departing from $Reply_m$. The *value deviation*, D_v, is the unsigned difference

between the data value in $Reply_m$ and the true value at l_i when $Reply_m$ enters the reply processing engine.

- The data path for a *cache hit* is much shorter than for a cache miss. This shorter data path is indicated by the dark grey arrow in Figure 2. If the cache is indexed by location, and a cache entry is present for a location l_i specified in a query, a reply can be generated using only the information in the tuple that corresponds to l_i Since the processing required to perform this cache lookup and generate a reply is relatively small, we assume that the system delay for a cache hit (S_d) and its associated cost are both zero. We also determine the value deviation for cache hits (D_v) in the same way as for cache misses.

We exploit spatial locality within sensor field data in the cache. Specifically, some caching and lookup policies allow cache "hits" in which the value at location l_i is approximated based on values $v_{i'}$ from neighboring location(s) $l_{i'} \in N(l_i)$. (Here $N(l_i)$ denotes the neighborhood of location l_i.) In the next section we develop and describe three such policies that implement what we call *approximate* lookups and queries. We compare these approximate policies with four *precise* lookup and query policies that only use information associated with location l_i to process queries that reference location l_i.

SENSOR FIELD RESOURCE MANAGEMENT

Caching and Lookup Policies

Our caching and lookup policies are designed to explore alternative techniques for increasing the effective cache hit ratio, thus conserving sensor field resources.

*Figure 2. Sensor Network Data Server (**SNDS**) or gateway with a cache*

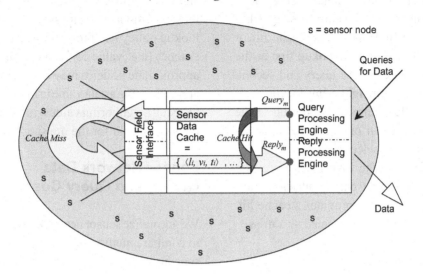

All of the caching and lookup policies we propose and evaluate incorporate an age threshold parameter T that specifies how long each entry is stored in the cache. We now describe all seven of our caching and lookup policies. *All hits, all misses, simple lookups* and *piggybacked queries* implement precise lookups and queries. On the other hand, *greedy age lookups, greedy distance lookups*, and *median-of-3 lookups* implement approximate lookups and queries.

- **All hits** (age threshold parameter $T = \infty$): In this policy cache entries are loaded into the cache but are never deleted, updated, or replaced.
- **All misses** (age parameter $T=0$): In this policy entries are not stored in the cache.
- **Simple lookups** (T): This caching policy results in a cache hit or cache miss based on a lookup at the location specified in each user query. If consecutive misses occur in the cache for the same location, this policy sends redundant queries into the sensor field. When a reply is received its value is loaded into the cache, stored for T seconds, and then deleted.

- **Piggybacked queries** (T): A cache hit or miss is determined only by a lookup at the location specified in the user query. If a query has already been issued to fill the cache at a particular location, subsequent queries block in a queue behind the original query and leverage the pending reply to fulfill multiple queries.
- **Greedy age lookups** (T): A cache hit or miss is determined by a lookup first at the location specified in the query and second by lookups at all neighboring locations. If there is more than one neighboring cache entry, the freshest (newest) cache entry is selected. As for piggybacked queries, if a query has already been issued to fill the cache at any of these locations, subsequent queries block in a queue behind the original query and leverage the pending reply to fulfill multiple queries. This is also true for the last two policies: *greedy distance lookups* and *median-of-3 lookups*.
- **Greedy distance lookups** (T): A cache hit or miss is determined by a lookup first at the location specified in the query and second by lookups at neighboring locations. If

there is more than one neighboring cache entry, the nearest cache entry is selected.

- **Median-of-3 lookups** (*T*). A cache hit or miss is determined by a lookup first at the location specified in the query and second by lookups at all neighboring locations. If there are at least three neighboring cache entries, the median of three randomly selected entries is selected as the value returned with a cache hit. If there are one or two neighboring cache entries, a randomly selected entry provides a cache hit. Otherwise, the query is treated as a miss.

By implementing blocking behind pending sensor field queries, four of these seven policies have an upper bound on the sensor field query rate, R_f. Specifically,

$$\max(R_f) = \frac{|\mathbf{N}|}{T}. \tag{1}$$

The four policies are piggybacked queries, median-of-3 lookups, and the two approximate greedy policies. In Equation (1), $|\mathbf{N}|$ is the number of distinct locations that can be specified in queries for sensor data.

The definition of a *cache hit* (vs. a *cache miss*) for approximate lookup and query policies has both similarities and differences with the traditional definition of a cache hit (or miss). One similarity is that the data value in the cache may be consistent with the value in the sensor field. This would always be true for a memory cache for a uniprocessor computer (Burroughs, 1964), or for a sequentially consistent multiprocessor (Fuller & Harbison, 1978). A related similarity is that the data value in the cache may be different from the value in the sensor field. This would also be the case, some of the time, in a distributed file system that caches blocks or files (Howard et al., 1988), or a web-based application that exploits web caches (Luotonen & Altis, 1994). These similarities are also shared with the precise lookup

and query policies. An important difference from these traditional caching systems is that the cache lookup policy itself might introduce some inconsistency (e.g., value deviation). This is because our approximate policies may emulate a cache hit by approximating a data value in the sensor field (e.g., by using neighboring values) rather than querying the sensor field for the missing value.

Sensor Network Data Quality and Query Cost

We normalize sensor network data quality in order to compare quality measurements from different sensing and mining systems, as well as for different system parameters (e.g., number of sensors, distance between sensors, etc.). We define data quality to be a linear combination of normalized *system delay* and normalized *value deviation* using a parameter *A*, which is the relative importance of delay when compared with value deviation. The expression that defines quality, denoted Q_n, is:

$$Q_n = A \frac{1}{(1+e^{-b})} + (1-A) \frac{1}{(1+e^{-c})} \tag{2}$$

where -*b* and -*c* are the exponents used to perform *softmax normalization* on delays and value deviations, and $0 \leq A \leq 1$. The exponents in Equation (2) are the *z* scores of their respective values and are therefore defined as follows:

$$-b = \frac{S_d - \mathrm{mean}(S_d)}{\mathrm{stddev}(S_d)}, \text{ and} \tag{3}$$

$$-c = \frac{D_v - \mathrm{mean}(D_v)}{\mathrm{stddev}(D_v)}. \tag{4}$$

Since small values of system delay (S_d) and value deviation (D_v) are both desirable, smaller values of Q_n, e.g., $0 < Q_n \ll 0.5$ imply better data quality, and larger values of Q_n correspond to worse quality.

We chose to use softmax normalization for the S_d and D_v values for several reasons:

- Softmax normalization works well on populations with a large dynamic range (e.g., some of our D_v values); and
- it is easy to use in practice since it requires that we know only the mean and standard deviation of our S_d and D_v populations.

Softmax normalization also yields transformed values that lie in the range [0,1]. Because of this property, and because of our definition of A, $0 \leq Q_n \leq 1$ and $Q_n = 0.5$ when the system delay and value deviation are simultaneously at their respective means. This type of normalization has been used by others in neural networks and data mining for pattern recognition and data classification (Bishop, 1995; Bridle, 1990; Han & Kamber, 2000). It has two other interesting properties that make it convenient for normalizing our system delays and value deviations:

- For finite values, softmax normalization reaches "softly" toward its maximum and minimum values of 0 and 1, never quite getting there (Rodriguez, 2004); and
- its transformed values are more or less linear in the middle range, and have nonlinearity at both ends that make it well suited for data values with distributions that have long tails (Rodriguez, 2004).

We considered four normalization methods for our system delays and value deviations:

- *min-max* normalization;
- *z-score* normalization;
- *sigmoidal* normalization; and
- *softmax* normalization.

We also thought it was important to have an expression for quality that is bounded for both our sensor network data and for data values from other sensor networks. Sigmoidal normalization and softmax normalization both have this property whereas min-max normalization and z-score normalization do not. This requirement makes median-based or percentile-based normalization methods undesirable as well. We ultimately chose softmax normalization over sigmoidal normalization because we found it convenient to conceptualize and graph non-negative values of quality. To provide some intuition for how softmax normalization varies between 0 and 1, Table 1 shows softmax-normalized values for ±6 standard deviations from the mean.

We use two different sensor field models in our research in order to generalize our results. The first model uses correlated random variables to simulate how the environment changes for 1000 sensor locations. This model gives us the flexibility to vary how the environment changes. The second model uses real-world trace data to drive how the environment changes. This trace data was taken from 54 light, temperature, and humidity sensors deployed in the Intel Berkeley Research lab over a five-week period (Deshpande et al., 2004).

Simulated Changes to the Environment

For simulated changes to the environment, the sensor field is a 3-dimensional field with rectangular planes on six faces. There is an 8-unit spacing between 10 sensors in the X-dimension, a 6-unit spacing for 10 sensors in the Y-dimension, and a 4-unit spacing for 10 sensors in the Z-dimension. Four base stations are placed on the X-Y plane. These four base stations are then connected to the sensor network data server that has the common cache. Sensors always communicate with their closest base station at a cost that incorporates free-space energy loss for each transmission (Raghunathan et al., 2002). Thus, the properties of each one-way communication to and from location l are as follows:

Table 1. Z scores and softmax-normalized values

Z score of data value	Softmax-normalized value
-6	0.0025
-5	0.0067
-4	0.0180
-3	0.0474
-2	0.1192
-1	0.2689
0	0.5
1	0.7311
2	0.8808
3	0.9526
4	0.9820
5	0.9933
6	0.9975

$$Cost_l = p \ r_{b'}^2 \mid \min(Cost_l) = 1 \text{ unit} \qquad (5)$$

where $r_{b'}$ is the distance between location l and its nearest base station b', and p is the normalization constant for the set of costs. In addition,

$$Delay_l = q \ r_{b'} \mid \min(Delay_l) = 1 \text{ second} \qquad (6)$$

where q is the normalization constant for the set of delays. We assume that all four base stations communicate with the **SNDS** containing the cache at zero cost, with zero delay, and using infinite bandwidth. Thus, the minimum cost to query a location in the sensor field is normalized to 2 units (1 for the query + 1 for the reply), and the maximum delay to query a location in the sensor field is 2 seconds (not including queuing delay). Finally, each base station is connected to the sensor field with an access link with a capacity of 25 queries per second.

Trace-Driven Changes to the Environment

For trace-driven changes to the environment, our second sensor field model has more than an order of magnitude fewer locations (54 instead of 1000). The sensors are arranged in a 2-dimensional field at the numbered locations in Figure 3, which is taken from (Deshpande et al., 2004). Each entry in the trace is from a Mica2Dot sensor, which senses humidity, temperature, light, and battery voltage. The trace contains over 2.2 million entries taken over more than five weeks in early 2004. This means that one location reads and records new sensor field values an average of about once every 1.33 seconds. We wanted to use the most dynamically changing of the sensor field values in our model to maximize the error in query accuracy. We therefore chose the value with the largest average difference between samples. This was light intensity, which is reported in Lux. A value of 1 Lux corresponds to moonlight, 400 Lux to a bright office, and 100,000 Lux to full sunlight.

Figure 3. Sensor field at the Intel Berkeley Research lab

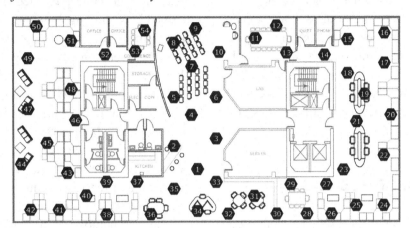

Four base stations are placed at the corners of the floor plan shown in Figure 3. As before, sensors always communicate with their closest base station. We further assume that the cost and delay of each one-way communication are given by Equations (5) and (6), respectively.

Query Workload Model

We use a query workload model that is well suited for real-world sensor network data mining applications that include monitoring and control functions. Many of these applications have a workload that includes a periodic arrival process of queries as well as a random arrival process. There are examples of query workloads that capture both of these components in the literature, e.g., (Intanagonwiwat et al., 2003; Jamieson et al., 2003; Wan et al., 2003). On the other hand, other researchers assume that queries either have exclusively periodic interarrival times (Lu et al., 2002; Madden et al., 2005) or random (usually exponential) interarrival times (Demers et al., 2003; Zhao & Govindan, 2003). We assume that the query workload for our applications consists of the superposition of two query processes: a polling component that slowly scans the sensor field at a fixed rate, and a random component that consists of queries to different locations in the sensor field. Within this random component it is equally likely that each location in the sensor field will be sampled. This workload model is similar to models used by others in (Intanagonwiwat et al., 2003; Jamieson et al., 2003; Wan et al., 2003). Specifically, our query workload is characterized by two parameters:

- τ = the period of the polling component of the query workload ($\tau > 0$); and
- λ = the average query arrival rate of a process that represents the random component of our workload.

For simulated changes to the environment, λ and τ are fixed: $\lambda = 81$ queries per second is used as the rate parameter to generate queries with exponentially distributed interarrival times with mean $1 / \lambda$. The parameter τ is set to $111.\overline{11}$ seconds so that the arrival rate for polling queries is 9 queries per second. When $\lambda = 81$ and $\tau = 111.\overline{11}$, the aggregate arrival rate for queries is $81 + 9 = 90$ queries per second. Since the total capacity of the sensor field access links is $4 \times 25 = 100$ queries per second, their average link utilization is 0.90 for "all miss" runs, and less for runs that include some cache hits.

For trace-driven changes to the environment, λ and τ are fixed for the results described in the next section: $\lambda = 0.81$ queries per second and τ = 600 seconds. This makes the average arrival rate for queries two orders of magnitude less than query rate for simulated changes, namely 0.9 queries per second. The total capacity of the sensor field access links is $4 \times 0.25 = 1$ query per second Thus, the average link utilization is also 0.90 for "all miss" runs.

DISCUSSION OF RESULTS

We wanted our simulated results to capture the fact that sensor field readings are correlated in both space and time. In our sensor field model, at time $t + 1$, the value at each location l is drawn from a normal distribution with mean

$$\mu_{l,t+1} = \frac{1}{3}\mu + \frac{1}{3}\mu_{l,t} + \frac{1}{3}\mu_{N(l),t}.$$

The long-term mean of this distribution is $\mu = 0$. The standard deviation $\sigma = 0.407514$, and the tails are truncated at minimum / maximum values of $\mu - 6\sigma / \mu + 6\sigma$. This standard deviation is the same as the standard deviation of the system end-to-end delays during a set of 20 runs without a cache for our 1000-node sensor network model. $N(l)$ denotes the neighbors of location l, and each neighboring location l' of l contributes to $\mu_{N(l),t}$ in proportion to

$$\mu_{l',t} \Big/ r_{l'}$$

where $r_{l'}$ is the distance between locations l and l'. The right hand side of this equation reflects the fact that sensor field data values are most often dependent on a long term mean, the previous value at the same location, and previous values at neighboring locations. This model for a changing environment is based on the model for correlated sensor network data developed by Jindal and Psounis (2006).

Each data value presented in our results is derived by averaging 20 simulation runs initialized with different seeds. Additional details of our experimental methodology are described in (Yates, 2006).

The odd-numbered figures, Figures 5, 7, 9 and 11 show results for light intensity in Lux measured over time in the Intel Berkeley lab data set. These results are for $T = 90$ seconds, and 0.9 queries per second. Note that because of Equation (1), the maximum sensor field query rate, $\max(R_f)$, is reduced to $54/90 = 0.6$ queries per second.

The even-numbered figures, Figures 4, 6, 8 and 10, are for correlated changes to the environment with both the age parameter, T, and the average query rate scaled for the more rapidly changing environment. Specifically, $\overline{T} = 8.88$ seconds and the average user query rate is 90 queries per second. Because of Equation (1), the maximum sensor field query rate $\max(R_f) = \frac{1000}{T} = 112.5$ queries per second.

We can draw two main conclusions from our experiments using the correlated and trace-driven models for how the environment changes. Results from these experiments appear in Figures 4 through 11.

1. There is a cost vs. quality trade-off for some data quality requirements but not others. For example, consider the results shown in Figures 4 and 5. Figure 5 shows cost versus quality for all seven caching and lookup policies, where $A = 0.1$ and the values at each location are changed according to the lab trace (Deshpande et al., 2004). At the smallest cost, we have a 100% cache hit ratio (labeled "All hits") that provides a quality of below 0.6 for zero cost. For the largest cost, we see that a 0% cache hit ratio (labeled "All misses") provides the third-best quality at a cost of approximately 19 units. Recall that for quality, smaller values indicate better quality. The remaining five caching and lookup policies provide a linear trade-off between cost and quality. Figure 4 also shows a trade-off between cost and quality for the same

Figure 4. Cost vs. quality for A = 0.1 and correlated changes over 1000 locations

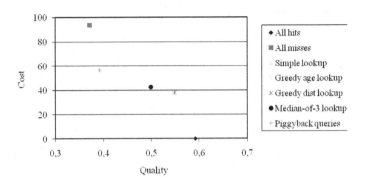

Figure 5. Cost vs. quality for A = 0.1 and trace-driven changes over 54 locations

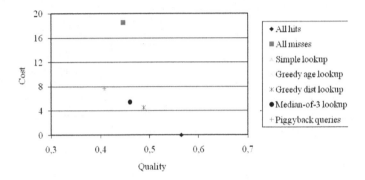

value of A and the same seven caching and lookup policies, but with changes to the environment now modeled by a series of values correlated in space and time. There are two observations worth noting when comparing these first two figures. First, the cost values in Figure 5 are less than in Figure 4 because the distances within the sensor field are smaller. Second, the trends are similar between these two figures, with the exception of the increase in quality of the "all misses" policy between Figure 4 and Figure 5. This worse "all misses" quality is due entirely to an increase in the normalized delay term in the right hand side of Equation (2). This can be verified by comparing the relative differences in delays between the policies, shown on the horizontal axes in Figure 6 and Figure 7.

It is reasonable to compare sensor network data quality between Figures 4 and 5 even though they show results from different data mining systems. This is because both systems use the same value of A ($A = 0.1$) and their underlying system delays and value deviations are normalized using Equation (2). Although these figures present important cost and quality results, it is also useful to examine additional performance metrics that are directly measured (e.g., the "raw" system delay and value deviation data). Figures 6 and 7 therefore show cost vs. delay for the different data mining systems. Similarly, Figures 8 and 9 show cost vs. value deviation for both systems.

We now examine system configurations for which delay is the more important component of quality in more detail. Figures 10 and 11 show

Figure 6. Cost vs. delay for A = 0.1 and correlated changes over 1000 locations

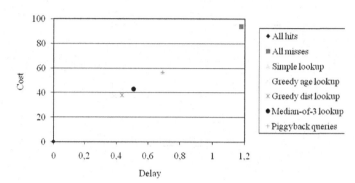

Figure 7. Cost vs. delay for A = 0.1 and trace-driven changes over 54 locations

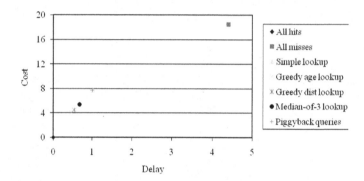

such configurations for a value of $A = 0.9$. The most remarkable result in these figures is that there is no trade-off between cost and quality when we significantly prioritize delay over value deviation. The two greedy caching and lookup policies have the best cost performance and the best quality performance for both models of changing the environment in Figures 10 and 11. Even though the "all hits" policy has the best absolute performance in these figures, we do not consider this a practical policy since it never updates the cache.

In studying Figures 4 through 11 it is interesting to understand which system variables depend on which system parameters. For example, cost, delay, and hit ratio values in these simulation results each depend on the following three variables:

- The caching and lookup policies themselves (including the value of T);
- the physical configuration of the sensor field; and
- the query arrival process.

Thus, a cost vs. delay or a cost vs. hit ratio graph is the same for different experiments in which these three variables are held constant. To see how cost and delay both increase with lower cache hit ratios, Table 2 shows the cache hit ratio for each of the (cost, delay) points in Figure 6. Similarly, Table 3 shows the hit ratio for each of the (cost, delay) points in Figure 7.

Value deviation depends on the same parameters listed above, and additionally on the manner in which the environment changes. Thus, cost

Figure 8. Cost vs. value deviation for A = 0.1 and correlated changes over 1000 locations

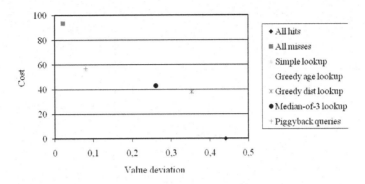

Figure 9. Cost vs. value deviation for A = 0.1 and trace-driven changes over 54 locations

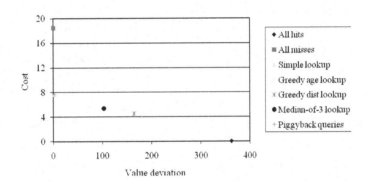

Table 2. Hit ratios, costs, and delays for $\overline{T} = 8.88$, 90 Queries per second, and correlated changes over 1000 locations

Policy	Hit ratio	Cost	Delay
All hits	1	0	0
All misses	0	94	1.18
Simple lookup	0.40	56	0.69
Greedy age lookup	0.62	37	0.39
Greedy distance lookup	0.60	38	0.44
Median-of-3 lookup	0.55	43	0.51
Piggyback queries	0.40	57	0.69

vs. value deviation graphs are the same when the policies, sensor field, query arrival process, and method for changing the environment are all identical. Figure 8 and Figure 9 show cost vs. value deviation results for correlated changes and trace-driven changes to the environment, respectively. The most interesting difference between the two figures is the overall increase in the dispersion

Figure 10. Cost vs. quality for A = 0.9 and correlated changes over 1000 locations

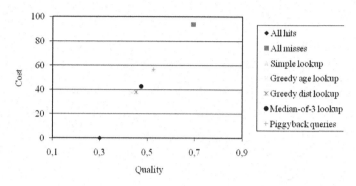

Figure 11. Cost vs. quality for A = 0.9 and TRACE-driven changes over 54 locations

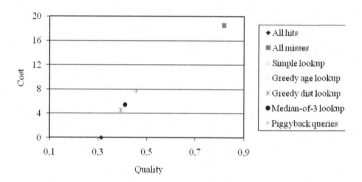

Table 3. Hit ratios, costs, and delays for T = 90, 0.9 queries per second, and trace-driven changes over 54 locations

Policy	Hit ratio	Cost	Delay
All hits	1	0	0
All misses	0	19	4.4
Simple lookup	0.59	7.7	1.0
Greedy age lookup	0.78	4.0	0.47
Greedy distance lookup	0.76	4.4	0.55
Median-of-3 lookup	0.71	5.4	0.68
Piggyback queries	0.59	7.7	1.0

of the value deviations in Figure 9 when compared with those in Figure 8. This is because the variation in sensor field values is much greater in the Intel Berkeley trace data than values that are drawn from our normal distribution with a time-dependent mean.

2. *Different lookup policies perform best depending on whether delay or value deviation is*

most important to the application. If data quality is more important to the application than cost, and value deviation is more important than delay, simple lookups and piggybacked queries provide the best performance. This can be seen in Figures 4 and 5. In both of these figures, simple lookups and piggybacked queries yield the best quality, other than the "all misses" policy for correlated changes. When value deviation is most important, the expense of taking a cache miss (by not computing an approximate value from neighboring values for these two policies) is worthwhile, since value deviation is deemed most important. If query cost is at a premium compared with quality, using greedy age lookups or greedy distance lookups is preferred. These two policies have the most favorable cost performance in both sensor field models, other than the "all hits" case.

If delay is more important to quality than value deviation, Figures 10 and 11 show that performing greedy age lookups or doing greedy distance lookups yields the best performance. This is true regardless of whether cost or quality is more important to the application. We again assume that the "all hits" case is not useful to realistic data mining applications. For these policies, getting the fast response time of a cache "hit" (which might be approximated from values at one or more neighboring locations) is worthwhile, since low delay is more important than a more accurate value.

The fact that different lookup policies perform best for different application requirements can be explained by examining the underlying delays and value deviations of the policies themselves. For example, consider the case where $A = 0.1$ and changes to the environment are driven by the lab traces. A value of $A = 0.1$ biases quality toward value deviation performance rather than delay performance. In this case, value deviation performance is significantly better when using precise lookups and queries, as shown in Figure 9. Figure 5 therefore shows that the data quality supported by the simple lookup and piggyback query policies is superior to the data quality supported by the greedy and median-of-3 lookup policies.

Now consider the case where $A = 0.9$ and changes to the environment are again driven by the lab trace data. A value of $A = 0.9$ biases quality toward delay performance rather than value deviation performance. In this case, both delay performance and cost performance are best for approximate lookups and queries, as shown in Table 3. Figure 11 thus shows that the query cost incurred for doing greedy age lookups or greedy distance lookups is superior to (i.e., less than) the query cost incurred by the other policies for quality that is also better.

It is helpful to summarize the cost and quality performance results presented above as follows:

- When value deviation is more important to quality than delay, there is a linear cost vs. quality trade-off. We obtain the best cost performance by implementing policies that approximate sensor values by using cached values from nearby locations. The best quality performance is achieved by policies that always query and cache the sensor field location specified in the user query.
- When delay is more important than value deviation, policies that approximate values using cached values from nearby locations provide the best cost performance as well as the best quality performance.
- These results hold for both simulated changes to the environment and trace-driven changes to the environment.

SENSOR NETWORK DATA MINING

Data mining is usually the most important step in a traditional knowledge discovery process, which consists of collecting and selecting data, preprocessing data, transforming data, mining data, and finally interpreting and evaluating patterns. Traditional data mining techniques include classification and prediction, cluster analysis, association rule mining, and outlier analysis, etc.

- *Classification* is the process of mapping data items into one of several predefined categories based on attribute values of the items. Examples of classification applications include fraud detection, computer and network intrusion detection, bank failure prediction, and image categorization. Classification is a type of supervised learning that consists of a training stage and a testing stage.
- *Clustering* is a type of unsupervised learning. It groups similar data items into clusters without prior knowledge of their class membership. The basic principle of clustering is to maximize intra-cluster similarity while minimizing inter-cluster similarity (Jain et al., 1999). Clustering has been used in a variety of applications including image segmentation, gene clustering, and document categorization.
- *Association rule mining* is the process of discovering frequently occurring item sets in a database. Association rule mining is often used in retail applications where the objective is to find which products are bought with what other products. An association is expressed as a rule, $X \Rightarrow Y$, indicating that item set X and item set Y occur together in the same transaction.

Although it has tremendous potential benefits to real-world applications, sensor network data mining brings new challenges to sensor network researchers and data mining researchers alike. Traditional data mining techniques are used to explore very large collections of "static" data that are stored on disk. Sensor network data mining, in contrast, requires exploration of "dynamic" data that change over time. This major difference lies in several areas, each of which poses unique challenges:

- Traditional data mining algorithms take a fixed set of data as input and extract previously unknown patterns as output. During the mining process the values of data items do not change. However, data gathered from sensor networks can change quickly as time passes, and thus must be managed as a set of *data streams* rather than a collection of data stored on disk. The data flows in Figure 1 are therefore best modeled as streams rather than one-time data transmissions. Research on mining data streams is still emerging as a new research area.
- Few data mining applications have *real-time processing requirements*. In contrast, a major goal in mining sensor network data is to identify and discover patterns in real time in order to monitor environments, detect changes as they are happening, and respond to events in a timely manner. Efficiency and scalability issues therefore become an important challenge. Thus, the distinction between the second data management phase (on-line processing) and the third data management phase (warehousing and mining) in Figure 1 is often blurred.
- Finally, in many traditional data mining applications only the values of the data items themselves are considered, and the spatial and temporal characteristics associated with data values are often ignored (except for specialized spatio-temporal mining applications). For sensor network data, however, the *spatial and temporal characteristics* associated with data values are almost always considered in order to correctly interpret the data.

Mining sensor network data also has different resource constraints than traditional data mining. Because sensor data are best modeled as data streams, these constraints are also common to all data stream processing applications. Data stream mining is aimed at "extracting knowledge structures represented in models and patterns in non-stopping streams of information" (Gaber et al., 2005, p. 18). Research problems in data stream mining have attracted attention in recent years because of important applications in different domains such as financial market monitoring, sensor network monitoring, security, web applications, etc. In this respect, sensor network data mining can be considered closely related to data stream mining.

Resource constraints specific to data stream mining pertain to data storage, computation, and communication. These constraints include but are not limited to (Babcock et al., 2002; Gaber et al., 2004, 2005):

- Data arrive rapidly and continuously, and traditional database management systems are not designed for loading such continuous streams;
- Memory requirements can be unbounded due to the continuous data flow;
- Because of real-world memory constraints, mining algorithms can only access data via a few linear scans and generally cannot perform repeated random access on data, which is often required by traditional data mining techniques;
- Not only do data values change over time, but the distributions resulting from different data-generating processes may also change, creating "concept drifts;" and
- Because mining algorithms cannot view the entire data set at once, it is usually difficult to generate results based on exact answers from queries. Some approximation techniques must be employed instead.

Facing these challenges and constraints, the data mining community has proposed a number of techniques and approaches to deal with new research questions raised by data stream mining. In the following subsections we review prior research on data stream and sensor network data mining, focusing on clustering, classification, and association rule mining. These studies suggest some promising directions for our future research, which is aimed at effectively mining wireless sensor network data that is acquired from sensor fields that explicitly manage resource consumption.

Clustering

As described earlier, clustering is an unsupervised machine learning technique that can be used in many situations. In sensor networks, for example, clustering can be used to group related sensor nodes and find cluster centers (Younis & Fahmy, 2004). These central nodes are then responsible for handling intra- and inter-cluster data transmission and exchange. Because only cluster centers, and not individual sensor nodes, are involved in communication between the sensor fields and external servers, communication costs are reduced and energy is saved.

One clustering problem is the so-called k-Median problem, which is defined as finding k centers among a set of nodes (or a sequence of nodes in the context of data streams) so as to minimize the total distance between individual nodes and their closest center node (Guha et al., 2000) given constraints such as time, space, and linear data scans. To address this problem, Guha et al. (2000) propose a constant-factor approximation algorithm that requires small space and only a single scan over the data stream. Based on a divide-and-conquer strategy, this algorithm examines data in a piecemeal fashion and then re-clusters the centers obtained. The final algorithm is a deterministic time, polylog(n)-approximation single-pass algorithm that uses small space, where n is the number of nodes processed. This algorithm has been improved by Babcock et al. (2003), who use an exponential histogram data structure to address the problem of merging distant clusters, and Charikar et al. (2003), who address the problem of increasing approximation factors by increasing the number of levels. Other researchers have also proposed various techniques for clustering data streams. Domingos and Hulten (2000, 2001), for example, proposed the VFML (Very Fast Machine Learning) approach and applied it to k-Means clustering and decision tree classification. Aggarwal (2003) proposed the CluStream algorithm which uses online and offline components to cluster streams of data. Both components deal with only a summarized version of data streams rather than individual data points. In the research of sensor network clustering, Younis and Fahmy (2004) have proposed a new energy-efficient approach to clustering sensor nodes in ad-hoc sensor networks. They present a HEED (Hybrid Energy-Efficient Distributed clustering) protocol that periodically finds clusters and their cluster centers according to available energy and other factors. This approach helps achieve the goals of prolonged network lifetime, scalability, and load balancing.

Classification

Classification is a type of supervised machine learning. Among the various classifiers, decision trees are the most widely studied classification technique in data stream mining. Usually, a decision tree is learned by first examining a set of training data that associates a category label with each data item based on its attribute values. The learned decision tree is then applied to the test data, whose items do not have labels, to assign labels to data items. Such a technique assumes that data are drawn from a sample with a stationary distribution. However, in the context of data streams, this assumption is often violated because the processes that generate the data may change continuously, creating a problem called "concept drift" (Schlimmer & Granger, 1986). Mining decision trees out of data streams must successfully deal with this problem in order to obtain correct classification results.

To address the concept drift problem, Hulten et al. (2001) propose a technique called Concept-adaptive Very Fast Decision Trees (CVFDT) to learn decision trees from high-speed, time-varying data streams. CVFDT keeps a sliding window of new examples (training data) and adds subtrees to the main tree when the old subtrees seem to be out-of-date. Old subtrees are totally replaced by new subtrees when the new subtrees become more accurate. CVFDT has been tested against web page request streams and has shown promising performance. Similarly, Last (2002) proposes a classifier that can adjust to concept drift. This algorithm keeps a sliding widow of the most recent examples and the classification model is rebuilt based on data present in the sliding window. Error rates indicating concept drift are used as a guide to the frequency of model rebuilding and the change in window size. Other examples of data stream classifiers include CluStream (Aggarwal et al., 2004) and AWSOM (Papdimitriou et al., 2003).

Association Rule Mining

Compared with clustering and classification, association rule mining has not been widely studied in the context of data stream mining. In sensor network data mining, association rule mining has been proposed to extract associations between sensors so that it is easy to find a set of sensors that can report events in the same time interval (Boukerche & Samarah, 2007; Loo et al., 2005). One goal of such mining is to find patterns, i.e., associated sensors, that deliver data together sufficiently often (e.g., greater than 90% of the time). However, because of resource constraints which are present when processing stream data, this problem is far more challenging than mining association rules in traditional data mining applications. To overcome these constraints, Boukerche and Samarah (2007) propose a new data structure called PLT (Positional Lexicographic Tree) in order to access and manipulate data effectively and efficiently. Their mining algorithm, based on a pattern growth approach, is able to find frequent association patterns from the sensor network data organized in a PLT. Association rule mining has also been used to estimate missing sensor readings (Halatchev & Gruenwald, 2005) and extract chronological patterns between sensors (Boukerche & Samarah, 2006).

FUTURE WORK

The main focus of our future research is to explore how different data mining techniques perform when wireless sensor network data is acquired and processed, considering cost performance, quality performance, and resource constraints. One research question is to ascertain whether a cost versus quality trade-off affects the results of data mining and if so, how, when, and why. Another possible direction is to design and develop new mining algorithms that can successfully address the various resource constraints of sensor network data mining while managing cost and quality performance.

CONCLUSION

Sensor network data mining requires bridging the gap between low-level data that is acquired in sensor fields and high-level knowledge that is useful to real-world applications. We describe a new approach to sensor field resource management, which is necessary but not sufficient to bridge this gap. While sensor network data is being gathered from sensor fields, the remaining stages in the data mining process need to produce accurate and timely output based on dynamically changing data. We have discussed the challenges and problems that arise within this data mining process and pointed to some directions for future research.

REFERENCES

Aggarwal, C. (2003). A framework for diagnosing changes in evolving data streams. In *Proceedings of the ACM SIGMOD Conference*, San Diego, California, June 2003.

Aggarwal, C., Han, J., Wang, J., & Yu, P. S. (2004). On demand classification of data streams. In *Proceedings of ACM International Conference on Knowledge Discovery and Data Mining*, Seattle, Washington, August 2004.

Babcock, B., Babu, S., Datar, M., Motwani, R., & Widom, J. (2002). Models and issues in data stream systems. In *Proceedings of the 21st Symposium on Principles of Database Systems (PODS)*, Madison, Wisconsin, June 2002.

Babcock, B., Datar, M., Motwani, R., & O'Callaghan, L. (2003). Maintaining variance and k-medians over data stream windows. In *Proceedings of the 22nd Symposium on Principles of Database Systems (PODS)*, San Diego, California, June 2003.

Bishop, C. M. (1995). Neural Networks for Pattern Recognition. Oxford, England: Oxford University Press.

Boukerche, A., & Samarah, S. (2006). A novel data mining technique for extracting events and inter knowledge based information from wireless sensor networks. In *Proceedings of IEEE Conference on Local Computer Networks*, Tampa, Florida, November 2006.

Boukerche, A., & Samarah, S. (2007). A new representation structure for mining association rules from wireless sensor networks. In *Proceedings of IEEE Wireless Communications and Networking*, Hong Kong, China, March 2007.

Boulis, A., Ganeriwal, S., & Srivastava, M. B. (2003). Aggregation in sensor networks: An energy-accuracy tradeoff. In *IEEE Workshop on Sensor Network Protocols and Applications (SNPA)*, Anchorage, Alaska, May 2003.

Bridle, J. S. (1990). Probabilistic interpretation of feed-forward classification network outputs, with relationships to statistical pattern recognition. In *Neurocomputing: Algorithms, Architecture and Applications*. Berlin, Germany: Springer-Verlag.

Burroughs Corporation. (1964). *Burroughs B5500 Information Processing Systems Reference Manual*. Burroughs Corporation, Detroit, Michigan, USA.

Charikar, M., O'Callaghan, L., & Panigrahy, R. (2003). Better streaming algorithms for clustering problems. In *Proceedings of 35th ACM Symposium on Theory of Computing (STOC)*, San Diego, California, June 2003.

Chu, D., Popa, L., Tavakoli, A., Hellerstein, J. M., Levis, P., Shenker, S., & Stoica, I. (2007). The design and implementation of a declarative sensor network system. In *Proceedings of ACM Conference on Embedded Networked Sensor Systems (SenSys)* (pp. 175-188), Sydney, Australia, November 2007.

Demers, A., Gehrke, J., Rajaraman, R., Trigoni, N., & Yao, Y. (2003). The Cougar project: A work-in-progress report. *SIGMOD Record, 32*(4), 53–59. doi:10.1145/959060.959070

Deshpande, A., Guestrin, C., Madden, S., Hellerstein, J. M., & Hong, W. (2004). Model-driven data acquisition in sensor networks. In *Proceedings of the International Conference on Very Large Data Bases (VLDB)* (pp. 588-599), Toronto, Canada.

Domingos, P., & Hulten, G. (2000). Mining high-speed data streams. In *Proceedings of the ACM International Conference on Knowledge Discovery and Data Mining* (pp.71-80).

Domingos, P., & Hulten, G. (2001). A general method for scaling up machine learning algorithms and its application to clustering. In *Proceedings of the Eighteenth International Conference on Machine Learning* (pp. 106-113), Williamstown, Massachusetts.

Fayyad, U. M., Piatetsky-Shapiro, G., & Smyth, P. (1996). From data mining to knowledge discovery: An overview. In U.M. Fayyad, G. Piatetsky-Shapiro, P. Smyth & R. Uthurusamy (Eds.), *Advances in Knowledge Discovery and Data Mining*. Menlo Park, CA, USA: AAAI Press/ The MIT Press.

Fuller, S. H., & Harbison, S. P. (1978). *The C.mmp multiprocessor* (Technical Report CMU-CS-78-146). Pittsburgh, Pennsylvania, USA: Department of Computer Science, Carnegie-Mellon University.

Gaber, M. M., Krishnaswamy, S., & Zaslavsky, A. (2004). Ubiquitous Data Stream Mining, *Current Research and Future Directions Workshop* held in conjunction with *The Eighth Pacific-Asia Conference on Knowledge Discovery and Data Mining*, Sydney, Australia.

Gaber, M. M., Zaslavsky, A., & Krishnaswamy, S. (2005). Mining data streams: A review. *SIGMOD Record, 34*(2), 18–26. doi:10.1145/1083784.1083789

Gibbons, P. B., Karp, B., Ke, Y., Nath, S., & Seshan, S. (2003). IrisNet: An architecture for a world-wide sensor web. *IEEE Pervasive Computing / IEEE Computer Society [and] IEEE Communications Society, 2*(4), 22–33. doi:10.1109/MPRV.2003.1251166

Gnawali, O., Greenstein, B., Jang, K.-Y., Joki, A., Paek, J., Vieira, M., et al. (2006). The Tenet architecture for tiered sensor networks. In *ACM Conference on Embedded Networked Sensor Systems (SenSys)* (pp. 153-166).

Guha, S., Mishra, N., Motwani, R., & O'Callaghan, L. (2000). Clustering data streams. In *Proceedings of the Annual Symposium on Foundations of Computer Science (FOCS)*, Redondo Beach, California, November 2000.

Gummadi, R., Gnawali, O., & Govindan, R. (2005). Macro-programming wireless sensor networks using Kairos. In *International Conference on Distributed Computing in Sensor Systems (DCOSS)*, Marina del Rey, California, June 2005.

Halatchev, M., & Gruenwald, L. (2005). Estimating missing values in related sensor data streams. In *11th International Conference on Management of Data* (COMAD), Goa, India, January 2005.

Han, J., & Kamber, M. (2000). *Data Mining: Concepts and Techniques*. San Francisco, California, USA: Morgan Kaufmann Publishers.

Howard, J. H., Kazar, M. L., Menees, S. G., Nichols, D. A., Satyanarayanan, M., Sidebotham, R. N., & West, M. J. (1998). Scale and performance in a distributed file system. *ACM Transactions on Computer Systems, 6*(1), 51–81. doi:10.1145/35037.35059

Hu, W., Misra, A., & Shorey, R. (2006). CAPS: Energy-efficient processing of continuous aggregate queries in sensor networks. In *IEEE International Conference on Pervasive Computing and Communications (PerCom)* (pp. 190-199).

Hulten, G., Spencer, L., & Domingos, P. (2001). Mining time-changing data streams. In *Proceedings of the ACM SIGKDD Conference*, San Francisco, California, August 2001.

Intanagonwiwat, C., Govindan, R., Estrin, D., Heidemann, J., & Silva, F. (2003). Directed diffusion for wireless sensor networking. *IEEE/ACM Transactions on Networking, 11*(1), 2–16.

Jain, A. K., Murty, M. N., & Flynn, P. J. (1999). Data clustering: A review. *ACM Computing Surveys, 31*(3), 264–323. doi:10.1145/331499.331504

Jamieson, K., Balakrishnan, H., & Tay, Y. C. (2003). *Sift: a MAC protocol for event-driven wireless sensor networks* (Technical Report 894). Cambridge, Massachusetts, USA: Laboratory for Computer Science, Massachusetts Institute of Technology.

Jindal, A., & Psounis, K. (2006). Modeling spatially correlated data in sensor networks. *ACM Transactions on Sensor Networks, 2*(4), 466–499. doi:10.1145/1218556.1218558

Last, M. (2002). Online classification of nonstationary data streams. *Intelligent Data Analysis, 6*(2), 129–147.

Loo, K. K., Tong, I., Kao, B., & Chenung, D. (2005). Online algorithms for mining inter-stream associations from large sensor networks. In *Proceedings of the Ninth Pacific-Asia Conference on Knowledge Discovery and Data Mining*, Hanoi, Vietnam, May 2005.

Lu, C., Blum, B. M., Abdelzaher, T. F., Stankovic, J. A., & He, T. (2002). RAP: A real-time communication architecture for large-scale wireless sensor networks. In *IEEE Real-Time and Embedded Technology and Applications Symposium* (pp. 55-66).

Luotonen, A., & Altis, K. (1994). World-wide web proxies. In *Selected papers of the First Conference on the World-Wide Web* (pp. 147-154), Amsterdam, The Netherlands, May 1994.

Madden, S., Franklin, M. J., Hellerstein, J. M., & Hong, W. (2005). TinyDB: An acquisitional query processing system for sensor networks. *ACM Transactions on Database Systems, 30*(1), 122–173. doi:10.1145/1061318.1061322

Newton, R., Morrisett, G., & Welsh, M. (2007). The Regiment macroprogramming system. In *International Conference on Information Processing in Sensor Networks (IPSN)* (pp. 489-498).

Papadimitriou, S., Faloutsos, C., & Brockwell, A. (2003). Adaptive, hands-off stream mining. In *Proceedings of the 29th International Conference on Very Large Data Bases (VLDB)* (pp. 560–571).

Raghunathan, V., Schurgers, C., Park, S., & Srivastava, M. B. (2002). Energy-aware wireless microsensor networks. *IEEE Signal Processing Magazine, 19*(2), 40–50. doi:10.1109/79.985679

Rodriguez, C. (2004). *A computational environment for data preprocessing in supervised classifications*. Unpublished master's thesis, University of Puerto Rico, Mayaguez, Puerto Rico.

Schlimmer, J. C., & Granger, R. H., Jr. (1986). Beyond incremental processing: Tracking concept drift. In *Proceedings of the 5th National Conference on Artificial Intelligence* (pp. 502-507), Philadelphia, Pennsylvania.

Schwetman, H. (2001). CSIM 19: A powerful tool for building systems models. In *Proceedings of the ACM Winter Simulation Conference* (pp. 250-255).

Schwetman, H. D. (1990). Introduction to process-oriented simulation and CSIM. In *Proceedings of the ACM Winter Simulation Conference* (pp. 154–157).

Sharaf, M. A., Beaver, J., Labrinidis, A., & Chrysanthis, P. K. (2004). Balancing energy efficiency and quality of aggregate data in sensor networks. *The VLDB Journal, 13*(4), 384–403. doi:10.1007/s00778-004-0138-0

Son, S.-H., Chiang, M., Kulkarni, S. R., & Schwartz, S. C. (2005). The value of clustering in distributed estimation for sensor networks. In *Proceedings of the IEEE International Conference on Wireless Networks, Communications, and Mobile Computing (WirelessCom)*, Maui, Hawaii, June 2005.

Tilak, S., Abu-Ghazaleh, N. B., & Heinzelman, W. (2002). Infrastructure tradeoffs for sensor networks. In *Proceedings of First ACM International Workshop on Wireless Sensor Networks & Applications* (pp. 49–58).

Wan, C.-Y., Eisenman, S. B., & Campbell, A. T. (2003). CODA: Congestion detection and avoidance in sensor networks. In *ACM Conference on Embedded Networked Sensor Systems (SenSys)* (pp. 266-279).

Yates, D. J. (2006). *Scalable data delivery for networked servers and wireless sensor networks*. Unpublished Ph.D. dissertation, Department of Computer Science, University of Massachusetts, Amherst, Massachusetts, USA.

Yates, D. J., Nahum, E., Kurose, J., & Shenoy, P. (2008). Data quality and query cost in pervasive sensing systems. In *IEEE International Conference on Pervasive Computing and Communications (PerCom)*, Hong Kong, China, March 2008.

Younis, O., & Fahmy, S. (2004). HEED: A hybrid, energy-efficient, distributed clustering approach for ad hoc sensor networks. *IEEE Transactions on Mobile Computing, 3*(4), 366–379. doi:10.1109/TMC.2004.41

Yu, Y., Krishnamachari, B., & Prasanna, V. K. (2004). Energy-latency tradeoffs for data gathering in wireless sensor networks. In *Proceedings of the IEEE Conference on Computer Communications (Infocom)*, Hong Kong, China, March 2004.

Zhao, J., & Govindan, R. (2003). Understanding packet delivery performance in dense wireless sensor networks. In *ACM Conference on Embedded Networked Sensor Systems (SenSys)* (pp. 1-13).

Section 6
Intelligent Techniques for Advanced Sensor Network Data Warehousing and Mining

Chapter 14
Event/Stream Processing for Advanced Applications[1]

Qingchun Jiang
Oracle Corporation, USA

Raman Adaikkalavan
Indiana University, USA

Sharma Chakravarthy
University of Texas, Arlington, USA

ABSTRACT

Event processing in the form of ECA rules has been researched extensively from the situation monitoring viewpoint to detect changes in a timely manner and to take appropriate actions. Several event specification languages and processing models have been developed, analyzed, and implemented. More recently, data stream processing has been receiving a lot of attention to deal with applications that generate large amounts of data in real-time at varying input rates and to compute functions over multiple streams that satisfy quality of service (QoS) requirements. A few systems based on the data stream processing model have been proposed to deal with change detection and situation monitoring. However, current data stream processing models lack the notion of composite event specification and computation, and they cannot be readily combined with event detection and rule specification, which are necessary and important for many applications. This chapter discusses a couple of representative scenarios that require both stream and event processing. The authors then summarize the similarities and differences between the event and data stream processing models. The comparison clearly indicates that for most of the applications considered for stream processing, event component is needed and is not currently supported. And conversely, earlier event processing systems assumed primitive (or simple) events triggered by DBMS and other applications, and did not consider computed events. By synthesizing these two and combining their strengths, the authors present an integrated model – one that will be better than the sum of its parts. The authors discuss the notion of a semantic window, which extends the current window concept for continuous queries, and stream modifiers in order to extend current stream computation model for complicated change detection. They further discuss the extension of event specification to include continuous queries. Finally, the authors demonstrate how one of the scenarios discussed earlier can be elegantly and effectively modeled using the integrated approach.

DOI: 10.4018/978-1-60566-328-9.ch014

INTRODUCTION

Event processing (Dayal et al., 1988; Schreier et al., 1991; Diaz, Paton, & Gray, 1991; Gehani, Jagadish, & Shmueli, 1992a; Gatziu & Dittrich, 1993; Kotz-Dittrich, 1993; Buchmann et al., 1993; Chakravarthy, Anwar, Maugis, & Mishra, 1994; Hanson, 1996; Lieuwen, Gehani, & Arlein, 1996; Seshadri, Livny, & Ramakrishnan, 1996; Engstrom, Berndtsson, & Lings, 1997; Dinn, Williams, & Paton, 1997) and lately data stream processing (Babu & Widom, 2001; Abadi et al., 2003; Chen et al., 2000; Madden & Franklin, 2002; Jiang & Chakravarthy, 2004a) have evolved independently based on situation monitoring application needs. Triggers have been successfully defined over relational databases and several event specification languages (Gehani, Jagadish, & Shmueli, 1992b, 1992c; Gatziu & Dittrich, 1993, 1994; Chakravarthy & Mishra, 1994; Roncancio, 1997; Adaikkalavan & Chakravarthy, 2003) for specifying composite events have been proposed. Different computation models (Gehani & Jagadish, 1991; Lieuwen et al., 1996; Gatziu & Dittrich, 1992, 1993; Engstrom et al., 1997; Buchmann et al., 1993; Chakravarthy et al., 1994; Dinn et al., 1997) for processing events, such as Petri nets (Gatziu & Dittrich, 1992, 1993), extended automata (Gehani & Jagadish, 1991; Lieuwen et al., 1996; Gehani et al., 1992c), and event graphs (Buchmann et al., 1993; Chakravarthy et al., 1994; Engstrom et al., 1997) – have been proposed and implemented. Various event consumption modes (Buchmann et al., 1993; Gatziu & Dittrich, 1992, 1993; Chakravarthy et al., 1994; Chakravarthy & Mishra, 1994) (or parameter contexts) have been explored. Similarly, data stream processing has received a lot of attention lately, and a number of issues – from architecture (Abadi et al., 2003; Madden & Franklin, 2002; Jiang & Chakravarthy, 2004a; Chen et al., 2000; Motwani et al., 2003) to quality of service (Tatbul et al., 2003; Babcock, Datar, & Motwani, 2004; Das, Gehrke, & Riedewald, 2003; Jiang & Chakravarthy, 2004b; Brian et al.,

2003; Carney et al., 2003) – have been explored. Although both of these topics seem different on the face of it, we argue that there are a number of similarities and some differences between them. Surprisingly, the computation model used for data stream processing is not very dissimilar from some of the event processing models (e.g., event graph), but with a different emphasis.

As many of the stream applications are based on sensor data, they invariably give rise to events on which some actions need to be taken. In other words, most of the stream applications seem to need not only computations on streams, but these computations generate interesting events (e.g., car accident detection and notification, network congestion control) and several such events may have to be composed, detected and monitored for taking appropriate actions. Currently, to the best of our knowledge, none of the work addresses the specification and computation of the above two threads of work. Our premise for this chapter is that although each one is useful in its own right, their combined expressiveness is critical for many applications of data stream processing. Hence there is a need for synthesizing the two into a more expressive model that combines the strengths of each one.

We use the running examples (discussed in Section 2) to explain the current limitations of each model, and the need for the integrated model.

Outline: Motivating examples and analysis of event and stream processing is discussed in Section 2. Integrated model is presented in Section 3. In Section 4, we demonstrate how one of the running example discussed in Section 2 is modeled using the integrated approach. In Section 5 we discuss the prototype implementation and experiments. Related work is discussed in Section 6 and Section 7 has conclusions.

Processing (including mining) of sensor data can be done in two ways: i) process archived data or ii) process data on the fly. Traditional techniques can be employed for processing archived data. The framework presented in this chapter addresses processing data on the fly as it comes in the form of one or more streams. Processing or mining of data streams is employed when quality of service (QoS) constraints are associated with the processing. In addition, this processing typically generates "interesting events" that needs to be correlated further for detecting appropriate situations (e.g., detection of fire or spreading of forest fire based on several sensor data streams). Another example of sensor networks is battlefield management where the data from each soldier is mined to detect the health or fitness of a soldier and correlated with others either to seek immediate help, or estimate group strength. The framework presented in this chapter accommodates both QoS satisfaction and allows for event correlation as the proposed architecture is a synergistic integration of stream and complex event processing.

Many sensor network application not only need to mine useful patterns from the data streams produced by sensors, but also need to perform pre-defined actions based on further processing of those patterns. In this chapter, we presented a framework in which users not only can mine patterns or events from sensor data streams by adding more mining-specific operators such as clustering and pattern changes, etc., they can also perform complex event processing based on the outputs from stream mining components. In addition, rules can also be defined on those events to take timely actions once that event is detected. The stream modifiers and semantic windows defined this chapter, along with other data stream management components such as scheduling, can help an user to express and compute complicated data mining operations over sensor streams. The event and rule processing extensions provided in this chapter can trigger pre-defined actions since most sensor stream applications need to detect changes or patterns and perform actions in a timely manner. These extensions can also help to further process mining results and provide feedback.

MOTIVATING EXAMPLES AND ANALYSIS

We have analyzed several real-world applications to understand the requirements and issues that need to be solved in order to have an end-to-end system. Network Fault Management application (Jiang, Adaikkalavan, & Chakravarthy, 2005) is summarized first. Linear Road Benchmark (Arasu et al., 2004) application is then discussed in detail and will be used as a running example in this chapter.

Example 1:*In telecommunication network management, Network Fault Management (NFM) is defined as the set of functions that: (a) detect, isolate, and correct malfunctions in a telecommunication network, (b) compensate for environmental changes, and (c) maintain and examine error logs, accept and act on error detection notifications, trace and identify faults, carry out sequence of diagnostic tests, correct faults, report error conditions, and localize and trace faults by examining and manipulating database information.*

A typical telecommunication network illustrated in Figure 1 is a multi-layered network, in which the bottom layer provides transport service through SDH/SONET networks. Above that, a PSTN switch network with a SS7 signaling network is used to provide traditional voice services, and an ATM network is used to provide Internet data service. Intelligent networks and other value-added networks can be added above the PSTN switch networks, and BGP/MPLS VPN network can be added above the ATM network. The NFM in such a multi-layered telecommunication network has been an interesting research problem (Baras, Li, & Mykoniatis, 1998; Bjerring, Lewis, & Thorarensen, 1996; Diaz-Caldera, Serrat-Fernandez, Berdekas, & Karayannis, 1999; Mountzia & Rodosek, 1999; Gambhir, Post, & Frisch, 1994; Frohlich & Nejdl, 1997; Medhi et al., 2001) in both industry and academia for a long time because of its high cost and complexity.

Figure 1. A typical telecomm network

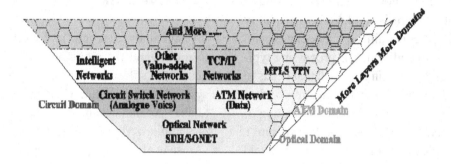

Currently, for each independent NFM system, due to the large volume of messages that are continuously reported by each network element (NE) and the complex message processing requirements, it is impossible to employ traditional database management system (DBMS) plus trigger mechanisms as the data processing paradigm for NFM. Current NFM systems have to hard code their data processing logic and specific monitoring rules (queries) in the system. As a result, various filters, pattern languages, regular expressions are employed to find their interesting alarm messages and group those messages into multiple subgroups based on various criteria. These subgroups are finally presented to experts to diagnose root causes or route to an event correlation system to identify causes automatically.

Based on our analysis, the techniques developed for stream processing are not sufficient for the above problem as the operators needed are different, the concept of window currently used in a stream processing system is inadequate, composite events cannot be defined over the outcome of stream processing, and number of events being generated cannot be curtailed by current approaches to event processing.

Example 2 (Car ADN):*In a car accident detection and notification system (adapted from the linear road benchmark (Arasu et al., 2004), each expressway in an urban area is modeled as a linear road, and is further divided into equal-length segments (e.g., 5 miles). Each registered vehicle on an express way is equipped with a sensor and reports its location periodically (say, every 30 seconds). Based on this location stream data, we want to detect a car accident in a near-real time manner. If a car reports the same location (or with speed zero mph) for four consecutive times, FOLLOWED BY at least one car in the same segment with a decrease in its speed by 30% during its four consecutive reports, then it is considered as a potential accident. Once an accident is detected, some actions may have to be taken immediately: i) notify the nearest police/ambulance control room about the car accident, ii) notify all the cars in 5 upstream segments about the accident, and iii) notify the toll station so that all cars that are blocked in the upstream for up to 20 minutes by the accident will not be tolled.*

Every car in the express way is assumed to report its location every 30 seconds forming the primary input data for the above example. The format of car location data stream (i.e., CarLocStr) is given below:

```
CarLocStr (timestamp, car_id, speed, exp_
way, lane, dir, x-pos)
```

CarSegStr is the car segment stream (or the input CarLocStr stream), but with the location of the car replaced by the segment corresponding to the location. Query shown below produces the CarSegStr from the CarLocStr stream.

```
SELECT timestamp, car_id, speed, exp_way,
lane, dir, (x-pos/5 miles) as seg FROM
CarLocStr;
```

Detecting an accident in the above CAR ADN example has three requirements:

(1) **IMMOBILITY:** Checking whether a car is at the same location for four consecutive time units (i.e., over a 2 minutes window, in our example, as the car reports its location every 30 seconds).

(2) **SPEED REDUCTION:** Finding whether there is at least one car that has reduced its speed by 30% or more during four consecutive time units.

(3) **SAME SEGMENT:** Determining whether the car that has reduced its speed (i.e., car identified in (2)) is in the same segment and it follows the car that is immobile (i.e., car identified in (1)).

Immobility of a car can be computed using CQs that are supported by the current data stream processing systems as shown below:

```
SELECT car_id, AVG(speed) as avg_speed
FROM CarLocStr [2 minutes sliding window]
GROUP BY car_id
HAVING avg_speed = 0;
```

With the current event and stream processing models using a non-procedural language[2], it is difficult or impossible to efficiently compute the speed reduction. Whether the cars that are found in requirements (1) and (2) are from the same segment can be readily determined in an event processing model using a sequence operator (Gatziu & Dittrich, 1993; Buchmann et al., 1993; Adaikkalavan & Chakravarthy, 2006). As the cars that are identified in requirement (3) can be separated by more than 4 time units, it requires an efficient, meaningful and less redundant approach to notifications. In other words, number of times the accident is reported should be kept to a minimum. This can be done efficiently using the current event processing models using the notion of contexts (e.g., recent context for this case), but not the current stream processing models. Although JOIN operator can be used to compute it in a roundabout manner, the number of notifications (or the number of times an event is raised) is not minimized.

The real-world examples we have analyzed clearly illustrate the need for stream processing followed by event processing to accomplish the task in an elegant manner[3]. In addition, the above notifications have to meet some Quality Of Service (QoS) requirements (see Table 1).

INTEGRATED ARCHITECTURE

Based on the above analysis it is clear that although there is some overlap, the class of applications for which they were developed were quite different and as a result some of the assumptions are also different. However, the monitoring applications based on very large amounts of raw data require combinations of the above technologies in a synergistic fashion. Below we present an integrated architecture for accomplishing that. This work is being continued to reconcile the differences between the two and have a single end-to-end framework that is applicable to stream processing applications, event processing applications, and their combination.

The integrated model is shown in Figure 2 and consists of four stages: 1) CQ processing stage used for computing CQs over data streams, 2) coupling stream output with event processing system, 3) event processing stage that is used for detecting events, and 4) rule processing stage that is used to check conditions, and to trigger predefined actions once events are detected. The seamless nature of our integrated model is due to a number of extensions (Jiang, Adaikkalavan, & Chakravarthy, 2007) and the compatibility of the

Table 1. Comparison of characteristics between event and stream processing work

Characteristics	Event Processing	Stream Processing
Plan model	• Event detection graphs	• Query tree/graphs
Input	• Event streams or event histories ordered by the time of occurrence. • Data sources are typically pre-determined with relatively low input rates.	• Data streams ordered in some way (by attribute or timestamp) and stored relations. • Data sources are mostly external with a wide range of input rates and bursty at times.
Output	• Simple or complex events as streams ordered by their occurrence timestamp. • When events are detected, a set of ECA rules can be triggered.	• Outputs forms another data stream • Also a stream either consumed by an application or is an *interesting* event
Duration of events/items in streams	• Event consumption modes (or parameter contexts) are used to determine when the events can be dropped or how they are combined with other events for detecting a composite/complex event.	• A window (such as time-based, tuple-based, and so) is defined for each data stream in queries. • Mainly for the purpose of unblocking operators. • The precise relationship between windows and contexts/modes is not yet established.
Algebra/Operators	• Mainly used to express and compose abstract events using primitive events and operators. • Both point- and interval-based semantics have been proposed. • Semantics typically based on time of occurrence and uses Allen's interval logic.	• Current stream operators used for continuous queries (CQs) are mostly modified relational operators. • CQ operators have input queues and internal windows (synopsis) in order to deal with highly bursty inputs and to convert blocking operators to non-blocking operators.
Computation Model	• Data flow computation models, Petri nets, and extended finite automata have been used. • Event computation models do not assume input queues and the notion of windows.	• Primarily data flow models are used for computation. Input queue(s) is (are) used for each operator. • Synopses are used at each operator for storing intermediate results needed for computing the operator for a window.
Best-Effort Vs. QoS	• The notion of QoS is not present. Although, there is some work on real-time events and event showers, they do not support any specific QoS requirements. • Event detection is typically synchronous; that is, whenever an event occurs it is detected or propagated to form a composite event. Thus, events are detected based on the best-effort method.	• QoS support is an integral part of data stream management systems. Typically, the following 3 QoS requirements are considered: tuple latency, maximum memory requirement, and smoothness of the throughput.
Optimization and Scheduling	• There is some work on rewriting event expressions and grouping common event sub-expressions Common event sub-expressions are grouped in order to reduce the overall response time and computation effort. • In general, event processing has not dealt with runtime optimizations.	• Significant amount of work for optimizing total memory requirements, tuple latency, and smoothness of throughput exists. Scheduling and load shedding algorithms have been proposed for data stream management systems.
Rules, Triggers, and Notification	• Rules in event processing systems describe what the underlying system should when an event occurs. • Existing event processing systems support dynamic enabling and disabling of rules. • Rule execution semantics have been proposed and analyzed. It includes rule processing granularity, instance/set oriented execution, iterative/recursive execution, conflict resolution, sequential/concurrent execution, coupling modes, confluence, and termination.	• Stream processing systems do not support high-level rule specification and processing explicitly. Several systems allow notification specification for a CQ or a CEP (complex event processing).

chosen event processing model (It will be difficult to integrate either the Petri net event processing model of SAMOS or the extended automata model of ODE with stream processing models) (i.e., an event detection graph) with the model used for stream processing.

Based on our analysis, synthesizing both the processing models requires the following issues

Figure 2. Four stage integration model

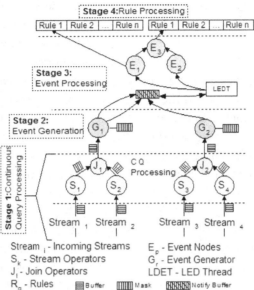

to be addressed: 1) handling highly bursty event streams (generated by the CQ processing stage) in event processing, 2) processing of events streams based on attributes and not solely on timestamp, 3) specification of events/event expressions, rules and CQs.

Both models have been enhanced to address the above mentioned issues: 1) Output of CQs is fed as inputs to the primitive events in the event processing stage. Continuous queries have been named, so that in the event processing stage the outputs of CQs can be used for detecting events. 2) Stream modifiers detect complex changes between tuples in a stream, 3) Masks are used to generate multiple event types from a CQ, 4) Semantic window enhances the expressiveness and computation efficiency of CQs, and allows the creation of more meaningful windows; For the event processing model, 5) Event operators have been enhanced by introducing input queue(s) for each operator, which makes it possible to handle highly bursty outputs from CQ processing stage and take advantages of the techniques (i.e., scheduling strategies, load shedding) developed for stream processing model. 6) Event expres-

sions have also been enhanced in such a way that primitive events can process event streams based on event attributes, and not only on timestamp. Finally, 7) Extended SQL like language allowing user to specify events/event expressions, rules and CQs together.

Below we summarize on why and how these enhancements are carried out. In Section 4 we show how these enhancements are used to solve the Car ADN example discussed in Section 2.

Continuous Query Processing

This stage processes normal CQs where it takes streams as inputs and gives computed continuous streams. The scheduling algorithms and QoS delivery mechanisms (i.e., load shedding techniques) along with other techniques developed for stream processing model can be applied directly. In many cases, final results of stream computations need to be viewed as interested data points where primitive events can be computed for defining situations that use multiple streams and composite events. A CQ may give rise to multiple events based on the attribute values of the output stream. In Figure

2, operators S_1, S_2, and J_1 form a CQ. Similarly, operators S_3, S_4 and J_2 form a CQ. This stage supports more complicated computations required by many stream applications:

- **Named Continuous Queries:** Many computations over streaming data are difficult to express and to be computed as a single CQ. In order to express computations clearly, CQs are named. It can be used to define primitive events, and can be used in defining other CQs.

```
CREATE CQ CQName AS (Normal CQ create
statements)
```

The name of a CQ is analogous to the name of a table in a DBMS and a named CQ has the same scope and usage as a table. The queue (buffer) associated with each operator in a CQ supports the output of a named CQ to be fed into the input queue of another named CQ. A named CQ is defined by using the CREATE CQ statement shown above. However, the FROM clause in a named CQ can use any previously defined CQs through their unique names. The Meta information of a named CQ is maintained in a CQ dictionary in the system. The Meta information includes the *query name*, its *input sources*, all *output attributes* ordered by their order in final output tuples, and its *output destination(s)*. Events can be specified by using named CQs and in addition provide conditions on attributes to generate multiple event types. If the output destination is to an application, it can be in the form of a named pipe, socket, or an output queue/buffer. A default output destination is defined in the system simply as a sink if there is no destination associated with this CQ. A CQ with a sink as its destination can be disabled in the system until a meaningful destination is associated to it. A named CQ can output its final results to multiple destinations.

- **Semantic Window:** Current types of windows available in stream processing systems are tuple-based, time-based, partitioned and attribute-based windows. The major problem with these types of windows is that they cannot express more meaningful windows. On the other hand, semantic window allows the specification of the number of tuples based on a computation. Semantic window computations can be carried out through well-developed and highly optimized SQL query processing engines. The main function of a semantic window is to determine which tuples should be in the current window by performing deletion of existing tuples and addition of new tuples. Before defining a semantic window, we need to identify the scope of data that a semantic window can access to perform computations. Obviously, all tuples in the current window, new tuples, and static relations, if needed, are fully accessible by that window. All data accessible by a semantic window are referred as *semantic window input data* (*SWID*).

Definition 1 (Semantic Window) *A semantic window is defined as a finite portion of historical tuples at any time point, which satisfy a* semantic window condition (*SWC*), *by computing SWC over SWID. The SWC can be any arbitrary condition over SWID.* However, to simplify the way to express a *SWC*, we use the CHECK statement shown below:

```
Stream [ CHECK      logical Expression
        SELECT      a1, a2, ..., an
        FROM SWID
        WHERE       Conditions
        GROUP BY    Attributes
        HAVING      Conditions ]
```

All the clauses used in the above statement (i.e., SELECT, FROM, WHERE, GROUP BY, and HAVING) have the same semantics and usage as in the standard SQL. However, only *SWID*, include all tuples in the current window, new tuples, and static relations can appear in FROM clause. $a1, a2, ..., an$ are the attribute names (or alias after applying aggregate functions) from *SWID*. The CHECK clause is a logical expression that consists of the attribute names used in the SELECT statement, relational and logical operators, and parentheses. The CHECK clause is the last clause to be evaluated in the SQL statement and returns a Boolean value. The CHECK clause requires only one row from the FROM clause after applying other clauses. If more than one row is returned, only the oldest one in current window is used to evaluate the CHECK clause. Semantic window can be expressed using current SQL statements (i.e., SELECT-FROM-WHERE statement) over *SWID* with little effort needed for its implementation. We can also take advantage of the well-developed SQL query processing engine and its optimization techniques[4] provided in stream processing model to compute semantic windows with little effort to modify current data stream systems.

- **Stream Modifiers:** A family of operators that can be used to compute the changes of one or more attributes over a window, thus reducing the number of tuples before they are sent to the event generator. It is defined as a function to compute the changes (i.e., relative change of an attribute) between two consecutive tuples of its input stream. It is denoted by

$$M(<t_1, t_2, ..., t_i> [, \text{P} <pseudo>][, O|N <v_1, v_2, ..., v_j>])$$

where M is the modifier function that computes a particular kind of change. The i-tuple $<t_1, t_2, ..., t_i>$ is the parameter required by the modifier function *M*. The following P $<pseudo>$ defines a pseudo value for the M function in order to prevent underflow. The following j-tuple element

is called the untouched attribute that needs to be output without any change. The $O|N$ part is called modifier profile, which determines whether the oldest values or the latest values of the j-tuple that needs to be output. If O is specified, the oldest values are output or the latest values are output if N is specified. Both untouched attributes and modifier profile are optional. A family of stream modifiers specific to applications can be defined using the above definitions. Output stream of a CQ is given as input to stream modifiers *SM1* and *SM2*. Once the change computations are performed, stream modifiers give their output as inputs to the event generator. Some of the stream modifiers are; ADiff() to detect absolute changes over two consecutive states, RDiff() to detect relative changes over two consecutive states, and ASlope() to compute the slope ratio of two attributes over two consecutive states. Below we explain RDiff() which is used to detect the relative changes over two consecutive states. It returns relative change of the values of attribute s_1, s_2, and the values of a subset of attributes given in $O|N <>$ profile. It is formally defined for case N as follows:

$$RDiff(<s_1>, \text{P} <pseudo> [, \text{N} <v_1, v_2, ..., v_j>]) =$$
$$< \frac{s_1^{i+1} - s_1^i + pseudo}{s_1^i + pseudo} [, v_1^{i+1}, v_2^{i+1}, ..., v_j^{i+1}] >$$

These enhancements not only greatly improve the ability of stream processing model to compute more complicated computation requirements through semantic windows and named CQs, but also improve the ability to compute more accurate final results in a more efficient way through stream modifiers and semantic windows. However, none of the enhancements affect the operator semantics, scheduling algorithms, QoS delivery mechanisms, and other components proposed for stream data processing.

Event Processing

Below we discuss two limitations of current event processing systems. Event detection graphs (or EDGs) in the current event processing systems do not have input queues/buffer for event operators as the input rate of an event stream is not assumed to be very high and highly bursty. Thus, in our integrated model, input queues/buffers are added to event operator nodes (shown in Figure 2) to handle the highly bursty input generated by the CQs from the CQ processing stage.

In a traditional event processing system, primitive events can be either class or instance level, but both of them are based on *timestamps*. Instance level events play an important role for events generated by stream processing, but with the dynamic nature of incoming streams it is difficult or *impossible* to determine the instance level events ahead of time. The example discussed below highlights the limitations of the current event operators that operate solely on timestamp.

Consider the CAR ADN (from Section 2) example. Event *Eimm* represents IMMOBILITY and event *Edec* represents SPEED REDUCTION. Event *Eacc* represents the accident and is detected when an event *Eimm* happens before event *Edec*. In addition, *Eacc* is detected only when events *Eimm* and *Edec* are generated by cars from the SAME SEGMENT.

```
CQ1 <timestamp, car_id, speed, exp_way,
lane, dir, seg_id> CQ2 <timestamp, car_
id, speed, exp_way, lane, dir, seg_id,
decrease_in_speed>
```

Stream CarSegStr (Section 2) sends inputs to the named continuous queries CQ1 and CQ2. CQ1 checks the car for IMMOBILITY and CQ2 checks for SPEED REDUCTION. Attributes of both CQ1 and CQ2 are shown above. We define events *Eimm* and *Edec* on CQ1 and CQ2, respectively.

```
Eimm <9.00 am, 1, 0 mph, EW1, 3, NW,
104> Edec <9.03 am, 2, 40 mph, EW1, 1,
NW, 109, 45%> Edec <9.04 am, 5, 20 mph,
EW1, 4, NW, 104, 40%>
```

From the above, *Eimm* occurs at 9.00 am and *Edec* occurs at 9.03 am and 9.04 am. *Eacc* is detected when *Eimm* precedes *Edec* in time. From the above tuples, two accidents are detected; car_id 1 and car_id 2, and car_id 1 and car_id 5. Thus, it is evident that in current event processing systems, the important condition that both the cars should be from the SAME SEGMENT is checked only after the event *Eacc* is detected. This introduces a high overhead on the event computation as there can be many unnecessary detection of event *Eacc* with nature of data stream applications.

The above example can be modeled using instance level events, but all the instances of a class should be predefined (or known previously). This may be impossible in a system where the data streams' attribute values are dynamic. Even if the values are predefined, they require large number of event nodes, which introduce high computation and memory overhead. Hence, event processing needs to be burdened less to support *efficient* detection. We have generalized event expression computation, so that attribute conditions are checked *before* the events are detected. This generalized expression allows both primitive and composite event nodes to detect events based on MASKS or attribute-based constraints. MASKS are pushed to the event generator node with primitive events (i.e., for leaf nodes in EDG) and are pushed into the event operator nodes (i.e., internal nodes in EDG) for other events. For instance, in Figure 2 MASKS corresponding to CQ with J_1 as the root node is pushed to event generator node $G1$. Thus, when multiple events are defined on the same CQ but with different MASKS, all of them are pushed to the corresponding event generator node.

```
CREATE EVENT    Ename
SELECT          A1, A2, …, An
```

```
MASK          Conditions
FROM          ES | EX
```

Users can specify events based on CQs (for primitive events) or on Events using the CREATE EVENT statement shown above.

- CREATE EVENT creates a named event *Ename*
- SELECT selects attributes *A1, A2, ..., An*
- MASK applies conditions on the attributes
- *ES* is a named CQ or a CREATE CQ statement
- *EX* is an event expression that combines more than one event using event operators

Coupling Event and Stream Processing

The local event detector (LED) has a common *notify buffer* (or event processor buffer) into which all events that are raised are queued. A single queue is necessary as events are detected and raised by different components of the system (CQs in this case) and they need to be processed using their time of occurrence. Briefly, a new operator is added to every stream query at the root if an event is associated with that CQ. This operator can take any number of MASKS and for each MASK, a different event tuple/object is created and sent to the notify buffer. This operator is activated only when an ECA rule associated with that CQ is enabled. This operator is similar to the select operator except that when it generates an event, it invokes an API of LED to queue that event in the notify buffer. CQs output data streams in the form of tuples and *event generator* operator nodes are attached to the root node of the CQ. As shown in Figure 2, nodes *J*1 and *J*2 are attached to event generator nodes *G*1 and *G*2. In addition, nodes *G*1 and *G*2 are also associated with MASKS. Thus, stream tuples from *J*1 and *J*2 are converted to events by nodes *G*1 and *G*2. Stream modifiers specific to applications can be defined using the above definitions. Output stream of a CQ is given

as input to stream modifiers *SM*1 and *SM*2. Once the change computations are performed, stream modifiers give their output as inputs to the event generator.

Rule Processing

The rule system is responsible for triggering pre-defined actions. A rule is used to trigger predefined actions once its associated event is detected. In our integrated model rules can be specified and created using the CREATE RULE statement as shown below.

```
CREATE RULE    Rname [, CM, CT, P]
ON             Ename
R CONDITION    Begin; (Simple or Complex
Condition); End;
R ACTION       Begin; (Simple or Complex
Action); End;
```

As shown above, CREATE RULE creates the rule *Rname* along with its properties such as coupling mode *CM* (e.g., immediate, deferred), consumption mode or context *CT* (e.g., recent, continuous) and priority *P* (e.g., 1, 2 where 1 is the highest) a positive integer used to set rule priority. ON specifies the event *Ename* associated with the rule and it can be replaced by the CREATE EVENT statement. In addition a rule also contains conditions associated with the rule and actions to be performed when conditions results are true. Conditions on attributes act as event mask. Other conditions that are pertinent to the rule, and those that are complex (i.e., any arbitrary condition such as average, standard deviations, PL/SQL code etc.,) are specified in the rule condition.

LINEAR ROAD BENCHMARK USING INTEGRATED ARCHITECTURE

In this section we will show how the Car ADN Example discussed in section 2 is modeled using the integrated model discussed above.

CQ for CAR ADN Example

The following IMMOBILE (CQ1) and DE-CREASE (CQ2) queries can be used to find all cars that stay at the same location and the cars whose speed has decreased by 30% within the last 2 minutes using the extensions described so far.

```
CREATE CQ  CQ1 AS
SELECT     RDiff (<speed> as C_speed, p
<0.01>, N<car_id, location, timestamp>)
FROM       CarLocStr [
           CHECK CWtime -NTtime <= 2
           SELECT MIN(CW.timestamp) AS
CWtime, MIN(N T .timestamp) AS NTtime
           FROM CW, N T ROUP BY CW.car_id
]
WHERE      C_speed = 0.0
CREATE CQ  CQ2 AS
SELECT     RDiff(<speed> as C_speed,
p<0.01>, N<car_id, location, timestamp>)
FROM       CarLocStr [
           CHECK CWtime -NTtime <= 2
           SELECT MIN(CW.timestamp) AS
CWtime, MIN(N T .timestamp) AS NTtime
           FROM CW, N T ROUP BY CW.car_id
]
WHERE      C_speed <= -30
```

Both the above defined queries CQ1 and CQ2 have the same semantic window, which is defined based on all old tuples in the window and the current tuple. The CHECK clause is evaluated after the projection list for each car, which is grouped by car_id. For each car, the PROJECT clause outputs CWtime and NTtime, the CHECK clause is used to force the difference between oldest timestamp

of each car in old window and the timestamp of current tuple is less than or equal to 2 minutes. If the difference is larger than 2 minutes, the tuple with the oldest timestamp in old window is deleted. The process is repeated until the oldest timestamp of each car in the window is less than or equal to 2 minutes. In CQ1, WHERE clause selects the cars that are immobile and in CQ2 cars with C_speed reduced by 30 are selected. The main SELECT statement selects any car that currently is in the defined semantic window and computes the speed changes using our previously defined stream modifier RDiff.

Events for CAR ADN Example

Below we show how we model the CAR ADN example (Section 2) using multiple (or nested) CREATE EVENT statements.

In our integrated model, event generator nodes are created on CQ1 and CQ2 so that events can be raised whenever there is an output from CQ1 (i.e., immobility) or CQ2 (speed decrease). Event *Eimm* is created from CQ1. It has two attributes car_id and seg_id denoting the car and segment numbers. Whenever there is a car that is immobile an event is generated by the event generator and propagated to the event processing stage.

```
CREATE EVENT    Eimm
SELECT          CQ1.car_id, CQ1.seg_id
FROM            CQ1
```

Event *Edec* is created from CQ2. This event has two attributes car_id and seg_id denoting the car and segment numbers. This event is detected whenever there is a car that reduces its speed.

```
CREATE EVENT    Edec
SELECT          CQ2.car_id, CQ2.seg_id
FROM            CQ2
```

Event *Eacc* detects the SAME SEGMENT condition mentioned in Section 2. This event determines whether the car that has reduced its speed is in the same segment and follows the car that is immobile. Event expression $_EX$ for the accident is *Eacc = Eimm* SEQUENCE *Edec*, where SEQUENCE is an event operator that is detected when the first event precedes the second event in time. In addition, MASK specifies that cars should be from the SAME SEGMENT and is checked in the SEQUENCE operator node.

```
CREATE EVENT    Eacc
SELECT          Eimm.car_id, Edec.car_id
                Eimm.seg_id, Eimm.time-
stamp
MASK            Eimm.seg_id = Edec.seg_id
FROM            Eimm SEQUENCE Edec
```

If we need to check cars only from a particular segment (seg_id) then a MASK condition can be associated with events *Eimm* and *Edec*.

Rules for Car ADN Example

When an event corresponding to an accident *Eacc* is detected, various types of life saving actions are required to be performed. Creation of event *Eacc* is shown in section 4.2. Rule corresponding to the CAR ADN example is shown below.

Whenever the event *Eacc* is detected it triggers the rule *AccidentNotify* in immediate coupling mode and Recent context. The rule takes four different actions by invoking four different function calls: 1) alert police control room (PCR), 2) alert ambulance control room (ACR), 3) inform upstream cars (UpSSeg), and 4) inform toll station (TollSt).

```
CREATE RULE    AccidentNotify, IMMEDIATE,
               RECENT, 1
ON             EVENT Eacc
R CONDITION    Begin; (true); End;
R ACTION       Begin;
               PCR(Eacc.seg_id, Eacc.Eimm.
car_id, Eacc.Edec.car_id, Eacc.time-
```
```
stamp);
               ACR(Eacc.seg_id, Eacc.
timestamp, Eacc.Eimm.car_id, Eacc.Edec.
car_id);
               UpSSeg(Eacc.seg_id, Eacc.
timestamp);
               TollSt(Eacc.seg_id, Eacc.
timestamp);
               End;
```

PROTOTYPE IMPLEMENTATION

Four stage integration model (also called EStream) discussed in the previous sections, is implemented by integrating the Local Event Detector into the MavStream server. Both the systems are home-grown, implemented in Java and run in the same address space. Integration model implementation consists of an extended Continuous Event Query (CEQ) input processor, an instantiator, a query/operator scheduler, a query processor, an event generator, a rule and event manager, an event detector, a runtime optimizer and a load shedder. The user submits the CEQ event generator (or EG) with each query. The ECA part of CEQ is given to the rule and event manager, which generates the computational model for the events. The rule and event manager then defines rules on the event nodes specified in the CEQ. The query scheduler schedules the query which is executed by the query processor. At runtime the event generator is responsible for raising events which are enqueued in the LED buffer as event objects. The event detector consumes the event objects from the LED buffer and detects the corresponding events. For each detected event, the conditions defined on it are checked and actions are taken if the corresponding conditions evaluate to true. The runtime optimizer monitors the QoS and if the user defined performance metric is not met then it dynamically changes the scheduling strategy associated with the stream computational model.

Input Processor

The input processor accepts the CEQs from the user as an input file. The input file is parsed by Input File Parser which splits the information of CQs and the events and rules. The CQ definitions are given to the query plan generator for generating query plans and the event and rule definitions are given to the event container to be temporarily stored before they are defined. Query plan is an object generated by the system for a CQ given by the user. Once the query plans are instantiated and the event generator operators are attached to each query, the event container can be accessed to define events and rules. In this section, we have explained the implementation of CEQ input file, the query plan generator and the event container data structure.

Continuous Event Query (CEQ)

The CEQ provides the capability of defining events, rules and continuous queries. CEQ supports CQs to be defined as future queries with start and end times. CEQ specifications are such that any definition can be given alone as well as together. To provide this capability the definitions for rules and events should contain enough information to uniquely specify the CQ on which events and rules should be created. It also provides the capability to modify the masks associated with events and deleting rules associated with events. The specification of CQs, events, masks and rules are provided to the system through an input file. The user can give his input in the form of an ASCII file which will be parsed and analyzed by EStream server. Eventually this input file will be generated by a Graphical User Interface. Following is the list of headers which can be defined in the input file. Each header has a format in which the input should be defined. The input file can have multiple definitions for the same header in a single file.

- **CQ definition:** The CQ's are defined by giving the stream operators and the information required for instantiating each stream operator. The CQ specification also requires the user to give the association between operators to form a query tree.
- **Event info:** Primitive events are attached to a query name to uniquely identify the query with which the respective events are associated. Each event can be defined with an optional mask. In case a composite event is defined then the query name should not be given and the operator and event names on which the composite event is created have to be defined.
- **Rule info:** Rules for the events have to be defined under the Rule Info header. A rule has a rule name associated with it, a condition and an action. The condition and action are predefined methods which could be associated with the event.
- **Modify mask:** The modify mask option can be used to modify a mask defined on the event continuous query. The mask can be modified by defining the input to uniquely identify the previous mask and giving a new mask condition. The server updates the mask with the new condition defined.
- **Delete rule:** This option allows the user to delete some conditions and actions associated with an event (primitive or composite). The system will drop the rule associated with the event. This becomes useful when there is more than one rule defined on the event and at a later stage one wants to drop a rule.

Query Plan Generator

Query plan generator generates a query plan and gives it to the server. Each operator definition is populated in a data structure called operatorData. The operatorData is wrapped in an OperatorNode that has references to the parent and child opera-tors. The query plan stores the operatorNode for the root operator and is given to the server for instan-tiation. Since the OperatorNode has references, the server can access all the OperatorNodes of the query. The query plan generator is implemented as the QueryPlanGenerator class.

Event Container

The event container temporarily stores the infor-mation regarding events and rules until they can be created, following the creation of the CQ. This is implemented in the ECADefinitionContainer data structure. The ECADefinitionContainer can be described as

- CompositeEventInfo: This HashTable stores the information of composite events defined by the user.
- PrimitiveEventInfo: This HashTable stores the information of primitive events defined by the user. Optimization is done by map-ping the primitive event definitions based on the query name on which the events have to be defined. This design avoids the event generator to be locked multiple times when more than one events are defined on a query which is already executing.
- Rule info: This HashTable stores all the rule definitions given by the user.

Rule and Event Manager

This is responsible for creation of the events and rules. Since both LED and event and rule manager are in the same address space, the APIs of LED can be called for creation of event nodes and associating rules on the event nodes created. It accesses the event container for getting the definitions of the events and rules and creates the event nodes for the ECA part of the CEQ defined by the user. Creation of the events is done in an

order such that primitive events are created first then composite events and finally the rules are defined on them.

Query Processor

The query processor has the implementation of all operators. The event generation for each CQ is done by the event generator operator which is attached as the root operator of each query after it is instantiated. The event generator operator executes in the query processor. Stream tuples are fed to the event generator operator where they are compared against available masks and then converted into event objects. Attributes of the stream tuples are inserted as event attributes and event objects are put into the LED buffer. This is implemented by making extensions to the query processor and the instantiator by implementing the event generator operator and associating it before every query is scheduled.

Event Generator Operator and Masks

The main issues for the event generator operator were dynamic addition of masks and persisting the query to event mapping for evaluating masks using the condition evaluator and generating event. The hashtable for storing the masks and the cases for addition of masks are described below.

- **MaskAndEventHandle HashTable:** Stores the mask condition and eventHan-dle key value pair, which is accessed to ob-tain the reference of the event nodes when a mask is evaluated to be true.
- **Adding of events and Masks:** There may be various cases for the addition of events and masks. It should be noted that in the EDG model, the case where a single mask will generate two events will never oc-cur for the same event generator. This is because the event node can be shared to

create complex events instead of creating another event with the same mask. If new conditions and actions are to be added then a separate rule is defined on the same event node.

○ The event does not have a mask defined: Here the eventHandle is added with a mask condition as "true". The eventHandle associated with the event is added to the MaskAndEventHandle HashTable with a true mask condition.

○ Mask is to be modified: The user can also modify mask at the operator execution time by giving event name and the new mask. This operator will lock itself and access the mask for the corresponding event name and update the HashTables.

Experiments

The experiment below shows the effect of the implementation of masks for a CEQ. The experiment was run on a machine with a single Xeon processor, 2.4GHz, 1GB RAM and Red Hat Linux 8.0 as the operating system. The data set for performance evaluation is a modified version of the dataset used by the Stanford Stream project. The data is stored in our database that is modified

to generate synthetic data stream. This synthetic data stream is fed to this system using the feeder module, where the delays between tuples follow Poisson distribution.

In this experiment, a CEQ is executed with and without masks to observe the performance difference. Time difference at which the rule associated with the query evaluates to true and the time at which the event was generated is measured and is named as the action execution latency. The experiment is run for a datasets of 2000 tuples, 10,000 tuples and 30,000 tuples. The data rate is 100 tuples/sec. The scheduling strategy used is path capacity scheduling (Jiang, & Chakravarthy, 2004b). The buffer is unbounded. The query that is evaluated is

```
CREATE CQ      AUTOMATEDMONITOR AS
      SELECT * from CarLocStr
      WHERE          carId > 100
      CREATE EVENT   "ResidentialSpeed-
ingTicket" on AUTOMATEDMONITOR
      MASK           "true"
      CREATE RULE    "SpeedingTicket"
      CONDITION      Speed > 30
      ACTION         "CalculateTicket-
BasedOnZone"
```

The query named AUTOMATEDMONITOR is to monitor the speed of cars which are public

Figure 3. Events generated with and without masks

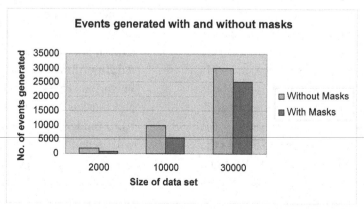

cars with CarId > 100 to be within the speed limit in the Residential area. If the speed of the car increases by 30mph then a ticket event "ResidentialSpeedingTicket" will be generated and the owner will be mailed the ticket.

In Figure 3 and Figure 4, we see that with the application of masks the number of events generated is considerably reduced thus reducing the traffic in the LED buffer. This also prevents the event detector to consume event objects from the LED buffer and drop them at the rule evaluation time. With the application of masks only those events whose rules evaluate to true are generated. This result in a decrease of the average event execution latency for various datasets and can be seen from the Figure 4. For more details on implementation, please refer to (Garg 2005).

RELATED WORK

Our work is closely related to the two threads of work, event processing (Dayal et al., 1988; Schreier et al., 1991; Diaz, Paton, & Gray, 1991; Gehani, Jagadish, & Shmueli, 1992a; Gatziu & Dittrich, 1993; Kotz-Dittrich, 1993; Buchmann et al., 1993; Chakravarthy, Anwar, Maugis, & Mishra, 1994; Hanson, 1996; Lieuwen, Gehani, & Arlein, 1996; Seshadri, Livny, & Ramakrishnan, 1996; Engstrom, Berndtsson, & Lings, 1997; Dinn,

Williams, & Paton, 1997) and stream processing (Babu & Widom, 2001; Abadi et al., 2003; Chen et al., 2000; Madden & Franklin, 2002; Jiang & Chakravarthy, 2004a). In this section we will discuss the systems which have tried to do a similar integration as the four stage integration model. For discussion on the systems developed on these two independent streams of works with respect to the limitations and capabilities required for their integration, please refer to (Garg 2005; Jiang, Adaikkalavan, & Chakravarthy, 2004). To the best of our knowledge there is no such system that does a complete integration hence we will discuss systems which are based on the concept of detecting events over data streams.

HiFi

HiFi generates simple events out of receptor data at its Edges and provides the functionality of complex event processing on these Edges (Cooper et al., 2004; Rizvi et al., 2005). It addresses the issue for generating the simple events by Virtual Devices (Rizvi et al., 2005), which interact with the heterogeneous sensors to produce application level simple events. Then complex event processing can be done on these simple events to correlate into a sophisticated application level event. An application of this system to a library scenario is also described. Although this system is a step in

Figure 4. Average action execution latency

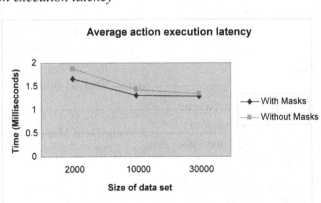

the right direction for the detection of events over sensor data, it does not define and detect events over stream queries. The events detected at Edges are simple events and cannot be defined over the result of the data preprocessed by a Continuous Query (Arasu, Babu, & Widom, 2003). Example of events that could be detected by this system are simple events such as

- "The book with ID 10 is on the shelf"
- "The person with ID 7 is leaving the library"
- "The book with ID 4 is being checked out".

Tiny DB

TinyDB has Event Based Queries (Madden et al., 2002, 2003), which is processing of events over stream queries. They address the need for event processing over stream processing is essential. The aim of implementing event here was to save power of the microprocessors as it is a push based mechanism which saves the process of constantly polling for events. The events are initiated by low-level lying operating system events. Events are interrupt lines, which are raised on the processor or sensor readings going above or below some threshold. Events have to be explicitly defined and then registered with the Query. This research is a step towards providing event capability for stream processing but does not integrate a complete event processing system. It lacks the capability of complex event processing and rule specification.

Financial Applications

Financial applications like Streambase (www. streambase.com), Apama (www.apama.com), GemStone (www.gemstone.com), etc., are systems which are used for mining for patters in data streams and raising events when some financial scenarios are detected. They also provide the ca-

pability of complex event processing over simple events. These systems are developed for a specific application domain and do not allow the user to evaluate stream queries on sensor data using the database operators. It provides a dashboard to the user to define financial scenarios to be mined in the input streams. It detects patterns and process events but does not span detection of events over events detected for other queries.

CONCLUSION

Synergistic integration of event processing and stream processing cater a large class of advanced applications. In this chapter we explained the need for the integration using advance applications. We then discussed the integrated architecture. Finally, we showed how the integrated architecture supports the linear road benchmark application.

REFERENCES

Abadi, D. (2003, Aug.). Aurora: A New Model and Architecture for Data Stream Management. *The VLDB Journal, 12*(2). doi:10.1007/s00778-003-0095-z

Adaikkalavan, R., & Chakravarthy, S. (2003, Sep.). SnoopIB: Interval-Based Event Specification and Detection for Active Databases. In *Proc. of ADBIS*.

Adaikkalavan, R., & Chakravarthy, S. (2006, Oct.). SnoopIB: Interval-Based Event Specification and Detection for Active Databases. *Data & Knowledge Engineering, 59*(1), 139–165. doi:10.1016/j. datak.2005.07.009

Arasu, A., et al. (2004, Sep.). Linear Road: A Stream Data Management Benchmark. In *Proc. of VLDB*.

Arasu, A., Babu, S., & Widom, J. (2003, Oct.). *The CQL Continuous Query Language: Semantic Foundations and Query Execution* (Tech. Rep. Nos. 2003-67, Stanford University).

Babcock, B., Datar, M., & Motwani, R. (2004, Mar.). Load Shedding for Aggregation Queries over Data Streams. In *Proc. of ICDE*.

Babu, S., & Widom, J. (2001, Sep.). Continuous Queries over Data Streams. *SIGMOD Record, 30*(3), 109–120. doi:10.1145/603867.603884

Baras, J., Li, H., & Mykoniatis, G. (1998, Apr.). *Integrated, Distributed Fault Management for Communication Networks* (Tech. Rep. Nos. CS-TR 98-10, University of Maryland).

Bjerring, L. H., Lewis, D., & Thorarensen, I. (1996). Inter-Domain Service Management of Broadband Virtual Private Networks. *Journal of Network and Systems Management, 4*(4), 355–373. doi:10.1007/BF02283160

Brian, B., et al. (2003). Chain: Operator Scheduling for Memory Minimization in Stream Systems. In *Proc. of ACM SIGMOD*.

Buchmann, A. P., et al. (1993). Rules in an Open System: The REACH Rule System. In Proc. of *Rules in Database Systems Workshop*.

Carney, D., et al. (2003, Sep.). Operator Scheduling in a Data Stream Manager. In *Proc. of VLDB*.

Chakravarthy, S., Anwar, E., Maugis, L., & Mishra, D. (1994). Design of Sentinel: An Object-Oriented DBMS with Event-Based Rules. *Information and Software Technology, 36*(9), 559–568. doi:10.1016/0950-5849(94)90101-5

Chakravarthy, S., & Mishra, D. (1994). Snoop: An Expressive Event Specification Language for Active Databases. *Data & Knowledge Engineering, 14*(10), 1–26. doi:10.1016/0169-023X(94)90006-X

Chen, J., et al. (2000). NiagaraCQ: A Scalable Continuous Query System for Internet Databases. In *Proc. of SIGMOD*.

Cooper, O., et al. (2004). HiFi: A Unified Architecture for High Fan-in Systems. In *Proc. of VLDB* (p. 1357 - 1360).

Das, A., Gehrke, J., & Riedewald, M. (2003). Approximate Join Processing over Data Streams. In *Proc. of SIGMOD*.

Dayal, U. (1988, Mar.). The HiPAC Project: Combining Active Databases and Timing Constraints. *SIGMOD Record, 17*(1), 51–70. doi:10.1145/44203.44208

Diaz, O., Paton, N., & Gray, P. (1991, Sep.). Rule Management in Object-Oriented Databases: A Unified Approach. In *Proc. of VLDB*.

Diaz-Caldera, R., Serrat-Fernandez, J., Berdekas, K., & Karayannis, F. (1999). An Approach to the Cooperative Management of Multitechnology Networks. *Communications Magazine, IEEE, 37*(5), 119–125. doi:10.1109/35.762867

Dinn, A., Williams, M. H., & Paton, N. W. (1997). ROCK & ROLL: A Deductive Object-Oriented Database with Active and Spatial Extensions. In *Proc. of ICDE*.

Engstrom, H., Berndtsson, M., & Lings, B. (1997). *ACOOD essentials* (Tech. Rep. HS-IDA-TR-97-010). University of Skovde.

Frohlich, P., & Nejdl, W. (1997, Jan.). Model-based Alarm Correlation in Cellular Phone Networks. In *Proc. of the international symposium on modeling, analysis, and simulation of computer and telecommunications systems (MASCOTS)*.

Gambhir, D., Post, M., & Frisch, I. (1994). A Framework for Adding Real-Time Distributed Software Fault Detection and Isolation to SNMP-based Systems Management. *Journal of Network and Systems Management, 2*(3). doi:10.1007/BF02139365

Garg, V. (2005, Dec.). *EStream: An Integration of Event and Stream Processing* (Master's Thesis, ITLab, The University of Texas at Arlington).

Gatziu, S., & Dittrich, K. R. (1992, Dec.). SAMOS: An Active, Object-Oriented Database System. *IEEE Quarterly Bulletin on Data Engineering*, *15*(1-4), 23–26.

Gatziu, S., & Dittrich, K. R. (1993, Sep.). Events in an Object-Oriented Database System. In *Proc. of Rules in database systems*.

Gatziu, S., & Dittrich, K. R. (1994, Feb.). Detecting Composite Events in Active Databases using Petri Nets. In *Proc. of workshop on research issues in data engineering*.

Gehani, N. H., & Jagadish, H. V. (1991). Ode as an Active Database: Constraints and Triggers. In *Proc of VLDB*.

Gehani, N. H., Jagadish, H. V., & Shmueli, O. (1992a, Dec.). *COMPOSE: A System For Composite Event Specification and Detection* (Tech. Rep.). AT&T Bell Laboratories.

Gehani, N. H., Jagadish, H. V., & Shmueli, O. (1992b). Composite Event Specification in Active Databases: Model & Implementation. In *Proc. of VLDB* (p. 327-338).

Gehani, N. H., Jagadish, H. V., & Shmueli, O. (1992c, June). Event Specification in an Object-Oriented Database. In *Proc. of SIGMOD* (p. 81-90). San Diego, CA.

Hanson, E. N. (1996). The Design and Implementation of the Ariel Active Database Rule System. *IEEE TKDE, 8*(1).

Jiang, Q., Adaikkalavan, R., & Chakravarthy, S. (2004). *Towards an Integrated Model for Event and Stream Processing* (Tech. Rep. Nos. CSE-2004-10, ITLab, The University of Texas at Arlington).

Jiang, Q., Adaikkalavan, R., & Chakravarthy, S. (2005, Apr.). NFM[i]: An Inter-domain Network Fault Management System. In *Proc. of ICDE* (pp. 1036-1047). Tokyo, Japan.

Jiang, Q., Adaikkalavan, R., & Chakravarthy, S. (2007, Jul.). MavEStream: Synergistic Integration of Stream and Event Processing. In *IEEE International Workshop on Data Stream Processing*.

Jiang, Q., & Chakravarthy, S. (2004a, Mar.). Data Stream Management System for MavHome. In *Proc. of ACM SAC*.

Jiang, Q., & Chakravarthy, S. (2004b, Jul.). Scheduling Strategies for Processing Continuous Queries over Streams. In *Proc. of BNCOD*.

Kotz-Dittrich, A. (1993, Mar.). Adding Active Functionality to an Object-Oriented Database System - A Layered Approach. In *Proc. of the Conference on Database Systems in Office, Technique and Science*.

Lieuwen, D. L., Gehani, N. H., & Arlein, R. (1996, Mar.). The Ode Active Database: Trigger Semantics and Implementation. In *Proc of ICDE* (pp. 412-420).

Madden, S., et al. (2002, Dec.). TAG: A Tiny Aggregation Service for Ad-Hoc Sensor Networks. In *Proc. of OSDI*.

Madden, S., et al. (2003). The Design of an Acquisitional Query Processor for Sensor Networks. In *Proc. of SIGMOD*.

Madden, S., & Franklin, M. J. (2002). Fjording the Stream: An Architecture for Queries over Streaming Sensor Data. In *Proc. of ICDE*.

Medhi, D., et al. (2001, May.). A Network Management Framework for Multi-Layered Network Survivability: An Overview. In *IEEE/IFIP conf. on integrated network management* (pp. 293-296).

Motwani, R., et al. (2003, Jan.). Query Processing, Resource Management, and Approximation in a Data Stream Management System. In *Proc. of CIDR*.

Mountzia, M. A., & Rodosek, G. D. (1999). Using the Concept of Intelligent Agents in Fault Management of Distributed Services. *Journal of Network and Systems Management*, 7(4). doi:10.1023/A:1018739932618

Rizvi, S., et al. (2005). Events on the Edge. In *Proc. of SIGMOD* (pp. 885-887).

Roncancio, C. (1997). Toward Duration-Based, Constrained and Dynamic Event Types. In *Second International Workshop on Active, Real-Time, and Temporal Database Systems* (LNCS 1553, pp. 176-193).

Schreier, U., et al. (1991). Alert: An Architecture for Transforming a Passive DBMS into an Active DBMS. In *Proc. of VLDB*.

Seshadri, P., Livny, M., & Ramakrishnan, R. (1996). The Design and Implementation of a Sequence Database System. In *Proc. of VLDB* (p. 99-110).

Tatbul, N., et al. (2003, Sep.). Load Shedding in a Data Stream Manager. In *Proc. of VLDB*.

ENDNOTES

[1] This work was supported, in part, by NSF grants IIS-0326505, EIA-0216500, and IIS 0534611.

[2] Models that are based on procedures may compute this, but they are more difficult to use than those models that are based on non-procedural languages (i.e., SQL). In this presentation we consider the latter one.

[3] Although the literature from which the example is taken seems to indicate that event processing can be combined into stream processing directly, we believe that it will be a clumsy way to approach this problem and other complex problems as well. Semantically, stream and event processing play different roles and hence need to be brought together in a synergistic way that preserves their individual semantics. Furthermore, specification of events as continuous queries needs to be expressive and flexible.

[4] We do not elaborate on implementation of semantic window and its optimization as SQL-based optimizations are well-known and can be directly used/adapted.

Chapter 15
A Survey of Dynamic Key Management Schemes in Sensor Networks

Biswajit Panja
Morehead State University, USA

Sanjay Kumar Madria
Missouri University of Science and Technology, USA

ABSTRACT

In sensor networks, the large numbers of tiny sensor nodes communicate remotely or locally among themselves to accomplish a wide range of applications. However, such a network poses serious security protocol design challenges due to ad hoc nature of the communication and the presence of constraints such as limited energy, slower processor speed and small memory size. To secure such a wireless network, the efficient key management techniques are important as existing techniques from mobile ad hoc networks assume resource-equipped nodes. There are some recent security protocols that have been proposed for sensor networks and some of them have also been implemented in a real environment. This chapter provides an overview of research in the area of key management for sensor networks mainly focused on using a cluster head based architecture. First we provide a review of the existing security protocols based on private/public key cryptography, Kerberos, Digital signatures and IP security. Next, the authors investigate some of the existing work on key management protocols for sensor networks along with their advantages and disadvantages. Finally, some new approaches for providing key management, cluster head security and dynamic key computations are explored.

INTRODUCTION

Sensor networks are used in complex environments for large-scale, real-time data processing and their foreseeable applications in military and in monitoring critical infrastructures such as bridges,

water resources etc. A sensor network consists of a collection of sensor nodes with computation, communication, and sensing capabilities that spread across a geographical area. They run on low power batteries, and thus their capabilities are limited. The sensor network is an ad hoc wireless network with nodes having limited hardware and software capabilities. For example, Table 1 shows

DOI: 10.4018/978-1-60566-328-9.ch015

Table 1. Hardware of Mica2 sensor nodes

Processor Speed	7.3728Mhz
Program Memory (Flash)	128Kb
Variable Memory (RAM)	4Kb
On chip storage (EPROM)	4Kb
Off chip storage (Flash)	4Mb
Digital IO	48
Analog to digital converters	8 10 bit
Radio Frequency communications rate	38.4kbit/sec
Radio power requirements	16mA transmit 9mA receive
Operating system	TinyOS

the hardware configuration of crossbow mica2 (http://www.xbow.com/Products/Wireless_Sensor_Networks.htm) sensor nodes. In addition, their limited computing power, bandwidth and memory size restrict the types of data processing algorithms to be used, and intermediate results that can be stored on the sensor nodes (Carman, Kruus, & Matt, 2000; Chan, Perrig, & Song, 2003; Du, Deng, Han, Chen, & Varshney, 2004; Eschenauer & Gligor, 2002).

Once deployed, sensor nodes may be moveable. They have sensing capabilities such as seismic, acoustic, magnetic, and infrared (Carman, Kruus, & Matt, 2000) . The architecture of the sensor network depends on the application environment.

Raw data is processed locally, and data fusion is used to aggregate the number of data streams from multiple nodes to conserve the most limiting factor in a sensor node's life expectancy, its battery capacity. Energy conservation is an important issue at node as well as at network level since it is not possible to recharge or change the battery in many applications. Figure 1 shows the architecture of a most common sensor network.

In a sensor network, there is a node designated as the cluster head (Intanagonwiwat, Govindan, & Estrin, 2000; Subramanian & Katz, 2000) responsible for communicating among its group members and with other cluster heads. The other functions of cluster heads are to aggregating data,

Figure 1. Sensor network architecture

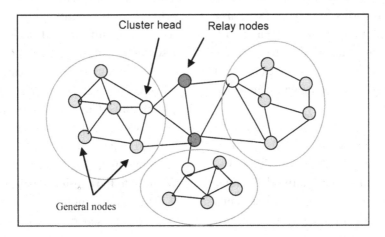

assigning identification to the sensor nodes in its cluster, accommodating newly joined nodes, and taking care of failure nodes etc. Sensor nodes communicate with other clusters through their cluster heads. Relay sensor nodes are responsible for passing information from one cluster to the other. By minimizing resource consumption, sensor networks can be made reliable. It should be self-organizing with both centralized and decentralized control. It must be robust and survivable, and should be able to handle failures. Significant amounts of research have been done to conserve the energy (Intanagonwiwat, Govindan, & Estrin, 2000).

Fault-tolerant, scalable, and reliable sensor networks have become important area of research (Carman, Kruus, & Matt, 2000; Du, Deng, Han, Chen, & Varshney, 2004) due to highly fragile and dynamic nature of the network. It is difficult to provide confidentiality and authentication in sensor networks because of high failure rate, limited energy and frequently changing network behavior. It is also difficult to prevent an adversary sensor node from compromising because of untraceable sensor nodes and less physical protection.

Security is one of the most important issues in sensor networks as it is more vulnerable to attack than a wired network due to use of RF channel which is not secure. In a wireless network, attack can be an active or passive. Active attacks involve some modification of the data stream or creation of a false stream. Passive attacks are like eavesdropping on transmissions. It may be expensive or unrealistic to deploy an intrusion free network. Network security can be handled by using a less expensive method, where the local attack may not be handled but the security of the whole network can be maintained.

The constraints in a sensor network security protocol are (Carman, Kruus, & Matt, 2000; Du, Deng, Han, Chen, & Varshney, 2004; Eschenauer & Gligor, 2002; The Network Simulator - ns-2, n.d.; http://pf.itd.nrl.navy.mil/projects.php?name=nrlsensorsim) battery power/energy, computational energy consumption, communication energy consumption, re-chargeability, sleep patterns, transmission range, memory, tamper protection, and unattended operations. The networking constraints are data rate, channel error rate, latency, limited pre-configuration, unreliable communication, frequent route changes, unknown recipients, population density, etc.

Symmetric cryptography in Sensor Net relies on a shared secret key between two parties to enable secure communication. Asymmetric cryptography, on the other hand, employs two different keys, a secret one and a public one. The public key is used for encryption and can be published. The secret private key is used for decryption. From computational point of view, asymmetric cryptography requires orders of magnitude more resources than symmetric and therefore, some authors argue for using symmetric keys in Sensor Net.

In wireless networks, attacks (Eschenauer & Gligor, 2002; Du, Deng, Han, & Varshney, 2003; Kong, Zerfos, Luo, Lu, & Zhang, 2001) could range from deleting messages, injecting erroneous messages to impersonate a node. One of the reasons is the poor physical protection of the network. The attacks can be not only from outside but also from within the network by a compromised node. An insider node may not forward the message. A node may reply the message to take control of the network. Secure routing (Yi, Naldurg, & Kravets, 2001) is also an issue in a dynamically changing network as false routing information can be generated by compromised nodes. Other problems associated with the wireless communication are trust relationship between neighbors. Different schemes have been developed (Eschenauer & Gligor, 2002; Chan & Rogers Sr., 2004) to protect the network such as using pre-deployed security infrastructure, using independent Security Agents (SA), and installing extra facilities in the network to mitigate the routing misbehavior.

The requirements for developing security protocols in sensor networks are confidentiality, authenticity, integrity, freshness, scalability, availability, self-organization and flexibility. To maintain confidentiality, messages, keys should be protected. For authenticity, the sender and receiver nodes should have keys for encryption/decryption and it should not be disclosed to unauthorized nodes. For integrity, the unauthorized nodes should not be able to participate in computing a shared key. To achieve freshness, the re-computation of keys is necessary. The security protocol must be able to handle large networks. The keys must be available to authorized nodes. The keying protocol should work after the self-organization of the networks.

The key management in sensor networks is to generate and provide keys for secure communication. If we use traditional cryptography for security then every sensor node should have a {private, public} key pair. Computing {private, public} key pair is not energy efficient (Chan, Perrig, & Song, 2003; Du, Deng, Han, Chen, & Varshney, 2004; Eschenauer & Gligor, 2002; Du, Deng, Han, & Varshney, 2003) in sensor networks because of the resource prune nodes and the computationally heavy algorithms used in these applications. Group key management in sensor networks is to provide a security framework for creating the cryptographic groups. It provides mechanisms to decide group security policies, performs access control, and eliminates compromised nodes. In a group key management scheme, a single key is computed for each group. Every sensor node in a group should have the same key for encryption and decryption of messages. For example, if there are four different groups with a different group key then a member of a group say A will not able to decrypt messages in other group say B, C or D unless it has all the group keys. If group members can detect any compromised nodes then they need to eliminate that node from the group by re-computing the group key immediately. There are two approaches (Steiner, Tsudik, & Waidner, 2000; Kim, Perrig, & Tsudik, 2000) for group key distribution; centralized and decentralized.

In the centralized approach, a key is distributed among a group of sensor nodes by a Trusted Third Party (TTP). In the decentralized method, a group member is selected for distributing the key to other members. The centralized approach is simple but more vulnerable to attacks. On the other hand, TTP provides a single point of failure. In the centralized approach to compute a group key, a centralized key server needs to be present for every subset of the groups. It is not practical to implement a TTP in a dynamic environment such as a sensor network. Decentralized key distribution approach is more appropriate for sensor networks.

Our motivation for exploring research issues related to key management techniques come from the following facts:

- Computing the {private, public} key pair in every node is impractical because of high energy consumption and in applications like military the battery can not be replaced or re-charged.
- If TTP is used to share a master secret key with a group of sensor nodes then the security of the whole group will be compromised in case an attacker can forge the key.
- Securing cluster heads is more important than securing general nodes as the cluster head aggregates data or makes decisions on behalf of a group.
- The failure rate of sensor nodes is higher because of less available energy.
- Storing pre-deployed keys is impractical because of memory size constraint as too many keys would be needed by each node to establish a secure communication path.

In this paper, we review some of the existing approaches for key management in sensor networks with respect to their merits and demerits. We have critically compared those with respect to some important parameters. We observed that issues such as confidentiality, authenticity and integrity of nodes in sensor networks are achieved using pre-deployed keys. However, some of these schemes (Zhu, Setia, & Jajodia, 2003; Ye, Luo, Lu, & Zhang, 2004; Malan, Welsh, & Smith, 2004) need to store large number of keys not feasible in sensor network due to the restricted memory size in motes (http://www.xbow.com/Products/Wireless_Sensor_Networks.htm; Karlof, Sastry, & Wagner, 2004). Matt et al. (Carman, Kruus, & Matt, 2000) showed that the total number of pre-deployed keys that are to be generated in groups of sensor nodes of different sizes increases exponentially with the network size. On the other hand, too few keys increase the possibility of insecure network as nodes may not be able to communicate to other authentic nodes. Thus, the key size and dynamic generation of keys will play an important role in securing communication in sensor networks. Towards this goal, we have also explored some new ideas for dynamic key management and tentatively outlined some of the schemes for the readers benefit.

Note that sensor data mining is an important area of research; however, data mining and data warehousing results can be affected by incorrect or compromised data. Therefore, a security mechanism is required in order to collect correct data for the analysis. In addition, privacy preserving data mining seeks is to safeguard sensitive information from unsolicited or unsanctioned disclosure. Most traditional data mining techniques analyze and model the dataset statistically, while privacy preservation is to protect against disclosure of individual data generated by sensors. The key management approach with respect to privacy preserving data mining assumes that the data is stored at several sensors, who agree to disclose the result of a certain data mining computation performed jointly over their data. The parties engage in a cryptographic protocol exchange messages encrypted to make some operations efficient while others computationally intractable. Classical works in secure multiparty computation show that any function computable in polynomial time is also securely computable in polynomial time by n parties, each holding one argument, under quite broad assumptions regarding how much the parties trust each other. The key cryptographic constructs often used include homomorphic and commutative encryption functions, secure multiparty scalar product and polynomial computation.

The rest of this review paper is organized as follows. Section 2 is the review for cryptography. Section 3 is the detailed review of work done for key management in sensor networks. Section 4 proposes requirement of group key management and different approaches for key management. Section 5 concludes the paper.

OVERVIEW OF CRYPTOGRAPHY

A security attack is an action that compromises the data integrity (Stallings, n.d.). Different security mechanisms have been developed to detect or prevent security attacks. There are different types of security attacks; interruption, interception, modification, and fabrication. Interruption occurs when somebody blocks the flow of information. By intercepting, an attacker listens to the messages which are modified before reaching the destination. In fabrication an unauthentic party inserts spurious messages. The attacks can be further divided into two categories: they are passive and active attacks. The examples of passive attacks could be eavesdropping or monitoring and traffic analysis. The examples of active attacks are modification of messages, denial of service, replay etc. The main security goals are confidentiality, integrity and authenticity. Confidentiality is for protecting data; authenticity is to make sure that the source and destination are authentic. Integrity is applied to the stream of messages to make sure that the data has not been altered. The other available security services are nonrepudiation and access control. Nonrepudiation guarantees that senders and receivers are authentic. Access control limits the access to the system. The methods for protection are encryption, software controls, hardware controls, different policies and physical controls. Example of software controls could be like giving different types of access to a database such as based on their roles. Hardware controls can be achieved by using a smart card. Policies like frequent change of password can be used to protect systems.

Conventional encryption uses one key which is also called a symmetric key. There are five components there in this scheme; plaintext, encryption algorithm, secret key, cipher text and decryption algorithm. The plain text is converted to cipher text using the encryption algorithm and the secret key. The decryption algorithm takes the cipher text and secret key to get the plaintext. This scheme depends on the secrecy of the key but not on the algorithm. The commonly used algorithms are feistel cipher structure, DES (Data encryption standard), DEA (Data encryption algorithm), TDEA (Triple DEA), Blowfish, etc.

Message Authentication Code (MAC) is used to verify the source of the message. It requires that the destination should be able to verify that the message came from the expected system, the message has not been altered in the transition and it was not delayed in the network. This helps to guarantee that an unauthentic user has not altered the message. The conventional approach for the message authentication is sharing the same secret key between the sender and the receiver. Message authentication can be done using a tag which does not use encryption. The tag is generated on the sender side and attached to the message. Once the receiver receives the message he can verify the tag. If somebody modifies the message during transition then the tag changes and the receiver will know that. $MACM = F(KAB, M)$ where KAB is the secret key shared between source A and destination B and M is the message. One-way hash functions can be used to generate the MAC. The shared key is used to encrypt/decrypt the MAC. It can be sent without encryption, in that case the MAC is created using a shared key between sender

and the receiver $MDM = H(SAB \| M)$. MAC can also be generated using SHA-1 secure hash function, MD5, RIPEMD-160 and HMAC.

In public key cryptosystem, two keys are used for authentication. There are three different types of application, encryption/decryption, digital signatures and key exchange. In encryption/decryption the sender encrypts the message using a public key and the receiver decrypts it using a private key. In a digital signature the sender sends the message with a signature which is generated using the private key of the sender. Two parties cooperate to generate a common key in the key exchange method. Two widely used algorithms for public key cryptosystem are RSA and Diffie-Hellman. In the RSA algorithm the cipher text is $C = Me \bmod n$, and message is $M = Cd \bmod n = Med \bmod n$. Both the sender and the receiver know n and e but only the receiver knows d. The public key in this algorithm is $KU = \{e, n\}$ and the private key is $KR = \{d, n\}$.

The digital signature is created using the private key of the sender and the receiver can verify it using the public key of the sender. As the private key is only known to the sender other parties can not generate the same digital signature. This helps the receiver to verify whether the message is coming from the expected sender. A trusted third party is used for providing digital certificate. It consists of the public key and ID of an authentic user which is signed by certification authority (CA). One widely used certificate is X.509.

Kerberos is an application level authentication protocol. It handles three types of threats, they are: a user may say that he/she is a different user, alters the network address of a work station and the replay attack. It has a centralized server which handles the entire authentication. The terms are: C = Client, AS = authentication server, V = server, IDc = identifier of user on C, IDv = identifier of V, Pc = password of user on C, ADc = network address of C, Kv = secret encryption key shared by AS an V, TS = timestamp, $\|$ = concatenation. An authentication dialog could be $C \rightarrow AS$: IDc

|| Pc || IDv, AS → C:Ticket, C→ V:IDc || Ticket, Ticket = EKv[IDc || Pc || IDv]. Version 4 has the following steps:

Obtaining ticket:

(1) C → AS:IDc || IDtgs ||TS1
(2) AS → C:EKc [Kc,tgs|| IDtgs || TS2 || Lifetime2 || Tickettgs]

Obtaining service:

(3) C → TGS:IDv ||Tickettgs ||Authenticatorc
(4) TGS → C:EKc [Kc,v|| IDv Authentication|| TS4 || Ticketv]

Authentication:

(5) C → V:Ticketv || Authenticatorc
(6) V → C:EKc,v[TS5 +1]

Another application level protocol is X.509 which is used to provide certificate to the users. It is created by signing the public key of the users by a certification authority. The certificate is revoked if a user is compromised. Pretty good privacy (PGP) and S/MIME are used for electronic mail security. PGP has five operational steps. They are: authentication, confidentiality, compression, email compatibility and segmentation. SHA-1 and RSA are used for authentication, IDEA or TDEA are used for confidentiality, ZIP is used for compression and radix-64 conversion is used for compatibility.

IP security is used to create secure access of different system or network through Internet. This is implemented below the transport layer protocol and is transparent to users. It guarantees that the routing is authentic. Security association (SA) is used to provide authentication and confidentiality. SA relates the sender and the receiver by using the security parameter index, the IP destination address and the security protocol identifier.

PRE-DISTRIBUTION AND GROUP KEY MANAGEMENT SCHEMES

This section provides the brief overview of the work done on security and key management in sensor networks. Below we describe some of the work in this direction and also provide some advantages and disadvantages.

Perrig et al. (Perrig, Szewczyk, Wen, Culler, & Tygar, 2001) developed a security protocol called SPIN (security protocols for sensor networks). Two concepts have been presented in this protocol. They are μTESLA (micro version of timed, efficient, streaming, loss-tolerant authentication protocol) and SNEP (secure network encryption protocol). They setup a network in which the sensor nodes communicate with a base station using RF channel. The basic properties of security such as confidentiality, data freshness and data authentication are maintained in this protocol. SNEP provides two party data authentication, data confidentiality, freshness and integrity whereas μTESLA takes care of authentication for broadcasted data.

The advantages of SNEP are: (1) It has low communication overhead. (2) It uses a counter. (3) It even has semantic security. (4) DES-CBC chaining algorithm is used to maintain the data confidentiality. MAC is used to keep message unaltered; and the counter is used to maintain the sequence of messages. The encrypted data:

$$E = \{D\}(kencr, c), M = MAC(kmac, c|\{D\}(kencr,c))$$

Message from A->B: $\{D\}(kencr, c)$
$MAC(kmac, c|\{D\}(kencr,c))$

where D is the data, C is the counter (initialization vector IV of DES-CBC) and kencr is the key. The counter value will never be the same as it is incremented after each message. So the encrypted message is different for the same data.

μTESLA works with delayed disclosure of symmetric keys as shown in Figure 2. The sender

Figure 2. Key association with packet in SPIN

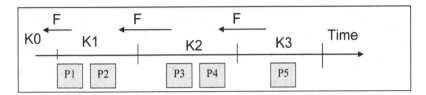

chooses a key Kn from a key chain and uses a one-way function to compute the other keys based on the formula Ki = F(Ki+1). It divides the broadcast time interval and associates each key with an interval. A MAC is computed using the key and the packet. The receiver node stores the message with MAC in buffer. The key is sent from the sender after a certain delay. The receiver knows the schedule for disclosing the key. In this method if some keys are lost they can still be recovered by applying the function multiple times. For example K0 can be computed from K2 by applying the function twice. K0=(F(F(K2)).

Advantages of SPIN: SPIN uses fewer numbers of keys (occupies less memory) for security. It uses symmetric key and does not use private/public key cryptography. The computation of symmetric key consumes less energy. After choosing one key it can compute the other keys based on a function.

Disadvantages of SPIN: If an intruder can know the disclosing time of the key from the sender then it can compute other keys. It is based on one-to-one communication specifically from the base station to the general sensor nodes. However, the scheme may need more resources if it is implemented among sensor only. Failure of the nodes has not been considered in their approach.

Eschenauer et al. (Eschenauer & Gligor, 2002) presented a key management scheme which has selective distribution and revocation of keys in sensor nodes. Their scheme is based on the probabilistic distribution of the key in which they guarantee that two neighboring nodes will have at least one common key in their key ring. This key is used by the neighboring nodes to encrypt/decrypt messages. The three steps followed in this method are: key pre-distribution, shared-key discovery, and path-key establishment. In key pre-distribution phase a large pool of keys (217 – 220) and their key identifiers are generated. Then k keys out of P are chosen to generate a key ring and given to the sensor nodes.

In shared-key discovery phase the sensor nodes find their neighbors who share at least one common key from their key ring. The nodes broadcast the identifiers of the keys within their communication range. The neighboring nodes if finds a common identifier creates a link with the nodes. Thus, the nodes create paths by using the common key with the neighbors. The figure 3 shows how the links are created among the neighbors. It also shows that if the source node wants to send some message to the destination node then it encrypts it with K1 then node A decrypts it with K1 and encrypts it with K9 and so on. The path is source-A-B-C-D-E-Destination.

Advantages: The neighboring node can set up a secure path if they share a key. It needs very few keys to find a secure path. In the paper, the authors proved that 75 keys are needed from 10,000 keys in order to have sharing key probability of 0.5.

Disadvantages: The disadvantage of this technique is that it uses pre-deployed keys which are comparatively easy to forge than a dynamic key. To make the probability higher than 0.75, it needs to have many more common keys with the neighbors. It is not possible to store many keys as the sensor nodes have very less memory. To have the highest probability of key sharing all the nodes needs to have a master key. If an attacker

Figure 3. Discovering neighbor using common key

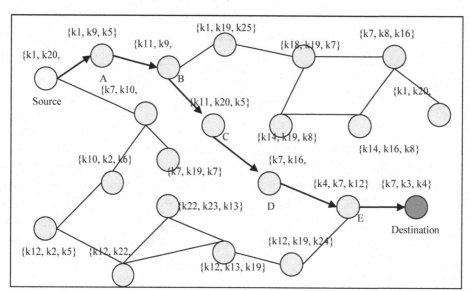

can forge the master key then the security of the whole network is compromised. To do the key freshness no new method has been proposed.

Liu and Ning (2003) proposed a special scheme for sensor networks where a pool of randomly generated bivariate polynomials is used to establish pairwise keys between sensors. So the nodes select different polynomials from the polynomial pool, in which each polynomial with different ID. In the direct key establishment phase, sensor nodes exchange the ID of polynomials to find shared polynomials, and so establish the pairwise key. Compared with previous methods, this scheme is more secure and sensors can be added dynamically without having to contact the previously deployed sensors. A grid-based scheme is also proposed in (Liu & Ning, 2003), where the polynomials are arranged in a grid. The set-up server assigns each sensor in the network to a unique intersection in the grid. This grid-based method is described as low communication overhead, and is more intrusion tolerant than previous schemes.

The location information can be used to improve the polynomial-based key pre-distribution schemes, as proposed in (Liu & Ning, 2005), called the closest polynomial scheme. This scheme combines the expected locations of sensor nodes with the random subset assignment scheme in (Liu & Ning, 2003), and allows tradeoff between the security against node captures and the probability of establishing direct keys with a given memory constraint. In all the polynomial-based key pre-distribution schemes, the essential computation in sensor nodes lies in the evaluation of a t-degree polynomial.

Zhu et al. (Zhu, Setia, & Jajodia, 2003) developed a key management protocol called LEAP (security mechanism for large scale distributed sensor networks). It has four types of keys: individual key, group key, cluster key and pairwise shared key. The individual key is shared with the base station. The base station can perform secure communication with the nodes using this key. The group key is used by the whole group and the base station. It is used for broadcasting. The cluster key is shared by two nodes from different groups. It is used to have secure individual communication with other group's nodes. The pairwise key is shared between immediate neighbors. The individual key Kum is generated in node u using a random function and master key Ksm, where Kum = f Ksm(u). Each node is uniquely identified

Table 2. Kerberos protocol computational energy consumption

Processor	Node A	Node B	KDC
MIPS R4000	0.0081	0.0075	0.0052
SA-1110 "StrongARM"	0.015	0.014	0.0097
Z-180	1.7	1.6	1.1
MC68328 "DragonBall"	0.091	0.098	0.072

in this scheme. A pairwise key Kuv is computed in u with the help of key of v Kv and a function Kuv = f Kv (u). For establishing cluster keys, the node u generates random key Kuc and encrypts it with a pairwise key for neighbors v1, v2, ...,vm. So u -> vi: (Kuc) kuvi . A pre-deployed group key is used for every group.

Advantages: Each node has different types of keys to communicate with a neighbor or cluster head or with a node from another group which makes the network secure. No new key is needed after the deployment.

Disadvantages: Each sensor node needs to store many keys which is difficult due to limited memory constraint. Thus, the communication overhead is high for the key establishment phases. If the keys are lost then no method was given to get a new pairwise key. The master key is the bottleneck as all other keys are derived from it. It also provides digital certificate which is computationally expensive.

Matt et al. in NAI LAB report (Carman, Kruus, & Matt, 2000) explored a different security and key management protocol. They show the difference of power consumption with different processor and security protocols. Table 2 (Carman, Kruus, & Matt, 2000) shows the Kerberos protocol's computational energy consumption, and Table 3 shows sensor node communication energy consumption for different hops in the sensor network for Kerberos protocol.

NAI Labs (Carman, Kruus, & Matt, 2000) has developed a novel key management protocol which is designed for the distributed sensor network environment. They have developed Identity-Based Symmetric Keying and the Rich Uncle protocol. Their main goal was to overcome energy constraints in sensor networks. Their research focused on energy consumption in key management, evaluating protocols based on energy consumption in system, at average and individual sensor nodes.

Advantages: Different security protocols (Kerberos, Otway-Rees, Group keying, etc) have been implemented using the actual sensor nodes. Energy consumption has been evaluated for different protocol with respect to different processors and hops in the network.

Disadvantages: The report does not explain the organization of the network which is the basic for proposing a security protocol. Since the re-keying is not done and therefore, if it follows the same steps for re-keying as for the first time keying then it is not computationally efficient.

Table 3. Sensor node communication energy consumption

Number of Hops	Communication Energy Consumption (mJ)
1	52
2	88
4	130
8	220

Du et al. (Du, Deng, Han, & Varshney, 2003) model a scheme which uses node deployment knowledge to provide key management in sensor networks. It is assumed that sensor nodes are deployed in groups. If N is the number of deployed sensor nodes, they can be divided it into T x N groups. They specify an index (i, j) and points (xi, yj) associated with the group Gi,j .

Advantages: The neighboring nodes can share a key from a small key chain as the nodes know the probable neighbors at the time of key pre-distribution. It avoids having a master key.

Disadvantages: The simultaneous dropped nodes from a helicopter may not be neighbors after the self-organization process, and therefore, the assumption that nodes know their probable neighbors may not be valid. Also, it is not explained how to refresh the key.

Kong et al. (Kong, Zerfos, Luo, Lu, & Zhang, 2001) explained the security of mobile ad-hoc networks in which the secret share of the key is distributed among the hosts so that no certification authority is needed to implement the robustness. The proposed scheme is based on one hop network with K nodes. Instead of a TTP or CA, K nodes jointly provide the system security. This means a coalition of K neighbors serve as the CA. The number of intruders has to be K in a group in order to get the whole key and attack the system. Secret share updates are used to maintain the freshness of the key. By periodically changing the key, the system achieves more resistance from the intruder or eavesdropper. Certificate based public key infrastructure is used in this paper. Two entities can exchange messages without alteration by the intruder. Certification services help the entities protecting the messages. The basic security requirements confidentiality, data integrity and authentication are maintained here. For certification service the idea of threshold secret sharing (Shamir, 1979) is used. In that scheme no particular entity in the network holds the whole key; instead they hold any part of the key. Using this concept, SK can be divided among the enti-

ties, each entity vi holds secret share pvi. Each entity also has another pair of {sk, pk}.

Advantages/Disadvantages: The above proposed (Kong, Zerfos, Luo, Lu, & Zhang, 2001) approach is mainly for mobile ad hoc networks. Here we analyze to see if the scheme can work efficiently in sensor networks. Threshold secret sharing and secret updates are for a limited number of secret share holders, but a sensor network is large in size. Threshold secret sharing does not handle network partitioning, join/leave operation, or wireless channel error. It cannot be applied to dynamically changing network. In large scale sensor network, if the centralized certification provider is used for issuing certificate and renewal then maintenance become an issue.

After reviewing the above work we observe that two approaches can be applied in large sensor networks. In the centralized approach, there will be a CA for certification of services. If the CA fails then the whole network becomes unprotected as it becomes a single point of failure. Also, if anyone attacks the CA (Amir et al, 2000; Steiner, Tsudik, & Waidner, 2000) then it can get the key for that group. If the network is divided in groups, then it is possible that each group has a different group-key for secure communication. In the hierarchical approach, collaboration of each node in a group can be used to form a CA. Different access points can be used for providing different levels of security. RSA can be used in the base station for providing the key pair {SK, PK} to the CA, where SK is the secret key and PK is the public key. SK is used for signing certificates of the entities in the network and the sign is verified by the public key PK.

Steiner et al. propose (Steiner, Tsudik, & Waidner, 2000) a group key agreement protocol which uses Diffie-Hellman key exchange. A distributed key agreement has been considered rather than a centralized key approach. They proposed a fault tolerant key management protocol and a different group agreement protocol. In centralized group key distribution, one key server generates keys and distributes them in a group. TGDH (Amir et al, 2000; Steiner, Tsudik, & Waidner, 2000; Kim, Perrig, & Tsudik, 2000) provides two interesting features: self clustering and cascaded key management operation. The basic requirements for group key agreement are: key freshness, group key secrecy, forward secrecy, backward secrecy, and key independence. Key freshness guarantees that a key is new. In group key secrecy, it is computationally infeasible for a passive adversary to discover any group key. When a member wants to join a particular group, it broadcasts a join request message. When the current members receive the message they first decide the insertion point. The right shallowest node is the first option for insertion. The tree is updated by adding an extra node or by deleting it. The tree is kept balanced and so is the height.

The implementation of TGDH protocols in sensor networks requires robust hardware. Sensor networks have limitations on bandwidth and computational power of individual nodes. It may be possible that the super nodes or cluster heads can be configured initially with more robust hardware and memory. Though EVS (Extended Virtual Synchrony) (Amir et al, 2000) and VS (View Synchrony) (Amir et al, 2000) have good communication semantics, they may not be implemented in sensor networks as they need tight clock synchronization.

For military data access policies (Bell & La Padula, 1973) multi level security is needed. It requires that only authorized personnel will be able to access the sensitive information. There are four types of security level: unclassified, confidential, secret, and top-secret. In this policy there is no read up and no write down, but one can read and write in the same level. There are Information leakage problems using Trojan horse. Using Trojan horse, one can create subject S that has access to some highly secret object O. Then, subject S' does not have access to O, but would like to create a new object O'. Then it grants S

write access to O'. Then, it grants S' read access to O', and then copies O to O'. The Bell & La-Padula (1973) model helps for proper accessing of privileged data with the goal to provide access control for military systems. It is not sufficient to use the sensitivity labels to classify objects. Every object is associated with a set of compartments (e.g. crypto, nuclear, biological, reconnaissance, etc.). People are also classified according to their security clearance for each given compartment.

From the work of Steiner et al. (2000) and Bell & LaPadula (1973) we conclude that Subset-difference based approach can be used to identify the intruder nodes in the hierarchical sensor network. In this approach, the cluster head transmits a message to a larger group. The steps are:

- Scheme Initiation – a method to assign secret information to different nodes.
- The broadcast algorithm – given a message M and a set of R nodes to be revoked, output a cipher text message to broadcast to all.
- A decryption algorithm – a non-revoked node should produce M from cipher text. Decryption should be based on the current message and the secret information.

Rogers et al. (Chan & Rogers Sr., 2004) proposed a model in which the key discovery is done using privacy homomorphism and Chinese remainder theorem. The pre-distribution of the keys are done before deploying the network. Instead of sending the index associated with the key to find the neighbor who is sharing at least one key from the key chain, Modified Rivest's Scheme (MRS) is used and the keys are sent in encrypted format. The main properties of privacy homomorphism theorem are:

(i) **Additive:** If x and y are plain text then the encryption using key k would be Ek(x) and Ek(y). Using additive property Ek(x+y) can be computed from Add(Ek(x), Ek(y)). For computing Ek(x+y) we do not need to know the plain text x and y.

(ii) **Scalar Multiplicative:** Given Ek(x) and constant t we can compute Ek(t.x) from Multi(Ek(x), t).

For finding a common key MRS is used with the following steps:

(i) **Choosing the prime numbers and the keys:** Assume primes p and q and n=pq. Key r1=s1 and r2=s2.

(ii) **Encrypting messages:** Break down m into l parts and apply the encryption function Ep,q,r,s(m) = (m1r1 mod p, m1s1 mod q…..)

(iii) **Decrypting messages:** (x1r1-1 mod p, y1s1-1 mod q…)

(iv) **Encrypted data processing:** Ek(a+b) = (((x1+u1) mod n, (y1+v1) mod n, …)) and Ek(t.a)=((tx1, ty1),….,(txl, tyl)) (mod n)

Advantages: The path discovery for secure communication is done without knowing the actual keys. The distributed key selection is used to have a probability>0.5 for sharing a common key with the neighboring nodes.

Disadvantages: It needs to do much processing for key discovery which is not practical in sensor networks. RSA used similar approach as of MRS for computing public/private key pair which is also computationally expensive.

Sun et al. (Sun & Liu, K.J.R., 2004) proposed a model for providing access control in the form of different groups. For providing data to a particular group they form a distributed network in a tree structure and each subtree can have different keys depending on the application. The users can have a separate private key for inter-subtree communication. For building the key graph they follow three steps: i) associate leafs with their parents. ii) associate the upper level parents. iii) connect all the sub trees.

Advantages: Each level in the network can have different access policies. The proposed scheme is very useful for military application where No Read Up and No Write Down policy is used.

Disadvantages: Different operations (join, merge, leave partition) of groups are not handled. The computational cost will be high for a large number of nodes in a group. It is not clear how to get the optimum number of nodes in a group and the optimum key size.

Malan et al. (Malan, Welsh, & Smith, 2004) proposed a key distribution scheme for sensor networks (EccM) based on elliptic curve cryptography. Elliptic curve cryptography is an approach of public-key cryptography based on the elliptic curves developed by Koblitz et al (Miller, 1986). In EccM each node computes a {private, public} key pair with the help of another node. This scheme has been implemented for Mica2 motes using TinyOs.

Advantage: It uses 163 bit key which can take 5.8×10^{42} years if an intruder tries to decrypt it using exhaustive key search at a decryption rate of $10^{6} decrytion / \mu \sec$. Thus, the scheme is robust against attacks.

Disadvantage: This scheme is not scalable (with respect to memory, communication overhead and energy consumption) with increasing number of nodes in the network which in turn creates many key pairs too.

Ye et al. (Ye, Luo, Lu, & Zhang, 2004) proposed a scheme to prevent injecting false data by the intruders. In this approach multiple sensors generate a report. One of nodes from a group of nodes who are sensing similar data act as a Center of Stimulus (COS). The COS merges the report from its neighboring nodes. The report generated in COS is in the form {LE, t, E}, where LE: the location of the event, t: the time of detection, E: the type of event sent to all detecting nodes. Each node i computes a MAC Mi =(Ki, LE||t||E) and sends back {i, Mi} to the COS. The COS generates the final report {LE, t, E, i1, Mi1, i2, Mi2, …, iT,

MiT}. The report is sent to the receiver node. For a report to be legitimate it should contain T MACs of distinct categories and T key indices of distinct partitions. Each node in a route checks whether T{ij, Mij} tuples exist and whether T key indices {ij, $1 \leq j \leq T$} belong to T distinct partitions. If it has a key $K \in$ {Kij, $1 \leq j \leq T$}, it computes M = (K, LE||t||E). When Mij = M, it means if the MAC is verified then the packet is forwarded else it is dropped otherwise. If intermediate nodes do not have any key in {Kij, $1 \leq j \leq T$} then they forward the packet.

Advantages: False data can be detected in the route before it reaches to the sink node. It saves the bandwidth of the network by attaching index information with each packet, so that the nodes can know if certain packets are for them. The communication overhead in the network is reduced by using bloom filters.

Disadvantages: If the nodes in a route do not share a common key of the report then the packets need to reach to the sink node before it can be verified. The security of the sink node is not considered. The sink node could be the bottleneck of the network as it shares key with every nodes in the network.

Yu and Guan (2005) presented a group-based key pre-distribution scheme using sensor deployment knowledge based on (Yu & Guan, 2005). This scheme distributes secret information instead of secret keys in sensor nodes to generate pairwise keys for nodes. Because most neighbors of a node are from its own group and neighboring groups, this scheme assigns each group a distinct secret matrix and makes neighboring groups share some other secret matrices, so that pairwise keys can be efficiently generated for neighboring nodes. To use the location information, sensor field is divided into hexagonal grids. Accordingly, sensor nodes are divided into groups each of which is deployed into a grid. By using location knowledge, the resulting scheme has a high degree of connectivity and low memory requirement. At the same time it shows a good resilience against node capture.

MiniSec (Luk, Mezzour, Perrig, A., Gligor, V.D, 2007) is a secure communication architecture for wireless sensor networks. It learns from the design of TinySec and ZigBee to create a communication protocol that has both high security and low energy consumption. It was created to be used with the Telos motes and has been fully integrated. They use OCB (Offset CodeBook), Bloom filters, and loose time synchronization. OCB is able to convert a text to a ciphertext of the same length not including the tag. Bloom filter's is a data structure which can quickly test to see if items are part of a set. MiniSec combines the better of two popular security programs and uses a counter to prevent replay attacks. Unlike its predecessors, it only sends the last bits of the counter to prevent the need for counter resynchronization. It only has to resynchronize if several packets are dropped. They chose Skipjack to be there block cipher and a block size of 64. They chose this number largely because OCB needs the nounce to be as big as the block size. They set the tag length to be 32 bits. The energy consumption is affect able through changing the amount of the counter that gets set so that you can find the perfect balance to prevent the need for counter resynchronization. In MiniSec-B it is used to securely communicate via broadcast. To limit the amount of memory needed they take away the individual counter for each node. It is replaced with a sliding window approach to time synchronization. They use two finite periods of time. They use two to lower the chance of dropping good messages because of the latency. However, it also increases the chances of an attacker being able to penetrate the network. To fix this vulnerability they added an internal counter which is reset at the beginning of each time period to save on memory. To implement MiniSec they change TinyOS to meet their needs by rewriting part of the network stack. MiniSec adds three bytes to the standard TinyOS packet. MiniSec-U uses OCB and Skipjack with 80 bit encryption keys for security. It also sends the last bit of the counter. MinSec uses time synchronization and two periods of time for which it will accept packets along with a counter and bloom filters.

MobiWorp (Khalil, Bagchi, Shroff, 2006) is a countermeasure for mobile networks that mitigates wormhole attacks (attacks on networks where nodes lie about locations so that messages are routed through them so that they can drop or read the messages). They have developed several tools including a way of preventing a node from claiming to exist at more than one position, schemes to detect and diagnose wormhole attacks in mobile networks, as well as to isolate malicious nodes. They also analyze the efficiency of their tools. For their work they make several assumptions about the network like nodes know their current and future locations using GPS or location discovery algorithms and beacon nodes. The network is loosely time synchronized. The network contains a Central Authority without resource constraints while each node can identify the CA's public key and are forced to identify themselves with the CA. They use designated nodes to monitor the traffic of its neighbors as a way to detect malicious nodes. These guard nodes watch for irregularities in communication between nodes and add to an internal counter every time it is observed. Once the counter reaches a certain threshold they send out an alert telling nodes in the network that there is a malicious node. The suspected node can only send messages originating from it. To integrate a node in the network using MobiWorp the CA sends a new node its public key so that it may convince the other nodes of its physical location. Each node keeps a list of nodes within two hops of them which they call the neighbor list. The CA keeps a master list which monitors the level of suspicion of malicious nodes in the network. To prevent the problem that might arise if a node is unable to communicate with the CA, they create the Selfish Move Protocol which only allows nodes to send and receive their own traffic when they

are on the move or unable to perform and Node-to-CA Handshake. They also give the ANUM(the public key given to nodes to identify themselves as valid to other nodes in their neighborhood) an expiration so that nodes have to check it with the CA. This expiration is started shortly after the node begins to move. To help the fact that in a network with a large number of moving nodes there would be a large amount of disconnected nodes they introduce a protocol that requires the nodes to tell the CA where they are going and at what speed they will be moving.

Parno et al. (Parno, Luk, Gaustad, & Perrig, 2006) set out to create a new secure routing protocol from the ground up. They begin by explaining their approach by giving network addresses to nodes in the network. They also explain the setup of how information will be routed through the network. They use a recursive algorithm to give nodes unique addresses in the network. They do so by starting with single nodes and then combining them with other nodes into groups and then combining those groups into bigger groups until the entire network is covered. To begin routing they run through a discover process in which they send their id and network authentication out to other nodes. When the other nodes confirm this information they add them to their neighbor list and integrate themselves to form a routing table. They create a process to handle the case when a node dies, and a vital link from one node to reach a certain part of the network is broken so that they can find a new route if one exists to that portion of the network. They send a message to their neighbors asking for a way to reach the area of the network they wish to go. If their neighbors don't know an alternate route then they send the same question to their neighbors, and so on. They offer solutions to the problem of adding new nodes to the network. Either by have some sort of time synchronization and executing the recursive algorithm to group or by having the main station send out a request to run the grouping algorithm. In its simplest form they send information from group to group until it is forwarded to the destination. They also allow the sender to choose a path which it determines as the best route. They allow route it to its destination but keep it away from areas of the network which are suspect for malicious behavior. This also allows it to send information along the shortest path possible. The routing tables are created and the ones which are closest to certain parts of a network in comparison to a certain node would be at the top of the list for the node to send anything to that area of the network. They introduce three ways of keeping the routing secure. The first is a Grouping Verification Tree (GVT). This makes checks against the updated the grouping of the network. They use it to ensure no malicious influence over grouping. To prevent a node from claiming multiple network addresses at the end of the grouping, they make each node announce its network address. They also use a system in which once a node detects a malicious node it sends a network wide message stating its identity and the malicious nodes identity. This results in revoking both nodes from the network.

In Yin & Madria (2008), a secure routing protocol against Wormhole attacks in sensor networks (SeRWA) is proposed. SeRWA protocol avoids using any special hardware such as the directional antenna and the precise synchronized clock to detect a wormhole. Moreover, it provides a real secure route against the wormhole attack. Simulation results show that SeRWA protocol only has very small false positives for wormhole detection during the neighbor discovery process (less than 10%). The average energy usage at each node for SeRWA protocol during the neighbor discovery and route discovery is below 25 mJ, which is much lower than the available energy (15 kJ) at each node. The cost analysis shows that SeRWA protocol only needs small memory usage at each node (below 14 KB if each node has 20 neighbors), which is suitable for the sensor network.

In Panja, Madria, & Bhargava (2008), a scheme called role-based access in sensor networks (RBASH) which provides role-based multilevel

security in sensor networks is given. Each group is organized in such a way that they can have different roles based on the context and thus, can provide or have different levels of accesses. RBASH provides the desired security level based on the application need. The multilevel security is based on assigned keys to different nodes at different levels. To achieve this goal, we organize the network using Hasse diagram then compute the key for each individual node and extend it further to construct the key for a group. Based on experimental observations, we concluded that RBASH is energy and communication efficient in providing security compared to some other protocols which provides uniform security for all the nodes (Table 4).

GROUP KEY MANAGEMENT SCHEMES

The military information system uses a dynamic hierarchical sensor environment in the battle field. There system is combined with multi partner's information systems in a hierarchical structure to achieve effective results. Military data access policies needs multi level security (Bell & La Padula, 1973) which requires that only authorized entities will be able to access the sensitive information. Thus, to control information access, only authorized individuals share the cryptographic keys to decode the disseminated information. To maintain the forward confidentiality, the group key needs to be re-keyed to make sure that departing members will not be able to decode the future messages. There are four types of security levels: first is unclassified, the second is confidential, the third is secret, and the fourth is top-secret. In this policy, there is no read up and no write down, but one can read and write at the same level. Group key management is appropriate for hierarchical military environment as it can implement different access control policies at each level. Each group of sensor nodes will have a different group key.

Table 4. Comparison of key management protocols

	Symmetric key	Security implementation	Base station can be used	Security scheme is implemented in TinyOS	Less key	
TinySec (Karlof, Sastry, & Wagner, 2004)	Symmetric key		Base station can be used	Security scheme is implemented in TinyOS	Less key	New packet format is implemented with respect to initialization vector.
Ye et al. (2004)	Symmetric key	Pre-deployed	Base station not used	False data is filtered by using MAC generated from group of nodes	Too many keys	The number of keys increases with the increase of number of nodes in a group.
EccM (Malan, Welsh, & Smith, 2004)	Public/Private key	Pre-deployed	Base station not used	Elliptic cure cryptography is used to compute the keys	Less key	ECC is used to implement security in TinyOS. It is to show the use of asymmetric key in sensors.
MRS (Roger et al) (2005)	Symmetric/ Asymmetric key	Dynamic key	Base station used	Privacy homomorphism and Modified Rivest's Scheme is used	Less key	The similar method of RSA is used except, path can be discovered without knowing the plain text.
Secret share (kong et al) (2001)	Symmetric/ Asymmetric key	Dynamic key	Base station used	TTP generates public/private key and uses secret share	Less key	Private/Public key and certification are used which are not energy efficient.
Deployment knowledge based (Du, Deng, Han, & Varshney, 2003)	Symmetric key	Pre-deployed	No base station	Key is distributed using same probability	Too many keys	This scheme does not work if nodes do not know the possible neighbors.
Pair-wise key (Eschenauer & Gligor, 2002; Liu & Ning, 2003; Liu & Ning, 2005)	Symmetric key	Pre-deployed	No base station	Key is distributed using probability distribution function	Too many keys	Key chains are implemented which is impractical for memory
TGDH (Steiner, Tsudik, & Waidner, 2000)	Group key	Dynamic key	No base station	Group key computation using keys from the group	One key per group	Used for wired network
NAI Report (Carman, Kruus, & Matt, 2000)	Symmetric/ Asymmetric key	Both pre-deployed and dynamic key	With/without base station for performance analysis	Traditional cryptography algorithms are explained with experimental evaluation	Too many keys	Rich Uncle protocol is developed for sensor networks
LEAP (Zhu, Setia, & Jajodia, 2003)	Symmetric key	Pre-deployed	Base station used	Master key is used to get the other keys	Too many keys	Impractical for sensor nodes
SPIN (Perrig, Szewczyk, Wen, Culler, & Tygar, 2001)	Symmetric key	Pre-deployed	Base station used	Previous key can be computed using a function	Less number of keys	Sensor hardware constrain taken into account

The members of a particular group are assumed to be at the only one level.

It is difficult to manage cryptographic keys for large and dynamically changing sensor networks. When a node leaves or joins a group, the group key must be changed. The group member should be able to compute the key consuming using less computational power. The communication cost inside the group and among the groups should be very low. Re-keying should be fast enough, so that disruptive membership events (join, leave, merge, partition) can be handled.

Here we explore some of our ideas for a group key management scheme for a hierarchical sensor network. In this scheme, every sensor node in a group contributes its partial key for computing the group key. The partial key can be a unique random number. Two entities are important; they are initiator and the leader. The key computation can start by the contribution of partial key from the initiator node. Every node contributes its partial key. One of the sensor nodes will act as a leader. It will compute the final group key using all the partial keys. There is no fixed rule for selecting the group initiator or leader. It is difficult to manage cryptographic keys for large dynamically sensor groups because of scalability. Dividing the sensor network in different groups in hierarchical fashion helps managing cryptographic keys, as each group will be having a unique group key. When a sensor node leaves or joins, the group key must be changed to make sure that newly joined nodes have the key, and departed nodes do not have the key.

We incorporate two important concepts in our approach: 1) Key trees 2) Diffie-Hellman key exchange protocol (Diffie, W., & Hellman, M.E., 1979). Key tree is used for computing and updating group keys and Diffie-Hellman protocol is used for computing the group key. Our approach uses modified Tree Based Group Diffie-Hellman (TGDH) (Amir et al, 2000) protocol for adapting it to the sensor network. We have formulated a new group key management scheme which will work efficiently in sensor networks. The TGDH protocol suite has many useful features and will be able to handle cascaded key management operations. It also provides self-clustering.

Proposed Approaches for Key Management

In this section, we propose different research issues for key management under the following assumption:

1. Each group of sensor nodes has one cluster head and it is resource-rich in terms of it computational ability, storage capacity, and battery power compared to general sensor nodes.
2. A cluster head communicates with cluster heads of other group for sensitive decision-making.
3. It is a self-reconfigurable sensor network.

Securing Cluster Heads

The research challenge is to secure the cluster head against threats such as tampering, and eavesdropping (see Figure 4). Security of the cluster head is important because it is responsible for communicating with other cluster heads on behalf of its group. A notable example to show the importance of the cluster head security is as follows. Assume that there is a group of sensors in region A rooted by cluster heads. The cluster head of the region A wants to communicate with the cluster head of the region B. If both the cluster heads are using private/public key cryptography then they are having {private, public} key pair for authentication of data and secure communication. According to this cryptography technique, if the cluster head of region A wants to send data to the cluster head of region B then it should encrypt data using the public key of cluster head B. A cluster head of region B decrypts the data using its own private key. This encryption/decryption method

Figure 4. Cluster head security

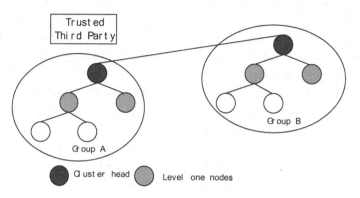

has a problem when applied to sensor network. In this approach if an attacker can take control of the cluster head then it can get the private key and decode the encoded data sent by other cluster heads. We present a technique, in which instead of keeping the private key in the cluster head, we split the key (as partial keys) using a secret function and distribute those among the sensor nodes in the hierarchy. In this model, when a cluster head wants to decode some data which is encrypted by other cluster head, then it must accumulate the partial-keys to rebuild the private key. In this model an attacker will not be able to get the private key just by attacking the cluster head, and attacking multiple sensor nodes would be too tedious. The advantage of this approach in providing secure communication out of a cluster head is that an adversary will not be able to get hold of the function by attacking the cluster head alone. This is because instead of storing the function in the cluster head, sub-functions are generated and distributed among the sensor nodes in each group. After the distribution of the sub-functions, these functions are discarded from each cluster head. When a cluster head wants to communicate with the base station it accumulates sub-functions to compute the function to compute the key used to encrypt/decrypt messages sent and received to and from the base station. For updating the functions, the sub-functions are updated using the information provided by the base station. Note

that a key can also be split by the cluster head and distributed as partial keys among the group members since the cluster head can split a key using Lagrange's interpolation and a function. However, the advantages of splitting the function by the cluster head into different sub-functions over splitting the keys is that the updating the sub-functions consumes less computation and communication overhead compared to updating the sub-keys because the re-generating the keys is done from the scratch and updating the key is done by an updating parameter.

Fault Tolerant Key Recovery using Lagrange's Interpolation

In the previous sub-section, we proposed a model for cluster head security in which we will split the private key of the cluster head and distribute it among the sensor nodes. The cluster head will accumulate the partial keys to get its private key in order to decode encrypted data. The cluster head will not be able to compute the private key if it does not get all the partial keys. However, the partial keys can be lost because of sensor node failures. To rectify this problem, we propose a key recovery technique in which the private key can be computed using a certain number of the partial keys. We are showing two possible steps for private key distribution and recovery.

Figure 5. Intra-cluster key computation

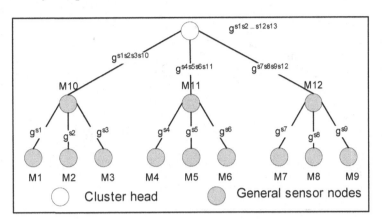

Private Key distribution: The certification authority (trusted third party) will generate a key pair {SK, PK} for each cluster head, where SK is a private key and PK is a public key. Each cluster head will have different key pair {SK, PK}. The private key SK will be split (as partial keys) and distributed among the sensor nodes. The sensor nodes which are having the partial keys can jointly act as certification authority.

Private Key recovery: The Lagrange's interpolation (Kong, Zerfos, Luo, Lu, & Zhang, 2001) can be used to accumulate the key in case of failure of nodes which are having the partial keys. According to Lagrange's interpolation, if the scheme is (k, n) then the private key SK can be divided into n parts. If we can get k number of partial keys from n then we can recover the original private key SK, here k<n.

Hierarchical Key Computation Scheme

In this task, we plan to compute the group key using bottom-up approach. We propose a modification to the multi party Diffie-Hellman protocol to accommodate it for sensor networks. For doing so, some of the group members are chosen as initiator and one member as a leader. Starting from the initiator sensor node every sensor node contributes their partial keys for computing the group key. Partial key can be a random number. The leader node accumulates all the partial keys for the computation of the group key. We will choose the leaf level sensor nodes as initiator and the cluster head as the leader. The group key can be computed with or without using a blind factor.

Group key computation without using blinding factor: For group key computation without using blinding factor we can use the following approach. As the leaf nodes work as initiators, they first broadcast their partial keys. The parent nodes of the leaf nodes get the partial keys and they in turn add their partial keys. As it is a bottom up approach, the nodes above leaf nodes broadcast their partial keys along with the leaf levels' partial keys and so on. Finally, the cluster head will have all the partial keys, and it will calculate the group key using its partial key contribution. After that, the cluster head broadcasts the group key.

A symmetric key will be used to encrypt and decrypt the partial keys. This is to make sure that unauthentic nodes cannot decrypt the partial keys. The identification of the nodes is used with the encrypted partial keys. The sensor nodes can then authenticate nodes before decrypting the partial keys, as the parent nodes need the partial keys of their children and they do not need the partial

Figure 6. Inter-cluster key computation

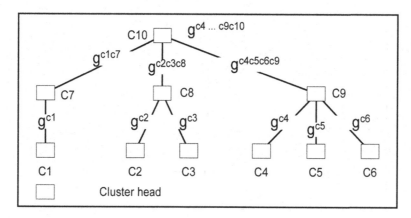

keys from other sensor nodes. The other group members cannot compute the group key because they cannot get the partial key of the cluster head since the cluster head do not broadcasts its partial key.

In Figure 5, the leaf nodes are M1, M2, ..., M9. M1, the initiator sensor M1 computes the partial key gS1 and broadcasts it. The parent node M10 of M1 gets the partial keys from its children. Here, g is a generator of the multiplicative group ZP* (i.e. the set {1, 2... p-1}, p is the prime) and S1 is a randomly chosen secret number for member M1. Likewise, the member M2 computes gS2 and broadcasts it, and the parent M10 gets the partial key. In this way, the member M10 receives g S1S2S3, and raises power by S10 to get the intermediate key (IK). Here, gS10 is the partial key contribution of M10.

We next discuss two types of group keys; the intra-group key and the inter-group key. The intra-group key is used by group of sensor nodes led by a cluster head, and inter-group key is used by a group of cluster heads.

The intermediate keys in M10, M11, and M12 are IK1 = g S1 S2 S3 S10, IK2 = g S4 S5 S6 S11, and IK3 = g S7 S8 S9 S12 respectively. The intermediate keys are encrypted using a symmetric key. All the sensor nodes in a group have the symmetric key. The cluster head calculates

the group key K, using IK1, IK2, and IK3 and its contribution gs13.

K = g S1 S2 S3 S10 S4 S5 S6 S11 S7 S8 S9 S12s13 [Intra- group key]

Next it encrypts the group key using the symmetric key. The authentic nodes, which have the symmetric key, can decrypt the group key. The cluster head broadcasts the group key to its group members so that every sensor node in that group gets the group key. This group key is called intra-group key, and is used for encryption/decryption inside the group of sensor nodes.

For inter-group encryption/decryption, a different group key is computed. This group key is unknown to the general sensor nodes. Figure 6 shows the computation of inter-group key. The intermediate key in C7, C8, C9 are IK1intra = g C1C7, IK2intra = g C2C3C8, IK3intra = g C4C5C6C9

The head of the cluster heads (HCH) computes this group key C using intermediate keys IK1inter, IK2inter, IK3inter, and its contribution gc10

C = g C1C7 C2C3C8 C4C5C6C9c10 [Inter-group key]

The HCH broadcasts the intra-group key to the cluster heads. The cluster heads use this group key for encryption/decryption of messages among themselves.

Group key computation using blinding factor: We can compute the group key using blinding factor . The advantage of using blinding factor is that an attacker will not be able to get the group key when cluster head broadcasts the group key. In Figure 6, we show that the intermediate keys are IK1 = g S1 S2 S3 S10, IK2 = g S4 S5 S6 S11 and IK3 = g S7 S8 S9 S12. After computation of IK1, IK2, IK3 the parent nodes M10, M11, M12 broadcast the intermediate keys. The children of M10, M11, M12 are interested in those keys, as they need to remove their contribution from the IK. Then, they insert randomly chosen blinding factor B. The keys after inserting blinding factor are as follows.

IKB1 = g B1 S2 S3 S10, IKB2 = g S1 B2 S3 S10, and IKB9 = g S7 S8 B9 S12. The cluster head gets the broadcasted keys IKB1,…, IKB9. The cluster head computes the group key K, using IKB1…IKB9 and its contribution gs13.

K = g B1 S2 S3 S10 S4 S5 S6 S11 S7 S8 S9 S12s13

After the group key computation, the cluster head broadcasts the group key with the blind factor. Now the authentic sensor node can recognize its blind factor. Each member unblinds its blinding factor that it received from the cluster head. It reinserts its original contribution Si (i = 1...n) for getting the group key. The same method is used to compute the inter-group key. A symmetric key is used for encryption and decryption of partial keys.

Dynamic Partial Key Computation

In this subsection, we propose a partial key computation technique. These partial keys will be used for the group key computation method shown in the previous sub section. Initially the sensor nodes at non-leaf level do not have their

partial keys. The approach to calculate the partial keys is as follows. The parents of the leaf nodes calculate their partial keys using a function. The arguments of the function are the partial keys of their children. Then, using bottom up approach all non-leaf sensor nodes can calculate their partial keys.

If the tree is binary then the function for calculating the key K is f (K<l+1, 2v> K<l+1, 2v+1>). Where "l" is the level in the tree and "v" is the position of the node from left. For example in Figure 7 if we want to calculate the K<2, 0>, then we need the function f(K<3, 0> K< 3, 1>). The computation of K<2, 1>, is not possible using f(K<3, 2> K< 3, 3>) as children are having different parents. If we use the function f(K<l+1, 2v+1> K<l+1, 2v+2>) then we are able to calculate K<2, 1>. From this we observe that the level does not change, but the position changes depends on the number of children.

In a non-binary tree, for calculating a key in the parent node, it needs to count the siblings in the left part of the sub-tree then it calculates m where m = siblings - 2. So, the function would be f(K<l+1, 2v+m> K<l+1, 2v+1+m>).

Group Key Computation using Elliptic Curve Polynomial

In this subsection, we propose an Energy and Communication Efficient Group key management (ECEG) protocol for sensor networks. We have discussed the EccM scheme (Malan, Welsh, & Smith, 2004) in Section 3, which is based on Elliptic Curve Cryptography (Miller, 1986). However, this scheme is not scalable with respect to memory, communication overhead and energy consumption. ECEG will elect the cluster head dynamically based on the energy left and will reduce the communication overhead by processing key management related operations (computing key chain, broadcasting points of elliptic curve) in the base station. The scheme will compute and update the keys dynamically using polynomials of

Figure 7. Dynamic partial keys computation

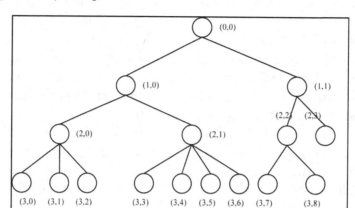

elliptic curve. ECEG will provide keys which can be used for encryption/decryption of messages. To prevent the failure of entire network due to a single compromised node, ECEG will provide the key only to the authentic nodes and updates it before the predefined expiration time. The key will be computed from the key chain and elliptic curve is same for each group hence denoted as a group key. We propose a scheme in which the memory, communication overhead and energy consumption can be reduced by performing some of the steps of the elliptic curve cryptography in the base station. Involving the base station for computing the key chains and later the points of the elliptic curve for computing and updating the key is suitable here as it does not have the energy or memory constraints. We do not propose use symmetric key from the base station because it is not cost effective in terms of communication and bandwidth to broadcast a new key from the base station before the expiration of previous key. Oppose to that, using ECEG nodes can compute a new key from the broadcasted message from the base station. This message contains a new point which can be used by the function and the key chain.

Four types of logical levels of nodes can be considered in this protocol. They are, general sensor node, covering sensor, cluster head and the base station. The covering sensors are a subset of general sensor nodes which covers the region of other sensors. Each group of nodes will have a cluster head responsible for communicating with the base station. For computing and updating the group key it will follow these steps:

1. Pre-distributing the key chain
2. Secure communication between nodes and the base station.
3. Elliptic curve point computation at the base station for computing group keys in the clusters
4. Key updating

CONCLUSION

This paper presents an overview of the key management schemes in sensor network environment to provide security. None of the protocols discussed in the paper maintains all the requirements of sensor network security as explained earlier. For example, SPIN is implemented for one hop network but in real world scenario the sensor network is always multi-hop and self-organized network since the sensor nodes fail often so it may not be possible for some nodes to connect to the base station. LEAP uses a master key to form the individual node's key or the cluster key. A sensor network can be compromised if an intruder can forge the master key. Most of the described schemes use pre-deployed keys, where the number of keys increases with the increase in network size to guarantee that neighboring nodes share a key. This is impractical because of limited memory constraint. Also, it may not be possible to change the key if it has been hard coded. Some of the problems of pre-deployed keys can be handled by using dynamic keys. Dynamic keys can be computed after deployment of the network. The two most important advantages of dynamic keys over pre-deployed keys are, senor nodes need not store too many keys, and the dynamic key may not be compromised by multiple decoding attempts by the intruder because of its frequently changing behavior. Since sensor networks work in groups, providing a group key would be a good idea. Each group can form a trust relationship among nodes. Also, using a group key management method, multi level security can be achieved as different users can have different levels of access in the hierarchy. Thus, there is a need to develop secure and robust key management protocols for the demanding requirements of sensor networks which will help on processing correct data for applications such as data mining in sensor networks.

REFERENCES

Amir, Y., Ateniese, G., Hasse, D., Kim, Y., Nita-Rotaru, C., Schlossnagle, T., Schultz, J., Stanton, J., & Tsudik, G. (2000). Secure group communication in asynchronous networks with failures: Integration and experiments. ICDCS 2000.

Bell, D. E., & La Padula, L. J. (1973). Secure Computer Systems: Mathematical Foundations (Technical Report ESD-TR-73-278). Hanscom AFB, Bedford, Mass.

Carman, D.W., Kruus, P.S., &.Matt, B. J. (2000). Constraints and approaches for distributed sensor network security (NAI Labs Technical Report #00-010).

Chan, A. C. F., & Rogers, E. S., Sr. (2004). Distributed Symmetric Key Management for Mobile Ad hoc Networks. IEEE INFOCOM.

Chan, H., Perrig, A., & Song, D. (2003). Random key predistribution schemes for sensor networks. In IEEE Symposium on Security and Privacy, Berkeley, California, May 11-14 2003.

Deb, B., Bhatnagar, S., & Nath, B. (2002). A Topology Discovery Algorithm for Sensor Networks with Applications to Network Management. IEEE CAS workshop.

Diffie, W., & Hellman, M. E. (1979). Privacy and authentication. An introduction to cryptography. *Proceedings of the IEEE*, *67*(3), 397–427. doi:10.1109/PROC.1979.11256

Du, W., Deng, J., Han, Y. S., Chen, S., & Varshney, P. (2004). A Key Management Scheme for Wireless Sensor Networks Using Deployment Knowledge. IEEE INFOCOM.

Du, W., Deng, J., Han, Y. S., & Varshney, P. (2003). A Pairwise Key Pre-distribution Scheme for Wireless Sensor Networks. In Proceedings of the 10th ACM Conference on Computer and Communications Security (CCS), Washington DC, October 27-31, 2003.

Eschenauer, L., & Gligor, V.D. (2002). A key-management scheme for distributed sensor networks. In Proceedings of the 9th ACM conferenceon Computer and communications security, Washington, DC, USA, November 18-22 2002.

Heinzelman, W., Chandrakasan, A., & Balakrishnan, H. (2000). Energy-Efficient Communication Protocol for Wireless Microsensor Networks. In Proceedings of the 33rd Hawaii International Conference on System Sciences (HICSS '00).

Intanagonwiwat, C., Govindan, R., & Estrin, D. (2000). Directed diffusion: A scalable and robust communication paradigm for sensor networks. In Proceedings of the Sixth Annual International Conference on Mobile Computing and Networking (pp. 56-67). Boston, MA: ACM Press.

Karlof, C., Sastry, N., & Wagner, C. (2004). TinySec: A Link Layer Security Architecture for Wireless Sensor Networks. In Proceedings of the Second ACM Conference on Embedded Networked Sensor Systems (SenSys 2004).

Khalil, I., Bagchi, S., & Shroff, N. B. (2006). MOBIWORP: Mitigation of the Wormhole Attack in Mobile Multihop Wireless Networks. Securecomm.

Kim, Y., Perrig, A., & Tsudik, G. (2000). Simple and fault-tolerant key agreement for dynamic collaborative groups. In ACM CCS 2000.

Kong, P., Zerfos, P., Luo, H., Lu, S., & Zhang, L. (2001). Providing Robust and Ubiquitous Security Support for Mobile Ad-Hoc Networks. International Conference on Network Protocols (ICNP).

Liu, D., & Ning, P. (2003). Establishing pairwise keys in distributed sensor networks. In Proceedings of 10th ACM Conference on Computer and Communications Security (CCS'03) (pp. 41-47). Washington DC: ACM Press.

Liu, D., & Ning, P. (2005). Improving key pre-distribution with deployment knowledge in static sensor networks. *ACM Transactions on Sensor Networks*, *1*(2), 204–239. doi:10.1145/1105688.1105691

Luk, M., Mezzour, G., Perrig, A., & Gligor, V. D. (2007). MiniSec: A Secure Sensor Network Communication Architecture. In Proceedings of IEEE International Conference on Information Processing in Sensor Networks (IPSN).

Malan, D., Welsh, M., & Smith, M. (2004). A Public-Key Infrastructure for Key Distribution in TinyOS Based on Elliptic Curve Cryptography. IEEE SECON 2004.

Miller, V.S. (1986). Use of Elliptic Curves in Cryptography, Advances in Cryptology – CRYPTO '85 Proceedings (LNCS 218, pp. 417-426).

Panja, B., Madria, S. K., & Bhargava, B. K. (2008). A role-based access in a hierarchical sensor network architecture to provide multilevel security. *Computer Communications*, *31*(4), 793–806. doi:10.1016/j.comcom.2007.10.036

Parno, B., Luk, M., Gaustad, E., & Perrig, A. (2006). Secure Sensor Network Routing: A Clean-Slate Approach. In Proceedings of the 2nd Conference on Future Networking Technologies (2006).

Perkins, C. (1994). Highly Dynamic Destination-Sequenced Distance-Vector Routing (DSDV) for Mobile Computers. ACM SIGCOMM'94 Conference on Communications Architectures, Protocols and Applications.

Perkins, C. E., & Royer, E. M. (1999). Ad hoc On-Demand Distance Vector Routing. In Proceedings of the 2nd IEEE Workshop on Mobile Computing Systems and Applications, New Orleans, LA, February 1999 (pp. 90-100).

Perrig, A., Szewczyk, R., Wen, V., Culler, D., & Tygar, J. D. (2001). SPINS: Security protocols for sensor networks. In Proceedings of Mobicom.

Shamir, A. (1979). How to Share a Secret. *Communications of the ACM*, *22*(11), 612–613. doi:10.1145/359168.359176

Stallings, W. (n.d.). Network Security Essentials Applications and Standards.

Steiner, M., Tsudik, G., & Waidner, M. (2000). Key Agreement in Dynamic Peer Groups. *IEEE Transactions on Parallel and Distributed Systems*, *11*(8), 769–780. doi:10.1109/71.877936

Subramanian, L., & Katz, R. H. (2000). An Architecture for Building Self-Configurable Systems. IEEE MobiHoc.

Sun, Y., & Liu, K. J. R. (2004). Scalable Hierarchical Access Control in Secure Group Communications. IEEE INFOCOM.

The Network Simulator - ns-2 (n.d.). Retrieved from http://www.isi.edu/nsnam/ns/

Ye, F., Luo, H., Lu, S., & Zhang, L. (2004). Statistical En-route Filtering of Injected False Data in Sensor Networks. INFOCOM 2004.

Yi, S., Naldurg, P., & Kravets, R. (2001). A Security-Aware Routing Protocol for Wireless Ad Hoc Networks. Poster presentation, ACM Symposium on Mobile Ad Hoc Networking & Computing (Mobihoc 2001), Long Beach, California, October, 2001

Yin, J., & Madria, S. (2008). SeRWA: A Secure Routing Protocol against Wormhole Attacks in Sensor Networks. Ad Hoc Networks.

Yu, Z., & Guan, Y. (2005). A robust group-based key management scheme for wireless sensor networks. In Proceedings of IEEE Wireless Communications and Networking Conference (WCNC 2005). New Orleans, LA USA: IEEE

Zhu, S., Setia, S., & Jajodia, S. (2003). LEAP: efficient security mechanisms for large-scale distributed sensor networks. In Proceedings of the 10th ACM conference on Computer and communication security

Compilation of References

Abadi, D. (2003, Aug.). Aurora: A new model and architecture for data stream management. *The VLDB Journal, 12*(2). doi:10.1007/s00778-003-0095-z

Abadi, D. J., Madden, S., & Lindner, W. (2005). Reed: robust, efficient filtering and event detection in sensor networks. In *Proceedings of the 31ˢᵗ International Conference on Very Large Data Bases* (pp. 769-780). VLDB Endowment.

Acharya, S., Gibbons, P. B., Poosala, V., & Ramaswamy, S. (1999). Join synopses for approximate query answering. In *Proceedings of 19ᵗʰ ACM Symposium on Principles of Database Systems* (pp. 275-286).

Adaikkalavan, R., & Chakravarthy, S. (2003, Sep.). SnoopIB: Interval-based event specification and detection for active databases. In *Proceedings of ADBIS*.

Adaikkalavan, R., & Chakravarthy, S. (2006, Oct.). SnoopIB: Interval-based event specification and detection for active databases. *Data & Knowledge Engineering, 59*(1), 139–165. doi:10.1016/j.datak.2005.07.009

Adjiman, P., Chatalic, P., & Goasdou, F. Rousset, M., & Simon, L. (2005). Somewhere in the Semantic Web. *3ʳᵈ Workshop on Principles and Practice of Semantic Web Reasoning* (LNCS, 3703, pp. 1-16).

Agarwal, S., Agrawal, R., Deshpande, P. M., Gupta, A., Naughton, J. F., Ramakrishnan, R., & Sarawagi, S. (1996). On the computation of multidimensional aggregates. In *Proceedings of the 22ⁿᵈ International Conference on Very Large Data Bases* (pp. 506-521).

Aggarwal, C. (2003, June). A framework for diagnosing changes in evolving data streams. In *Proceedings of the ACM SIGMOD Conference*, San Diego, California.

Aggarwal, C. C., & Yu, P. S. (2001). Outlier detection for high dimensional data, In *Proceedings of the ACM SIGMOD Conference on Management of Data* (pp. 37-47).

Aggarwal, C. C., Han, J., Wang, J., & Yu, P. S. (2003). A framework for clustering evolving data streams. In *VLDB 2003, Proceedings of 29th International Conference on Very Large Data Bases* (pp. 81–92). Morgan Kaufmann.

Aggarwal, C. C., Han, J., Wang, J., & Yu, P. S. (2004, August). On demand classification of data streams. In *Proceedings of the Tenth ACM SIGKDD International Conference on Knowledge Discovery and Data Mining* (pp. 503-508), Seattle, WA.

Agrawal, R., & Srikant, R. (1994). Fast algorithms for mining association rules. In *Proceedings of the 20th International Conference on Very Large Data Bases* (pp. 487-499).

Agrawal, R., & Srikant, R. (1995). Mining sequential patterns. In *ICDE'95* (pp. 3-14).

Agrawal, R., & Srikant, R. (2000). Privacy-preserving data mining. In *Proceedings of the 2000 ACM SIGMOD Conference on Management of Data*. New York: ACM.

Agrawal, R., Gehrke, J., Gunopulos, D., & Raghavan, P. (1998). Automatic subspace clustering of high dimensional data for data mining applications. In *Proceedings of the ACM-SIGMOD International Conference on Management of Data* (pp. 94–105). Seattle, WA: ACM Press.

Ahmed, N., Kanhere, S. S., & Jha, S. (2005). The holes problem in wireless sensor networks: a survey. *SIGMOBILE Mob. Comput. Commun. Rev.*

Akyildiz, I. F., Pompili, D., & Melodia, T. (2005). Underwater acoustic sensor networks: Research challenges . *Ad Hoc Networks, 3*, 257–279. doi:10.1016/j.adhoc.2005.01.001

Akyildiz, I. F., Su, W., Sankarasubramaniam, Y., & Cayirci, E. (2002). Wireless sensor networks: a survey. *International Journal of Computer Networks, 38*(4), 393–422. doi:10.1016/S1389-1286(01)00302-4

Akyildiz, I., Su, W., Sankarasubramaniam, Y., & Cayirci, E. (2002). A survey on sensor networks. *IEEE Communications Magazine, 40*(8), 102–114. doi:10.1109/MCOM.2002.1024422

Albert, R., & Barabási, A.-L. (2002). Statistical mechanics of complex networks. *Reviews of Modern Physics, 74*, 47–97. doi:10.1103/RevModPhys.74.47

Albert, R., Jeong, H., & Barabási, A.-L. (1999). Diameter of the world-wide web. *Nature, 401*, 130. doi:10.1038/43601

Albert, R., Jeong, H., & Barabási, A.-L. (2000). Error and attack tolerance of complex networks. *Nature, 406*, 378–382. doi:10.1038/35019019

Alert System. (2007). Retrieved from http://www.alertsystems.org

Amir, Y., Ateniese, G., Hasse, D., Kim, Y., Nita-Rotaru, C., Schlossnagle, T., Schultz, J., Stanton, J., & Tsudik, G. (2000). Secure group communication in asynchronous networks with failures: Integration and experiments. ICDCS 2000.

Ananthakrishna, R., Das, A., Gehrke, J., Korn, F., Muthukrishnan, S., & Srivastava, D. (2003). Efficient approximation of correlated sums on data streams. *IEEE Transactions on Knowledge and Data Engineering, 15*(3), 569–572. doi:10.1109/TKDE.2003.1198391

Angiulli, F., & Pizzuti, C. (2002). Fast outlier detection in high dimensional spaces. *In Proceedings of Pacific-Asia Conference on Knowledge Discovery and Data Mining* (pp. 15-26).

Ankerst, M., Breunig, M., Kriegel, H.-P., & Sander, J. (1999). OPTICS: Ordering points to identify the clustering structure. In *ACM SIGMOD Record, Proceedings of the 1999 ACM SIGMOD International Conference on Management of Data, 28*(2), 49-60.

Antoniou, T., Chatzigiannakis, I., Mylonas, G., Nikoletseas, S., & Boukerche, A. (2004). A new energy efficient and fault-tolerant protocol for data propagation in smart dust networks using varying transmission range. In *Annual Simulation Symposium (ANSS)*. ACM/IEEE.

Apiletti, D., Baralis, E., & Bruno, G. G., & Cerquitelli, T. (2006). IGUANA: Individuation of global unsafe anomalies and alarm activation. In *3rd International IEEE Conference on Intelligent Systems,* (pp. 267-272). IEEE Press.

Arasu, A., Babu, S., & Widom, J. (2003, Oct.). *The CQL continuous query language: Semantic foundations and query execution* .(Tech. Rep. Nos. 2003-67) CA: Stanford University.

Arasu, A., et al. (2004, September). Linear road: A stream data management benchmark. In *Proceedings of VLDB.*

Armstrong, M. P. (2002). Geographic information technologies and their potentially erosive effects on personal privacy. *Studies in the Social Sciences, 27*, 19–28. doi:10.1016/S0165-4896(01)00085-3

Armstrong, M. P., & Ruggles, A. J. (2005). Geographic information technologies and personal privacy. *Cartographica, 40*, 4.

Associated Press. (2005). Tracking cell phones for real-time traffic data. Retrieved from http://www.wired.com/news/wireless/0,1382,69227,00.html

Avnur, R., & Hellerstein, J. M. (2000). Eddies: Continuously adaptive query processing. In *Proceedings of the 2000 ACM International Conference on Management of Data* (pp. 261-272).

Babcock, B., Datar, M., & Motwani, R. (2004, Mar.). Load shedding for aggregation queries over data streams. In *Proceedings of ICDE.*

Babcock, B., Babu, S., Datar, M., Motwani, R., & Widom, J. (2002). Models and issues in data stream systems. In *Proceedings of the Twenty-First ACM SIGMOD-SIGACT-SIGART Symposium on Principles of Database Systems* (pp. 1-16). New York: ACM.

Babcock, B., Babu, S., Datar, M., Motwani, R., & Widom, J. (2002, June). Models and issues in data stream systems. In *Proceedings of the 21st Symposium on Principles of Database Systems (PODS),* Madison, Wisconsin.

Babcock, B., Datar, M., Motwani, R., & O'Callaghan, L. (2003, June). Maintaining variance and k-medians over data stream windows. In *Proceedings of the 22nd Symposium on Principles of Database Systems (PODS),* San Diego, California.

Babu, S., & Widom, J. (2001). Continuous queries over data streams. *SIGMOD Record, 30*(3), 109–120. doi:10.1145/603867.603884

Barabási, A. L. (2002). *Linked: The new science of networks.* New York: Penguin.

Barabási, A.-L., & Albert, R. (1999). Emergence of scaling in random networks. *Science, 286,* 509–512. doi:10.1126/science.286.5439.509

Baralis, E., Cerquitelli, T., & D'Elia, V. (2007). Modeling a sensor network by means of clustering. In *18th International Conference on Database and Expert Systems Applications,* (pp. 177-181). IEEE Press.

Baras, J., Li, H., & Mykoniatis, G. (1998, Apr.). *Integrated, distributed fault management for communication networks.* (Tech. Rep. Nos. CS-TR 98-10, University of Maryland).

Barbará, D. (2002). Requirements for clustering data streams. *SIGKDD Explorations (Special Issue on Online, Interactive, and Anytime Data Mining), 3*(2), 23–27.

Barbará, D., & Chen, P. (2000). Using the fractal dimension to cluster datasets. In *Proceedings of the Sixth ACM-SIGKDD International Conference on Knowledge Discovery and Data Mining,* pages 260–264. ACM Press.

Barbara, D., Domeniconi, C., & Rogers, J. P. (2006) Detecting outliers using transduction and statistical significance testing. In *Proceedings of the ACM SIGKDD conference on Knowledge Discovery and Data Mining* (pp. 55-64).

Barbarà, D., Du Mouchel, W., Faloutsos, C., Haas, P. J., Hellerstein, J. M., & Ioannidis, Y. E. (1997). The New Jersey data reduction report. *A Quarterly Bulletin of the Computer Society of the IEEE Technical Committee on Data Engineering, 20*(4), 3–45.

Barnett, V., & Lewis, T. (1994). *Outliers in statistical data* (3rd ed.). New York: John Wiley & Sons.

Barrat, A., & Weigt, W. (2000). On the properties of small-world networks. *The European Physical Journal B, 13,* 547–560. doi:10.1007/s100510050067

Barriere, L., Fraigniaud, P., Narayanan, L., & Opatrny, J. (2003). Robust position-based routing in wireless ad hoc networks with irregular transmission ranges. *Wireless Communications and Mobile Computing.*

Bash, B. A., Byers, J. W., & Considine, J. (2004). Approximately uniform random sampling in sensor networks. In *DMSN* (pp. 32-39). ACM Press.

Bay, S., & Schwabacher, M. (2003). Mining distance-based outliers in near linear time with randomization and a simple pruning rule. In *Proceedings of the ACM SIGMOD Conference on Knowledge Discovery and Data* (pp. 29-38).

Bell, D. E., & La Padula, L. J. (1973). Secure computer systems: Mathematical foundations (Technical Report ESD-TR-73-278). Hanscom AFB, Bedford, Mass.

Bellahsène, Z., & Roantree, M. (2004). Querying distributed data in a super-peer based architecture. *15th International Conference on Database and Expert System Applications* (LNCS 3180, pp. 296-305).

Benaloh, J. (1994). Dense probabilistic encryption. In *Proceedings of the Workshop on Selected Areas of Cryptography* (pp. 120-128).

Bentley, J. L. (1975). Multidimensional binary search trees used for associative searching. *Communications of the ACM, 18,* 509517. doi:10.1145/361002.361007

Bentley, J. L., & Friedman, J. H. (1979). Data structures for range searching. *ACM Computing Surveys, 11,* 397409. doi:10.1145/356789.356797

Berchtold, S., Böhm, C., & Kriegel, H.-P. (1998). The pyramid-technique: Towards breaking the curse of dimensionality. In *Proceedings of the 1998 ACM International Conference on Management of Data* (pp. 142-153).

Beringer, J., & Hüllermeier, E. (2006). Online clustering of parallel data streams. *Data & Knowledge Engineering, 58*(2), 180–204. doi:10.1016/j.datak.2005.05.009

Bern, M., & Eppstein, D. (1996). Approximation algorithms for NP-hard problems. Chapter 8 in *Approximation algorithms for geometric problems* (pp. 296-345). PWS Publishing Company.

Berthold, M., & Hand, D. (1999). *Intelligent data analysis - An introduction.* Springer Verlag.

Beutel, J. (2005). Location management in wireless sensor networks. In *Handbook of Sensor Networks: Compact Wireless and Wired Systems.* CRC Press.

Beyer, K., & Ramakrishnan, R. (1999). Bottom-up computation of sparse, iceberg cubes. In *Proceedings of the 1999 ACM International Conference on Management of Data* (pp. 359-370).

Bicking, C., & Gryna, F. M., Jr. (1979). Quality control handbook. In J. M. Juran, F. M. Gryna, Jr., & R. Bingham, Jr. (Eds.), (pp. 23-1-23-35). New York: McGraw Hill.

Birant, D., & Kut, A. (2006). Spatio-temporal outlier detection in large database. *Journal of Computing and Information Technology, 14*(4), 291–298.

Bishop, C. M. (1995). *Neural networks for pattern recognition.* UK: Oxford University Press.

Bjerring, L. H., Lewis, D., & Thorarensen, I. (1996). Inter-domain service management of broadband virtual private networks. *Journal of Network and Systems Management, 4*(4), 355–373. doi:10.1007/BF02283160

Boccaletti, S., Latora, V., Moreno, Y., Chavez, M., & Hwang, D.-U. (2006). Complex networks: Structure and dynamics. *Physics Reports, 424*, 175–308. doi:10.1016/j.physrep.2005.10.009

Bohm, C., Faloutsos, C., & Plant, C. (2008). Outlier-robust clustering using independent components. In *Proceedings of the ACM SIGMOD Conference on Management of Data* (pp. 185-198).

Bonnet, P., Gehrke, J., & Seshadri, P. (2000). Querying the physical world . *IEEE Personal Communications, 7*(5), 10–15. doi:10.1109/98.878531

Bonnet, P., Gehrke, J., & Seshadri, P. (2001). Towards sensor database systems. In *Proceedings of 2nd International Conference on Mobile Data Management* (pp. 3-14).

Bordwell, B., & Thompson, K. (1997). *Film art: An introduction.* McGraw-Hill.

Bose, P., Morin, P., Stojmenovic, I., & Urrutia, J. (1999). Routing with guaranteed delivery in ad hoc wireless networks. In *Discrete algorithms and methods for mobile computing and communications.*

Boukerche, A., & Samarah, S. (2006, November). A novel data mining technique for extracting events and inter knowledge based information from wireless sensor networks. In *Proceedings of IEEE Conference on Local Computer Networks*, Tampa, Florida.

Boukerche, A., & Samarah, S. (2007, March). A new representation structure for mining association rules from wireless sensor networks. In *Proceedings of IEEE Wireless Communications and Networking*, Hong Kong, China.

Boulis, A., Ganeriwal, S., & Srivastava, M. B. (2003, May). Aggregation in sensor networks: An energy-accuracy tradeoff. In *IEEE Workshop on Sensor Network Protocols and Applications (SNPA)*, Anchorage, Alaska.

Bradley, P., Fayyad, U., & Reina, C. (1998). Scaling clustering algorithms to large databases. In *Proceedings of the Fourth International Conference on Knowledge Discovery and Data Mining* (pp. 9-15). AAAI Press.

Brand, M. (2006). Fast low-rank modifications of the thin singular value decomposition. *Linear Algebra and Its Applications, 415*(1), 20–30. doi:10.1016/j.laa.2005.07.021

Brandes, U. (2001). A faster algorithm for betweenness centrality. *The Journal of Mathematical Sociology, 25*, 163–177.

Breunig, M. M., Kriegel, H. P., Ng, R. T., & Sander, J. (2000). LOF: Identifying density-based local outliers. In *Proceedings of the ACM SIGMOD Conference on Management of Data* (pp. 93-104).

Brian, B., et al. (2003). Chain: Operator scheduling for memory minimization in stream systems. In *Proc. of ACM SIGMOD.*

Bridle, J. S. (1990). Probabilistic interpretation of feed-forward classification network outputs, with relationships to statistical pattern recognition. In *Neurocomputing: Algorithms, architecture and applications.* Berlin: Springer-Verlag.

Brin, S., & Page, L. (1998). The anatomy of a large-scale hypertextual Web search engine. *Computer Networks and ISDN Systems, 30*(1-7), 107-117.

Buccafurri, F., Furfaro, F., Saccà, D., & Sirangelo, C. (2003). A quad-tree based multiresolution approach for two-dimensional summary data. In *Proceedings of the 15th IEEE International Conference on Scientific, Statistical Database Management* (pp. 127-137).

Buchmann, A. P., et al. (1993). Rules in an open system: The REACH rule system. In *Proceedings of Rules in Database Systems Workshop.*

Budalakoti, S., Cruz, S., Srivastava, A. N., Akella, R., & Turkov, E. (2006). *Anomaly detection in large sets of high-dimensional symbol sequences.* The United States: California, National Aeronautics and Space Administration.

Burrell, J., Brooke, T., & Beckwith, R. (2004). Vineyard computing: Sensor networks in agricultural production. *IEEE Pervasive Computing / IEEE Computer Society [and] IEEE Communications Society, 1*(3), 38-45. doi:10.1109/MPRV.2004.1269130

Burroughs Corporation. (1964). *Burroughs B5500 information processing systems reference manual.* Burroughs Corporation, Detroit, Michigan, USA.

Cai, Y. D., Clutterx, D., Papex, G., Han, J., Welgex, M., & Auvilx, L. (2004). MAIDS: Mining alarming incidents from data streams. In *Proceedings of the 2004 ACM International Conference on Management of Data* (pp. 919-920).

Camara, K., & Jungert, E. (2007). A visual query language for dynamic processes applied to a scenario driven environment. *Journal of Visual language and Computation. Special Issue on Human-GIS Interaction, 18*(3), 315–338.

Campos, M. M., & Milenova, B. L. (2005). Creation and deployment of data mining-based intrusion detection systems in Oracle Database 10g. *In Proceedings of the 2005 International Conference on Machine Learning and Applications* (pp. 105-112).

Candia, J., González, M. C., Wang, P., Schoenharl, T., Madey, G., & Barabási, A.-L. (2008, June). Uncovering individual and collective human dynamics from mobile phone records. *Journal of Physics A. Mathematical and Theoretical, 41,* 224015. doi:10.1088/1751-8113/41/22/224015

Carman, D.W., Kruus, P.S., & Matt, B. J. (2000). Constraints and approaches for distributed sensor network security (NAI Labs Technical Report #00-010).

Carney, D., et al. (2003, Sep.). Operator scheduling in a data stream manager. In *Proceedings of VLDB.*

Carrara, W. H., Majewski, R. M., & Goodman, R. S. (1995). *Spotlight synthetic aperture radar: Signal processing algorithms.* Artech House.

Cauley, L. (2006). *NSA has massive database of Americans' phone calls.* USA Today.

Chakrabarti, D., Kumar, R., & Tomkins, A. (2006). Evolutionary clustering. In *KDD* (pp. 554–560).

Chakrabarti, K., Porkaew, K., & Mehrotra, S. (2000Feb. 28-March 3). Efficient query refinement in multimedia databases. In *Proceedings of the 16th International Conference on Data Engineering.* San Diego, California.

Chakravarthy, S., & Mishra, D. (1994). Snoop: An expressive event specification language for active databases. *Data & Knowledge Engineering, 14*(10), 1–26. doi:10.1016/0169-023X(94)90006-X

Chakravarthy, S., Anwar, E., Maugis, L., & Mishra, D. (1994). Design of sentinel: An object-oriented DBMS with event-based rules. *Information and Software Technology, 36*(9), 559–568. doi:10.1016/0950-5849(94)90101-5

Chan, A. C. F., & Rogers, E. S., Sr. (2004). Distributed symmetric key management for mobile ad hoc networks. IEEE INFOCOM.

Chan, H., Luk, M., & Perrig, A. (2005). Using clustering information for sensor network localization. In *First IEEE International Conference on Distributed Computing in Sensor Systems* (pp. 109–125).

Chan, H., Perrig, A., & Song, D. (2003, May 11-14). Random key predistribution schemes for sensor networks. In *IEEE Symposium on Security and Privacy*, Berkeley, California.

Chandola, V., Banerjee, A., & Kumar, V. (2007). *Outlier detection: A survey* (Tech. Rep.). University of Minnesota.

Chang, S.-K., & Jungert, E. (1996). *Symbolic projection for image information retrieval and spatial reasoning.* London: Academic Press.

Chang, S.-K., & Jungert, E. (1998). A spatial/temporal query language for multiple data sources in a heterogeneous information system environment. *International Journal of Cooperative Information Systems, 7*(2-3), 167–186. doi:10.1142/S021884309800009X

Chang, S.-K., Costagliola, G., & Jungert, E. (2002). Multi-sensor information fusion by query refinement. *Recent Advances in Visual Information Systems* (LNCS 2314, pp. 1-11).

Chang, S.-K., Costagliola, G., Jungert, E., & Orciuoli, F. (2004). Querying distributed multimedia databases and data sources for sensor data fusion. *IEEE Transactions on Multimedia, 6*(5), 687–672. doi:10.1109/TMM.2004.834862

Chang, S.-K., Dai, W., Hughes, S., Lakkavaram, P., & Li, X. (2002, September 26-28). Evolutionary query processing, fusion and visualization. In *Proceedings of the 8th International Conference on Distributed Multimedia Systems.* San Francisco, California, (pp. 677-686).

Chang, S.-K., Jungert, E., & Li, X. (2006). A progressive query language and interactive reasoner for information fusion. *Journal of Information Fusion, 8*(1), 70–83. doi:10.1016/j.inffus.2005.09.004

Charikar, M., Chekuri, C., Feder, T., & Motwani, R. (1997) Incremental clustering and dynamic information retrieval. In *Proceedings of the 29th Annual ACM Symposium on Theory of Computing.*

Charikar, M., O'Callaghan, L., & Panigrahy, R. (2003). Better streaming algorithms for clustering problems. In *Proceedings of 35th ACM Symposium on Theory of Computing (STOC)*, San Diego, California, June 2003.

Chatterjea, S., & Havinga, P. (2008). An adaptive and autonomous sensor sampling frequency control scheme for energy-efficient data acquisition in wireless sensor networks. *4th IEEE International Conference on Distributed Computing in Sensor Systems,* DCOSS 2008.

Chatzigiannakis, I., Dimitriou, T., Nikoletseas, S., & Spirakis, P. (2004). A probabilistic forwarding protocol for efficient data propagation in sensor networks. In *European Wireless (EW) Conference on Mobility and Wireless Systems beyond 3G.*

Chatzigiannakis, I., Dimitriou, T., Nikoletseas, S., & Spirakis, P. (2006). A probabilistic forwarding protocol for efficient data propagation in sensor networks. *Journal of Ad hoc Networks.*

Chatzigiannakis, I., Kinalis, A., & Nikoletseas, S. (2006). Efficient and robust data dissemination using limited extra network knowledge. In *IEEE Conference on Distributed Computing in Sensor Systems (DCOSS).*

Chatzigiannakis, I., Mylonas, G., & Nikoletseas, S. (2006). Modeling and evaluation of the effect of obstacles on the performance of wireless sensor networks. In *Annual Simulation Symposium (ANSS).* ACM/IEEE.

Chatzigiannakis, I., Nikoletseas, S., & Spirakis, P. (2002). Smart dust protocols for local detection and propagation. In *Principles of Mobile Computing (POMC).* ACM.

Chatzigiannakis, I., Nikoletseas, S., & Spirakis, P. (2005). Smart dust protocols for local detection and propagation. *Journal of Mobile Networks (MONET)*.

Chaudhuri, S., & Dayal, U. (1997). An overview of data warehousing and OLAP technology. *SIGMOD Record*, *26*(1), 65–74. doi:10.1145/248603.248616

Chawathe, S., Krishnamurthy, V., Ramachandran, S., & Sarma, S. (2004). Managing RFID data. In *VLDB'04*.

Chen, J., et al. (2000). NiagaraCQ: A scalable continuous query system for Internet databases. In *Proceedings of SIGMOD*.

Chen, Y., & Tu, L. (2007). Density-based clustering for real-time stream data. In *Proceedings of the 13th ACM SIGKDD International Conference on Knowledge Discovery and Data Mining* (pp. 133–142).

Chen, Y., Dong, G., Han, J., Wah, B. W., & Wang, J. (2002). Multi-dimensional regression analysis of time-series data streams. In *Proceedings of the 28th International Conference on Very Large Data Bases* (pp. 323-334).

Cheng, T., & Li, Z. (2006). A multiscale approach for spatio-temporal outlier detection. *Transactions in Geographic Information System*, *10*(2), 253–263.

Cheu, E. Y., Keongg, C., & Zhou, Z. (2004). On the two-level hybrid clustering algorithm. In *International Conference on Artificial Intelligence in Science and Technology* (pp. 138-142).

Chipman, H., & Tibshirani, R. (2006). Hybrid hierarchical clustering with applications to microarray data. *Biostatistics (Oxford, England)*, *7*(2), 286–301. doi:10.1093/biostatistics/kxj007

Chiu, A. L., & Fu, A. W. (2003). Enhancements on local outlier detection. In *Proceedings of International Database Engineering and Applications Symposium* (pp. 298-307).

Chu, D., Deshpande, A., Hellerstein, J., & Hong, W. (2006). Approximate data collection in sensor networks using probabilistic models. In *Proc. of the 2006 Intl. Conf. on Data Engineering*.

Chu, D., Popa, L., Tavakoli, A., Hellerstein, J. M., Levis, P., Shenker, S., & Stoica, I. (2007, November). The design and implementation of a declarative sensor network system. In *Proceedings of ACM Conference on Embedded Networked Sensor Systems (SenSys)* (pp. 175-188). Sydney, Australia.

Chung, R. (1997). *Spectral graph theory* (Vol. 92). CBMS Regional Conference Series in Mathematics.

Clifton, C., Kantarcioglu, M., & Vaidya, J. (2002). Defining privacy for data mining. In *Proceedings of the National Science Foundation Workshop on Next Generation Data Mining*.

Coble, J., Cook, D. J., & Holder, L. B. (2006). Structure discovery in sequentially-connected data streams. *International Journal of Artificial Intelligence Tools*, *15*(6), 917–944. doi:10.1142/S0218213006003041

Collard, S. B., & Lugo-Fernández, A. (1999). *Coastal upwelling and mass mortalities of fishes and invertebrates in the northeastern Gulf of Mexico during spring and summer 1998* (OCS Study MMS 99-0049). New Orleans, LA: U.S. Department of the Interior, Minerals Management Service, Gulf of Mexico OCS Region. Retrieved August 4, 2008, from http://www.gomr.mms.gov/PI/PDFImages/ESPIS/3/3207.pdf

Colliat, G. (1996). OLAP, relational, and multidimensional database systems. *SIGMOD Record*, *25*(3), 64–69. doi:10.1145/234889.234901

Cook, D. J., & Holder, L. B. (1994). Substructure discovery using minimum description length and background knowledge. *Journal of Artificial Intelligence Research*, *1*, 231–255.

Cooper, O., et al. (2004). HiFi: A unified architecture for high fan-in systems. In *Proceedings of VLDB* (p. 1357 - 1360).

Cormode, G., Muthukrishnan, S., & Zhuang, W. (2007). Conquering the divide: Continuous clustering of distributed data streams. In *Proceedings of the 23rd International Conference on Data Engineering (ICDE 2007)* (pp. 1036–1045).

Cortes, C., Fisher, K., Pregibon, D., Rogers, A., & Smith, F. (2000). Hancock: A language for extracting signatures from data streams. In *Proceedings of the 6th ACM International Conference on Knowledge Discovery and Data Mining* (pp. 9-17).

Cortes, C., Pregibon, D., & Volinsky, C. (2003). Computational methods for dynamic graphs. *Journal of Computational and Graphical Statistics, 12*, 950–970. doi:10.1198/1061860032742

Cristescu, R., Beferull-Lozano, B., & Vetterli, M. (2004). On network correlated data gathering. In *Proceedings IEEE INFOCOM 2004*.

Cui, J.-H., Kong, J., Gerla, M., & Zhou, S. (2006). The challenges of building mobile underwater wireless networks for aquatic applications . *IEEE Network, 20*(3), 12-18. doi:10.1109/MNET.2006.1637927

Culler, D. E., & Mulder, H. (2004). Smart sensors to network the world. *Scientific American.*

Cummings, J. A. (1994). Global and regional ocean thermal analysis systems at Fleet Numerical Meteorology and Oceanography Center. In *OCEANS '94, 'Oceans Engineering for Today's Technology and Tomorrow's Preservation.' Proceedings, Vol. 3,* (pp.III/75-III/81). Brest, France: IEEE Press.

Cuzzocrea, A. (2005). Overcoming limitations of approximate query answering in OLAP. In *Proceedings of the 9ᵗʰ IEEE International Database Engineering, Applications Symposium* (pp. 200-209).

Dai, B.-R., Huang, J.-W., Yeh, M.-Y., & Chen, M.-S. (2006). Adaptive clustering for multiple evolving streams. *IEEE Transactions on Knowledge and Data Engineering, 18*(9), 1166–1180. doi:10.1109/TKDE.2006.137

Damiani, M. L., & Spaccapietra, S. (2006). Spatial data warehouse modelling. In *Processing and managing complex data for decision support* (pp. 21-27).

Dancyger, K. (2002). *The Technique of Film and Video Editing. History, Theory and Practice.* Focal Press.

Das, A., Gehrke, J., & Riedewald, M. (2003). Approximate Join Processing over Data Streams. In *Proc. of SIGMOD.*

Datta, S., Bhaduri, K., Giannella, C., Wolff, R., & Kargupta, H. (2006). Distributed data mining in peer-to-peer networks. *IEEE Internet Computing, 10*(4), 18–26. doi:10.1109/MIC.2006.74

Davidson, I., & Ravi, S. S. (2005). Distributed pre-processing of data on networks of Berkeley motes using non-parametric EM. In *Proc. of SIAM SDM Workshop on Data Mining in Sensor Networks* (pp. 17-27).

Dayal, U. (1988, Mar.). The HiPAC Project: Combining Active Databases and Timing Constraints. *SIGMOD Record, 17*(1), 51–70. doi:10.1145/44203.44208

Deb, B., Bhatnagar, S., & Nath, B. (2002). A Topology Discovery Algorithm for Sensor Networks with Applications to Network Management. IEEE CAS workshop.

Deligiannakis, A., Kotidis, Y., & Roussopoulos, N. (2003). *Data reduction techniques for sensor networks.* (Technical Report CS-TR-4512). UM Computer Science Department.

Deligiannakis, A., Kotidis, Y., & Roussopoulos, N. (2004). Compressing historical information in sensor networks. In *Proceedings of the ACM International Conference on Management of Data* (pp. 527-538).

Deligiannakis, A., Kotidis, Y., & Roussopoulos, N. (2004). Hierarchical In-network data aggregation with quality guarantees. In *Proceedings of 9th International Conference on Extending Database Technology* (pp. 658-675).

Demers, A., Gehrke, J., Rajaraman, R., Trigoni, N., & Yao, Y. (2003). The Cougar project: A work-in-progress report. *SIGMOD Record, 32*(4), 53–59. doi:10.1145/959060.959070

Dempster, A., Laird, N., & Rubin, D. (1977). Maximum likelihood from incomplete data via the EM algorithm. *Journal of the Royal Statistical Society. Series A (General), 39*(1), 1–38.

Deshpande, A., Guestrin, C., & Madden, S. (2005). Using probabilistic models for data management in acquisitional environments. In *Proc. Biennial Conf. on Innovative Data Sys. Res* (pp. 317-328).

Deshpande, A., Guestrin, C., Madden, S., & Hong, W. (2005). Exploiting correlated attributes in acqusitional query processing. In *ICDE*, 2005.

Deshpande, A., Guestrin, C., Madden, S., Hellerstein, J. M., & Hong, W. (2004). Model-driven data acquisition in sensor networks. In *Proceedings of the International Conference on Very Large Data Bases (VLDB)* (pp. 588-599), Toronto, Canada.

Dhillon, I. S., & Modha, D. S. (1999). A data-clustering algorithm on distributed memory multiprocessors. *Large-scale parallel data mining* (pp. 245-260). Springer.

Diaz, O., Paton, N., & Gray, P. (1991, Sep.). Rule management in object-oriented databases: A unified approach. In *Proceedings of International Conference on Very Large Data Bases VLDB.*

Diaz-Caldera, R., Serrat-Fernandez, J., Berdekas, K., & Karayannis, F. (1999). An approach to the cooperative management of multitechnology networks. *Communications Magazine, IEEE, 37*(5), 119–125. doi:10.1109/35.762867

Diffie, W., & Hellman, M. E. (1979). Privacy and authentication. An introduction to cryptography. *Proceedings of the IEEE, 67*(3), 397–427. doi:10.1109/PROC.1979.11256

Dinn, A., Williams, M. H., & Paton, N. W. (1997). ROCK & ROLL: A Deductive Object-Oriented Database with Active and Spatial Extensions. In *Proc. of ICDE.*

Dobra, A., Garofalakis, M., Gehrke, J., & Rastogi, R. (2002). Processing Complex aggregate queries over data streams. In *Proceedings of the 2002 ACM International Conference on Management of Data* (pp. 61-72).

Domingos, P., & Hulten, G. (2000). Mining high-speed data streams. In *Proceedings of the 6th ACM International Conference on Knowledge Discovery, Data Mining* (pp. 71-80).

Domingos, P., & Hulten, G. (2001). A general method for scaling up machine learning algorithms and its application to clustering. In *Proceedings of the Eighteenth International Conference on Machine Learning* (pp. 106-113), Williamstown, Massachusetts.

Du, W., Deng, J., Han, Y. S., & Varshney, P. (2003, October 27-31). A pairwise key pre-distribution scheme for wireless sensor networks. In *Proceedings of the 10th ACM Conference on Computer and Communications Security (CCS)*, Washington DC.

Du, W., Deng, J., Han, Y. S., Chen, S., & Varshney, P. (2004). A key management scheme for wireless sensor networks using deployment knowledge. IEEE INFOCOM.

Dunne, L. E., Brady, S., Smyth, B., & Diamond, D. (2005). Initial development and testing of a novel foam-based pressure sensor for wearable sensing. *Journal of Neuroengineering and Rehabilitation, 2*(1), 4. doi:10.1186/1743-0003-2-4

Dy, J. G., & Brodley, C. E. (2004). Feature selection for unsupervised learning. *Journal of Machine Learning Research, 5*, 845–889.

Engstrom, H., Berndtsson, M., & Lings, B. (1997). *ACOOD essentials* (Tech. Rep. HS-IDA-TR-97-010). University of Skovde.

Eschenauer, L., & Gligor, V. D. (2002, November 18-22). A key-management scheme for distributed sensor networks. In *Proceedings of the 9th ACM conference on Computer and Communications Security*, Washington, DC.

Eskin, E. (2000). Anomaly detection over noisy data using learned probability distributions. In *Proceedings of International Conference on Machine Learning* (pp. 222-262).

Ester, M., Kriegel, H. P., Sander, J., & Xu, X. (1996). A density-based algorithm for discovering clusters in large spatial databases with noise. In *Proceedings of the 2nd International Conference on Knowledge Discovery and Data Mining (KDD)*, (pp. 226-231). Portland, OR: AAAI Press.

Ester, M., Kriegel, H. P., Sander, J., Wimmer, H. P., & Xu, X. (1998). Incremental clustering for mining in a data warehousing environment. In *Proceedings of the 24th International Conference on Very Large Data Bases* (pp. 323-333).

Estrin, D., Govindan, R., Heidemann, J., & Kumar, S. (1999). Next century challenges: Scalable coordination in sensor networks. In *Mobile computing and networking*. ACM.

European Parliament and Council of the European Union. (1995). Directive 95/64/EC of the European Parliament and of the Council of 24 October 1995. *Official Journal of the European Communities*, *281*, 30–51.

Evans, A. C. (1981). European data protection law. *The American Journal of Comparative Law*, *29*, 571–582. doi:10.2307/839754

Fan, H., Zaiane, O. R., Foss, A., & Wu, J. (2006). A nonparametric outlier detection for effectively discovering top-n outliers from engineering data. In *Proceedings of Pacific-Asia Conference on Knowledge Discovery and Data Mining (KDD)*, (pp. 557-566).

Fang, M., Shivakumar, N., Garcia-Molina, H., Motwani, R., & Ullman, J. D. (1998). Computing iceberg queries efficiently. In *Proceedings of the 24th International Conference on Very Large Data Bases (VLDB)* (pp. 299-310).

Fang, Q., Gao, J., & Guibas, L. (2006). Locating and bypassing holes in sensor networks. *Mobile Networks and Applications*.

Fayyad, U. M., Piatetsky-Shapiro, G., & Smyth, P. (1996). From data mining to knowledge discovery: An overview. In U.M. Fayyad, G. Piatetsky-Shapiro, P. Smyth & R. Uthurusamy (Eds.), *Advances in knowledge discovery and data mining*. Menlo Park, CA: AAAI Press/The MIT Press.

Ferrer, F., Aguilar, J., & Riquelme, J. (2005). Incremental rule learning and border examples selection from numerical data streams. *Journal of Universal Computer Science*, *11*(8), 1426–1439.

Finkenzeller, K. (2003). *RFID Handbook: Fundamentals and applications in contactless smart cards and identification*. John Wiley and Sons.

Finn, G. G. (1987). *Routing and addressing problems in large metropolitan-scale internetworks* (Tech. Rep.). Information Sciences Institute.

Fisher, D. H. (1987). Knowledge acquisition via incremental conceptual clustering. *Machine Learning*, *2*(2), 139–172.

Folkesson, M., Grönwall, C., & Jungert, E. (2006). A fusion approach to coarse-to-fine target recognition. In B. V. Dasarathy (Ed.), *Proceedings of SPIE -6242- Multisensor, Multisource Information Fusion: Architectures, Algorithms, and Applications*.

Foss, A., & Zaïane, O. (2002). A parameterless method for efficiently discovering clusters of arbitrary shape in large datasets. In *Proceedings of International Conference on Data Mining* (pp. 179-186*)*.

Freeman, L. (1977). A set of measures of centrality based on betweenness. *Sociometry*, *40*, 35–41. doi:10.2307/3033543

Freeman, L. (1979). Centrality in social networks: conceptual clarifications. *Social Networks*, *1*, 215–239. doi:10.1016/0378-8733(78)90021-7

Frey, H., & Gorgen, D. (2006). Geographical cluster-based routing in sensing-covered networks. *IEEE Transactions on Parallel and Distributed Systems*, *17*(9), 899–911. doi:10.1109/TPDS.2006.124

Friedman, A., Schuster, A., & Wolff, R. (2006). *k*-Anonymous decision tree induction. In J. Fürnkranz, T. Scheffer, & M. Spiliopoulou (Eds.), *Proceedings of the 10th European Conference on Principles and Practice of Knowledge Discovery in Databases*.

Frohlich, P., & Nejdl, W. (1997, Jan.). Model-based alarm correlation in cellular phone networks. In *Proc. of the International Symposium On Modeling, Analysis, And Simulation Of Computer And Telecommunications Systems (MASCOTS)*.

Fu, J., & Yu, X. (2006). Rotorcraft acoustic noise estimation and outlier detection. In *Proceedings of International Joint Conference on Neural Networks* (pp. 4401-4405).

Fuller, S. H., & Harbison, S. P. (1978). *The C.mmp multiprocessor* (Technical Report CMU-CS-78-146). Pittsburgh, Pennsylvania: Department of Computer Science, Carnegie-Mellon University.

Gaber, M. M. (2007). Data stream processing in sensor networks. In J. Gama & M. M. Gaber (Eds.), *Learning from data streams processing techniques in sensor network* (pp. 41-48). Berlin-Heidelberg: Springer.

Gaber, M. M., Krishnaswamy, S., & Zaslavsky, A. (2004). Ubiquitous data stream mining, *Current Research and Future Directions Workshop* held in conjunction with *The Eighth Pacific-Asia Conference on Knowledge Discovery and Data Mining*, Sydney, Australia.

Gaber, M. M., Zaslavsky, A., & Krishnaswamy, S. (2005). Mining data streams: A review. *SIGMOD Record, 34*(2), 18–26. doi:10.1145/1083784.1083789

Gaede, V., & Günther, O. (1998). Data structures for range searching. *ACM Computing Surveys, 30*, 170–231. doi:10.1145/280277.280279

Gama, J., & Rodrigues, P. P. (2007). Data stream processing. In J. Gama & M.M. Gaber (Eds.), *Learning from Data Streams - Processing Techniques in Sensor Networks* (pp. 25–39). Springer Verlag.

Gama, J., & Rodrigues, P. P. (2007). Stream-based electricity load forecast. In J. N., Kok, J. Koronacki, R.L. de Mantaras, S. Matwin, D.Mladenic, & A. Skowron (Eds.), *Proceedings of the 11th European Conference on Principles and Practice of Knowledge Discovery in Databases (PKDD 2007)* (LNAI 4702, pp. 446-453). Warsaw, Poland. Springer Verlag.

Gama, J., Fernandes, R., & Rocha, R. (2006). Decision trees for mining data streams. *Intelligent Data Analysis, 10*(1), 23–45.

Gama, J., Medas, P., Castillo, G., & Rodrigues, P. P. (2004). Learning with drift detection. In A. L. C. Bazzan,& S. Labidi (Eds.), *Proceedings of the 17th Brazilian Symposium on Artificial Intelligence (SBIA 2004)* (LNCS 3171, pp. 286-295). São Luiz, Maranhão, Brazil. Springer Verlag.

Gambhir, D., Post, M., & Frisch, I. (1994). A framework for adding real-time distributed software fault detection and isolation to SNMP-based systems management. *Journal of Network and Systems Management, 2*(3). doi:10.1007/BF02139365

Ganesan, P., Yang, B., & Garcia-Molina, H. (2004). One torus to rule them all: multi-dimensional queries in P2P systems. In *Proceedings of the 7th International Workshop on the Web and Databases* (pp. 19-24). New York: ACM.

Ganti, V., Li Lee, M., & Ramakrishnan, R. (2000). ICICLES: Self-tuning samples for approximate query answering. In *Proceedings of 26th International Conference on Very Large Data Bases* (pp. 176-187).

Garg, V. (2005, Dec.). *EStream: An integration of event and stream processing* (Master's Thesis, ITLab, The University of Texas at Arlington).

Garofalakis, M., Gehrke, J., & Rastogi, R. (2002). Querying and mining data streams: You only get one look a tutorial. In *Proceedings of the 2002 ACM SIGMOD International Conference on Management of Data*.

Gatziu, S., & Dittrich, K. R. (1992, Dec.). SAMOS: An active, object-oriented database system. *IEEE Quarterly Bulletin on Data Engineering, 15*(1-4), 23–26.

Gatziu, S., & Dittrich, K. R. (1993, Sep.). Events in an object-oriented database system. In *Proc. of Rules in database systems*.

Gatziu, S., & Dittrich, K. R. (1994, Feb.). Detecting composite events in active databases using Petri Nets. In *Proc. of Workshop On Research Issues In Data Engineering*.

Gay, D., Levis, P., von Behren, R., Welsh, M., Brewer, E., & Culler, D. (2003). The nesC Language: A holistic approach to network embedded systems. In *ACM SIGPLAN Conference on Programming Language Design and Implementation (PLDI)* (pp. 1-11). ACM Press.

Gehani, N. H., & Jagadish, H. V. (1991). Ode as an active database: Constraints and triggers. In *Proc of VLDB*.

Gehani, N. H., Jagadish, H. V., & Shmueli, O. (1992, Dec.). *COMPOSE: A system for composite event specification and detection* (Tech. Rep.). AT&T Bell Laboratories.

Gehani, N. H., Jagadish, H. V., & Shmueli, O. (1992). Composite event specification in active databases: Model & implementation. In *Proc. of VLDB* (p. 327-338).

Gehani, N. H., Jagadish, H. V., & Shmueli, O. (1992, June). Event specification in an object-oriented database. In *Proc. of SIGMOD* (p. 81-90). San Diego, CA.

Gehrke, J., & Madden, S. (2004). Query processing in sensor networks. *IEEE Pervasive Computing / IEEE Computer Society [and] IEEE Communications Society, 3*(1), 46–55. doi:10.1109/MPRV.2004.1269131

Getoor, L., & Diehl, C. P. (2005). Link mining: A survey. *SIGKDD Explorations Newsletter, 7*(2), 3–12. doi:10.1145/1117454.1117456

Ghoting, A., Parthasarathy, S., & Otey, M. (2006). Fast mining of distance-based outliers in high dimensional datasets. *In Proceedings of the SIAM International Conference on Data Mining* (pp. 608-612).

Giannella, C., Han, J., Pei, J., Yan, X., & Yu, P. (2003). Mining frequent patterns in data streams at multiple time granularities. In *NSF Workshop on Next Generation Data Mining*.

Giannotti, F., Nanni, M., Pedreschi, D., & Pinelli, F. (2007). Trajectory pattern mining. In *KDD'07* (pp. 330-339). ACM.

Gibbons, P. B., Karp, B., Ke, Y., Nath, S., & Seshan, S. (2003). IrisNet: An architecture for a world-wide sensor Web. *IEEE Pervasive Computing / IEEE Computer Society [and] IEEE Communications Society, 2*(4), 22–33. doi:10.1109/MPRV.2003.1251166

Gilbert, A. C., Kotidis, Y., Muthukrishnan, S., & Strauss, M. J. (2001). Surfing wavelets on streams: One-pass summaries for approximate aggregate queries. In *Proceedings of 27th International Conference on Very Large Data Bases* (pp. 79-88).

Gnawali, O., Greenstein, B., Jang, K.-Y., Joki, A., Paek, J., Vieira, M., et al. (2006). The Tenet architecture for tiered sensor networks. In *ACM Conference on Embedded Networked Sensor Systems (SenSys)* (pp. 153-166).

Golab, L., & Özsu, M. T. (2003). Issues in data stream management. *SIGMOD Record, 32*(2), 5–14. doi:10.1145/776985.776986

Gonzalez, H., Han, J., & Li, X. (2006). Flowcube: constructuing RFID flowcubes for multi-dimensional analysis of commodity flows. In *VLDB'06*, Seoul, Korea.

Gonzalez, H., Han, J., & Li, X. (2006). Mining compressed commodity workflows from massive RFID data sets. In *CIKM'06*, Virginia.

Gonzalez, H., Han, J., & Shen, X. (2007). Cost-conscious cleaning of massive RFID data sets. In *ICDE'07*, Istanbul, Turkey.

Gonzalez, H., Han, J., Li, X., & Klabjan, D. (2006). Warehousing and analyzing massive RFID data sets. In *Proceedings of the 22nd IEEE International Conference on Data Engineering* (pp. 83-93). Atlanta, GA.

González, M. C., & Barabási, A.-L. (2007). Complex networks: From data to models. *Nature Physics, 3,* 224–225. doi:10.1038/nphys581

González, M. C., Hidalgo, C. A., & Barabási, A.-L. (2008). Understanding individual mobility patterns. *Nature, 453,* 479–482. doi:10.1038/nature06958

Gonzalez, T. F. (1985). Clustering to minimize the maximum inter-cluster distance. *Theoretical Computer Science, 38,* 293–306. doi:10.1016/0304-3975(85)90224-5

Grafe, G. (1993). Query evaluation techniques for large databases. *ACM Computing Surveys, 25*(2), 73–170. doi:10.1145/152610.152611

Gray, J., Chaudhuri, S., Bosworth, A., Layman, A., Reichart, D., & Venkatrao, M. (1997). Data cube: A relational aggregation operator generalizing group-by, cross-tab and sub-totals. *Data Mining and Knowledge Discovery, 1*(1), 29–54. doi:10.1023/A:1009726021843

Greenstein, B., Estrin, D., Govindan, R., Ratnasamy, S., & Shenker, S. (2003). DIFS: A distributed index for features in sensor networks. *Ad Hoc Networks*.

Grubbs & Frank. (1969). Procedures for detecting outlying observations in samples. *Technometrics, 11*(1), 1–21. doi:10.2307/1266761

Guha, S., Meyerson, A., Mishra, N., Motwani, R., & O'Callaghan, L. (2003). Clustering data streams: Theory and practice. *IEEE Transactions on Knowledge and Data Engineering*, *15*(3), 515–528. doi:10.1109/TKDE.2003.1198387

Guha, S., Mishra, N., Motwani, R., & O'Callaghan, L. (2000). Clustering data streams. In *Proceedings of the Annual Symposium on Foundations of Computer Science (FOCS)*, Redondo Beach, California, November 2000.

Guha, S., Rastogi, R., & Shim, K. (1998). CURE: An efficient clustering algorithm for large databases. In L. M. Haas & A. Tiwary, A., editors, *Proceedings of the 1998 ACM-SIGMOD International Conference on Management of Data* (pp. 73–84). ACM Press.

Gummadi, R., Gnawali, O., & Govindan, R. (2005, June). Macro-programming wireless sensor networks using Kairos. In *International Conference on Distributed Computing in Sensor Systems (DCOSS)*, Marina del Rey, California.

Gupta, H., Zhou, Z., Das, S.R., & Gu, Q. (2006). Connected sensor cover: Self-organization of sensor networks for efficient query execution. *IEEE/ACM Transactions on Networking*, *14*(1), 55-67.

Güting, R. H., & Schneider, M. (2005). *Moving object databases*. Morgan Kaufman.

Halatchev, M., & Gruenwald, L. (2005). Estimating missing values in related sensor data streams. In *11th International Conference on Management of Data* (COMAD), Goa, India, January 2005.

Halkidi, M., Batistakis, Y., & Varzirgiannis, M. (2001). On clustering validation techniques. *Journal of Intelligent Information Systems*, *17*(2-3), 107–145. doi:10.1023/A:1012801612483

Hall, D. L., & Llinas, J. (Eds.). (2001). *Handbook of multisensor data fusion*. New York: CRC Press.

Han, J., & Kamber, M. (2000). *Data Mining: Concepts and Techniques*. San Francisco: Morgan Kaufmann Publishers.

Han, J., & Kamber, M. (2006). *Data mining: Concepts and techniques*. The Morgan Kaufmann Series in Data Management Systems, Morgan Kaufmann Publishers.

Han, J., Chen, Y., Dong, G., Pei, J., Wah, B. W., Wang, J., & Cai, Y. D. (2005). Stream cube: An architecture for multi-dimensional analysis of data streams. *Distributed and Parallel Databases*, *18*(2), 173–197. doi:10.1007/s10619-005-3296-1

Han, J., Pei, J., Dong, G., & Wang, K. (2001). Efficient computation of iceberg cubes with complex measures. In *Proceedings of the 2001 ACM International Conference on Management of Data* (pp. 1-12).

Han, J., Stefanovic, N., & Kopersky, K. (1998). Selective materialization: An efficient method for spatial data cube construction. In *PAKDD'98* (pp. 144-158).

Hanson, E. N. (1996). The design and implementation of the Ariel active database rule system. *IEEE TKDE, 8*(1).

Hardin, J., & Rocke, D. M. (2004). Outlier detection in the multiple cluster setting using the minimum covariance determinant estimator. *Journal of Computational Statistics and Data Analysis*, *44*, 625–638. doi:10.1016/S0167-9473(02)00280-3

Harinarayan, V., Rajaraman, A., & Ullman, J. D. (1996). Implementing data cubes efficiently. In *SIGMOD'96*.

Harkins, S., He, H., Willams, G. J., & Baster, R. A. (2002). Outlier detection using replicator neural networks. In *Proceedings of International Conference on Data Warehousing and Knowledge Discovery* (pp. 170-180).

Hartigan, J. A. (1975). *Clustering algorithms*. New York: Wiley.

Hartigan, J. A., & Wong, M. (1979). Algorithm AS136: A k-means clustering algorithm. *Applied Statistics*, *28*, 100–108. doi:10.2307/2346830

Hawkins, D. M. (1980). *Identification of outliers*. London: Chapman and Hall.

He, T., Krishnamurthy, S., Stankovic, J., Abdelzaher, T., Luo, L., Stoleru, R., et al. (2004). Energy-efficient surveillance system using wireless sensor networks. *2nd International conference on Mobile systems, applications, and services* (pp. 270-283).

He, Z., Deng, S., & Xu, X. (2005). An optimization model for outlier detection in categorical data. In *Proceedings of International Conference on Intelligent Computing* (pp. 400-409).

He, Z., Xu, X., & Deng, S. (2003). Discovering cluster based local outliers. *Official Publication of Pattern Recognition Letters, 24*(9-10), 1651–1660.

He, Z., Xu, X., & Deng, S. (2003). Outlier detection over data streams. In *Proceedings of International Conference for Young Computer Scientists. Harbin, China.*

Heinzelman, W., Chandrakasan, A., & Balakrishnan, H. (2000). Energy-Efficient Communication Protocol for Wireless Microsensor Networks. In Proceedings of the 33rd Hawaii International Conference on System Sciences (HICSS '00).

Hellsten, H., Ulander, L. M. H., Gustavsson, A., & Larsson, B. (1996). Development of VHF CARABAS-II SAR. In . *Proceedings of the Radar Sensor Technology SPIE., 2747,* 48–60.

Hemmerling, A. (1989). *Labyrinth Problems: Labyrinth-Searching Abilities of Automata.* B.G. Teubner, Leipzig.

Henzinger, M. R., Raghavan, P., & Rajagopalan, S. (1998). Computing on data streams (Technical Report 1998-011). Digital Systems Research Center.

Hettich, S., & Bay, S. D. (1999). The UCI KDD archive. Irvine, CA: University of California, Department of Information and Computer Science. Retrieved July 22, 2008, from http://kdd.ics.uci.edu/databases/el_nino/tao-all2.dat.gz

Hightower, J., & Borriello, G. (2001). Location systems for ubiquitous computing. *Computer.*

Hinneburg, A., & Keim, D. A. (1998). An efficient approach to clustering in large multimedia databases with noise. In R. Agrawal, P. Stolorz, & G. Piatetsky-Shapiro (Eds.), *Proceedings, The Fourth International Conference on Knowledge Discovery and Data Mining* (pp. 58-65). Menlo Park, CA: AAAI Press.

Ho, C.-T., Agrawal, R., Megiddo, N., & Srikant, R. (1997). Range queries in OLAP data cubes. In *Proceedings of the 1997 ACM International Conference on Management of Data* (pp. 73-88).

Hodge, V. J., & Austin, J. (2003). A survey of outlier detection methodologies. *International Journal of Artificial Intelligence Review, 22,* 85–126.

Hoeffding, W. (1963). Probability inequalities for sums of bounded random variables. *Journal of the American Statistical Association, 58*(301), 13–30. doi:10.2307/2282952

Horney, T., Ahlberg, J., Jungert, E., Folkesson, M., Silvervarg, K., Lantz, F., et al. (2004, April). An information system for target recognition. In *Proceedings of the SPIE Conference on Defense and Security, Orlando, Florida,* (pp. 12-16).

Horney, T., Holmberg, M., Silvervarg, K., & Brännström, M. (2006, Sept.3-6). MOSART Research Testbed. *IEEE International Conference on Multisensor Fusion and Integration for Intelligent Systems,* (pp. 225-229).

Howard, J. H., Kazar, M. L., Menees, S. G., Nichols, D. A., Satyanarayanan, M., Sidebotham, R. N., & West, M. J. (1998). Scale and performance in a distributed file system. *ACM Transactions on Computer Systems, 6*(1), 51–81. doi:10.1145/35037.35059

Hu, T., & Sung, S. Y. (2003). Detecting pattern-based outliers. *Official Publication of Pattern Recognition Letters, 24*(16), 3059–3068. doi:10.1016/S0167-8655(03)00165-X

Hu, W., Misra, A., & Shorey, R. (2006). CAPS: Energy-efficient processing of continuous aggregate queries in sensor networks. In *IEEE International Conference on Pervasive Computing and Communications (PerCom)* (pp. 190-199).

Hulten, G., Spencer, L., & Domingos, P. (2001, August). Mining time-changing data streams. In *Proceedings of the Seventh ACM SIGKDD International Conference on Knowledge Discovery and Data Mining*, San Francisco, California.ACM Press.

Ibriq, J., & Mahgoub, I. (2004). Cluster-based routing in wireless sensor networks: Issues and challenges. In *International Symposium on Performance Evaluation of Computer and Telecommunication Systems* (pp. 759-766).

Ihler, A., Hutchins, J., & Smyth, P. (2006). Adaptive event detection with time-varying poisson processes. In *Proceedings of the 12th ACM SIGKDD International Conference on Knowledge Discovery and Data Mining.* New York: ACM.

Ihler, A., Hutchins, J., & Smyth, P. (2007). Learning to detect events with markov-modulated poisson processes. *ACM Transactions on Knowledge Discovery from Data, 1*(3).

Intanagonwiwat, C., Estrin, D., Govindan, R., & Heidemann, J. S. (2002). Impact of network density on data aggregation in wireless sensor networks. In *Proceedings of the 22nd IEEE International Conference on Distributed Computing Systems* (pp. 457-458).

Intanagonwiwat, C., Govindan, R., & Estrin, D. (2000). Directed diffusion: A scalable and robust communication paradigm for sensor networks. In Proceedings of the Sixth Annual International Conference on Mobile Computing and Networking (pp. 56-67). Boston: ACM Press.

Intanagonwiwat, C., Govindan, R., Estrin, D., Heidemann, J., & Silva, F. (2003). Directed diffusion for wireless sensor networking. *IEEE/ACM Transactions on Networking, 11*, 2-16.

Ives, Z. G., Levy, A. Y., & Weld, D. S. (2000). Efficient evaluation of regular path expressions on streaming XML data (Technical Report UW-CSE-2000-05-02). University of Washington.

Jain, A. K., & Dubes, R. C. (1988). *Algorithms for clustering data*. Englewood Cliffs, NJ: Prentice-Hall.

Jain, A. K., Murty, M. N., & Flynn, P. J. (1999, September). Data clustering: A review. *ACM Computing Surveys, 31*(3), 264–323. doi:10.1145/331499.331504

Jamieson, K., Balakrishnan, H., & Tay, Y. C. (2003). *Sift: A MAC protocol for event-driven wireless sensor networks* (Technical Report 894). Cambridge, Massachusetts: Laboratory for Computer Science, Massachusetts Institute of Technology.

Janakiram, D., Mallikarjuna, A., Reddy, V., & Kumar, P. (2006). Outlier detection in wireless sensor networks using Bayesian belief networks. In *Proceedings of Communication System Software and Middleware* (pp. 1-6).

Januzaj, E., Kriegel, H. P., & Pfeifle, M. (2003). Towards effective and efficient distributed clustering. *Workshop on Clustering Large Data Sets (ICDM2003).*

Jeffery, S. R., Alonso, G., Franklin, M. J., Hong, W., & Widom, J. (2006). Declarative support for sensor data cleaning. In *Proceedings of International Conference on Pervasive Computing* (pp. 83-100).

Jeffery, S. R., Garofalakis, M., & Franklin, M. J. (2006). Adaptive cleaning for RFID data streams. In *VLDB'06*, Seoul, Korea.

Jeffrey, S. R., Alonso, G., Franklin, M., Hong, W., & Widom, J. (2006). A pipelined framework for online cleaning of sensor data streams. In *ICDE'06*, Atlanta, Georgia.

Jensen, F. V. (1996). *An introduction to Bayesian networks.* New York: Springer Verlag.

Jiang, M. F., Tseng, S. S., & Su, C. M. (2001). Tw-phase clustering process for outliers detection. *Official Publication of Pattern Recognition Letters, 22*(6-7), 691–700. doi:10.1016/S0167-8655(00)00131-8

Jiang, N., (2007). *A data imputation model in sensor databases* (LNCS 4782, pp. 86-96).

Jiang, Q., & Chakravarthy, S. (2004, Mar.). Data Stream Management System for MavHome. In *Proc. of ACM SAC.*

Jiang, Q., & Chakravarthy, S. (2004, Jul.). Scheduling Strategies for Processing Continuous Queries over Streams. In *Proc. of BNCOD*.

Jiang, Q., Adaikkalavan, R., & Chakravarthy, S. (2004). *Towards an Integrated Model for Event and Stream Processing* (Tech. Rep. Nos. CSE-2004-10, ITLab, The University of Texas at Arlington).

Jiang, Q., Adaikkalavan, R., & Chakravarthy, S. (2005, Apr.). NFM^i: An Inter-domain Network Fault Management System. In *Proc. of ICDE* (pp. 1036-1047). Tokyo, Japan.

Jiang, Q., Adaikkalavan, R., & Chakravarthy, S. (2007, Jul.). MavEStream: Synergistic Integration of Stream and Event Processing. In *IEEE International Workshop on Data Stream Processing*.

Jin, W., Tung, A. K. H., & Han, J. (2001). Mining top-n local outliers in large databases. In *Proceedings of the ACM SIGMOD Conference on Knowledge Discovery and Data* (pp. 293-298).

Jindal, A., & Psounis, K. (2006). Modeling spatially correlated data in sensor networks. *ACM Transactions on Sensor Networks, 2*(4), 466–499. doi:10.1145/1218556.1218558

Johnson, E., & Kargupta, H. (1999). Collective, hierarchical clustering from distributed heterogeneous data. In M. Zaki & C. Ho (Eds.), *Large-Scale Parallel KDD Systems*, Volume 1759 of Lecture Notes in Computer Science (pp. 221-244). Berlin/Heidelberg, Germany: Springer.

Johnson, T., Kwok, I., & Ng, R. T. (1998). Fast computation of 2-dimensional depth contours. In *Proceedings of the ACM SIGMOD Conference on Knowledge Discovery and Data* (pp. 224-228).

Juang, B. H., & Rabiner, L. R. (1990). The segmental K-Means algorithm for estimating parameters of hidden Markov models. *IEEE Transactions on Acoustics, Speech, and Signal Processing, 9*(38), 1639–1641. doi:10.1109/29.60082

Jungert, E. (1999). A qualitative approach to reasoning about objects in motion based on symbolic projection. In *Proceedings of the Conference on Multimedia Databases and Image Communication (MDIC'99). Salerno, Italy, October 4-5* (pp. 89-100).

Jungert, E., Söderman, U., Ahlberg, S., Hörling, P., Lantz, F., & Neider, G. (1999, April 7-8). Generation of high resolution terrain elevation models for synthetic environments using laser-radar data. In *Proceedings of SPIE Modeling, Simulation and Visualization for Real and Virtual Environments. Orlando, Florida,* (pp. 12-20).

Kalnis, P., Ng, W., Ooi, B., Papadias, D., & Tan, K. (2002). An adaptive peer-to-peer network for distributed caching of OLAP results. In SIGMOD Record, *ACM SIGMOD International Conference on the Management of Data, 31*(2), 25-36. ACM Press.

Kantarcioglu, M., Jin, J., & Clifton, C. (2004). When do data mining results violate privacy? In *Proceedings of the 10th ACM SIGKDD International Conference on Knowledge Discovery and Data Mining* (pp. 599-604). New York: ACM Press.

Kargupta, H., & Chan, P. (Eds.). (2000). *Distributed and parallel data mining*. Menlo Park, CA / Cambridge, MA: AAAI Press / MIT Press.

Kargupta, H., & Sivakumar, K. (2004). Existential pleasures of distributed data mining. In *Data mining: Next generation challenges and future directions*. AAAI/MIT Press.

Kargupta, H., Huang, W., Sivakumar, K., & Johnson, E. L. (2001). Distributed clustering using collective principal component analysis. *Knowledge and Information Systems, 3*, 422–448. doi:10.1007/PL00011677

Kargupta, H., Park, B.-H., Hershberger, D., & Johnson, E. (2000). Collective data mining: a new perspective toward distributed data mining. In H. Kargupta & P. Chan (Eds.), *Advances in distributed and parallel knowledge discovery* (pp. 131-174). Menlo Park, CA / Cambridge, MA: AAAI Press / MIT Press.

Karl, H., & Willig, A. (2005). *Protocols and Architectures for Wireless Sensor Networks*, chapter 10.1.2 Aspects of topology control algorithms. Wiley.

Karlof, C., Sastry, N., & Wagner, C. (2004). TinySec: A Link Layer Security Architecture for Wireless Sensor Networks. In Proceedings of the Second ACM Conference on Embedded Networked Sensor Systems (SenSys 2004).

Karp, B., & Kung, H. (2000). GPSR: Greedy perimeter stateless routing for wireless networks. In *MobiCom '00: 6th Annual International Conference on Mobile Computing and Networking*, (pp. 243-254). New York: ACM.

Kawashima, H., Hirota, Y., Satake, S., & Imai, M. (2006). MeT: A Real World Oriented Metadata Management System for Semantic Sensor Networks. *3rd International Workshop on Data Management for Sensor Networks (DMSN)* (pp. 13-18).

Khalil, I., Bagchi, S., & Shroff, N. B. (2006). MOBIWORP: Mitigation of the Wormhole Attack in Mobile Multihop Wireless Networks. Securecomm.

Kim, S., & Cho, S. (2006). Prototype based outlier detection. In *Proceedings of International Joint Conference on Neural Networks* (pp. 820-826).

Kim, Y., Perrig, A., & Tsudik, G. (2000). Simple and fault-tolerant key agreement for dynamic collaborative groups. In ACM CCS 2000.

Kim, Y.-J., Govindan, R., Karp, B., & Shenker, S. (2005). Geographic routing made practical. In *Networked Systems Design & Implementation*.

Kim, Y.-J., Govindan, R., Karp, B., & Shenker, S. (2005). On the pitfalls of geographic face routing. In *Foundations of mobile computing*.

Kim, Y.-J., Govindan, R., Karp, B., & Shenker, S. (2006). Lazy cross-link removal for geographic routing. In *Embedded Networked Sensor Systems*.

Klampanos, I. A., & Jose, J. M. (2004). An architecture for information retrieval over semi-collaborating peer-to-peer networks. In *Proceedings of the 2004 ACM Symposium on Applied Computing* (pp. 1078-1083). New York: ACM.

Klampanos, I. A., Jose, J. M., & van Rijsbergen, C. J. K. (2006). Single-pass clustering for peer-to-peer information retrieval: The effect of document ordering. In *INFOSCALE '06. Proceedings of the First International Conference on Scalable Information Systems*. New York: ACM.

Klusch, M., Lodi, S., & Moro, G. (2003). Distributed clustering based on sampling local density estimates. In *Proceedings of the 19th International Joint Conference on Artificial Intelligence* (pp. 485-490). Acapulco, Mexico: AAAI Press.

Knorr, E., & Ng, R. (1998). Algorithms for mining distance-based outliers in large data sets. *International Journal of Very Large Data Bases* (pp. 392-403).

Kollios, G., Gunopulos, D., Koudas, N., & Berchtold, S. (2003). Efficient biased sampling for approximate clustering and outlier detection in large data sets. *International Journal of Knowledge and Data Engineering, 15*(5), 1170–1187. doi:10.1109/TKDE.2003.1232271

Kong, P., Zerfos, P., Luo, H., Lu, S., & Zhang, L. (2001). Providing Robust and Ubiquitous Security Support for Mobile Ad-Hoc Networks. International Conference on Network Protocols (ICNP).

Koontz, W. L. G., Narendra, P. M., & Fukunaga, K. (1976). A graph-theoretic approach to nonparametric cluster analysis. *IEEE Transactions on Computers, C-25,* 936–944. doi:10.1109/TC.1976.1674719

Kothuri, R., Hanckel, R., & Yalamanchi, A. (2008). Using Oracle Extensibility Framework for Supporting Temporal and Spatio-Temporal Applications. In *Proceedings of the fifteenth International Symposium on Temporal Representation and Reasoning* (pp. 15-18).

Kotidis, Y. (2005). Snapshot queries: Towards data-centric sensor networks. In *Proc. of the 2005 Intl. Conf. on Data Engineering* (pp. 131-142).

Kotidis, Y. (2006). Processing proximity queries in sensor networks. *3rd International Workshop on Data Management for Sensor Networks (DMSN)* (pp. 1-6).

Kotz-Dittrich, A. (1993, Mar.). Adding Active Functionality to an Object-Oriented Database System - A Layered Approach. In *Proc. of the Conference on Database Systems in Office, Technique and Science.*

Kou, Y., Lu, C., & Chen, D. (2006). Spatial weighted outlier detection. In *Proceedings of SIAM International Conference on Data Mining* (pp. 613-617).

Kranakis, E., Singh, H., & Urrutia, J. (1999). Compass routing on geometric networks. In *Canadian Conference on Computational Geometry.*

Kudela, R., Pitcher, G., Probyn, T., Figueiras, F., Moita, T., & Trainer, V. (2005). Harmful algal blooms in coastal upwelling systems. *Oceanography (Washington, D.C.), 18*(2), 184–197.

Kuhn, F., Wattenhofer, R., & Zollinger, A. (2003). Worst-case optimal and average-case efficient geometric ad-hoc routing. In *Mobile ad hoc Networking & Computing.*

Kuhn, F., Wattenhofer, R., Zhang, Y., & Zollinger, A. (2003). Geometric ad-hoc routing: of theory and practice. In *Principles of Distributed Computing.*

Kuruvila, J., Nayak, A., & Stojmenovic, I. (2006). Progress and location based localized power aware routing for ad hoc and sensor wireless networks. *International Journal of Distributed Sensor Networks.*

Last, M. (2002). Online classification of nonstationary data streams. *Intelligent Data Analysis, 6*(2), 129–147.

Laurikkala, J., Juhola, M., & Kentala, E. (2000). Informal identification of outliers in medical data. In *Proceedings of International Workshop on Intelligent Data Analysis in Medicine and Pharmacology.*

Lazarevic, A., Ozgur, A., Ertoz, L., Srivastava, J., & Kumar, V. (2003). A comparative study of anomaly detection schemes in network intrusion detection. In *Proceedings of SIAM Conference on Data Mining.*

Lehane, B., & O'Connor, N. E. (2004). Action sequence detection in motion pictures. *The International Workshop on Multidisciplinary Image, Video, and Audio Retrieval and Mining.*

Lehane, B., & O'Connor, N. E. (2006). Movie indexing via event detection. *7th International Workshop on Image Analysis for Multimedia Interactive Services.*

Lehane, B., O'Connor, N. E., & Murphy, N. (2005). Dialogue scene detection in movies. *International Conference on Image and Video Retrieval (CIVR)* (pp. 286-296).

Lehane, B., O'Connor, N. E., Smeaton, A. F., & Lee, H. (2006). A system for event-based film browsing. *TIDSE 2006, 3rd International Conference on Technologies for Interactive Digital Storytelling and Entertainment* (LNCS 4326, pp. 334-345).

Leone, P., Moraru, L., Powell, O., & Rolim, J. (2006). A localization algorithm for wireless ad-hoc sensor networks with traffic overhead minimization by emission inhibition. In *Algorithmic Aspects of Wireless Sensor Networks (ALGOSENSORS).*

Leong, B., Liskov, B., & Morris, R. (2006). Geographic routing without planarization. In *NSDI'06: Proceedings of the 3rd conference on 3rd Symposium on Networked Systems Design & Implementation* (pp. 25–25). Berkeley, CA. USENIX Association.

Li, M., Lee, G., Lee, W.-C., & Sivasubramaniam, A. (2006). PENS: An algorithm for density-based clustering in peer-to-peer systems. In *INFOSCALE '06. Proceedings of the First International Conference on Scalable Information Systems.* New York: ACM.

Li, S., & Kernighan, B. (1971). An effective heuristic algorithm for the TSP. *Operations Research, 21,* 498–516.

Li, X., & Chang, S.-K. (2003). An interactive visual query interface on spatial/temporal data. In *Proceedings of the 10th international conference on Distributed Multimedia Systems. San Francisco, September 8-10, 2003* (pp. 257-262).

Li, X., Han, J., & Gonzalez, H. (2004). High-dimensional OLAP: A minimal cubing approach. In *Proceedings of the 30th International Conference on Very Large Data Bases* (pp. 528-539).

Li, X., Kim, Y., Govindan, R., & Hong, W. (2003). Multidimensional range queries in sensor networks. In *SenSys '03: Proceedings of the 1st International Conference on Embedded Networked Sensor Systems* (pp. 63-75). New York: ACM.

Lieuwen, D. L., Gehani, N. H., & Arlein, R. (1996, Mar.). The Ode Active Database: Trigger Semantics and Implementation. In *Proc of ICDE* (pp. 412-420).

Lindell, Y., & Pinkas, B. (2002). Privacy preserving data mining. *Journal of Cryptology*, *15*(3), 177–206. doi:10.1007/s00145-001-0019-2

Liu, D., & Ning, P. (2003). Establishing pairwise keys in distributed sensor networks. In Proceedings of 10th ACM Conference on Computer and Communications Security (CCS'03) (pp. 41-47). Washington DC: ACM Press.

Liu, D., & Ning, P. (2005). Improving key pre-distribution with deployment knowledge in static sensor networks. *ACM Transactions on Sensor Networks*, *1*(2), 204–239. doi:10.1145/1105688.1105691

Lodi, S., Moro, G., & Sartori, C. (in press). Distributed data clustering in multi-dimensional peer-to-peer networks. In H. T. Shen & A. Bouguettaya (Eds.), Conferences in Research and Practice in Information Technology (CRPIT): Vol. 103. Proceedings of the Twenty-First Australasian Database Conference (ADC2010). Brisbane, Australia: Australian Computer Society.

Loo, K. K., Tong, I., Kao, B., & Chenung, D. (2005). Online algorithms for mining inter-stream associations from large sensor networks. In *Proceedings of the Ninth Pacific-Asia Conference on Knowledge Discovery and Data Mining*, Hanoi, Vietnam, May 2005.

Lopez, I., Snodgrass, R., & Moon, B. (2005). Spatiotemporal aggregate computation: A survey. *IEEE TKDE*, *17*(2), 271–286.

Lopez, X., & Das, S. (2007). *Semantic data integration for the enterprise*. Retrieved from http://www.oracle.com/technology/tech/semantic_technologies/pdf/semantic11g_dataint_twp.pdf

Lu, C. T., Chen, D., & Kou, Y. (2003). Algorithms for spatial outlier detection. In *Proceedings of International Conference on Data Mining* (pp. 597-600).

Lu, C. T., Chen, D., & Kou, Y. (2003). Detecting spatial outliers with multiple attributes. *In Proceedings of International Conference on Tools with Artificial Intelligence* (pp. 122-128).

Lu, C., Blum, B. M., Abdelzaher, T. F., Stankovic, J. A., & He, T. (2002). RAP: A real-time communication architecture for large-scale wireless sensor networks. In *IEEE Real-Time and Embedded Technology and Applications Symposium* (pp. 55-66).

Luk, M., Mezzour, G., Perrig, A., & Gligor, V. D. (2007). MiniSec: A Secure Sensor Network Communication Architecture. In Proceedings of IEEE International Conference on Information Processing in Sensor Networks (IPSN).

Luo, X., Dong, M., & Huang, Y. (2006). On distributed fault-tolerant detection in wireless sensor networks. *IEEE Transactions on Computers*, *55*(1), 58–70. doi:10.1109/TC.2006.13

Luotonen, A., & Altis, K. (1994). World-wide web proxies. In *Selected papers of the First Conference on the World-Wide Web* (pp. 147-154), Amsterdam, The Netherlands, May 1994.

Ma, X., Yang, D., Tang, S., Luo, Q., Zhang, D., & Li, S. (2004). Online mining in sensor networks. In *Proceedings of International conference on network and parallel computing* (pp. 544-550).

Madden, S., & Franklin, M. J. (2002). Fjording the stream: An architecture for queries over streaming sensor data. In *Proceedings of the 18th IEEE International Conference on Data Engineering* (pp. 555-566).

Madden, S., & Hellerstein, J. M. (2002). Distributing queries over low-power wireless sensor networks. In *Proceedings of the 2002 ACM International Conference on Management of Data* (pp. 622).

Madden, S., Franklin, M. J., & Hellerstein, J. M. (2002). TAG: A Tiny AGgregation service for ad-hoc sensor networks. *ACM SIGOPS Operating Systems Review*, *36*, 131–146. doi:10.1145/844128.844142

Madden, S., Franklin, M. J., Hellerstein, J. M., & Hong, W. (2003). The design of an acquisitional query processor for sensor networks. In *Proceedings of the 2003 ACM SIGMOD international conference on Management of data* (pp. 491-502). ACM Press.

Madden, S., Franklin, M. J., Hellerstein, J. M., & Hong, W. (2005). TinyDB: An acquisitional query processing system for sensor networks. *ACM Transactions on Database Systems, 30*(1), 122–173. doi:10.1145/1061318.1061322

Madden, S., Szewczyk, R., Franklin, M. J., & Culler, M. J. (2002). Supporting aggregate queries over ad-hoc wireless sensor networks. In *Proceedings of the 4th IEEE Workshop on Mobile Computing and Systems, Applications* (pp. 49-58).

Madey, G. R., Barabási, A.-L., Chawla, N. V., Gonzalez, M., Hachen, D., Lantz, B., et al. (2007). Enhanced situational awareness: Application of DDDAS concepts to emergency and disaster management. In Y. Shi, G. D. van Albada, J. Dongarra, & P. M. A. Sloot (Eds.), *Proceedings of the International Conference on Computational Science* (Vol. 4487, pp. 1090-1097). Berlin, Germany: Springer.

Madey, G. R., Szabó, G., & Barabási, A.-L. (2006). WIPER: The integrated wireless phone based emergency response system. In V. N. Alexandrov, G. D. val Albada, P. M. A. Sloot, & J. Dongarra (Eds.), *Proceedings of the International Conference on Computational Science* (Vol. 3993, pp. 417-424). Berlin, Germany: Springer-Verlag.

Malan, D., Welsh, M., & Smith, M. (2004). A Public-Key Infrastructure for Key Distribution in TinyOS Based on Elliptic Curve Cryptography. IEEE SECON 2004.

Malerba, D., Appice, A., & Ceci, M. (2004). A data mining query language for knowledge discovery in a geographical information system. In *Database Support for Data Mining Applications* (pp. 95-116).

Manjunath, B. S., Salember, P., & Sikora, T. (2002). *Introduction to MPEG-7, Multimedia content description language.* John Wiley and Sons Ltd.

Manku, G. S., & Motwani, R. (2002). Approximate frequency counts over data streams. In *Proceedings of the 28th International Conference on Very Large Data Bases* (pp. 346-357).

Marchant, P., Briseboi, A., Bédard, Y., & Edwards, G. (2004). Implementation and evaluation of a hypercube-based method for spatiotemporal exploration and analysis. *ISPRS Journal of Photogrammetry and Remote Sensing, 59*, 6–20. doi:10.1016/j.isprsjprs.2003.12.002

Margaritis, D., Faloutsos, C., & Thrun, S. (2001). NetCube: A scalable tool for fast data mining and compression. In *Proceedings of the 27th International Conference on Very Large Data Bases* (pp. 311-320).

Markos, M., & Singh, S. (2003). Novelty detection: A review-part 1: statistical approaches. *International Journal of Signal Processing, 83*, 2481–2497. doi:10.1016/j.sigpro.2003.07.018

McLachlan, G., & Krishnan, T. (1997). *The EM algorithm and extensions.* Wiley series in probability and statistics, John Wiley and Sons.

Medhi, D., et al. (2001,May.). A Network Management Framework for Multi-Layered Network Survivability: An Overview. In *IEEE/IFIP conf. on integrated network management* (pp. 293-296).

Merugu, S., & Ghosh, J. (2003). Privacy-preserving distributed clustering using generative models. In X. Wu, A. Tuzhilin, & J. Shavlik (Eds.), *Proceedings of the 3rd IEEE International Conference on Data Mining.* Los Alamitos, CA: IEEE Computer Society.

Milenova, B. L., & Campos, M. M. (2002). O-Cluster: Scalable clustering of large high-dimensional data sets. In *Proceedings of the IEEE International Conference on Data Mining* (pp. 290-297).

Milenova, B. L., & Campos, M. M. (2005). Mining high-dimensional data for information fusion: A database-centric approach. In *Proceedings of the 2005 International Conference on Information Fusion.*

Milenova, B. L., Yarmus, J., & Campos, M. M. (2005). SVM in Oracle Database 10g: Removing the barriers to widespread adoption of support vector machines. In *Proceedings of the 31st International Conference on Very Large Data Bases* (pp. 1152-1163).

Miller, V.S. (1986). Use of Elliptic Curves in Cryptography, Advances in Cryptology – CRYPTO '85 Proceedings (LNCS 218, pp. 417-426).

Milojicic, D. S., Kalogeraki, V., Lukose, R., Nagaraja, K., Pruyne, J., Richard, B., et al. (2002). *Peer-to-peer computing* (Technical Report HPL-2002-57R1). HP Lab.

Mitra, P., Murthy, C., & Pal, S. K. (2002). Unsupervised feature selection using feature similarity. *IEEE Transactions on Pattern Analysis and Machine Intelligence, 24*(3), 301–312. doi:10.1109/34.990133

Mitzenmacher, M., & Upfal, E. (2005). *Probability and computing: Randomized algorithms and probabilistic analysis*. Cambridge, UK: Cambridge University Press.

Monteiro, R. S., Zimbrao, G., Schwarz, H., Mitschang, B., & de Souza, J. M. (2005). Building the data warehouse of frequent itemsets in the dwfist approach. In *ISMIS* (pp. 294-303).

Monti, G., & Moro, G. (2008). Multidimensional range query and load balancing in wireless ad hoc and sensor networks. In K.Wehrle, W. Kellerer, S. K. Singhal, & R. Steinmetz (Eds.), *Eighth International Conference on Peer-to-Peer Computing* (pp. 205-214). Los Alamitos, CA: IEEE Computer Society.

Monti, G., & Moro, G. (2009). Self-organization and local learning methods for improving the applicability and efficiency of data-centric sensor networks. *Sixth International ICST Conference on Heterogeneous Networking for Quality, Reliability, Security and Robustness, QShine/AAA-IDEA 2009*, (LNICST 22, pp. 627-643). Berlin/Heidelberg, Germany: Springer.

Monti, G., Moro, G., & Lodi, S. (2007). W*-Grid: A robust decentralized cross-layer infrastructure for routing and multi-dimensional data management in wireless ad-hoc sensor networks. In M. Hauswirth, A. Montresor, N. Shahmehri, K. Wehrle, & A. Wierzbicki, *Seventh IEEE International Conference on Peer-to-Peer Computing* (pp. 159-166). Los Alamitos, CA: IEEE Computer Society.

Monti, G., Moro, G., & Sartori, C. (2006). W^R-Grid: A scalable cross-layer infrastructure for routing, multi-dimensional data management and replication in wireless sensor networks. In G. Min, B. Di Martino, L.T.Yang, M. Guo, & G. Ruenger (Eds.), *Frontiers of High Performance Computing and Networking – ISPA 2006 Workshops* (LNCS 4331, pp. 377-386). Berlin/Heidelberg, Germany: Springer.

Moor, J. H. (1997). Towards a theory of privacy in the information age. *ACM SIGCAS Computers and Society, 27*(3), 27–32. doi:10.1145/270858.270866

Moraru, L., Leone, P., Nikoletseas, S., & Rolim, J. D. P. (2007). Near optimal geographic routing with obstacle avoidance in wireless sensor networks by fast-converging trust-based algorithms. In *Q2SWinet '07: Proceedings of the 3rd ACM workshop on QoS and security for wireless and mobile networks* (pp. 31–38). New York: ACM.

Moreira, A., & Santos, M. Y. (2005). Enhancing a user context by real-time clustering mobile trajectories. In *International Conference on Information Technology: Coding and Computing (ITCC'05)* (Vol. 2, p. 836). Los Alamitos, CA, USA: IEEE Computer Society.

Moro, G., & Monti, G. (2006). W-Grid: A cross-layer infrastructure for multi-dimensional indexing, querying and routing in ad-hoc and sensor networks. In A. Montresor, A. Wierzbicki, & N. Shahmehri (Eds.), *IEEE Int. Conference on P2P Computing* (pp. 210-220). Los Alamitos, CA: IEEE Computer Society.

Moro, G., & Ouksel, A. M. (2003). G-Grid: A class of scalable and self-organizing data structures for multi-dimensional querying and content routing in P2P networks. *Agents and Peer-to-Peer Computing* (LNCS 2872, pp. 123-137). Berlin/Heidelberg, Germany: Springer.

Motwani, R., et al. (2003, Jan.). Query Processing, Resource Management, and Approximation in a Data Stream Management System. In *Proc. of CIDR*.

Mountzia, M. A., & Rodosek, G. D. (1999). Using the Concept of Intelligent Agents in Fault Management of Distributed Services. *Journal of Network and Systems Management, 7*(4). doi:10.1023/A:1018739932618

Murray, S. D. (Ed.). (2001). U.S.-EU "Safe Harbor" Data Privacy Arrangement. *The American Journal of International Law, 91*, 169–169.

Muthukrishnan, S., Shah, R., & Vitter, J. S. (2004). Mining deviants in time series data streams. In *Proceedings of International Conference on Scientific and Statistical Database Management*.

Nakashima, E. (2007). *Cellphone tracking powers on request*. Washington Post.

Nascimento, S., Casimiro, H., Sousa, F. M., & Boutov, D. (2005). Applicability of fuzzy clustering for the identification of upwelling areas on sea surface temperature images. In B. Mirkin & G. Magoulas (Eds.), *Proceedings of the 2005 UK Workshop on Computational Intelligence* (pp. 143–148). London, United Kingdom. Retrieved August 6, 2008, from http://www.dcs.bbk.ac.uk/ukci05/ukci05proceedings.pdf

National Research Council. (2007). *Putting people on the map: Protecting confidentiality with linked social-spatial data* (M. P. Gutmann & P. C. Stern, Eds.). Washington, D.C.: The National Academies Press.

Nejdl, W., et al. (2002). EDUTELLA: a P2P networking infrastructure based on RDF. *11th International World Wide Web Conference* (pp. 604-615). ACM Press.

Newman, M. (2003). The structure and function of complex networks. *SIAM Review, 45*(2), 167–256. doi:10.1137/S003614450342480

Newman, M. E. J. (2001). Clustering and preferential attachment in growing networks. *Physical Review E: Statistical, Nonlinear, and Soft Matter Physics, 64*(025102).

Newman, M. E. J. (2002). Assortative mixing in networks. *Physical Review Letters, 89*(208701).

Newman, M. E. J. (2004). Detecting community structure in networks. *The European Physical Journal B, 38*, 321–330. doi:10.1140/epjb/e2004-00124-y

Newman, M., Barabási, A.-L., & Watts, D. J. (Eds.). (2006). *The structure and dynamics of networks*. Princeton, NJ: Princeton University Press.

Newton, R., Morrisett, G., & Welsh, M. (2007). The Regiment macroprogramming system. In *International Conference on Information Processing in Sensor Networks (IPSN)* (pp. 489-498).

Nikoletseas, S., & Powell, O. (2008). Obstacle avoidance algorithms in wireless sensor networks. *Encyclopedia of Algorithms*. Springer.

Noble, C. C., & Cook, D. (2003). Graph-based anomaly detection. In *Proceedings of the 9th ACM SIGKDD International Conference on Knowledge Discovery and Data Mining*.

Nowak, R. D. (2003). Distributed EM algorithms for density estimation and clustering in sensor networks. *IEEE Transactions on Signal Processing, 51*, 2245–2253. doi:10.1109/TSP.2003.814623

O'Callaghan, L., Meyerson, A., Motwani, R., Mishra, N., & Guha, S. (2002). Streaming-data algorithms for high-quality clustering. In *Proceedings of the Eighteenth Annual IEEE International Conference on Data Engineering* (pp. 685–696). IEEE Computer Society.

O'Connor, M., Bellahsene, Z., & Roantree, M. (2005). An Extended Preorder Index for Optimising XPath Expressions. *3rd XML Database Symposium* (LNCS 3671, pp 114-128).

Oliver, M. J., Glenn, S., Kohut, J. T., Irwin, A. J., Schofield, O. M., & Moline, M. A. (2004). Bioinformatic approaches for objective detection of water masses on continental shelves. *Journal of Geophysical Research, 109*.

Onnela, J.-P., Saramäki, J., Hyvönen, J., Szabó, G., de Menezes, M. A., & Kaski, K. (2007a). Analysis of a large-scale weighted network of one-to-one human communication. *New Journal of Physics, 9*, 179. doi:10.1088/1367-2630/9/6/179

Onnela, J.-P., Saramäki, J., Hyvövnen, J., Szabó, G., Lazer, D., & Kaski, K. (2007b). Structure and tie strengths in mobile communication networks. *Proceedings of the National Academy of Sciences of the United States of America, 104*(18), 7332–7336. doi:10.1073/pnas.0610245104

Orlando, S., Orsini, R., Raffaetà, A., Roncato, A., & Silvestri, C. (2007). Trajectory data warehouses: Design and implementation issues. *Journal of Computing Science and Engineering, 1*(2), 240–261.

Otey, M. E., Ghoting, A., & Parthasarathy, S. (2006). Fast distributed outlier detection in mixed-attribute data sets. *International Journal of Data Mining and Knowledge Discovery, 12*(2-3), 203–228. doi:10.1007/s10618-005-0014-6

Paillier, P. (1999). Public-key cryptosystems based on composite degree residuosity classes. In *Advances in Cryptology -Eeurocrypt '99 Proceedings* (LNCS 1592, pp. 223-238).

Palpanas, T., Papadopoulos, D., Kalogeraki, V., & Gunopulos, D. (2003). Distributed deviation detection in sensor networks. In *Proceedings of the ACM SIGMOD Conference on Management of Data* (pp. 77-82).

Panatier, Y. (1996) *Variowin: Software for spatial data analysis in 2D.* New York: Springer-Verlag Berlin Heidelberg.

Panja, B., Madria, S. K., & Bhargava, B. K. (2008). A role-based access in a hierarchical sensor network architecture to provide multilevel security. *Computer Communications, 31*(4), 793–806. doi:10.1016/j.comcom.2007.10.036

Papadias, D., Tao, Y., Kalnis, P., & Zhang, J. (2002). Indexing spatio-temporal data warehouses. In *ICDE'02* (p. 166-175).

Papadimitriou, S., Faloutsos, C., & Brockwell, A. (2003). Adaptive, hands-off stream mining. In *Proceedings of the 29th International Conference on Very Large Data Bases (VLDB)* (pp. 560–571).

Papadimitriou, S., Kitagawa, H., Gibbons, P. B., & Faloutsos, C. (2003). LOCI: fast outlier detection using the local correlation integral. In *Proceedings of International Conference on Data Engineering* (pp. 315-326).

Park, B. H., & Kargupta, H. (2003). Distributed data mining. In N. Ye (Ed.), *The handbook of data mining* (pp. 341-348). Lawrence Erlbaum Associates.

Park, N. H., & Lee, W. S. (2004). Statistical grid-based clustering over data streams. *SIGMOD Record, 33*(1), 32–37. doi:10.1145/974121.974127

Parno, B., Luk, M., Gaustad, E., & Perrig, A. (2006). Secure Sensor Network Routing: A Clean-Slate Approach. In Proceedings of the 2nd Conference on Future Networking Technologies (2006).

Pawling, A., Chawla, N. V., & Madey, G. (2007). Anomaly detection in a mobile communication network. *Computational & Mathematical Organization Theory, 13*(4), 407–422. doi:10.1007/s10588-007-9018-7

Pawling, A., Schoenharl, T., Yan, P., & Madey, G. (2008). WIPER: An emergency response system. In *Proceedings of the 5th International Information Systems for Crisis Response and Management Conference.*

Pedersen, R., & Jul, E. (Eds.). (2005). *First International Workshop on Data Mining in Sensor Networks.* Retrieved April 18, 2007, from http://www.siam.org/meetings/sdm05/sdm-Sensor-Networks.zip

Pedersen, T., & Tryfona, N. (2001). Pre-aggregation in spatial data warehouses. In *SSTD'01* (Vol. 2121, pp. 460-480).

Perkins, C. (1994). Highly Dynamic Destination-Sequenced Distance-Vector Routing (DSDV) for Mobile Computers. ACM SIGCOMM'94 Conference on Communications Architectures, Protocols and Applications.

Perkins, C. E., & Royer, E. M. (1999). Ad hoc On-Demand Distance Vector Routing. In Proceedings of the 2nd IEEE Workshop on Mobile Computing Systems and Applications, New Orleans, LA, February 1999 (pp. 90-100).

Perrig, A., Szewczyk, R., Wen, V., Culler, D., & Tygar, J. D. (2001). SPINS: Security protocols for sensor networks. In Proceedings of Mobicom.

Petroveskiy, M. I. (2003). Outlier detection algorithms in data mining system. *Journal of Programming and Computer Software, 29*(4), 228–237. doi:10.1023/A:1024974810270

Pfoser, D., Jensen, C. S., & Theodoridis, Y. (2000). Novel approaches in query processing for moving object trajectories. In *VLDB'00* (pp. 395-406).

Pokrajac, D., Lazarevic, A., & Latechi, L. J. (2007). Incremental local outlier detection for data streams. In *Proceedings of IEEE Symposium on Computational Intelligence and Data Mining* (pp. 504-515).

Powell, O., & Nikoletseas, S. (2007a). Simple and efficient geographic routing around obstacles for wireless sensor networks. In *WEA 6th Workshop on Experimental Algorithms*, Rome, Italy. Springer Verlag.

Powell, O., & Nikoletseas, S. (2007b). *Geographic routing around obstacles in wireless sensor networks* (Technical report). Computing Research Repository (CoRR).

Preparata, F., & Shamos, M. (1988). *Computational geometry: An introduction*. New York: Springer-Verlag.

Qiao, L., Agrawal, D., & El Abbadi, A. (2002). RHist: Adaptive summarization over continuous data streams. In *Proceedings of the 11th ACM International Conference on Information and Knowledge Management* (pp. 469-476).

Queensland Department of Natural Resources and Mines. Environmental Protection Agency, Queensland Health, Department of Primary Industries, & Local Governments Association of Queensland (2002). *Queensland Harmful Algal Bloom Response Plan*. Version 1. Retrieved from http://www.nrw.qld.gov.au/water/blue_green/pdf/multi_agency_hab_plan_v1.pdf

Rabaey, J. M., Ammer, M. J., da Silva, J. L., Patel, D., & Roundy, S. (2000). Picoradio supports ad hoc ultra-low power wireless networking. *Computer*.

Raghunathan, V., Schurgers, C., Park, S., & Srivastava, M. B. (2002). Energy-aware wireless microsensor networks. *IEEE Signal Processing Magazine*, *19*(2), 40–50. doi:10.1109/79.985679

Ramaswamy, S., Rastogi, R., & Shim, K. (2000). Efficient algorithms for mining outliers from large data sets. In *Proceedings of the ACM SIGMOD Conference on Management of Data* (pp. 427-438).

Rand, W. M. (1971). Objective criteria for the evaluation of clustering methods. *Journal of the American Statistical Association*, *66*, 846–850. doi:10.2307/2284239

Rao, A., Ratnasamy, S., Papadimitriou, C., Shenker, S., & Stoica, I. (2003). Geographic routing without location information. In *Mobile Computing and Networking*.

Rao, J., Doraiswamy, S., Thakar, H., & Colby, L. (2006). A deferred cleansing method for RFID data analytics. In *VLDB'06*, Seoul, Korea.

Ratnasamy, S., Francis, P., Handley, M., Karp, R., & Schenker, S. (2001). A scalable content-addressable network. In *Proceedings of the 2001 Conference on Applications, Technologies, Architectures, and Protocols for Computer Communications* (pp. 161–172). New York, NY: ACM.

Ratnasamy, S., Karp, B., Shenker, S., Estrin, D., Govindan, R., & Yin, L. (2003). Data-centric storage in sensornets with GHT, a geographic hash table. *Mobile Networks and Applications*, *8*, 427–442. doi:10.1023/A:1024591915518

Ratnasamy, S., Karp, B., Yin, L., Yu, F., Estrin, D., Govindan, R., & Shenker, S. (2002). GHT: a geographic hash table for data-centric storage. In *Wireless Sensor Networks and Applications*. ACM.

Ren, D., Rahal, I., & Perrizo, W. (2004). A vertical outlier detection algorithm with clusters as by-product. In *Proceedings of International Conference on Tools with Artificial Intelligence* (pp. 22-29).

Rigaux, P., & Scholl, M. A. V. (2001). *Spatial database: With application to GIS*. Morgan Kaufmann.

Rivest, S., Bédard, Y., & Marchand, P. (2001). Towards better support for spatial decision making: Defining the characteristics of spatial on-line analytical processing (SOLAP). *Geomatica*, *55*(4), 539–555.

Rizvi, S., et al. (2005). Events on the Edge. In *Proc. of SIGMOD* (pp. 885-887).

Rodrigues, P. P., & Gama, J. (2007). Clustering techniques in sensor networks. In J. Gama & M. M. Gaber, (Eds.), *Learning from Data Streams - Processing Techniques in Sensor Networks* (pp. 125–142). Springer Verlag.

Rodrigues, P. P., & Gama, J. (2008). Dense pixel visualization for mobile sensor data mining. In *Proceedings of the 2nd International Workshop on Knowledge Discovery from Sensor Data* (pp. 50–57). ACM Press.

Rodrigues, P. P., Gama, J., & Lopes, L. (2008a). Clustering distributed sensor data streams. In W. Daelemans, B. Goethals & K. Morik (Eds.), *Proceedings of the European Conference on Machine Learning and Knowledge Discovery in Databases (ECMLPKDD 2008)* (LNAI 5212, pp. 282–297) Antwerpen, Belgium: Springer Verlag.

Rodrigues, P. P., Gama, J., & Lopes, L. (2009). Requirements for clustering streaming sensors. In A. R. Ganguly, J. Gama, O. A. Omitaomu, M. M. Gaber, & R. R. Vatsavai (Eds.), *Knowledge Discovery from Sensor Data* (pp. 33–51). CRC Press.

Rodrigues, P. P., Gama, J., & Pedroso, J. P. (2008b). Hierarchical clustering of time-series data streams. *IEEE Transactions on Knowledge and Data Engineering, 20*(5), 615–627. doi:10.1109/TKDE.2007.190727

Rodriguez, C. (2004). *A computational environment for data preprocessing in supervised classifications.* Unpublished master's thesis, University of Puerto Rico, Mayaguez, Puerto Rico.

Roncancio, C. (1997). Toward Duration-Based, Constrained and Dynamic Event Types. In *Second International Workshop on Active, Real-Time, and Temporal Database Systems* (LNCS 1553, pp. 176-193).

Rose, K., Malaika, S., & Schloss, R. (2006). Virtual XML: A toolbox and use cases for the XML World View. *IBM Systems Journal, 45*(2), 411–424.

Rothwell, S., Lehane, B., Chan, C. H., Smeaton, A. F., O'Connor, N. E., Jones, G. J. F., & Diamond, D. (2006). The CDVPlex Biometric Cinema: Sensing Physiological Responses to Emotional Stimuli in Film. *Pervasive 2006 - the 4th International Conference on Pervasive Computing.*

Rousseeuw, P. J., & Leroy, A. M. (1996). *Robust regression and outlier detection.* John Wiley and Sons.

Ruts, I., & Rousseeuw, P. (1996). Computing depth contours of bivariate point clouds. *Journal of Computational Statistics and data . Analysis, 23*, 153–168.

Sarma, S., Brock, D. L., & Ashton, K. (2000). *The networked physical world* [White paper]. MIT Auto-ID Center.

Saxon Project. (2007). Retrieved from http://saxon.sourceforge.net

Schlimmer, J. C., & Granger, R. H., Jr. (1986). Beyond incremental processing: Tracking concept drift. In *Proceedings of the 5th National Conference on Artificial Intelligence* (pp. 502-507), Philadelphia, Pennsylvania.

Schoenharl, T., & Madey, G. (2008). Evaluation of measurement techniques for the validation of agent-based simulations agains streaming data. In M. Bubak, G.D. van Albada, J. Dongarra, & P.M.A. Sloot (Eds.), Proceedings of the *International Conference on Computational Science* (Vol. 5103, pp. 6-15). Heidelberg, Germany: Springer.

Schoenharl, T., Bravo, R., & Madey, G. (2006). WIPER: Leveraging the cell phone network for emergency response. *International Journal of Intelligent Control and Systems, 11*(4), 209–216.

Schoenharl, T., Madey, G., Szabó, G., & Barabási, A.-L. (2006). WIPER: A multi-agent system for emergency response. In B. van de Walle & M. Turoff (Eds.), *Proceedings of the 3rd International Information Systems for Crisis Response and Management Conference.*

Schofield, O., Grzymski, J., Paul Bissett, W., Kirkpatrick, G. J., Millie, D. F., & Moline, M. (1999). Optical monitoring and forecasting systems for harmful algal blooms: Possibility or pipe dream? *Journal of Phycology, 35*, 1477–1496. doi:10.1046/j.1529-8817.1999.3561477.x

Schölkopf, B., Platt, J., Shawe-Taylor, J., Smola, A. J., & Williamson, R. C. (2001). Estimating the support of a high dimensional distribution. *Journal of Neural Computation, 13*(7), 1443–1471. doi:10.1162/089976601750264965

Schreier, U., et al. (1991). Alert: An Architecture for Transforming a Passive DBMS into an Active DBMS. In *Proc. of VLDB.*

Schwetman, H. (2001). CSIM 19: A powerful tool for building systems models. In *Proceedings of the ACM Winter Simulation Conference* (pp. 250-255).

Schwetman, H. D. (1990). Introduction to process-oriented simulation and CSIM. In *Proceedings of the ACM Winter Simulation Conference* (pp. 154–157).

Scott, D. W. (1992). *Multivariate density estimation. Theory, practice, and visualization.* New York: Wiley.

Seada, K., Helmy, A., & Govindan, R. (2001). On the effect of localization errors on geographic face routing in sensor networks. In *Information Processing in Sensor Networks.*

Serving, X. M. L. Project (2007). Retrieved from http://servingxml.sourceforge.net

Seshadri, P., Livny, M., & Ramakrishnan, R. (1996). The Design and Implementation of a Sequence Database System. In *Proc. of VLDB* (p. 99-110).

Shamir, A. (1979). How to Share a Secret. *Communications of the ACM, 22*(11), 612–613. doi:10.1145/359168.359176

Sharaf, M. A., Beaver, J., Labrinidis, A., & Chrysanthis, P. K. (2004). Balancing energy efficiency and quality of aggregate data in sensor networks. *The VLDB Journal, 13*(4), 384–403. doi:10.1007/s00778-004-0138-0

Shekhar, S., & Chawla, S. (2003). *Spatial databases: A tour*. Prentice Hall.

Shekhar, S., Lu, C. T., & Zhang, P. (2001). A unified approach to spatial outliers detection. *International Journal of GeoInformatica, 7*(2), 139–166. doi:10.1023/A:1023455925009

Shekhar, S., Lu, C., Tan, X., Chawla, S., & Vatsavai, R. (2001). Map cube: A visualization tool for spatial data warehouse. In H. J. Miller & J. Han (Eds.), *Geographic data mining and knowledge discovery*. Taylor and Francis.

Sherrill, D. M., Moy, M. L., Reilly, J. J., & Bonato, P. (2005). Using hierarchical clustering methods to classify motor activities of copd patients from wearable sensor data. *Journal of Neuroengineering and Rehabilitation, 2*(16).

Silberstein, A., Braynard, R., & Yang, J. (2006). Constraint chaining: On energy-efficient continuous monitoring in sensor networks. *SIGMOD '06: Proceedings of the 2006 ACM SIGMOD international conference on Management of data* (pp. 157-168). New York: ACM

Silverman, B. W. (1986). *Density estimation for statistics and data analysis*. London: Chapman and Hall.

Silvervarg, K., & Jungert, E. (2004, Sept. 8-10). Visual specification of spatial/temporal queries in a sensor data independent information system. In *Proceedings of the 10ᵗʰ International Conf, on Distributed Multimedia Systems. San Francisco, California*, (pp. 263-268).

Silvervarg, K., & Jungert, E. (2005, September 5-7). Uncertain topological relations for mobile point objects in terrain. In *Proceedings of 11ᵗʰ International Conference on Distributed Mutimedia Systems, Banff, Canada*, (pp. 40-45).

Silvervarg, K., & Jungert, E. (2005). A visual query language for uncertain spatial and temporal data. In *Proceedings of the Conference on Visual Information systems (VISUAL). Amsterdam, The Netherlands* (pp. 163-176).

Silvervarg, K., & Jungert, E. (2006). A scenario driven decision support system. In *Proceedings of the eleventh International Conference on Distributed Multimedia Systems. Grand Canyon, USA, August 30 – September 1, 2006* (pp. 187-192).

Silvestri, C., & Orlando, S. (2005). Distributed approximate mining of frequent patterns. In *SAC'05* (p. 529-536). ACM.

Silvestri, C., & Orlando, S. (2007). Approximate mining of frequent patterns on streams. *Int. Journal of Intelligent Data Analysis, 11*(1), 49–73.

Smith, P., & Hobbs, L. (2008). *Comparing materialized views and analytic workspaces in Oracle Database 11g*. Retrieved from http://www.oracle.com/technology/products/bi/db/11g/pdf/comparision_aw_mv_11g_twp.pdf

Solove, D. J. (2002). Conceptualizing privacy. *California Law Review, 90*, 1088–1155. doi:10.2307/3481326

Solove, D. J. (2007). "I've got nothing to hide" and other misunderstandings of privacy. *The San Diego Law Review, 44*.

Son, S.-H., Chiang, M., Kulkarni, S. R., & Schwartz, S. C. (2005). The value of clustering in distributed estimation for sensor networks. In *Proceedings of the IEEE International Conference on Wireless Networks, Communications, and Mobile Computing (WirelessCom)*, Maui, Hawaii, June 2005.

Sozer, E. M., Stojanovic, M., & Proakis, J. G. (2000). Underwater acoustic networks. *IEEE Journal of Oceanic Engineering, 25*, 72–83. doi:10.1109/48.820738

Spiliopoulou, M., Ntoutsi, I., Theodoridis, Y., & Schult, R. (2006). Monic: modeling and monitoring cluster transitions. In *KDD* (pp. 706–711).

Stackowiak, R., Rayman, J., & Greenwald, R. (2007). *Oracle data warehousing and business intelligence solutions.* Indianapolis, IN: Wiley Publishing Inc.

Stallings, W. (n.d.). Network Security Essentials Applications and Standards.

Stauffer, D., & Aharony, A. (1992). *Introduction to Percolation Theory.* (2nd ed.) London: Taylor and Francis.

Steinbach, M., Karypis, G., & Kumar, V. (2000). A comparison of document clustering techniques. In *Proceedings of the ACM SIGKDD Workshop on Text Mining.* Retrieved from http://www.cs.cmu.edu/~dunja/KDDpapers/

Steiner, M., Tsudik, G., & Waidner, M. (2000). Key Agreement in Dynamic Peer Groups. *IEEE Transactions on Parallel and Distributed Systems, 11*(8), 769–780. doi:10.1109/71.877936

Stinson, D. R. (2006). *Cryptography: Theory and practice.* Boca Raton, FL: Chipman & Hall.

Stojmenovic, I. (2002). Position-based routing in ad hoc networks. *Communications Magazine.*

Stojmenovic, I., & Datta, S. (2004). Power and cost aware localized routing with guaranteed delivery in unit graph based ad hoc networks. *Wireless Communications and Mobile Computing.*

Stojmenovic, I., & Lin, X. (2001). Loop-free hybrid single-path/flooding routing algorithms with guaranteed delivery for wireless networks. *IEEE Transactions on Parallel and Distributed Systems, 12*(10). doi:10.1109/71.963415

Stonebraker, M. (1975). Implementation of integrity constraints and views by query modification. In *Proceedings of the 1975 ACM SIGMOD international conference on Management of data San Jose. California, 1975* (pp. 65–78).

Subramaniam, S., Palpanas, T., Papadopoulos, D., Kalogeraki, V., & Gunopulos, D. (2006). Online outlier detection in sensor data using non-parametric models. *International Journal of Very Large Data Bases* (pp. 187-198).

Subramanian, L., & Katz, R. H. (2000). An Architecture for Building Self-Configurable Systems. IEEE MobiHoc.

Sun, J.-Z., & Sauvola, J. (2006). Towards advanced modeling techniques for wireless sensor networks. In *Proceedings of the 1st International Symposium on Pervasive Computing and Applications* (pp. 133–138). IEEE Computer Society.

Sun, P. (2006). *Outlier detection in high dimensional, spatial and sequential data sets.* Doctoral dissertation, University of Sydney, Sydney.

Sun, P., & Chawla, S. (2004). On local spatial outliers. In *Proceedings of International Conference on Data Mining* (pp. 209-216).

Sun, P., Chawla, S., & Arunasalam, B. (2006). Mining for outliers in sequential databases. In *Proceedings of SIAM International Conference on Data Mining* (pp. 94-105).

Sun, Y., & Liu, K. J. R. (2004). Scalable hierarchical access control in secure group communications. *IEEE INFOCOM.*

Surdeanu, M., Turmo, J., & Ageno, A. (2005). A hybrid unsupervised approach for document clustering. In *Proceedings of the 11th ACM SIGKDD International Conference on Knowledge Discovery and Data Mining.*

Sweeney, L. (1997). Weaving technology and policy together to maintain confidentiality. *The Journal of Law, Medicine & Ethics, 25,* 98–110. doi:10.1111/j.1748-720X.1997.tb01885.x

Sykacek, P. (1997) Equivalent error bars for neural network classifiers trained by bayesian inference. In *Proceedings of European Symposium on Artificial Neural Networks* (pp. 121-126).

Szewczyk, R., Mainwaring, A., Polastre, J., & Culler, D. (2004). An analysis of a large scale habitat monitoring application. In *Proceedings of the 2nd international conference on Embedded networked sensor systems* (pp. 214-226). ACM Press.

Takagi, H., & Kleinrock, L. (1984). Optimal transmission ranges for randomly distributed packet radio terminals. *Communications, IEEE Transactions on [legacy, pre - 1988], 32*(3), 246–257.

Tamminen, M. (1984). Comment on quad- and oc-trees. *Communications of the ACM, 27,* 248–249. doi:10.1145/357994.358026

Tan, P. N., Steinbach, M., & Kumar, V. (2006). *Introduction to data mining.* Addison-Wesley.

Tan, P.-N., Steinbach, M., Kumar, V., Potter, C., Klooster, S., & Torregrosa, A. (2001). Finding spatio-temporal patterns in earth science data. In *KDD Workshop on Temporal Data Mining.*

Tao, Y., & Papadias, D. (2005). Historical spatio-temporal aggregation. *ACM TOIS, 23,* 61–102. doi:10.1145/1055709.1055713

Tao, Y., Kollios, G., Considine, J., Li, F., & Papadias, D. (2004). Spatio-temporal aggregation using sketches. In *ICDE'04* (p. 214-225). IEEE.

Taomir, B., & Rothkrantz, L. (2005). Crisis management using mobile ad-hoc wireless networks. In *Proceedings of the 2nd International Information Systems for Crisis Response Management Conference.*

Tasoulis, D. K., & Vrahatis, M. N. (2004). Unsupervised distributed clustering. In M. H. Hamza (Ed.), *IASTED International Conference on Parallel and Distributed Computing and Networks* (pp. 347-351). Innsbruck, Austria: ACTA Press.

Tatbul, N., et al. (2003, Sep.). Load Shedding in a Data Stream Manager. In *Proc. of VLDB.*

Tax, D. M. J., & Duin, R. P. W. (1999). Support vector domain description. *Official Publication of Pattern Recognition Letters, 20,* 1191–1199. doi:10.1016/S0167-8655(99)00087-2

The Network Simulator - ns-2 (n.d.). Retrieved from http://www.isi.edu/nsnam/ns/

Thomas, M., Andoh-Baidoo, F., & George, S. (2005). Evresponse–moving beyond traditional emergency response notification. In *Proceedings of the 11th Americas Conference on Information Systems.*

Tilak, S., Abu-Ghazaleh, N. B., & Heinzelman, W. (2002). Infrastructure tradeoffs for sensor networks. In *Proceedings of First ACM International Workshop on Wireless Sensor Networks & Applications* (pp. 49–58).

Tiny, O. S. (2000). The TinyOS Documentation Project. Retrieved from http://www.tinyos.org.

Trigoni, N., Yao, Y., Demers, A., Gehrke, J., & Rajaraman, R. (2005). Multi-query optimization for sensor networks. In *Proceedings of the First IEEE International Conference on Distributed Computing in Sensor Systems (DCOSS 2005).* Springer.

Tukey, J. (1997). *Exploratory data analysis.* Addison-Wesley.

Tulone, D., & Madden, S. (2006). An energy-efficient querying framework in sensor networks for detecting node similarities. In *Proceedings of the 9th ACM international symposium on Modeling analysis and simulation of wireless and mobile systems* (pp. 191-300).

Tulone, D., & Madden, S. (2006). PAQ: Time series forecasting for approximate query answering in sensor networks. *Paper presented at the 3rd European Workshop on Wireless Sensor Networks (EWSN01906),* Zurich, Switzerland.

Valero-Mora, P. M., Young, F. W., & Friendly, M. (2003). Visualizing categorical data in ViSta. *Journal of Computational Statistics & Data Analysis, 43,* 495–508. doi:10.1016/S0167-9473(02)00289-X

Vassiliadis, P., & Sellis, T. K. (1999). A survey of logical models for OLAP databases. *SIGMOD Record, 28*(4), 64–69. doi:10.1145/344816.344869

Vélez, B., Weiss, R., Sheldon, M. A., & Gifford, D. K. (1997). Fast and effective query refinement. In *Proceedings of the 20th ACM Conference on Research and Development in Information Retrieval (SIGIR97). Philadelphia, Pennsylvania.*

Wan, C.-Y., Eisenman, S. B., & Campbell, A. T. (2003). CODA: Congestion detection and avoidance in sensor networks. In *ACM Conference on Embedded Networked Sensor Systems (SenSys)* (pp. 266-279).

Wang, W., Yang, J., & Muntz, R. R. (1997). STING: A statistical information grid approach to spatial data mining. In M. Jarke, M. J. Carey, K. R. Dittrich, F. H. Lochovsky, P. Loucopoulos, & M. A. Jeusfeld (Eds.), *Proceedings of the Twenty-Third International Conference on Very Large Data Bases* (pp. 186–195). Athens, Greece. Morgan Kaufmann.

Warneke, B., Last, M., Liebowitz, B., & Pister, K. S. J. (2001). Smart dust: communicating with a cubic-millimeter computer. *Computer.*

Watts, D. J. (1999). *Small worlds.* NJ: Princeton University Press.

Watts, D. J., & Strogatz, S. H. (1998). Collective dynamics of 'small-world' networks. *Nature, 393,* 440–442. doi:10.1038/30918

Wei, L., Qian, W., Zhou, A., Jin, W., & Yu, J. X. (2003). HOT: Hypergraph-based outlier test for categorical data. In *Proceedings of Pacific-Asia Conference on Knowledge Discovery and Data Mining* (pp. 399-410).

Werner-Allen, G., Lorincz, K., Welsh, M., Marcillo, O., Johnson, J., Ruiz, M., & Lees, J. (2006). Deploying a wireless sensor network on an active volcano. *IEEE Internet Computing, 10*(2). doi:10.1109/MIC.2006.26

White, F. E. (1998, October). Managing data fusion systems in joint and coalition warfare. In *Proceedings of EuroFusion98 – International Conference on Data Fusion, Great Malvern, United Kingdom* (pp. 49-52).

Whitehouse, K., Zhao, F., & Liu, J. (2006). Semantic Streams: a Framework for Composable Semantic Interpretation of Sensor Data. *3rd European Workshop on Wireless Sensor Networks (EWSN)* (LNCS 3868, pp. 5-20).

Witkowski, A., Bellamkonda, S., Li, H., Liang, V., Sheng, L., Smith, W., et al. (2007). Continuous queries in Oracle. In *Proceedings of the 33rd International Conference on Very Large Data Bases* (pp. 1173-1184).

Worboys, M., & Duckham, M. (2004). *GIS: A computing perspective* (2nd ed.). CRC Press.

Xia, P., Chrysanthis, P., & Labrinidis, A. (2006). Similarity-aware query processing in sensor networks. *Parallel and Distributed Processing Symposium* (p. 8).

Xiao, L., & Ouksel, A. M. (2005). Tolerance of localization imprecision in efficiently managing mobile sensor databases. In U. Cetintemel & A. Labrinidis (Eds.), *Proceedings of the 4th ACM International Workshop on Data Engineering for Wireless and Mobile* (pp. 25- 32). New York: ACM.

Xin, D., Han, J., Cheng, H., & Li, X. (2006). Answering top-k queries with multi-dimensional selections: The ranking cube approach. In *Proceedings of the 32nd International Conference on Very Large Data Bases* (pp. 463-475).

Xing, G., Lu, C., Pless, R., & Huang, Q. (2004). On greedy geographic routing algorithms in sensing-covered networks. In *MobiHoc '04: Proceedings of the 5th ACM international symposium on Mobile ad hoc networking and computing* (pp. 31–42). New York: ACM Press.

Xu, X., Ester, M., Kriegel, H.-P., & Sander, J. (1998). A distribution-based clustering algorithm for mining in large spatial databases. In *Proceedings 14th International Conference on Data Engineering* (pp. 324-331). Los Alamitos, CA: IEEE Computer Society.

Xu, X., Jager, J., & Kriegel, H. P. (1999). A fast parallel clustering algorithm for large spatial databases. *Data Mining and Knowledge Discovery, 3*(3), 263–290. doi:10.1023/A:1009884809343

Yager, Fedrizzi & Kacprzyk (eds.) (1994). *Advances in Dempster-Shafer theory of evidence.* New York: Wiley & Sons.

Yamanishi, K., & Takeuchi, J. (2006). A unifying framework for detecting outliers and change points from non-stationary time series data. *International Journal of Knowledge and Data Enginnering, 18*(4), 482–492. doi:10.1109/TKDE.2006.1599387

Yan, P., Schoenharl, T., Pawling, A., & Madey, G. (2007). *Anomaly detection in the WIPER system using a Markov modulated Poisson process.* Working Paper. http://www.nd.edu/~dddas/Papers/MMPP.pdf

Yang, X., Ong, K. G., Dreschel, W. R., Zeng, K., Mungle, C. S., & Grimes, C. A. (2002). Design of a wireless sensor network for long-term, in-situ monitoring of an aqueous environment. *Sensors, 2*, 455–472. doi:10.3390/s21100455

Yao, Y., & Gehrke, J. (2003). Query processing in sensor networks. In *Proceedings of the 1st International Conference on Innovative Data Systems Research.* Retrieved from http://www-db.cs.wisc.edu/cidr/cidr2003/program/p21.pdf

Yates, D. J. (2006). *Scalable data delivery for networked servers and wireless sensor networks.* Unpublished Ph.D. dissertation, Department of Computer Science, University of Massachusetts, Amherst, Massachusetts, USA.

Yates, D. J., Nahum, E., Kurose, J., & Shenoy, P. (2008, March). Data quality and query cost in pervasive sensing systems. In *IEEE International Conference on Pervasive Computing and Communications (PerCom)*, Hong Kong, China.

Ye, F., Luo, H., Lu, S., & Zhang, L. (2004). Statistical en-route filtering of injected false data in sensor networks. *INFOCOM 2004.*

Ye, W., Heidemann, J., & Estrin, D. (2002). An energy-efficient MAC protocol for wireless sensor networks. In *INFOCOM.*

Yi, S., Naldurg, P., & Kravets, R. (2001, October*). A security-aware routing protocol for wireless ad hoc networks.* Poster presentation, ACM Symposium on Mobile Ad Hoc Networking & Computing (Mobihoc 2001), Long Beach, California.

Yin, J., & Madria, S. (2008). *SeRWA: A secure routing protocol against wormhole attacks in sensor networks.* Ad Hoc Networks.

Yoon, S., & Shahabi, C. (2007). The clustered aggregation (CAG) technique leveraging spatial and temporal correlations in wireless sensor networks. *ACM Transactions on Sensor Networks, 3*(1). Article 3.

Younis, O., & Fahmy, S. (2004). HEED: A hybrid, energy-efficient, distributed clustering approach for ad hoc sensor networks. *IEEE Transactions on Mobile Computing, 3*(4), 366–379. doi:10.1109/TMC.2004.41

Yu, D., Sheikholeslami, G., & Zhang, A. (2002). Findout: finding outliers in very large datasets. *Journal of Knowledge and Information Systems, 4*(3), 387–412. doi:10.1007/s101150200013

Yu, J. X., Qian, W., Lu, H., & Zhou, A. (2006). Finding centric local outliers in categorical/numerical spaces. *Journal of Knowledge Information System, 9*(3), 309–338. doi:10.1007/s10115-005-0197-6

Yu, Y., Krishnamachari, B., & Prasanna, V. K. (2004, March). Energy-latency tradeoffs for data gathering in wireless sensor networks. In *Proceedings of the IEEE Conference on Computer Communications (Infocom)*, Hong Kong, China.

Yu, Z., & Guan, Y. (2005). A robust group-based key management scheme for wireless sensor networks. In *Proceedings of IEEE Wireless Communications and Networking Conference (WCNC 2005)*. New Orleans, LA USA: IEEE

Zaki, M. J., & Ho, C.-T. (Eds.). (2000). *Large-scale parallel data mining* (LNCS 1759). Berlin/Heidelberg, Germany: Springer.

Zhang, D., Gunopulos, D., Tsotras, V. J., & Seeger, B. (2002). Temporal aggregation over data streams using multiple granularities. In *Proceedings of 8th International Conference on Extending Database Technology* (pp. 646-663).

Zhang, K., Torkkola, K., Li, H., Schreiner, C., Zhang, H., Gardner, M., & Zhao, Z. (2007). A context aware automatic traffic notification system for cell phones. In *27th International Conference on Distributed Computing Systems Workshops (ICDCSW '07)* (pp. 48–50). IEEE Computer Society.

Zhang, T., Ramakrishnan, R., & Livny, M. (1996). BIRCH: An efficient data clustering method for very large databases. In *Proceedings of the 1996 ACM SIGMOD International Conference on Management of Data* (pp. 103–114). ACM Press.

Zhang, Y., Meratnia, N., & Havinga, P. J. M. (2007). *A taxonomy framework for unsupervised outlier detection techniques for multi-type data sets.* The Netherlands: University of Twente, Technique Report.

Zhao, F., & Guibas, L. (2004a). Geographic energy-aware routing. In *Wireless Sensor Networks, an Information Processing Approach.* Elsevier.

Zhao, F., & Guibas, L. (2004b). *Wireless sensor networks, an information processing approach.* Elsevier.

Zhao, F., & Guibas, L. (2004c). Localization and localization services. In *Wireless Sensor Networks, and Information Processing Approach.* Elsevier.

Zhao, J., & Govindan, R. (2003). Understanding packet delivery performance in dense wireless sensor networks. In *ACM Conference on Embedded Networked Sensor Systems (SenSys)* (pp. 1-13).

Zhao, Y., Deshande, P. M., & Naughton, J. F. (1997). An array-based algorithm for simultaneous multidimensional aggregates. In *Proceedings of the 1997 ACM International Conference on Management of Data* (pp. 159-170).

Zhou, A., Cao, F., Yan, Y., Sha, C., & He, X. (2007). Distributed data stream clustering: A fast EM approach. In *Proceedings of the 23rd International Conference on Data Engineering* (pp. 736–745).

Zhou, G., He, T., Krishnamurthy, S., & Stankovic, J. A. (2004). Impact of radio irregularity on wireless sensor networks. In *Mobile systems, applications and services.*

Zhu, C., Kitagawa, H., & Faloutsos, C. (2005). Example-based robust outlier detection in high dimensional datasets. In *Proceedings of International Conference on Data Mining* (pp. 829-832).

Zhu, S., Setia, S., & Jajodia, S. (2003). LEAP: efficient security mechanisms for large-scale distributed sensor networks. In Proceedings of the 10th ACM conference on Computer and communication security

Zhu, Y., Vedantham, R., Park, S. J., & Sivakumar, R. (2008). A scalable correlation aware aggregation strategy for wireless sensor networks. *Information Fusion, 9,* 354–369. doi:10.1016/j.inffus.2006.09.002

About the Contributors

Alfredo Cuzzocrea is actually a Researcher at the Institute of High Performance Computing and Networking of the Italian National Research Council, Italy, and an Adjunct Professor at the Department of Electronics, Computer Science and Systems of the University of Calabria, Italy. His research interests include multidimensional data modelling and querying, data stream modelling and querying, data warehousing and OLAP, OLAM, XML data management, Web information systems modelling and engineering, knowledge representation and management models and techniques, Grid and P2P computing. He is author or co-author of more than 100 papers in referred international conferences (including EDBT, SSDBM, ISMIS, ADBIS, DEXA, DaWaK, DOLAP, IDEAS, SEKE, WISE, FQAS, SAC) and international journals (including DKE, JIIS, IJDWM, WIAS). He serves as program committee member of referred international conferences (including ICDM, SDM, PKDD, PAKDD, CIKM, ICDCS, ER, WISE, DASFAA, FQAS, SAC) and as review board member of referred international journals (including TODS, TKDE, TSMC, IS, DKE, JIIS, IPL, TPLP, COMPJ, DPDB, KAIS, INS, IJSEKE, FGCS). He also serves as PC Chair in several international conferences and as Guest Editor in international journals like JCSS, DKE, KAIS, IJBIDM, IJDMMM and JDIM.

* * *

Raman Adaikkalavan is an Assistant Professor in the Computer and Information Sciences Department at Indiana University South Bend. His current research interests include: complex event processing, event stream processing systems, information security, and access control models.

Vincent Andrieu worked as a research intern from the Institut National des Sciences Appliquées (INSA) in Toulouse, France. His research was conducted with the Interoperable Systems Group in Dublin City University building peer-to-peer information systems for managing sensor data. His focus was on optimizing peer-to-peer queries.

Elena Baralis is full professor at the Dipartimento di Automatica e Informatica of the Politecnico di Torino since January 2005. She holds a Master degree in Electrical Engineering and a Ph.D. in Computer Engineering, both from Politecnico di Torino. Her current research interests are in the field of databases, in particular data mining, sensor databases, and bioinformatics. She has published numerous papers on journal and conference proceedings. She has served on the program committees of several international conferences and workshops, among which IEEE ICDM, VLDB, ACM CIKM, DaWak, ACM SAC, PKDD. She has managed several Italian and EU research projects.

Karin Camara received her M.S. in Computer Science from the University of Linköping in 1997. In 2007 she received a licentiate degree from the same university. Currently she is working as a computer consultant pursuing her interest in making the contact between computers and humans work more smothly.

Marcos M. Campos is the Technology Development Manager for the Data Mining Technologies group at Oracle. Previously he was a Senior Scientist with Thinking Machines. He holds a B.S. in Physics, a M.A. in Regional Science, and a Ph.D. in Cognitive & Neural Systems. Over the years Dr. Campos has been working on transforming databases into easy to use analytical servers. He has created, led, and participated in the design and implementation of multiple in-database analytics projects, including: Oracle Personalization, a distributed real-time recommendation system, and Oracle Data Mining, for which he holds several patents. He is an active contributor to the machine learning community through technical papers and participation in industry standards.

Fabrice Camous holds a PhD in Computer Science from Dublin City University and a Higher Diploma in Computer Science from University College Dublin. He also holds a Licence in Economics and Linguistics from the University of Aix-Marseille II, and the University of Rouen, respectively. His research interests are in the area of Bio-medical Information Retrieval and the use of semantic networks to establish similarity. He is also interested in Health Informatics, including the implementation challenges of the Electronic Health Record, and the technologies involved in ambient Health Monitoring, such as sensor data management and stream processing.

Julián Candia has authored more than 25 peer-reviewed publications on a variety of topics, ranging from Particle and High Energy Physics to Condensed Matter and Statistical Physics. His Doctoral Thesis focused on the "Propagation of High Energy Galactic Cosmic Rays", which developed models that are standard references by present experimental and theoretical works in the field. His current research is aimed at the understanding of complex and nonequilibrium systems, with a particular interest in applying the network approach to large-scale data analysis and to the modeling of complex systems of interdisciplinary nature.

Tania Cerquitelli is a post-doctoral researcher in Computer Engineering at the Politecnico di Torino since January 2007. She holds a PhD degree and a Master degree in Computer Engineering, both from the Politecnico di Torino. She also obtained a Master degree in Computer Science from Universidad De Las Américas Puebla. Her research activities are focused on the integration of data mining algorithms into database system kernels and on the exploitation of data mining techniques to analyze streaming data collected by sensor networks. She served in the program committee of DS2ME'08 and IADIS'06.

Sharma Chakravarthy is a Professor in the Computer Science and Engineering Department at the University of Texas at Arlington. He has established the Information Technology Laboratory (or ITLab) and a Distributed and Parallel Computing Cluster (DPCC) at UTA. His current research interests include: enabling technologies for advanced information systems, mining/knowledge discovery (conventional, graph, and text), active capability, web change monitoring, stream data processing, document/email classification, and information security. Prior to joining UTA, he was with the CISE Department at the University of Florida, Gainesville. He also worked as a Computer Scientist at the Computer Corporation of America (CCA), Cambridge, MA and as a Member of Technical Staff at Xerox Advanced Information Technology (XAIT), Cambridge, MA. He has published over 115 papers in refereed International journals and conference proceedings. He has consulted for MCC, BBN, SRA Systems (India), Enigmatec, London, and RTI Inc. He has given tutorial on a number of database topics, such as active, real-time, distributed, object-oriented, and heterogeneous databases in North America, Europe, and Asia. He has given several invited and keynote addresses at universities and International conferences.

Shi-Kuo Chang received the B.S. degree from National Taiwan University in 1965. He received the M.S. and Ph.D. degrees from the University of California, Berkeley, in 1967 and 1969, respectively. He was a research scientist at IBM Watson Research Center from 1969 to 1975. From 1975 to 1982 he was Associate Professor and then Professor at the Department of Information Engineering, University of Illinois at Chicago. From 1982 to 1986 he was Professor and Chairman of the Department of Electrical and Computer Engineering, Illinois Institute of Technology. From 1986 to 1991 he was Professor and Chairman of the Department of Computer Science, University of Pittsburgh. He is currently Professor and Director of Center for Parallel, Distributed and Intelligent Systems, University of Pittsburgh. Dr. Chang is a Fellow of IEEE. He was a consultant to IBM, Bell Laboratories, Standard Oil, Honeywell, Naval Research Laboratory and Siemens. His research interests include distributed multimedia systems, visual information systems, visual languages and visual computing. Dr. Chang has published over two hundred and forty papers and wrote or edited twelve books.

Hong Cheng (www.se.cuhk.edu.hk/~hcheng), is an Assistant Professor in the Department of Systems Engineering and Engineering Management at CUHK. She received her Ph.D. in Computer Science from University of Illinois at Urbana-Champaign in 2008. Before her Ph.D., she received a B.S. and M.Phil. in Computer Science from Zhejiang University and Hong Kong University of Science and Technology, respectively. Her primary research interests are in the fields of data mining, machine learning and database systems. Her current research projects include: (1) discriminative frequent pattern-based classification and its applications, and (2) scalable frequent pattern mining algorithms. She has published over 20 research papers in international conferences, journals, and book chapter, including SIGMOD, VLDB, SIGKDD, ICDE, ICDM, SDM, ACM Transactions on KDD, and Data Mining and Knowledge Discovery, and received research papers awards at ICDE'07, SIGKDD'06 and SIGKDD'05.

Gennaro Costagliola received his Laurea degree in Computer Science magna cum laude from the University of Salerno, Italy, in 1987. From 1989 to 1993 he was a post-graduate and a visiting researcher at the University of Pittsburgh, PA, USA and then joined the Faculty of Science at the University of Salerno as an Assistant Professor. Since 2001 he has been a Full Professor of Computer Science at the University of Salerno. His research interests include heterogenous data sources, web technologies, e-learning, human-computer interaction, parsing technologies, and theory, implementation and applications of visual languages. He is a member of the Association for Computing Machinery and the IEEE Computer Society.

Vincenzo D'Elia is a Ph.D. student of the Database and Data Mining Group at the Dipartimento di Automatica e Informatica of the Politecnico di Torino since January 2007. He obtained a Master degree in Computer Engineering from Politecnico di Torino in November 2006. He is currently working in the field of data mining, and time series analysis. In particular, his activity is focused on the analysis of sensor network data to support effective data collection by means of data mining techniques. His research activities are also devoted to analyze physiological signals to study athlete conditions while they perform their activities.

Filippo Furfaro received his Ph.D. in Computer Science and System Engineering from University of Calabria, Italy, in 2003. Currently, he is an assistant professor at the Faculty of Engineering — University of Calabria, where he is member of the Department of Electronics, Computer and System Sciences. His research interests include database theory, logic programming, inconsistent data management, semi-structured data, XML query languages, compression techniques for multidimensional data, computation on grid and peer-to-peer environments.

João Gama is a researcher at LIAAD-INESC Porto LA, the Laboratory of Artificial Intelligence and Decision Support of the University of Porto. His main research interest is Learning from Data Streams. He has published several articles in change detection, learning decision trees from data streams, hierarchical clustering from streams, etc. Editor of special issues on Data Streams in Intelligent Data Analysis, J. Universal Computer Science, and New Generation Computing. Co-chair of ECML 2005 Porto, Portugal 2005, Conference chair of Discovery Science 2009, and of a series of Workshops on Knowledge Discovery in Data Streams, ECML 2004, Pisa, Italy, ECML 2005, Porto, Portugal, ICML 2006, Pittsburg, US, ECML 2006 Berlin, Germany, SAC2007, Korea, and the ACM Workshop on Knowledge Discovery from Sensor Data held in conjunction with ACM SIGKDD 2007 and ACM SIGKDD 2008. Together with M. Gaber edited the book Learning from Data Streams - Processing Techniques in Sensor Networks, published by Springer.

Hector Gonzalez is a research scientist at Google. He received his Ph.D. in Computer Science from University of Illinois at Urbana-Champaign in 2008. Before his Ph.D. he received a his B.Sc. degree in Computer Science from Universidad Eafit, where he graduated first in class, he also completed an M.B.A. at Harvard Business School, graduating with distinction in 1999. His primary research interests are in the areas of data mining, data warehousing, and database systems. His thesis, entitled "Mining Massive Moving Object Datasets: From RFID Data Flow Analysis to Traffic Mining" deals with warehousing, mining, and managing object tracking data from supply chain applications, to transportation engineering systems.

Jiawei Han is a Professor in the Department of Computer Science at the University of Illinois. He has been working on research into data mining, data warehousing, stream data mining, spatiotemporal and multimedia data mining, biological data mining, social network analysis, text and Web mining, and software bug mining, with over 350 conference and journal publications. He has chaired or served in over 100 program committees of international conferences and workshops and also served or is serving on the editorial boards for Data Mining and Knowledge Discovery, IEEE Transactions on Knowledge and Data Engineering, Journal of Computer Science and Technology, and Journal of Intelligent Information Systems. He is currently the founding Editor-in-Chief of ACM Transactions on Knowledge Discovery from Data (TKDD). Jiawei has received three IBM Faculty Awards, the Outstanding Contribution Award at the 2002 International Conference on Data Mining, ACM Service Award (1999) and ACM SIGKDD Innovation Award (2004), and IEEE Computer Society Technical Achievement Award (2005). He is an ACM Fellow (2004). His book "Data Mining: Concepts and Techniques" (Morgan Kaufmann) has been used popularly as a textbook.

Paul J.M. Havinga is an associate professor in the pervasive system group at the University of Twente in the Netherlands. He received his PhD on the thesis entitled 'Mobile Multimedia Systems' in 2000 and was awarded with the 'DOW Dissertation Energy Award' for this work. His research interests include large-scale, heterogeneous wireless systems, sensor networks, energy-efficient architectures and protocols, ubiquitous computing, and personal communication systems. This research has resulted in over 180 scientific publications in journals and conferences. He is the editor of several journals and magazines and regularly serves as program committee chair, member, and reviewer for conferences and workshops.

Qingchun Jiang is a principal member of technical staff at Oracle Corporation. His primary research interests include SQL query processing and optimization, continuous query processing, and data stream management.

Erland Jungert has a Ph. D. in Computer Science from Linköping University, Sweden, 1980. Currently he is Director of Compute Science Research at the Swedish Defence Research Agency (FOI), in Linköping, Sweden since 1987. Since 1998 he is also a part time professor in Geoinformatics at the Department of Computer and Information Science, at Linköping University. He has been visiting Associate Professor at the department of Electrical Engineering, at the Illinois Institute of Technology, Chicago, Ill. Dr. Jungert is co-author of one book on spatial reasoning and co-editor of two other books on Visual Languages and on Intelligent Database Systems. Furthermore, he is also associate editor of the journal of Visual Languages and Computing. His interests concern methods for spatial reasoning, query languages for sensor data sources. Lately, he has developed an interest for command and control systems for crisis management including techniques and methods for prevention of crises and antagonistic threat activities also including studies of methods for protection of establishments, buildings and other important objects.

Nicolas Legeay worked as a research intern from the Institut National des Sciences Appliquées (INSA) in Lyon, France. His research was conducted with the Interoperable Systems Group in Dublin City University building a semantic enrichment process to process multiple sensor stream formats. This research facilitated an automated enrichment process whereby human interaction was not necessary when processing new sensor streams.

Stefano Lodi received his PhD in Computer Science and Electronic Engineering from the University of Bologna, where he is currently associate professor and teaches courses in information systems and decision support systems. He has co-authored papers on knowledge representation, agent-based and peer-to-peer distributed data mining, peer-to-peer semantic networks, visual data mining, and stream data mining, published in international conference proceedings, book chapters, and journals. He has been a team member in Italian and international projects in knowledge representation and information systems. His research interests include knowledge discovery and data mining, peer-to-peer systems, sensor networks, biomedical data analysis, and knowledge representation.

Luís Lopes received his MSc and PhD in Computer Science from the University of Porto, Portugal, in 1995 and 1999, respectively. He is currently an Auxiliar Professor at the Department of Computer Science at the Faculty of Science of the University of Porto. His main research interests are in virtual machines, domain specific programming languages, embedded systems (mostly sensor networks) and distributed systems in general.

Gregory Madey is a Research Professor in the Department of Computer Science and Engineering at the University of Notre Dame, IN, USA. He received the Ph.D and M.S. degrees in operations research from Case Western Reserve University and the M.S. and B.S degrees in mathematics from Cleveland State University. His research includes topics in emergency management modeling and simulation, agent-based modeling and simulation, distributed systems, web-services, data mining and service oriented architectures, web portals for scientific collaboration, web mining, and scientific databases. He is a member of the ACM, AIS, IEEE Computer Society, Informs, and the Society for Computer Simulation.

Sanjay Kumar Madria received his Ph.D. in Computer Science from Indian Institute of Technology, Delhi, India in 1995. He is an Associate Professor, Department of Computer Science at the University of Missouri-Rolla, USA. Earlier he was Visiting Assistant Professor in the Department of Computer Science, Purdue University, West Lafayette, USA. He has published more than 120 Journal and conference papers in the areas of mobile and sensor data management. He has organized international conferences, workshops and presented tutorials in the areas of mobile computing. He has given invited talks and served as panellist in National Science Foundation, USA and Swedish Research Council. His research is supported by NSF, DOE, UMRB and industrial grants for over $1.7M. He was awarded JSPS fellowship in 2006 and University of Missouri-Rolla's Faculty Excellence award in 2007. He is IEEE Senior Member and IEEE and ACM Distinguished Speaker.

Elio Masciari received his PhD in computer science engineering at the University of Calabria, Italy. Currently he is a Researcher at the ICAR institute of the Italian National Research Council. His research activity is mainly focused on techniques for the analysis and mining of data, text and web mining, XML query languages and XML structural similarity. He is also involved in several research project in the data mining field.

Nirvana Meratnia has received her Phd in 2005 from the database group at the University of Twente in the Netherlands and since then is working there in the pervasive system group as a researcher. Her research interests are in the area of distributed data management in wireless sensor networks, smart and collaborative objects, ambient intelligence, and context-aware applications. She is actively involved as program committee member and reviewer for conferences and workshop.

Boriana Milenova is a Principal Member of the Technical Staff in the Data Mining group of Oracle Corporation. She has designed and implemented a variety of data mining algorithms as core components within the Oracle database, for which she holds several patents. Dr. Milenova has authored a number of technical papers and book chapters. She received her B.S. and M.S. degrees in Electrical Engineering from the Czech Technical University, Prague, Czech Republic, and her Ph.D. in Cognitive & Neural Systems from Boston University.

Gabriele Monti was born in 1978 in Cesena, Italy. He received his university degree in Statistics and Computing for Businesses in 2002, and his Ph.D. in 2007, from the University of Bologna. He is currently a researcher at DEIS, Department of Electronics, Computer Sciences and Systems. The focus of his research are wireless mobile ad-hoc networks and sensor networks. He has worked in particular on routing algorithms, data management and workload balancing in distributed environments. Since November 2007 he is working on the DORII European Project which is supported by the European Commission. The project aims to deploy e-Infrastructure for new scientific communities.

Gianluca Moro is research associate at the Department of Electronics, Computer Science and Systems of the University of Bologna in Cesena. He received the Laurea degree in Computer Science in 1994 and the Ph. D. in Computer Science Engineering in 1999. He has been teaching since 2002 Distributed Information Systems and Information Processing Systems at the Faculty of Engigeering and at the Faculty of Statistical Science respectively. His main research interests focus on data management and data mining in distributed systems such as peer-to- peer, grid, ad-hoc and sensor networks. He has served in program committees of more than thirty international conferences such as P2P Computing, Globecomm, Data Management in Grid, Distributed Computing Systems etc. Moro organized several editions of the workshop DBISP2P at VLDB and AP2PC at AAMAS and in 2008 he has been vice chair of ACM International Conference on Intelligent Agent Technology. He published more than fifty papers in several areas of databases and distributed systems both in ACM, IEEE, Springer conferences and in journals such as Information Systems. He is member of RINGRID and DORII European projects.

Sotiris Nikoletseas is a Professor at the Computer Engineering and Informatics Dept of Patras Univ., Greece, a Senior Researcher at CTI and Director of its SensorsLab. His research interests include: Algorithmic Techniques in Distributed Computing (focus on wireless sensor networks and ad-hoc mobile networks), Probabilistic Techniques and Random Graphs, Algorithmic Engineering and Large Scale Simulation. He has coauthored over 100 publications in international Journals and refereed Conferences, several Invited Chapters in Books by major publishers and a Book on the Probabilistic Method. He has served as the Program Committee Chair/General Chair of many Conferences (including ALGO-SENSORS, MSWIM, MOBIWAC, WMAN, WEA, IPDPS, DCOSS), and as Editor of Special Issues and Member of the Editorial Board of major Journals (like TCS, JEA, IJDSN, IEEE Transactions on Computers). He has co-initiated several international events (including ALGOSENSORS, DCOSS). He has delivered several invited talks and tutorials. He has participated/coordinated several externally funded European Union R&D Projects related to fundamental aspects of modern networks (including ALCOM-FT, FLAGS, CRESCCO, DELIS, AEOLUS).

Noel E. O'Connor graduated from Dublin City University (DCU) with a B.Eng. in Electronic Engineering (1992) and a PhD (1998), after working for 2 years as a research assistant for Teltec Ireland at DCU. He is currently an Associate Professor in the School of Electronic Engineering. He is a Principal Investigator in CLARITY: Centre for Sensor Web Technologies. CLARITY is a €16.4m research centre funded by Science Foundation Ireland. CLARITY is a partnership between Dublin City University and University College Dublin, supported by research at the Tyndall National Institute (TNI) Cork. Prof. O'Connor's early research was in the field of video compression which subsequently led to an interest in video object segmentation and tracking as well as other aspects of computer vision. Prof. O'Connor has published over 160 peer-reviewed publications, made 11 standards submissions, filed 5 patents and spun off a campus company, Aliope Ltd.

Salvatore Orlando is an associate professor at the Department of Computer Science, University Ca' Foscari of Venice. He received a laurea degree cum laude in Computer Science from the University of Pisa, and a PhD in Computer Science from the same University. Then he was a post-doc fellow of the HP laboratories, and a post-doc fellow of the University of Pisa. In 1994 he joined as an assistant professor the Ca' Foscari University of Venice, where since 2000 he has been an associate professor. His research interests include the design of efficient and scalable algorithms for various data and Web mining techniques, parallel/distributed data mining by exploiting cluster and Grid platforms, distributed and P2P systems for information discovery, parallel systems, parallel languages and programming environments. Salvatore Orlando has published over 80 papers on international journals and conferences on several subjects, in particular on parallel processing, and data and Web mining, and has served on the Program Committees of many international conferences.

Biswait Panja completed his Ph.D. from University of Missouri-Rolla. He pursued his research on Security in Sensor Networks under the supervision of Dr. Sanjay Kumar Madria. Currently he is working as an assistant professor of computer science at Morehead State University. While pursing PhD at University of Missouri-Rolla he received following three awards: Outstanding Graduate Teaching Assistant, Outstanding Graduate Research Showcase and Outstanding Computer Science Graduate student.

Alec Pawling received the B.S. in Science in Computer Science from Truman State University and M.S. in Computer Science and Engineering from the University of Notre Dame. He is currently a Ph.D. candidate at the University of Notre Dame. His research interests include data mining, data warehousing, complex networks, cyberinfrastructure, and virtual organizations.

Olivier Powell is a Lecturer at the Computer Science Department of the University of Geneva in Switzerland. He was previously a Swiss National Science Foundation research fellow at the Computer Science and Informatics Department of the University of Patras in Greece and a post-doctoral research associate at the TCSensor lab of the Computer Science Department of the University of Geneva. He received a Ph.D. degree in Computer Science from the University of Geneva in the field of Quantitative Complexity Theory and a Masters Degree in Mathematics from the same University.

Alessandra Raffaetà is an assistant professor in the Department of Computer Science of the University Ca' Foscari of Venezia. She graduated in Computer Science (Laurea in Scienze dell'Informazione) summa cum laude in 1994, from the University of Pisa and she took a PhD in Computer Science from the University of Pisa in 2000. Her research interests include design and formal semantics of programming languages, spatio-temporal reasoning and Data warehouses. Contact her at raffaeta@dsi.unive.it.

Mark Roantree holds a B.Sc. from Trinity College (Ireland), M.Sc. from Dublin City University (Ireland) and PhD from Napier University (Edinburgh). He is currently a senior lecturer at Dublin City University where he has been Principal Investigator on 7 major funding awards. Dr Roantree's current research interests are in optimizing XML databases, peer-to-peer information system infrastructures, sensor networks and distributed query processing. He is co-founder and leader of the Interoperable

Systems Group for 10 years, during which time he has published almost 60 peer-reviewed publications. He has reviewed papers for numerous journals and international conferences including the International Journal of Cooperative Information Systems, Transactions on Information Systems (TOIS), SIGMOD Record and the International Conferences on Extending Database Technologies (EDBT), and the largest Semantic Web Conference (ISWC).

Pedro Pereira Rodrigues received the BSc and the MSc degrees in Computer Science from the Faculty of Science of the University of Porto, Portugal, in 2003 and 2005, respectively. He is currently a PhD candidate at the same university, working as a researcher at the Artificial Intelligence and Decision Support Laboratory, on distributed clustering of streaming data from sensor networks. His research interests concentrate on machine learning and data mining from distributed data streams and reliability of predictive and clustering analysis in streaming environments, with applications on industry- and health-related domains.

Jose Rolim is Director of the Center Universitaire d'Informatique and Full Professor at the Department of Computer Science of the University of Geneva where he heads the Theoretical Computer Science and Sensor Lab (TCSensor Lab). He received his Ph.D. degree in Computer Science from the University of California, Los Angeles, in the area of formal languages and complexity. He has published many articles in the areas of theoretical computer science, distributed systems, randomization and computational complexity and leads two major national projects in the areas of Power Aware Computing and Games and Complexity. He also participates as a partner in two European Projects in the areas of Dynamic Systems and Foundations of Sensor Networks. Prof. Rolim participates in the editorial board of several journals and conferences and he is the Steering Committee Chair and General Chair of the IEEE Distributed Computing Conference in Sensor Systems. He has been Program Committee Chair of major conferences such as ICALP, IPDPS, RANDOM and served as Program Committee Member of all major conferences in theoretical computer science.

Alessandro Roncato is an assistant professor in the Department of Computer Science of the University Ca' Foscari of Venezia. He graduated in Computer Science (Laurea in Scienze dell'Informazione) summa cum laude in 1989, from the University of Udine and he took a PhD in Computer Science from the University of Pisa in 1995. His research interests include distributed computing, spatio-temporal Data warehouses. Contact him at roncato@dsi.unive.it.

Domenico Saccà was born in Catanzaro (Italy) on November 5, 1950, and he received a Doctoral degree in Engineering from the University of Rome on 1975. Since 1987, he is full professor of Computer Engineering at the University of Calabria. From 2002 to 2008, he was Director of the CNR (the Italian National Research Council) Research Institute ICAR (Institute for High Performance Computing and Networking), located in Rende (CS) and branches in Naples and Palermo. Previously, from 1995 on, he was director of the CNR Institute on System and Computer Sciences (Istituto per la Sistemistica e l'Informatica, ISI-CNR). In the past he was visiting scientist at IBM Laboratory of San Jose, at the Computer Science Department of UCLA and at the ICSI Institute of Berkeley; furthermore, he was a scientific consultant of MCC, Austin, and manager of the Research Division of CRAI. His current research interests focus on advanced issues on database such as: scheme integration in data warehousing, compressed representation of datacubes, workflow and process mining, logic-based database query languages. His list of publications contains more than 100 papers on journals (including Journal of the ACM, SIAM Journal on Computing, ACM Transactions on Database Systems, IEEE Transactions on Knowledge and Data Engineering, IEEE Transactions on Software Engineering, Theoretical Computer Science, Journal of Computer and System Sciences, Acta Informatica) and on proceedings of international conferences. He has been member of the program committees of several international conferences, director of international schools and seminars and leader of many national and international research projects.

Claudio Sartori is Full Professor in System for Information Processing, in the Department of Electronics, Computer Science and Systems, of the University of Bologna. He has been doing research since 1984 in the areas of databases, information systems, distributed and peer-to-peer systems, data mining. He participated to several Italian and international research projects and directed also applied research joint projects with the industry. He is co-author of more than 80 scientific papers, mostly in international journals and conferences. He is the director of the undergraduate program "Statistics sciences", in the University of Bologna. He is co- organizer, on a regular basis, of workshops in the area of peer-to- peer systems, both from the point of view of intelligent agents and that of information systems. He teaches courses in computer science at base and advanced level, including database, information systems, data mining.

Timothy Schoenharl received the bachelors degree in Mathematics from John Carroll University and the Masters of Engineering and Ph.D. in Computer Science from the University of Notre Dame. His research interests include Agent-Based Modeling and Simulation, Dynamic Data-Driven Application Systems and Multi-Agent Systems. He currently works for Amazon.com developing DDDAS for Supply Chain applications.

Claudio Silvestri received the MS degree in Computer Science summa cum laude from the University Ca' Foscari of Venezia, and the PhD in Computer Science from the same University in 2006. He is currently a research fellow at Department of Computer Science and Communication, University of Milan. His research interests include: algorithms for frequent pattern mining and techniques for approximate data mining on distributed data and data streams, position based secure access to information systems, privacy preservation in location based services, and spatio-temporal data disclosure. Dr. Claudio Silvestri is a member of the program committee of the Track on Data Mining of the ACM Symposium on Applied Computing since 2006.

Since 1997, **Alan F. Smeaton** has been a full Professor of Computing at DCU. He has been the Dean of Faculty of Computing and Mathematical Sciences, and Head of the School of Computing. He is the founding director of the Centre for Digital Video Processing, a University-designated research centre with over 40 full-time researchers. Alan's background is in information retrieval, initially on using natural language processing techniques for text retrieval. In the mid 1990s he moved to information retrieval from visual media – image and video. He now has interests in managing all kinds of information including personal media, image, video, and information derived from a variety of personal sensors. Alan is the Deputy Director of the SFI CSET CLARITY, a collaboration between DCU, UCD and the Tyndall Institute.

Jennifer Xu is an Assistant Professor of Computer Information Systems at Bentley College. Jennifer's research areas include knowledge discovery and data mining, knowledge management, social network analysis, information visualization, human-computer interaction, and electronic commerce. Jennifer's work has been published and presented at international symposiums and conferences, and appeared in refereed journals. She has also received awards for her papers "A framework for locating and analyzing hate groups in blogs" with Michael Chau, and "CrimeNet explorer: A framework for criminal network knowledge discovery" with Hsinchun Chen. She holds a PhD from the University of Arizona. She also holds an MSc and an MA from the University of Mississippi, and a BSc from the Beijing Institute of Technology.

Ping Yan received the B.S. (1994) and M.S. (1999) degrees in computer science from Nanjing University and earned the M.S.E. (2003) and Ph.D. (2006) degree in computer science and engineering from the University of Notre Dame. Her research interest includes image processing, computer vision, biometrics and object recognition. After she received her Ph.D, she spent a year as a Post-doctoral researcher at Notre Dame investigating approaches to anomaly detection in streaming data and evaluated the Markov-Modulated Poisson Process for modeling periodic data. She is currently working as senior software developer at Swift Biometrics, Inc on iris biometrics.

David Yates is an Assistant Professor of Computer Information Systems at Bentley College. David's research areas include computer networking, data communications, sensor networks, embedded systems, operating systems, and computer architecture. Before joining Bentley, David held research and academic positions at the University of Massachusetts and Boston University. His work has been published and presented at international symposiums and conferences, and appeared in refereed journals. In the corporate arena, he was a co-founder and vice president of software development at InfoLibria – a startup that grew to become a leading provider of hardware and software for building content distribution and delivery networks before it was acquired. With various colleagues, he holds several U.S. patents for processes and systems related to computer networking, content management, and mobile computing. He holds a PhD and MSc from the University of Massachusetts. He also holds a BSc from Tufts University.

Tianyi Wu received the BS degree in computer science from Fudan University, Shanghai, in 2005. He is currently working toward the PhD degree in the Department of Computer Science, University of Illinois at Urbana-Champaign. His research interests include data mining, OLAP query processing, and text data management. He is a member of the ACM.

Yang Zhang is currently a Ph.D. student in the pervasive system group at the University of Twente in the Netherlands. He received his B.S. and M.S. degrees in computer science and technology from the University of Jiangsu, China, in 2002 and 2004, respectively. He received his second M.S. degree in Telematics from the University of Twente, the Netherlands in 2006. His research interests include distributed data mining, outlier detection and event detection in sensor networks. He is involved as publicity co-chair and reviewer for conferences and workshop.

Index